Instructor's Resource Manual

for

Atkins and Jones's

Chemical Principles

The Quest for Insight

THIRD EDITION

THOMAS SPENCE
Loyola University New Orleans

and

LYNN VOGEL KOPLITZ
Loyola University New Orleans

W. H. Freeman and Company
New York

ISBN: 0-7167-0700-4

EAN: 9780716707004

Printed in the United States of America

Second printing

W. H. Freeman and Company
41 Madison Avenue
New York, NY 10010
Houndmills, Basingstoke RG21 6XS England

www.whfreeman.com

CONTENTS

Acknowledgments

We would like to express our gratitude to the many individuals who made substantial contributions to the successful creation of this solutions manual. We thank Loretta Jones and Peter Atkins for their helpful clarifications of the exercises in their textbook and for their encouragement and support. Ken Whitmire graciously provided files from the previous version of this manual. Patricia Christie, Xiaodong Zhang, Scott Kowaleski, Keith Ponchot, Kevin Simpson, and Steven Simpson carefully reviewed the solutions and made many helpful suggestions for improvement. Xiaodong Zhang and Brent Koplitz helped by solving several mathematically challenging problems. D. Andrew Knight drew hundreds of Lewis dot structures. The staff at W. H. Freeman, including Jessica Fiorillo, Amanda McCorquodale, Shawn Churchman, Jenness Crawford, and Amy Shaffer did their best to make sure this project was completed in a timely fashion. We are especially grateful to Elizabeth Spence who spent many hours making this manual presentable.

T.G.S. and L.V. K.

FUNDAMENTALS

A.2 The boiling point and melting point are physical properties. The flammability of acetone is a chemical property.

A.4 The red-brown color of copper is a physical property. The formation of copper from copper sulfide ores, the formation of copper oxide from copper, the formation of impure copper from copper oxide and carbon, and the purification of copper by electrolysis are chemical properties.

A.6 (a) extensive; (b) intensive; (c) intensive; (d) intensive

A.8
$$d = \frac{m}{V}$$
$$= \left(\frac{10.00\text{ g}}{9.87\text{ mL} - 8.75\text{ mL}}\right)\left(\frac{1\text{ mL}}{1\text{ cm}^3}\right)$$
$$= 8.93\text{ g}\cdot\text{cm}^{-3}$$

A.10
$$m = Vd$$
mass of lead, m_l, equals mass of redwood, m_r
$$m_l = V_l d_l = m_r = V_r d_r$$
$$V_l d_l = V_r d_r$$
$$V_l = \frac{V_r d_r}{d_l} = \frac{(375\text{ cm}^3)(0.38\text{ g}\cdot\text{cm}^{-3})}{(11.3\text{ g}\cdot\text{cm}^{-3})}$$
$$= 13\text{ cm}^3$$

A.12 The average of the four density determinations is
$$d_{avg} = \frac{(1.68 + 1.67 + 1.69 + 1.69)\text{ g}\cdot\text{cm}^{-3}}{4} = 1.68\text{ g}\cdot\text{cm}^{-3}$$

or $0.06\,\text{g}\cdot\text{cm}^{-3}$ below the accepted value. The standard deviation of these four values is

$$s=\sqrt{\frac{(1.68-1.68)^2+(1.67-1.68)^2+(1.69-1.68)^2+(1.69-1.68)^2}{3}}$$

$$=0.01\ \text{g}\cdot\text{cm}^{-3}$$

So the accuracy of the data, as measured by the difference between the average and an accepted value, is poor compared to the precision (reproducibility) of the data as measured by the standard deviation. The data is precise to within $0.01\ \text{g}\cdot\text{cm}^{-3}$ but only accurate to within 0.06 $\text{g}\cdot\text{cm}^{-3}$.

A.14 (a) $d=\dfrac{m}{V}$

$$=\left(\frac{2.0\times10^{-23}\ \text{g}}{\frac{4}{3}\pi(1.5\times10^{-5}\ \text{pm})^3}\right)\left(\frac{1\ \text{pm}}{1\times10^{-10}\ \text{cm}}\right)^3$$

$$=1.4\times10^{21}\ \text{g}\cdot\text{cm}^{-3}$$

(b) From A.14a, density of a carbon nucleus is $1.4\times10^{21}\ \text{g}\cdot\text{cm}^{-3}$

$$d=\frac{m}{V}=\frac{m}{\frac{4}{3}\pi r^3}$$

$$r=\sqrt[3]{\frac{3m}{4\pi d}}\qquad m=d_1V_1=d_1\left(\tfrac{4}{3}\pi r_1^3\right)$$

$$r=\sqrt[3]{\frac{3d_1\left(\frac{4}{3}\pi r_1^3\right)}{4\pi d}}$$

$$=\sqrt[3]{\frac{d_1r_1^3}{d}}$$

$$=\sqrt[3]{\frac{(5.5\ \text{g}\cdot\text{cm}^{-3})(6.4\times10^3\ \text{km})^3}{\left(1.4\times10^{21}\text{g}\cdot\text{cm}^{-3}\right)}\left(\frac{1\ \text{m}}{1\times10^{-3}\ \text{km}}\right)^3}$$

$$=1.0\ \text{m}$$

A.16 $E_K = \frac{1}{2}mv^2$

$$= \frac{1}{2}(2.5 \times 10^5 \text{ kg})(25 \text{ km} \cdot \text{s}^{-1})^2 \left(\frac{1000 \text{ m}}{1 \text{ km}}\right)^2$$

$$= 7.8 \times 10^{13} \text{ kg} \cdot \text{m}^2 \cdot \text{s}^{-2}$$

$$= 7.8 \times 10^{13} \text{ J}$$

A.18 $E_P = mgh$

$$= (0.51 \text{ kg})(9.81 \text{ m} \cdot \text{s}^{-2})(3.0 \text{ m})$$

$$= 15 \text{ kg} \cdot \text{m}^2 \cdot \text{s}^{-2}$$

$$= 15 \text{ J}$$

A.20 When the skier is at rest at the top of the slope, the kinetic energy is 0. The total energy is equal to the potential energy. At the bottom of the slope, the potential energy is 0; it has been converted into kinetic energy.

$$E = E_K + E_P$$

$$= E_P$$

$$= mgh$$

$$= (99.5 \text{ kg})(9.81 \text{ m} \cdot \text{s}^{-2})(244 \text{ m})$$

$$= 2.38 \times 10^5 \text{ kg} \cdot \text{m}^2 \cdot \text{s}^{-2}$$

$$= 2.38 \times 10^5 \text{ J}$$

$$= 238 \text{ kJ}$$

A.22 Since $R = R_E + h$,

$$E_p = \Delta E_p = \frac{-Gm_E m}{R_E + h} - \left(\frac{-Gm_E m}{R_E}\right)$$

$$= \frac{-Gm_E m}{R_E}\left(\frac{1}{1+\frac{h}{R_E}} - 1\right) = \frac{Gm_E m}{R_E}\left(1 - \frac{1}{1+\frac{h}{R_E}}\right)$$

If $x = \frac{h}{R_E}$, then $\frac{1}{1+\frac{h}{R_E}} = 1 - \frac{h}{R_E}$ from the expansion in x.

Then

$$E_p = \frac{Gm_E m}{R_E}\left(1 - \left(1 - \frac{h}{R_E}\right)\right) = \frac{Gm_E mh}{R_E^2} = mgh$$

$$\text{where } g = \frac{Gm_E}{R_E^2}$$

A.24 $F = ma$

$$= (9.1 \times 10^{-31} \text{ kg})(1.2 \times 10^{6} \text{ m} \cdot \text{s}^{-2})$$
$$= 1.1 \times 10^{-24} \text{ kg} \cdot \text{m} \cdot \text{s}^{-2}$$
$$= 1.1 \times 10^{-24} \text{ N}$$

A.26 $P = \frac{F}{A}$

units of force = N = $\text{kg} \cdot \text{m} \cdot \text{s}^{-2}$

units of area = m^2

$$\text{units of pressure} = \frac{\text{kg} \cdot \text{m} \cdot \text{s}^{-2}}{\text{m}^2}$$
$$= \text{kg} \cdot \text{m}^{-1} \cdot \text{s}^{-2}$$
$$= \text{Pa}$$

A.28 (See Appendix 1)

$$\text{energy} = \text{power} \cdot \text{time} = I^2 R \cdot \text{time} \qquad V = IR$$
$$= \frac{V}{R} \cdot I \cdot R \cdot \text{time} = V \cdot I \cdot \text{time} = \text{V} \cdot \left(\frac{\text{C}}{\text{s}}\right) \cdot \text{s}$$
$$= \text{V} \cdot \left(\frac{\text{J}}{\text{V}}\right) = \text{J}$$

A.30 First find the actual mass of the sample, then multiply by Earth gravity.

$$m_{planet} = \frac{5.70 \text{ lome}\left(\frac{1.3 \text{ N}}{1 \text{ lome}}\right)\left(\frac{1 \text{ kg} \cdot \text{m} \cdot \text{s}^{-2}}{1 \text{ N}}\right)\left(\frac{1 \text{ coburg}}{2.78 \text{ m} \cdot \text{s}^{-2}}\right)}{9.8 \text{ coburg}}$$

$$= 0.27 \text{ kg}$$

$$m_{Earth} g = (0.27 \text{ kg})(9.8 \text{ m} \cdot \text{s}^{-2}) = 2.7 \text{ N}$$

B.2 (a) Part one of Dalton's hypothesis – that all atoms of a given element are identical – has been disproved. (b) Mass spectrometry revealed that some atoms with the same atomic number have different masses.

B.4 In an evacuated tube, we would expect these particles to be electrons from the metal cathode (as discussed on p. F15). In that case, the ratio of charge to mass would be

$\frac{q}{m} = \frac{-1.60218 \times 10^{-19}\text{C}}{9.10939 \times 10^{-31}\text{kg}} = -1.759 \times 10^{11}\text{C} \cdot \text{kg}^{-1}$. Since this ratio for an electron is much larger than the one given for the "canal rays" in this helium-filled tube, we can reason that these particles are about 10,000 times as massive as an electron. It turns out that He^{+} has the right mass:

$$\frac{q}{m} = \frac{1.60\,218 \times 10^{-19}\text{C}}{(4.00)(1.66\,054 \times 10^{-27}\text{kg})} = 2.41 \times 10^{7}\text{C} \cdot \text{kg}^{-1}$$

B.6 (a) A deuterium atom has 1 proton, 1 neutron, and 1 electron.

(b) A ^{127}I atom has 53 protons, 74 neutrons, and 53 electrons.

(c) A ^{15}N atom has 7 protons, 8 neutrons, and 7 electrons.

(d) A ^{209}Bi atom has 83 protons, 126 neutrons, and 83 electrons.

B.8 (a) ^{175}Lu; (b) ^{118}Sn; (c) ^{6}Li

B.10 (a) ^{12}C, ^{13}C, and ^{14}C all have 6 protons and 6 electrons. (b) They have different numbers of neutrons: an atom of ^{12}C has 6 neutrons, an atom of ^{13}C has 7 neutrons, and an atom of ^{14}C has 8 neutrons.

B.12 (a) Atoms of ^{55}Mn, ^{56}Fe, and ^{58}Ni all have 30 neutrons. (b) They have different masses, numbers of protons, and numbers of electrons.

B.14 (a) In NH_3, the nitrogen atom contributes 7 protons, 7 neutrons, and 7 electrons. Each hydrogen atom contributes 1 proton and 1 electron. There are 10 protons, 7 neutrons, and 10 electrons total.

(b) $\text{mass of protons} = (10)(1.673 \times 10^{-24}\ \text{g}) = 1.673 \times 10^{-23}\ \text{g}$

$\text{mass of neutrons} = (7)(1.675 \times 10^{-24}\ \text{g}) = 1.172 \times 10^{-23}\ \text{g}$

$\text{mass of electrons} = (10)(9.109 \times 10^{-28}\ \text{g}) = 9.109 \times 10^{-27}\ \text{g}$

B.16 (a) Silicon is a Group 14 metalloid. (b) Strontium is a Group 2 metal. (c) Tin is a Group 14 metal. (d) Antimony is a Group 15 metalloid.

B.18 (a) Ag, metal; (b) Ge, metalloid; (c) Kr, nonmetal

B.20 lithium, Li, 3; sodium, Na, 11; potassium, K, 19; rubidium, Rb, 37; cesium, Cs, 55. All the alkali metals react with water to produce hydrogen. Lithium reacts gently; the reaction becomes more vigorous down the periodic table. The melting points decrease down the periodic table.

B.22 (a) Metals are malleable, conduct electricity, and have luster. (b) Nonmetals are neither malleable nor ductile and do not conduct electricity.

C.2 An element is made up of the same kinds of atoms. These atoms can bond together to form molecules, making a molecular substance that is still an element. One example is oxygen, O_2.

C.4 Since $x:y:z$ is 1:2:2 and $x = 3$, $C_xH_yN_z$ becomes $C_3H_6N_6$.

C.6 (a) Sulfur is a Group 16 nonmetal and will form 2− anions, S^{2-}. (b) Lithium is a Group 1 metal and will form 1+ cations, Li^+. (c) Phosphorus is a Group 15 nonmetal and will form 3− anions, P^{3-}. (d) Oxygen is a Group 16 nonmetal and will form 2− anions, O^{2-}.

C.8 (a) $^{107}Ag^+$ has 47 protons, 60 neutrons, and 46 electrons. (b) $^{140}Ce^{4+}$ has 58 protons, 82 neutrons, and 54 electrons. (c) $^{17}O^{2-}$ has 8 protons, 9 neutrons, and 10 electrons. (d) $^{40}Ca^{2+}$ has 20 protons, 20 neutrons, and 18 electrons.

C.10 (a) $^{24}Na^+$; (b) $^{27}Al^{3+}$; (c) $^{79}Se^{2-}$; (d) $^{52}Cr^{2+}$

C.12 (a) Sodium forms Na^+ ions. Phosphorus forms P^{3-} ions. Three sodium ions produce a charge of $3 \times (+1) = +3$. One phosphorus ion balances the charge. The formula for sodium phosphide is Na_3P. (b) Aluminum forms Al^{3+} ions and chloride forms Cl^- ions. Since three chloride ions are needed to balance the charge on the aluminum cation, the empirical formula for aluminum chloride is $AlCl_3$. (c) Strontium forms Sr^{2+}. Fluorine forms F^-. Since two fluoride ions will balance the charge on the strontium cation, the formula for strontium fluoride is SrF_2. (d) Lithium forms +1 cations and selenium forms −2 anions. Two Li^+ ions are needed to balance the charge on every one Se^{2-}, so the formula for lithium selenide is Li_2Se.

C.14 (a) Element E belongs to Group 16 (VI, VIA). (b) Selenium, Se.

C.16 (a) 1 + 1 + 8 = 10 protons and an equal number of electrons; 0 + 0 + 8 = 8 neutrons

(b) $10(1.67262 \times 10^{-27}\text{ kg}) = 1.67262 \times 10^{-26}\text{ kg}$ protons

$8(1.67439 \times 10^{-27}\text{ kg}) = 1.33994 \times 10^{-26}\text{ kg}$ neutrons

$10(9.10939 \times 10^{-31}\text{ kg}) = 9.10939 \times 10^{-20}\text{ kg}$ electrons

(c) $$\frac{1.33994 \times 10^{-26}\text{ kg}}{1.33994 \times 10^{-26}\text{ kg} + 1.67262 \times 10^{-26}\text{ kg} + 9.10939 \times 10^{-20}\text{ kg}}$$

$= 0.444649$

So about 44% of my body mass is due to neutrons.

C.18 (a) $Ca_3(PO_4)_2$

(b) $(NH_4)_2SO_4$

(c) Since two silver ions have the same charge as one magnesium ion, each silver ion carries a charge of +1.

(d) Since one lead ion has the same charge as two potassium ions, each lead ion carries a charge of +2.

D.2 (a) $Ba(OH)_2$; (b) $AlPO_4$; (c) $ZnCl_2$; (d) $FeCl_3$

D.4 (a) sodium bicarbonate or sodium hydrogen carbonate; (b) mercury(I) chloride; (c) sodium hydroxide; (d) zinc(II) oxide

D.6 (a) N_2O_4; (b) H_2S; (c) Cl_2O_7; (d) NI_3; (e) SO_2; (f) HF; (g) I_2Cl_6

D.8 (a) silicon dioxide; (b) silicon carbide; (c) dinitrogen oxide or nitrous oxide; (d) phosphorus(V) oxide; (e) carbon disulfide; (f) sulfur dioxide; (g) ammonia

D.10 (a) $HClO_4$; (b) HClO; (c) HIO; (d) HF; (e) H_3PO_3;

(f) HIO_4

D.12 (a) $CaCl_2$; (b) $Fe_2(SO_4)_3$; (c) NH_4I; (d) Li_2S; (e) Ca_3P_2

D.14 (a) chromium(III) chloride hexahydrate; (b) cobalt(II) nitrate hexahydrate; (c) thallium(I) chloride; (d) bromine chloride; (e) manganese(IV) oxide; (f) mercury(II) nitrate; (g) nickel(II) perchlorate; (h) calcium hypochlorite; (i) niobium(V) oxide

D.16 (a) methane; (b) fluoromethane; (c) bromomethane; (d) iodomethane

D.18 $BaTiO_3$; hydrogen sulfate(VI) or dihydrogen sulfate

D.20 E = Sr; strontium bromide; strontium oxide

D.22 Oxygen is present as a molecular anion in each case. (a) potassium superoxide, ionic, K^+, O_2^-; (b) sodium peroxide, ionic, 2 Na^+, O_2^{2-}; (c) cesium ozonide, ionic, Cs^+, O_3^-

D.24 (a) uranyl chloride; (b) sodium azide; (c) calcium oxalate; (d) vanadyl sulfide; (e) potassium dichromate; (f) lithium thiocyanate

D.26 (a) alcohol; (b) carboxylic acid; (c) fluorocarbon

E.2 (a) moles of sand $= \left(\dfrac{1000 \text{ tons}}{10^{12} \text{ grains}}\right)(6.022 \times 10^{23} \text{ grains} \cdot \text{mol}^{-1})(1 \text{ mol})$

$= 6 \times 10^{14}$ tons

(b) depth of sand $= \left(\dfrac{\text{volume of 1 mol of sand}}{\text{area of United States}}\right)$

$$= (1\ \text{mm}^3 \cdot \text{grain}^{-1}) \frac{\left(6.02\times10^{23}\,\text{grain}\cdot\text{mol}^{-1}\right)}{\left(3.6\times10^{6}\,\text{mi}^2\right)\left(\dfrac{1\times10^{6}\,\text{mm}}{0.6214\ \text{mi}}\right)}(1\ \text{mol})\left(\frac{1\ \text{m}}{1000\ \text{mm}}\right)$$

$= 65$ m

E.4 molar mass of $CuSO_4 \cdot 5H_2O = 249.68\ g \cdot mol^{-1}$

$$29.50\ \text{g CuSO}_4\cdot5\text{H}_2\text{O}\left(\frac{1\ \text{mol CuSO}_4\cdot5\text{H}_2\text{O}}{249.68\ \text{g CuSO}_4\cdot5\text{H}_2\text{O}}\right)\left(\frac{1\ \text{mol Cu}}{1\ \text{mol CuSO}_4\cdot5\text{H}_2\text{O}}\right)$$

$$\times\left(\frac{63.54\ \text{g Cu}}{1\ \text{mol Cu}}\right)$$

$= 7.507$ g Cu

E.6 molar mass of sulfur

$$= \left(\frac{95.0}{100}\right)(31.97\ \text{g}\cdot\text{mol}^{-1}) + \left(\frac{0.8}{100}\right)(32.97\ \text{g}\cdot\text{mol}^{-1})$$

$$+\left(\frac{4.2}{100}\right)(33.97\ \text{g}\cdot\text{mol}^{-1})$$

$$= 32.06\ \text{g}\cdot\text{mol}^{-1}\ \left(30\ \text{g}\cdot\text{mol}^{-1}\ \text{to 1 SF}\right)$$

E.8 molar mass of krypton

$$= \left(\frac{0.3}{100}\right)(77.92\ \text{g}\cdot\text{mol}^{-1}) + \left(\frac{2.3}{100}\right)(79.91\ \text{g}\cdot\text{mol}^{-1})$$

$$+\left(\frac{11.6}{100}\right)(81.91\ \text{g}\cdot\text{mol}^{-1}) + \left(\frac{11.5}{100}\right)(82.92\ \text{g}\cdot\text{mol}^{-1})$$

$$+\left(\frac{56.9}{100}\right)(83.91\ \text{g}\cdot\text{mol}^{-1}) + \left(\frac{17.4}{100}\right)(85.91\ \text{g}\cdot\text{mol}^{-1})$$

$$= 83.8\ \text{g}\cdot\text{mol}^{-1}\ \left(80\ \text{g}\cdot\text{mol}^{-1}\ \text{to 1 SF}\right)$$

E.10 (a) $\text{mass} = \left(\dfrac{5.68\times10^{15}\text{atoms}}{6.022\times10^{23}\text{atoms}\cdot\text{mol}^{-1}}\right)\left(200.59\ \text{g}\cdot\text{mol}^{-1}\right)\left(\dfrac{1\times10^{6}\ \mu\text{g}}{1\ \text{g}}\right)$

$= 1.89\ \mu\text{g}$

(b) $\text{mass} = \left(7.924\times10^{-9}\ \text{mol}\right)\left(178.49\ \text{g}\cdot\text{mol}^{-1}\right)\left(\dfrac{1\times10^{6}\ \mu\text{g}}{1\ \text{g}}\right) = 1.41\ \mu\text{g}$

(c) $\text{mass} = \left(3.49\ \mu\text{mol}\right)\left(\dfrac{1\ \text{mol}}{1\times10^{6}\ \mu\text{mol}}\right)\left(157.25\ \text{g}\cdot\text{mol}^{-1}\right)\left(\dfrac{1\times10^{6}\ \mu\text{g}}{1\ \text{g}}\right)$

$= 549\ \mu\text{g}$

(d) $\text{mass} = \left(\dfrac{6.29\times10^{24}\text{atoms}}{6.022\times10^{23}\text{atoms}\cdot\text{mol}^{-1}}\right)\left(121.75\ \text{g}\cdot\text{mol}^{-1}\right)\left(\dfrac{1\times10^{6}\ \mu\text{g}}{1\ \text{g}}\right)$

$= 1.27\times10^{9}\ \mu\text{g}$

E.12 (a) $m_{\text{Cu}} = \left(\dfrac{0.00735\ \text{g K}}{39.10\ \text{g}\cdot\text{mol}^{-1}\ \text{K}}\right)(63.54\ \text{g}\cdot\text{mol}^{-1}\ \text{Cu})$

$= 0.0119\ \text{g Cu or } 11.9\ \text{mg Cu}$

(b) $m_{\text{Cu}} = \left(\dfrac{0.00735\ \text{g Au}}{196.97\ \text{g}\cdot\text{mol}^{-1}\ \text{Au}}\right)(63.54\ \text{g}\cdot\text{mol}^{-1}\ \text{Cu})$

$= 0.00237\ \text{g Cu or } 2.37\ \text{mg Cu}$

E.14 (a) molar mass of $H_2O = 18.02\ \text{g}\cdot\text{mol}^{-1}$

$$n_{\text{H}_2\text{O}} = \left(\frac{1.80\ \text{kg}}{18.02\ \text{g}\cdot\text{mol}^{-1}}\right)\left(\frac{1000\ \text{g}}{1\ \text{kg}}\right) = 1.00\times10^{2}\ \text{mol}$$

$N_{\text{H}_2\text{O}} = (1.00\times10^{2}\ \text{mol})(6.022\times10^{23}\ \text{atoms}\cdot\text{mol}^{-1}) = 6.02\times10^{25}$ molecules

(b) molar mass of benzene $= 78.11\ \text{g}\cdot\text{mol}^{-1}$

$$n_{\text{benzene}} = \left(\frac{49\ \text{kg}}{78.11\ \text{g}\cdot\text{mol}^{-1}}\right)\left(\frac{1000\ \text{g}}{1\ \text{kg}}\right) = 6.3\times10^{2}\ \text{mol}$$

$N_{\text{benzene}} = (6.3\times10^{2}\ \text{mol})(6.022\times10^{23}\ \text{atoms}\cdot\text{mol}^{-1}) = 3.8\times10^{26}$ molecules

(c) molar mass of P atom $= 30.97\ \text{g}\cdot\text{mol}^{-1}$

$$n_P = \left(\frac{100.0 \text{ g}}{30.97 \text{ g}\cdot\text{mol}^{-1}}\right) = 3.229 \text{ mol P atoms}$$

$$N_P = (3.229 \text{ mol})(6.022\times10^{23} \text{ atoms}\cdot\text{mol}^{-1}) = 1.944\times10^{24} \text{ P atoms}$$

molar mass of P_4 molecule = 123.88 $\text{g}\cdot\text{mol}^{-1}$

$$n_{P_4} = \left(\frac{100.0 \text{ g}}{123.88 \text{ g}\cdot\text{mol}^{-1}}\right) = 0.8072 \text{ mol } P_4 \text{ molecules}$$

$$N_{P_4} = (0.8072 \text{ mol})(6.022\times10^{23} \text{ atoms}\cdot\text{mol}^{-1}) = 4.861\times10^{23} \text{ } P_4 \text{ molecules}$$

(d) molar mass of CO_2 = 44.01 $\text{g}\cdot\text{mol}^{-1}$

$$n_{CO_2} = \left(\frac{5.0 \text{ g}}{44.01 \text{ g}\cdot\text{mol}^{-1}}\right) = 0.11 \text{ mol}$$

$$N_{CO_2} = (0.11 \text{ mol})(6.022\times10^{23} \text{ atoms}\cdot\text{mol}^{-1}) = 6.6\times10^{22} \text{ molecules}$$

(e) molar mass of NO_2 = 46.01 $\text{g}\cdot\text{mol}^{-1}$

$$n_{NO_2} = \left(\frac{5.0 \text{ g}}{46.01 \text{ g}\cdot\text{mol}^{-1}}\right) = 0.11 \text{ mol}$$

$$N_{NO_2} = (0.11 \text{ mol})(6.022\times10^{23} \text{ atoms}\cdot\text{mol}^{-1}) = 6.6\times10^{22} \text{ molecules}$$

E.16 (a) molar mass of KCN = 65.12 $\text{g}\cdot\text{mol}^{-1}$

$$n_{CN^-} = \left(\frac{1.00 \text{ g KCN}}{65.12 \text{ g}\cdot\text{mol}^{-1} \text{ KCN}}\right)\left(\frac{1 \text{ mol } CN^-}{1 \text{ mol KCN}}\right) = 1.54\times10^{-2} \text{ mol}$$

(b) molar mass of H_2O = 18.02 $\text{g}\cdot\text{mol}^{-1}$

$$n_H = \left(\frac{2.00\times10^{-1} \text{ g } H_2O}{18.02 \text{ g}\cdot\text{mol}^{-1} \text{ } H_2O}\right)\left(\frac{2 \text{ mol H}}{1 \text{ mol } H_2O}\right) = 2.2\times10^{-2} \text{ mol}$$

(c) molar mass of $CaCO_3$ = 100.09 $\text{g}\cdot\text{mol}^{-1}$

$$n_{CaCO_3} = \left(\frac{5.00\times10^{3} \text{ g } CaCO_3}{100.09 \text{ g}\cdot\text{mol}^{-1} \text{ } CaCO_3}\right) = 5.00 \text{ mol}$$

(d) molar mass of $La_2(SO_4)_3\cdot 9\,H_2O$ = 728.14 $\text{g}\cdot\text{mol}^{-1}$

$$n_{CN^-} = \left(\frac{5.00 \text{ g } La_2(SO_4)_3\cdot 9\,H_2O}{728.14 \text{ g}\cdot\text{mol}^{-1} \text{ } La_2(SO_4)_3\cdot 9\,H_2O}\right)\left(\frac{9 \text{ mol } H_2O}{1 \text{ mol } La_2(SO_4)_3\cdot 9\,H_2O}\right)$$

$$= 6.18\times10^{-2} \text{ mol}$$

E.18 (a) molar mass of CaH_2 = $42.10\ g\cdot mol^{-1}$

$$\text{number of formula units} = \left(\frac{5.294\ g}{42.10\ g\cdot mol^{-1}}\right)(6.022\times10^{23}\ \text{formula units}\cdot mol^{-1})$$

$= 7.572\times10^{22}$ formula units

(b) molar mass of $NaBF_4 = 109.80\ g\cdot mol^{-1}$

$$\text{mass} = \left(\frac{6.25\times10^{24}\ \text{formula units}}{6.022\times10^{23}\ \text{formula units}\cdot mol^{-1}}\right)(109.80\ g\cdot mol^{-1}) = 1.14\times10^{3}\ g$$

$$\text{(c) moles of } CeI_3 = \left(\frac{9.54\times10^{21}\ \text{g formula units}}{6.022\times10^{23}\ \text{formula units}\cdot mol^{-1}}\right) = 1.58\times10^{-2}\ mol$$

E.20 (a) molar mass of $C_8H_8 = 114.22\ g\cdot mol^{-1}$

$$\text{mass of one octane molecule} = \frac{114.22\ g\cdot mol^{-1}}{6.022\times10^{23}\ \text{molecules}\cdot mol^{-1}}$$

$$= 1.897\times10^{-22}\ g\cdot \text{molecule}^{-1}$$

$$\text{(b)}\ N_{C_8H_8} = \frac{0.82\ g}{1.897\times10^{-22}\ g\cdot \text{molecule}^{-1}}$$

$= 4.3\times10^{21}$ molecules

E.22

$$?\ g\ CuSO_4 = 100.\ g\ CuSO_4\cdot 5\ H_2O\left(\frac{1\ mol\ CuSO_4\cdot 5\ H_2O}{249.70\ g\ CuSO_4\cdot 5\ H_2O}\right)$$

$$\times\left(\frac{1\ mol\ CuSO_4}{1\ mol\ CuSO_4\cdot 5\ H_2O}\right)\left(\frac{159.62\ g\ CuSO_4}{1\ mol\ CuSO_4}\right)$$

$= 63.9\ g\ CuSO_4$

E.24 molar mass $AuCl_3\cdot 2\ H_2O = 339.35\ g\cdot mol^{-1}$

$$\text{mass Au} = 35.35 \text{ g AuCl}_3 \cdot 2\text{ H}_2\text{O}\left(\frac{1 \text{ mol AuCl}_3 \cdot 2\text{ H}_2\text{O}}{339.35 \text{ g AuCl}_3 \cdot 2\text{ H}_2\text{O}}\right)$$

$$\times\left(\frac{1 \text{ mol Au}}{1 \text{ mol AuCl}_3 \cdot 2\text{ H}_2\text{O}}\right)\left(\frac{196.67 \text{ g Au}}{1 \text{ mol Au}}\right)$$

$$= 20.46 \text{ g}$$

E.26 $? \text{ atoms} = 25.5 \text{ g chloroform}\left(\frac{1.000 \text{ cm}^3}{1.492 \text{ g}}\right)\left(\frac{0.250 \text{ mol}}{20.0 \text{ mL}}\right)$

$$\times\left(\frac{6.022\times10^{23}\text{ molecules}}{1 \text{ mol}}\right)\left(\frac{5 \text{ atoms}}{\text{molecule}}\right)$$

$$= 6.43\times10^{23}\text{ atoms}$$

$$? \text{ g } C_2H_4Cl_2 = 6.43\times10^{23}\text{ atoms}\left(\frac{1 \text{ molecule}}{8 \text{ atoms}}\right)$$

$$\times\left(\frac{1 \text{ mol}}{6.022\times10^{23}\text{ molecules}}\right)\left(\frac{98.95 \text{ g}}{1 \text{ mol}}\right)$$

$$= 13.2 \text{ g } C_2H_4Cl_2$$

E.28 $? \text{ Si atoms} = (28.0000... \text{ g } ^{28}\text{Si})\left(\frac{1 \text{ atom}}{4.64567\times10^{-23}\text{g}}\right)$

$$= 6.027118\times10^{23} \text{ atoms} = 1 \text{ mole}$$

Since there are more atoms per mole, each mole will weigh more and all of the molar masses will be proportionately higher

by $\dfrac{6.02712\times10^{23} - 6.02214\times10^{23}}{6.02214\times10^{23}}\times100 = 0.0827\%$.

(a) $^{12}\text{C} \rightarrow 12.0000... \text{ g}\cdot\text{mol}^{-1}\times1.000827 = 12.0099 \text{ g}\cdot\text{mol}^{-1}$

(b) $\text{Cl} \rightarrow 35.45 \text{ g}\cdot\text{mol}^{-1}\times1.000827 = 35.48 \text{ g}\cdot\text{mol}^{-1}$

E.30 If x_1, x_2, and x_3 represent percentage composition of each of three isotopes having m_1, m_2, and m_3 respectively, then $x_1+x_2+x_3 = L = 100$ and $m_{mean} = \dfrac{x_1m_1+x_2m_2+x_3m_3}{100}$. The composition triangle given in the problem can be mapped onto a corresponding mass triangle such that each apex represents a composition of 100% of one of the isotopes and 0% of the other two. The central point inside the equilateral triangle represents a composition consisting of equal parts of each isotope, or $33.\overline{33}$ % each. Since the composition can vary over "the whole range of possibilities," a plot marking all of these composition, or mass, points would cover the whole triangle.

F.2 molar mass of $C_{14}H_{18}N_2O_5 = 294.30\ g \cdot mol^{-1}$

$$\%C = \frac{14 \times 12.01\ g \cdot mol^{-1}}{294.30\ g \cdot mol^{-1}} \times 100 = 57.13\%$$

$$\%H = \frac{18 \times 1.0079\ g \cdot mol^{-1}}{294.30\ g \cdot mol^{-1}} \times 100 = 6.1645\%$$

$$\%N = \frac{2 \times 14.01\ g \cdot mol^{-1}}{294.30\ g \cdot mol^{-1}} \times 100 = 9.520\%$$

$$\%O = \frac{5 \times 16.00\ g \cdot mol^{-1}}{294.30\ g \cdot mol^{-1}} \times 100 = 27.18\%$$

F.4 (a) M_2O_3 69.9% M For 100 g of compound, 69.9 g is M, 30.1 g is O.

$$\frac{?\ g\ M_2O_3}{mole\ M_2O_3} = \left(\frac{100\ g\ M_2O_3}{30.1\ g\ O}\right)\left(\frac{16.00\ g\ O}{1\ mol\ O}\right)\left(\frac{3\ mol\ O}{1\ mol\ M_2O_3}\right)$$

$$= 159\ g \cdot mol^{-1}\ M_2O_3$$

Therefore, 159 − 48 = 111 g/mol are due to M. Since there are 2 moles of M per mole of M_2O_3, the molar mass of M = 55.5 g/mol. That molar mass matches Fe. (b) iron (III) oxide

F.6 (a) For 100 g of talc,

$$\text{moles of Mg} = \frac{19.2\ g}{24.31\ g \cdot mol^{-1}} = 0.790\ mol$$

$$\text{moles of Si} = \frac{29.6 \text{ g}}{28.09 \text{ g} \cdot \text{mol}^{-1}} = 1.05 \text{ mol}$$

$$\text{moles of O} = \frac{42.2 \text{ g}}{16.00 \text{ g} \cdot \text{mol}^{-1}} = 2.64 \text{ mol}$$

$$\text{moles of H} = \frac{9.0 \text{ g}}{1.0079 \text{ g} \cdot \text{mol}^{-1}} = 8.9 \text{ mol}$$

Dividing each number by 0.790 mol gives a ratio of 1 Mg : 1.33 Si : 3.34 O : 11.3 H. Multiplying by 3 in order to get appropriate whole numbers gives the formula $Mg_3Si_4O_{10}H_{34}$.

(b) For 100 g of saccharin,

$$\text{moles of C} = \frac{45.89 \text{ g}}{12.01 \text{ g} \cdot \text{mol}^{-1}} = 3.821 \text{ mol}$$

$$\text{moles of H} = \frac{2.75 \text{ g}}{1.0079 \text{ g} \cdot \text{mol}^{-1}} = 2.73 \text{ mol}$$

$$\text{moles of N} = \frac{7.65 \text{ g}}{14.01 \text{ g} \cdot \text{mol}^{-1}} = 0.546 \text{ mol}$$

$$\text{moles of O} = \frac{26.20 \text{ g}}{16.00 \text{ g} \cdot \text{mol}^{-1}} = 1.638 \text{ mol}$$

$$\text{moles of S} = \frac{17.50 \text{ g}}{32.06 \text{ g} \cdot \text{mol}^{-1}} = 0.546 \text{ mol}$$

Dividing by 0.546 mol gives a ratio of 7.00 C : 5.00 H : 1.00 N : 3.00 O : 1.00 S. The empirical formula is $C_7H_5NO_3S$.

(c) For 100 g of salicylic acid,

$$\text{moles of C} = \frac{60.87 \text{ g}}{12.01 \text{ g} \cdot \text{mol}^{-1}} = 5.068 \text{ mol}$$

$$\text{moles of H} = \frac{4.38 \text{ g}}{1.0079 \text{ g} \cdot \text{mol}^{-1}} = 4.35 \text{ mol}$$

$$\text{moles of O} = \frac{34.75 \text{ g}}{16.00 \text{ g} \cdot \text{mol}^{-1}} = 2.172 \text{ mol}$$

Dividing by 2.172 mole gives a ratio of 2.333 C : 2.00 H : 1.00 O. Multiplying by 3 gives the empirical formula $C_7H_6O_3$.

F.8 moles of S $= \dfrac{4.69\ \text{g}}{32.06\ \text{g}\cdot\text{mol}^{-1}} = 0.146\ \text{mol}$

moles of F $= \dfrac{15.81\ \text{g} - 4.69\ \text{g}}{19.00\ \text{g}\cdot\text{mol}^{-1}} = 0.585\ \text{mol}$

Dividing by 0.146 mol gives a ratio of 4 F : 1 S. The formula is SF_4.

F.10 For 100 g of compound,

moles of C $= \dfrac{64.56\ \text{g}}{12.01\ \text{g}\cdot\text{mol}^{-1}} = 5.376\ \text{mol}$

moles of H $= \dfrac{5.42\ \text{g}}{1.0079\ \text{g}\cdot\text{mol}^{-1}} = 5.38\ \text{mol}$

moles of Fe $= \dfrac{30.02\ \text{g}}{55.85\ \text{g}\cdot\text{mol}^{-1}} = 0.5375\ \text{mol}$

Dividing by 0.5375 mol gives 10.00 C : 10.00 H : 1.00 Fe. The empirical formula is $FeC_{10}H_{10}$ with a molar mass of $186.03\ \text{g}\cdot\text{mol}^{-1}$.

F.12 For 100 g of Paclitaxel,

moles of C $= \dfrac{66.11\ \text{g}}{12.01\ \text{g}\cdot\text{mol}^{-1}} = 5.504\ \text{mol}$

moles of H $= \dfrac{6.02\ \text{g}}{1.0079\ \text{g}\cdot\text{mol}^{-1}} = 5.97\ \text{mol}$

moles of N $= \dfrac{1.64\ \text{g}}{14.01\ \text{g}\cdot\text{mol}^{-1}} = 0.117\ \text{mol}$

mass of O = 100 − 66.11 − 6.02 − 1.64 = 26.23 g

moles of O $= \dfrac{26.23\ \text{g}}{16.00\ \text{g}\cdot\text{mol}^{-1}} = 1.639\ \text{mol}$

Dividing by 0.117 mol gives a ratio of 47.0 C : 51.0 H : 1.00 N : 14.0 O. The empirical formula of Paclitaxel is $C_{47}H_{51}NO_{14}$.

F.14 For 1 mol of cacodyl,

$$\text{moles of C} = \frac{(0.2288)(209.96\text{ g})}{12.01\text{ g}\cdot\text{mol}^{-1}} = 4.00\text{ mol}$$

$$\text{moles of H} = \frac{(0.0576)(209.96\text{ g})}{1.0079\text{ g}\cdot\text{mol}^{-1}} = 12.00\text{ mol}$$

$$\text{moles of As} = \frac{(0.7136)(209.96\text{ g})}{74.92\text{ g}\cdot\text{mol}^{-1}} = 2.00\text{ mol}$$

The molecular formula is $C_4H_{12}As_2$.

F.16 For the 1.78 mg sample of didemnin-A,

$$\text{mmoles of C} = \frac{1.11\text{ mg}}{12.01\text{ g}\cdot\text{mol}^{-1}} = 0.0924\text{ mmol}$$

$$\text{mmoles of H} = \frac{0.148\text{ mg}}{1.0079\text{ g}\cdot\text{mol}^{-1}} = 0.147\text{ mmol}$$

$$\text{mmoles of N} = \frac{0.159\text{ mg}}{14.01\text{ g}\cdot\text{mol}^{-1}} = 0.0113\text{ mmol}$$

$$\text{mmoles of O} = \frac{0.363\text{ mg}}{16.00\text{ g}\cdot\text{mol}^{-1}} = 0.0227\text{ mmol}$$

Dividing by 0.0113 mmol gives 8.18 C : 13.0 H : 1.00 N : 2.00 O.

Multiplying by 7 gives the formula $C_{49}H_{78}N_6O_{12}$ with a molar mass of $942\ \text{g}\cdot\text{mol}^{-1}$.

F.18 We can write the formula for dolomite as $(Mg_xCa_y)CO_3$. Then

$$(Mg_xCa_y)CO_3(s) \rightarrow x\ MgO(s) + y\ CaO(s) + CO_2(g)$$

mass of carbon dioxide lost = 9.66 g – 4.84 g = 4.82 g

$$\text{moles } CO_2(g) \text{ lost} = 4.82\text{ g}\left(\frac{1\text{ mol}}{44.01\text{ g}}\right) = 0.110\text{ mol, so } x + y = 0.110.$$

It must also be true that the sum of the masses of the oxides add up to 4.84 g.

$$x \text{ mol MgO}\left(\frac{40.30 \text{ g MgO}}{1 \text{ mol MgO}}\right) + y \text{ mol CaO}\left(\frac{56.08 \text{ g CaO}}{1 \text{ mol CaO}}\right)$$

$= 4.84$ g mixed oxides

Rearranging $x + y = 0.110$ gives $x = 0.110 - y$. Substituting for x,

$$(0.110 - y)\ (40.30) + y\ (56.08) = 4.84$$

$$15.78y = 0.41$$

$$y = 0.026$$

and

$$x = 0.110 - 0.026 = 0.084$$

Then the original sample contained

$$\%\text{ mass MgCO}_3 = \left(\frac{0.084 \text{ mol MgCO}_3}{9.66 \text{ g sample}}\right)\left(\frac{84.31 \text{ g MgCO}_3}{1 \text{ mol MgCO}_3}\right) \times 100$$

$$= 73\%$$

G.2 Distillation could be used to separate water from dissolved salts.

G.4 (a) heterogeneous, decanting; (b) homogeneous, distillation;
(c) heterogeneous, dissolving followed by filtration and distillation

G.6 (a) molarity of $NaNO_3 = \dfrac{2.345 \text{ g}}{(85.00 \text{ g}\cdot\text{mol}^{-1})(0.200 \text{ L})} = 0.1379$ M $NaNO_3$

(b) molarity of $NaNO_3 = \dfrac{2.345 \text{ g}}{(85.00 \text{ g}\cdot\text{mol}^{-1})(0.250 \text{ L})} = 0.1104$ M $NaNO_3$

G.8 molarity of $Na_2CO_3 = \dfrac{7.112 \text{ g}}{(105.99 \text{ g}\cdot\text{mol}^{-1})(0.2500 \text{ L})} = 0.2684$ M Na_2CO_3

(a) $V = \dfrac{5.112\times 10^{-3} \text{ mol Na}_2\text{CO}_3}{0.2684 \text{ mol Na}_2\text{CO}_3\cdot\text{L}^{-1}} = 1.905\times 10^{-2} \text{ L} = 19.05 \text{ mL}$

(b) $V = \dfrac{3.451\times 10^{-3} \text{ mol CO}_3^{2-}}{0.2684 \text{ mol CO}_3^{2-}\cdot\text{L}^{-1}} = 1.286\times 10^{-2} \text{ L} = 12.86 \text{ mL}$

G.10 (a) $\text{molarity} = \dfrac{(1.345\ \text{mol}\cdot\text{L}^{-1})(0.012\ 56\ \text{L})}{(0.2500\ \text{L})} = 0.067\ 57\ \text{mol}\cdot\text{L}^{-1}$

(b) $\text{molarity} = \dfrac{(0.366\ \text{mol}\cdot\text{L}^{-1})(0.025\ 00\ \text{L})}{(0.125\ 00\ \text{L})} = 0.0732\ \text{mol}\cdot\text{L}^{-1}$

G.12 (a) $\text{moles transferred to the second flask} = \dfrac{(0.094\ \text{g})(2.00\ \text{mL})}{(249.68\ \text{g}\cdot\text{mol}^{-1})(500\ \text{mL})}$

$= 1.5\times10^{-6}\ \text{mol}$

$\text{new concentration} = \dfrac{1.5\times10^{-6}\ \text{mol}}{0.5000\ \text{L}} = 3.0\times10^{-6}\ \text{mol}\cdot\text{L}^{-1}$

(b) $\text{mass} = \left(249.68\ \text{g}\cdot\text{mol}^{-1}\right)(1.5\times10^{-6}\ \text{mol})$

$= 3.7\times10^{-4}\ \text{g or } 0.37\ \text{mg}$

G.14 (a) $\text{volume of } 15.0\ \text{M NH}_3 \text{ that must be diluted} = \dfrac{(1.25\ \text{mol}\cdot\text{L}^{-1})(500\ \text{mL})}{(15.0\ \text{mol}\cdot\text{L}^{-1})}$

$= 41.7\ \text{mL}$

(b) $\text{volume of } 15.0\ \text{M NH}_3 \text{ needed} = \dfrac{(0.32\ \text{mol}\cdot\text{L}^{-1})(15.0\ \text{L})}{(15.0\ \text{mol}\cdot\text{L}^{-1})}$

$= 0.32\ \text{L or } 3.2\times10^{2}\ \text{mL}$

G.16 (a) $\text{molarity of the first solution} = \dfrac{(0.661\ \text{g})}{(294.20\ \text{g}\cdot\text{mol}^{-1})\ (0.2500\ \text{L})}$

$= 8.99\times10^{-3}\ \text{mol}\cdot\text{L}^{-1}\ \text{K}_2\text{Cr}_2\text{O}_7(\text{aq})$

$\text{molarity of the second solution} = \dfrac{(8.99\times10^{-3}\ \text{mol}\cdot\text{L}^{-1})(1\times10^{-3}\ \text{L})}{(0.500\ \text{L})}$

$= 1.80\times10^{-5}\ \text{mol}\cdot\text{L}^{-1}\ \text{K}_2\text{Cr}_2\text{O}_7(\text{aq})$

$\text{molarity of the final solution} = \dfrac{(1.80\times10^{-5}\ \text{mol}\cdot\text{L}^{-1})(0.010\ \text{L})}{(0.250\ \text{L})}$

$= 7.20\times10^{-7}\ \text{mol}\cdot\text{L}^{-1}\ \text{K}_2\text{Cr}_2\text{O}_7(\text{aq})$

(b) $\text{mass} = (7.20\times10^{-7}\ \text{mol}\cdot\text{L}^{-1})(0.250\ \text{L})(294.2\ \text{g}\cdot\text{mol}^{-1}) = 5.3\times10^{-5}\ \text{g}$

G.18 (a) mass of NaOH = $(5.0\ \text{mol}\cdot\text{L}^{-1})(0.0750\ \text{L})(40.00\ \text{g}\cdot\text{mol}^{-1}) = 15\ \text{g}$

Add 15 g of NaOH to a 75.0-mL volumetric flask and dilute to the line.

(b) mass of $BaCl_2 = (0.21\ \text{mol}\cdot\text{L}^{-1})(5.0\ \text{L})(208.24\ \text{g}\cdot\text{mol}^{-1}) = 2.2\times10^2\ \text{g}$

Add 220 g of $BaCl_2$ to a 5.0-L flask and fill with water to the line.

(c)
$$\begin{aligned}\text{mass of } AgNO_3 &= (0.0340\ \text{mol}\cdot\text{L}^{-1})(0.300\ \text{L})(169.88\ \text{g}\cdot\text{mol}^{-1})\\ &= 1.73\ \text{g}\end{aligned}$$

Add 1.73 g of $AgNO_3$ to a 300-mL volumetric flask and fill with water to the line.

G.20 $10\ \text{mL} \times \dfrac{0.10\ \text{mol}}{1000\ \text{mL}} \times \dfrac{6.02214\times10^{23}\text{molecules}}{1\ \text{mol}} = 6.0\times10^{20}$ molecules are in the first 10 mL aliquot of the solution. The number of successive ten-fold dilutions, *n*, required to get down to just one molecule can be found from this relationship:

$$\begin{aligned}6.0\times10^{20}\text{molecules}\times(0.1)^n &= 1\ \text{molecule}\\ \log(6.0\times10^{20}) + n\log(0.1) &= \log(1)\\ 20.78 + n(-1) &= 0\\ n &= 20.78\end{aligned}$$

So after 20 such dilutions, less than 10 molecules are left. The last dilution should be by a factor of $(0.1)^{0.78} = 0.17 > 0.1$ in order to try to ensure that one molecule will still remain.

H.2 (a) $2\ AgNO_3(s) \rightarrow 2\ Ag(s) + 2\ NO_2(g) + O_2(g)$

(b) $P_2S_5(s) + 3\ PCl_5(s) \rightarrow 5\ PSCl_3(g)$

(c) $2\ BF_3(g) + 6\ NaH(s) \rightarrow B_2H_6(g) + 6\ NaF(s)$

(d) $2\ LnC_2(s) + 6\ H_2O(l) \rightarrow 2\ Ln(OH)_3(s) + 2\ C_2H_2(g) + H_2(g)$

H.4 (a) $2\ NiS(s) + 3\ O_2(g) \rightarrow 2\ NiO(s) + 2\ SO_2(g)$

(b) $SiO_2(s) + 3\ C(s) \xrightarrow{\Delta} SiC(s) + 2\ CO(g)$

(c) $3\ H_2(g) + N_2(g) \rightarrow 2\ NH_3(g)$

(d) $3\ Mg(s) + B_2O_3(s) \rightarrow 2\ B(s) + 3\ MgO(s)$

H.6 $2\ O_3(g) \rightarrow 3\ O_2(g)$

H.8 $4\ BF_3(g) + 3\ NaBH_4(s) \rightarrow 3\ NaBF_4(s) + 2\ B_2H_6(g)$

$B_2H_6(g) + 3\ O_2(g) \rightarrow B_2O_3(s) + 3\ H_2O(l)$

H.10 $5\ Sb_2O_3(s) + 2\ SbCl_3(s) \rightarrow 3\ Sb_4O_5Cl_2(s)$

H.12 $C_{14}H_{18}N_2O_5(s) + 16\ O_2(g) \rightarrow 14\ CO_2(g) + 9\ H_2O(l) + N_2(g)$

H.14 $C_9H_8O_4(s) + 9\ O_2(g) \rightarrow 9\ CO_2(g) + 4\ H_2O(l)$

H.16 $4\ NH_3(g) + 5\ O_2(g) \rightarrow 4\ NO(g) + 6\ H_2O(l)$

$2\ NO(g) + O_2(g) \rightarrow 2\ NO_2(g)$

$3\ NO_2(g) + H_2O(l) \rightarrow 2\ HNO_3(aq) + NO(g)$

H.18 (a), (b), and (d) are conserved overall in any chemical reaction.

I.2 (a) strong electrolyte; (b) strong electrolyte; (c) weak electrolyte

I.4 (a) soluble; (b) insoluble; (c) insoluble; (d) soluble

I.6 (a) $PbSO_4(s)$ is insoluble but to the extent that it does dissolve the species are $Pb^{2+}(aq)$ and $SO_4^{2-}(aq)$; (b) $K^+, CO_3{}^{2-}$; (c) $K^+, CrO_4{}^{2-}$;

(d) $Hg_2Cl_2(s)$ is insoluble but to the extent that it does dissolve the species are $Hg_2^{2+}(aq)$ and $Cl^-(aq)$.

I.8 (a) $2\ NH_4NO_3(aq) + CaCl_2(aq) \rightarrow 2\ NH_4Cl(aq) + Ca(NO_3)_2(aq)$

No precipitate is expected.

(b) $MgCO_3(s)$ is insoluble. It will remain undissolved and the $NaNO_3$ will dissolve.

(c) $Na_2SO_4(aq) + BaCl_2(aq) \rightarrow 2\ NaCl(aq) + BaSO_4(s)$

Barium sulfate, $BaSO_4$, will precipitate.

I.10 (a) No reaction.

(b) $2\ H_3PO_4(aq) + 2\ CaCl_2(aq) \rightarrow Ca_3(PO_4)_2(s) + 6\ HCl(aq)$

net ionic equation: $3\ Ca^{2+}(aq) + 2\ PO_4^{3-}(aq) \rightarrow Ca_3(PO_4)_2(s)$

spectator ions: H^+, Cl^-

(c) No reaction.

(d) $NiSO_4(aq) + (NH_4)_2CO_3(aq) \rightarrow NiCO_3(s) + (NH_4)_2SO_4(aq)$

net ionic equation: $Ni^{2+}(aq) + CO_3^{2-}(aq) \rightarrow NiCO_3(s)$

spectator ions: NH_4^+, SO_4^{2-}

(e) $2\ HNO_3(aq) + Ba(OH)_2(aq) \rightarrow Ba(NO_3)_2(aq) + 2\ H_2O(l)$

net ionic equation: $H^+(aq) + OH^-(aq) \rightarrow H_2O(l)$

spectator ions: Ba^{2+}, NO_3^-

I.12 (a) overall equation:

$2\ AgNO_3(aq) + Na_2CO_3(aq) \rightarrow Ag_2CO_3(s) + 2\ NaNO_3(aq)$

complete ionic equation:

$$2\ Ag^+(aq) + 2\ NO_3^-(aq) + 2\ Na^+(aq) + CO_3^{2-}(aq) \rightarrow Ag_2CO_3(s) + 2\ NO_3^-(aq) + 2\ Na^+(aq)$$

net ionic equation: $2\ Ag^+(aq) + CO_3^{2-}(aq) \rightarrow Ag_2CO_3(s)$

spectator ions: Na^+, NO_3^-

(b) overall equation: $Pb(NO_3)_2(aq) + 2\ KI(aq) \rightarrow PbI_2(s) + 2\ KNO_3(aq)$

complete ionic equation:

$Pb^{2+}(aq) + 2\ NO_3^-(aq) + 2\ K^+(aq) + 2\ I^-(aq) \rightarrow PbI_2(s) + 2\ NO_3^-(aq) + 2\ K^+(aq)$

net ionic equation: $Pb^{2+}(aq) + 2\ I^-(aq) \rightarrow PbI_2(s)$

spectator ions: K^+, NO_3^-

(c) overall equation: $Ba(OH)_2(aq) + H_2SO_4(aq) \rightarrow BaSO_4(s) + H_2O(l)$

complete ionic equation:

$Ba^{2+}(aq) + 2\ OH^-(aq) + 2\ H^+(aq) + SO_4^{2-}(aq) \rightarrow BaSO_4(s) + 2\ H_2O(l)$

net ionic equations: $Ba^{2+}(aq) + SO_4^{2-}(aq) \rightarrow BaSO_4(s)$

and $OH^-(aq) + H^+(aq) \rightarrow H_2O(l)$

no spectator ions

(d) overall equation:

$(NH_4)_2S(aq) + Cd(NO_3)_2(aq) \rightarrow 2\ NH_4NO_3(aq) + CdS(s)$

complete ionic equation:

$2\ NH_4^+(aq) + S^{2-}(aq) + Cd^{2+}(aq) + 2\ NO_3^-(aq)$

$\rightarrow 2\ NH_4^+(aq) + 2\ NO_3^-(aq) + CdS(s)$

net ionic equation: $Cd^{2+}(aq) + S^{2-}(aq) \rightarrow CdS(s)$

spectator ions: NH_4^+, NO_3^-

(e) overall equation:

$2\ KOH(aq) + CuCl_2(aq) \rightarrow Cu(OH)_2(s) + 2\ KCl(aq)$

complete ionic equation:

$2\ K^+(aq) + 2\ OH^-(aq) + Cu^{2+}(aq) + 2\ Cl^-(aq)$

$\rightarrow Cu(OH)_2(s) + 2\ K^+(aq) + 2\ Cl^-(aq)$

net ionic equation: $Cu^{2+}(aq) + 2\ OH^-(aq) \rightarrow Cu(OH)_2(s)$

spectator ions: K^+, Cl^-

I.14 (a) $3\ Pb^{2+}(aq) + 6\ NO_3^-(aq) + 6\ K^+(aq) + 2\ PO_4^{3-}(aq) \rightarrow$

$Pb_3(PO_4)_2(s) + 6\ K^+(aq) + 6\ NO_3^-(aq)$

net ionic equation: $3\ Pb^{2+}(aq) + 2\ PO_4^{3-}(aq) \rightarrow Pb_3(PO_4)_2(s)$

(b) $6\ K^+(aq) + 3\ S^{2-}(aq) + 2\ Bi^{3+}(aq) + 6\ NO_3^-(aq) \rightarrow$

$Bi_2S_3(s) + 6\ K^+(aq) + 6\ NO_3^-(aq)$

net ionic equation: $2\ Bi^{3+}(aq) + 3\ S^{2-}(aq) \rightarrow Bi_2S_3(s)$

(c) $2\ Na^+(aq) + C_2O_4^{2-}(aq) + Mn^{2+}(aq) + 2\ CH_3CO_2^-(aq) \rightarrow$

$MnC_2O_4(s) + 2\ Na^+(aq) + 2\ CH_3CO_2^-(aq)$

net ionic equation: $Mn^{2+}(aq) + C_2O_4^{2-}(aq) \rightarrow MnC_2O_4(s)$

I.16 (a) $AgNO_3(aq)$ and $Na_2SO_4(aq)$; (b) $MgSO_4(aq)$ and $KOH(aq)$;

(c) $Ca(NO_3)_2(aq)$ and $(NH_4)_2PO_4(aq)$

I.18 (a) $Pb^{2+}(aq) + CrO_4^{2-}(aq) \rightarrow PbCrO_4(s)$

(b) $Al^{3+}(aq) + PO_4^{3-}(aq) \rightarrow AlPO_4(s)$

(c) $Fe^{2+}(aq) + 2\ OH^-(aq) \rightarrow Fe(OH)_2(s)$

(d) $Pb(NO_3)_2$ and K_2CrO_4; K^+, NO_3^-

$Al(NO_3)_3$ and Li_3PO_4; Li^+, NO_3^-

$FeSO_4$ and NaOH; Na^+, SO_4^{2-}

I.20 No precipitate with Cl^- indicates no Ag^+. The white precipitate with SO_4^{2-} is $CaSO_4$, so Ca^{2+} is present. The black precipitate with the supernatant is HgS, indicating the presence of Hg^{2+}. Therefore we would report Ca^{2+} and Hg^{2+} as being present in the solution.

I.22 (a) $(NH_4)_3PO_4(aq) + 3\ CaCl_2(aq) \rightarrow Ca_3(PO_4)_2(s) + 6\ NH_4Cl(aq)$

net ionic equation: $3\ Ca^{2+} + 2\ PO_4^{3-} \rightarrow Ca_3(PO_4)_2(s)$

(b) $$\frac{?\text{ mol NH}_4^+}{1\text{ L solution}} = \left(\frac{25.0\text{ g (NH}_4)_3\text{PO}_4}{70.0\text{ mL solution}}\right)\left(\frac{1\text{ mol (NH}_4)_3\text{PO}_4}{149.095\text{ g (NH}_4)_3\text{PO}_4}\right)$$
$$\times\left(\frac{3\text{ mol NH}_4^+}{1\text{ mol (NH}_4)_3\text{PO}_4}\right)\left(\frac{1000\text{ mL}}{1\text{ L}}\right)$$
$$= 7.19\text{ M NH}_4^+$$

$$\frac{?\text{ mol Cl}^-}{1\text{ L solution}} = \left(\frac{50.0\text{ mL CaCl}_2\text{(aq)}}{70.0\text{ mL solution}}\right)\left(\frac{0.125\text{ mol CaCl}_2}{1000\text{ mL CaCl}_2\text{(aq)}}\right)$$
$$\times\left(\frac{2\text{ mol Cl}^-}{1\text{ mol CaCl}_2}\right)\left(\frac{1000\text{ mL solution}}{1\text{ L solution}}\right)$$
$$= 0.179\text{ M Cl}^-$$

I.24 $?\text{ mol CuI} = 15.75\text{ g CuI}\left(\frac{1\text{ mol CuI}}{190.45\text{ g CuI}}\right) = 0.08270\text{ mol CuI}$

Since the reaction was limited by the $CuNO_3$ and the stoichiometry is 1:1, there was an equivalent number of moles of $CuNO_3$ in the original reactant solution, or

$$\left(\frac{0.08270\text{ mol CuNO}_3}{50.0\text{ mL CuNO}_3\text{(aq)}}\right)\left(\frac{1000\text{ mL}}{1\text{ L}}\right) = 1.654\text{ M CuNO}_3\text{(aq)}$$

J.2 (a) acid; (b) base; (c) acid; (d) base; (e) acid

J.4 (a) overall equation: $H_3PO_4(aq) + 3\ KOH(aq) \rightarrow K_3PO_4(aq) + 3\ H_2O(l)$

total ionic equation:

$$3\ H^+(aq) + PO_4^{3-}(aq) + 3\ K^+(aq) + 3\ OH^-(aq)$$
$$\rightarrow 3\ K^+(aq) + PO_4^{3-}(aq) + 3\ H_2O(l)$$

net ionic equation: $3\ H^+(aq) + 3\ OH^-(aq) \rightarrow 3\ H_2O(l)$

(b) overall equation:

$$Ba(OH)_2(aq) + CH_3CO_2H(aq) \rightarrow Ba(CH_3CO_2)_2(aq) + 2\ H_2O(l)$$

total ionic equation:

$Ba^{2+}(aq) + 2\ OH^-(aq) + 2\ CH_3CO_2H(aq)$
$\rightarrow Ba^{2+}(aq) + 2\ CH_3CO_2^-(aq) + 2\ H_2O(l)$

net ionic equation:

$OH^-(aq) + CH_3CO_2H(aq) \rightarrow CH_3CO_2^-(aq) + H_2O(l)$

(c) overall equation:

$Mg(OH)_2(aq) + HClO_3(aq) \rightarrow Mg(ClO_3)_2(aq) + 2\ H_2O(l)$

total ionic equation:

$Mg^{2+}(aq) + 2\ OH^-(aq) + 2\ H^+(aq) + 2\ ClO_3^-(aq)$
$\rightarrow Mg^{2+}(aq) + 2\ ClO_3^-(aq) + 2\ H_2O(l)$

net ionic equation: $OH^-(aq) + H^+(aq) \rightarrow H_2O(l)$

J.6 (a) CH_3CO_2K, potassium acetate;

$CH_3CO_2H(aq) + K^+(aq) + OH^-(aq) \rightarrow K^+(aq) + CH_3CO_2^-(aq) + H_2O(l)$

(b) NH_4I, ammonium iodide;

$NH_3(aq) + H_3O^+(aq) + I^-(aq) \rightarrow NH_4^+(aq) + I^-(aq) + H_2O(l)$

(c) $BaSO_4$, barium sulfate;

$Ba^+(aq) + 2\ OH^-(aq) + H^+(aq) + HSO_4^-(aq) \rightarrow BaSO_4(s) + 2\ H_2O(l)$

(d) $NaCN$, sodium cyanide;

$HCN(aq) + Na^+(aq) + OH^-(aq) \rightarrow Na^+(aq) + CN^-(aq) + H_2O(l)$

J.8 (a) acid: $CH_3CO_2H(aq)$, base: $NH_3(aq)$; (b) acid: $HCl(aq)$, base: $(CH_3)_3N(aq)$; (c) acid: $H_2O(l)$, base: $O^{2-}(aq)$

J.10 (a) $43.64\text{ g P} \times \left(\dfrac{1\text{ mol P}}{30.97\text{ g P}}\right) = 1.409\text{ mol P}$

$$56.36\text{ g O} \times \left(\dfrac{1\text{ mol O}}{16.00\text{ g O}}\right) = 3.523\text{ mol O}$$

$$\frac{3.523}{1.409} = 2.5 = \text{mole ratio of O:P in compound A}$$

P_4O_{10} is the molecular formula with this stoichiometry that has the correct molar mass since $(4\times 30.97)+(10\times 16.00)=283.88\ \text{g}\cdot\text{mol}^{-1}$.

(b) There is a mistake in the information given in the problem. Hydrogen composes 3.087% of compound B (not 1.029%).

$$3.087\ \text{g H}\times\left(\frac{1\ \text{mol H}}{1.0079\ \text{g H}}\right)=3.063\ \text{mol H}$$

$$31.60\ \text{g P}\times\left(\frac{1\ \text{mol P}}{30.97\ \text{g P}}\right)=1.020\ \text{mol P}$$

$$65.31\ \text{g O}\times\left(\frac{1\ \text{mol O}}{16.00\ \text{g O}}\right)=4.082\ \text{mol O}$$

$$\frac{4.082}{1.020}=4.00=\text{mole ratio of O:P in compound B}$$

$$\frac{3.063}{1.020}=3.00=\text{mole ratio of H:P in compound B}$$

The molecular formula H_3PO_4 has the correct stoichiometry as well as the appropriate molar mass since

$(3\times 1.0079)+(1\times 30.97)+(4\times 16.00)=97.99\ \text{g}\cdot\text{mol}^{-1}$.

(c) for part (a): $P_4(s)+5\ O_5\rightarrow P_4O_{10}(s)$;

for part (b): $P_4O_{10}(s)+6\ H_2O(l)\rightarrow 4\ H_3PO_4(aq)$;

for part (c): $3\ Ca(OH)_2(aq)+2\ H_3PO_4(aq)\rightarrow Ca_3(PO_4)_2(aq)+6\ H_2O(l)$

J.12 (a) ? M

$$C_6H_5NH_3^+=\left(\frac{50.0\ \text{g}\ C_6H_5NH_3Cl}{150.0\ \text{mL}}\right)\left(\frac{1000\ \text{mL}}{\text{L}}\right)\left(\frac{1\ \text{mol}\ C_6H_5NH_3Cl}{129.45\ \text{g}\ C_6H_5NH_3Cl}\right)$$

$$=2.57\ \text{M}\ C_6H_5NH_3Cl=2.57\ \text{M}\ C_6H_5NH_3^+$$

(b) $C_6H_5NH_3^+ + H_2O(l)\rightarrow C_6H_5NH_2(aq)+H_3O^+$

$C_6H_5NH_3^+$	$H_2O(l)$	$C_6H_5NH_2(aq)$	H_3O^+
acid	base	conjugate base	conjugate acid

K.2 (a) $6\ Hg^{2+}(aq)+2\ Fe(s)\rightarrow 3\ Hg_2^{2+}(aq)+2\ Fe^{3+}(aq)$

(b) $Pt^{4+}(aq)+H_2(g)\rightarrow Pt^{2+}(aq)+2\ H^+(aq)$

(c) $2\ Al(s) + Fe_2O_3(s) \rightarrow 2\ Fe(s) + Al_2O_3(s)$

(d) $2\ La(s) + 3\ Br_2(l) \rightarrow 2\ LaBr_3(s)$

K.4 (a) +4; (b) +6; (c) +5; (d) +4; (e) +5; (f) +2

K.6 (a) +3; (b) +3; (c) +4; (d) +2; (e) +4

K.8 (a) $-\frac{1}{2}$; (b) -1; (c) -1; (d) -1; (e) $-\frac{1}{3}$

K.10 (a) $Cl_2(g)$ is reduced to $Cl^-(aq)$ so the oxidation number on Cl changes from 0 to 1−. $I^-(aq)$ is oxidized to $I_2(aq)$ with the oxidation number on I changing from 1− to 0. (b) $Cl_2(g)$ is both oxidized to OCl^- (oxidation number changes from 0 to 1+) and reduced to Cl^- (oxidation number changes from 0 to 1−). A reaction in which a single substance is both oxidized and reduced is known as a *disproportionation reaction.*
(c) The N atom in NO, which has a 2+ oxidation state, is oxidized up to 4+ in NO_2 while one O atom is reduced from 0 in O_3 to 2− in NO_2.

K.12 (a) $KBrO_3$ will be the stronger oxidizing agent. Br^{5+} will more readily accept electrons than Br^+. (b) MnO_4^- will be the better oxidizing agent because Mn^{7+} will more readily accept electrons than Mn^{2+}.

K.14 (a) Oxidizing agent: $Cr_2O_3(s)$; reducing agent: Al(s)

(b) Oxidizing agent: $N_2(g)$; reducing agent: Li(s)

(c) Oxidizing agent: $Ca_3(PO_4)_2(s)$; reducing agent: C(s)

K.16 $2\ NaCl(s) \rightarrow 2\ Na(s) + Cl_2(g)$

Na(s) is produced by reduction while $Cl_2(g)$ is produced by oxidation.

K.18 (a) redox reaction: oxidizing agent: $I_2O_5(s)$; reducing agent: CO(g)

(b) redox reaction: oxidizing agent: $I_2(aq)$; reducing agent: $S_2O_3^{2-}(aq)$

(c) precipitation reaction: $Ag^+(aq) + Br^-(aq) \rightarrow AgBr(s)$

(d) redox reaction: oxidizing agent: Mg(s); reducing agent: $UF_4(g)$

K.20

Group	IA	IIA	IIIA	IVA	VA	VIA	VIIA
	H +1,−1						
	Li +1	Be +2	B +3	C +4,-4	N +5,−3	O +2,−2	F −1
	Na +1	Mg +2	Al +3	Si +4,-4	P +5,−3	S +6,−2	Cl +7,−1
	K +1	Ca +2	Ga +3	Ge +4,-4	As +5,−3	Se +6,−2	Br +5,−1
	Rb +1	Sr +2	In +3	Sn +4,+2	Sb +5,−3	Te +6,−2	I +7,−1

Metals only exhibit positive maximum oxidation states while non-metals have both positive maximum states and negative minimum states. For the most part, the maximum oxidation number corresponds to the Group number, or the number of valence electrons. The minimum oxidation number for the non-metals corresponds to the number of electrons needed to complete the outermost electron shell.

K.22 (a) Silver goes from 0 as Ag to 1+ as Ag_2S, so it is oxidized when it is tarnished.

(b) $$? \text{ moles Ag} = 5.0 \text{ g } Ag_2S\left(\frac{1 \text{ mol } Ag_2S}{247.81 \text{ g } Ag_2S}\right)\left(\frac{2 \text{ mol Ag}}{1 \text{ mol } Ag_2S}\right)$$
$$= 0.040 \text{ mol Ag was oxidized}$$

(c) Since one mole of electrons is transferred for every Ag^+ formed, 0.040 moles of electrons were lost from the 0.040 moles of Ag atoms that were oxidized.

L.2 (a) $Ca_3(PO_4)_2(aq) + 3\ H_2SO_4(aq) \rightarrow 3\ CaSO_4(s) + 2\ H_3PO_4(aq)$

$$(200\text{ kg H}_2\text{SO}_4)\left(\frac{1000\text{ g}}{1\text{ kg}}\right)\left(\frac{1\text{ mol H}_2\text{SO}_4}{98.08\text{ g H}_2\text{SO}_4}\right)\left(\frac{2\text{ mol H}_3\text{PO}_4}{3\text{ mol H}_2\text{SO}_4}\right)$$
$$= 1.36\times10^3\text{ mol H}_3\text{PO}_4$$

(b) $$(200\text{ mol Ca}_3(\text{PO}_4)_2)\left(\frac{3\text{ mol CaSO}_4}{1\text{ mol Ca}_3(\text{PO}_4)_2}\right)\left(\frac{136.14\text{ g CaSO}_4}{1\text{ mol CaSO}_4}\right)$$
$$= 8.17\times10^4\text{ g CaSO}_4$$

L.4 (a) $B_2H_6(g) + 3\ O_2(l) \rightarrow 2\ HBO_2(g) + 2\ H_2O(l)$

$$(257\text{ g B}_2\text{H}_6)\left(\frac{1\text{ mol B}_2\text{H}_6}{27.67\text{ g B}_2\text{H}_6}\right)\left(\frac{3\text{ mol O}_2}{1\text{ mol B}_2\text{H}_6}\right)\left(\frac{32.00\text{ g O}_2}{1\text{ mol O}_2}\right) = 892\text{ g O}_2(l)$$

(b) $$(106\text{ g B}_2\text{H}_6)\left(\frac{1\text{ mol B}_2\text{H}_6}{27.67\text{ g B}_2\text{H}_6}\right)\left(\frac{2\text{ mol HBO}_2}{1\text{ mol B}_2\text{H}_6}\right)\left(\frac{43.82\text{ g HBO}_2}{1\text{ mol HBO}_2}\right)$$
$$= 336\text{ g HBO}_2(g)$$

L.6 (a) $4\ KO_2(s) + 2\ H_2O(l) \rightarrow 3\ O_2(g) + 4\ KOH(s)$

$$(115\text{ g O}_2)\left(\frac{1\text{ mol O}_2}{32.00\text{ g O}_2}\right)\left(\frac{4\text{ mol KO}_2}{3\text{ mol O}_2}\right)\left(\frac{71.10\text{ g KO}_2}{1\text{ mol KO}_2}\right) = 341\text{ g KO}_2$$

(b) $$(75.0\text{ g KO}_2)\left(\frac{1\text{ mol KO}_2}{71.10\text{ g KO}_2}\right)\left(\frac{4\text{ mol KOH}}{4\text{ mol KO}_2}\right)$$
$$\times\left(\frac{1\text{ mol CO}_2}{1\text{ mol KOH}}\right)\left(\frac{44.01\text{ g CO}_2}{1\text{ mol CO}_2}\right)$$
$$= 46.4\text{ g CO}_2$$

L.8 $CH_2O(s) + O_2(g) \rightarrow CO_2(g) + H_2O(l)$

volume of wood $= 12\times14\times25\text{ cm}^3 = 4.2\times10^3\text{cm}^3$

mass of wood $= (4.2\times10^3\text{cm}^3)\left(\frac{0.72\text{ g}}{1\text{ cm}^3}\right) = 3.0\times10^3\text{g oak}$

$$\text{moles of wood} = (3.0\times10^3 \text{g oak})\left(\frac{1 \text{ mol oak}}{30.03 \text{ g oak}}\right) = 1.0\times10^2 \text{ mol oak}$$

$$(1.0\times10^2 \text{ mol oak})\left(\frac{1 \text{ mol } H_2O}{1 \text{ mol } CH_2O}\right)\left(\frac{18.02 \text{ g } H_2O}{1 \text{ mol } H_2O}\right) = 1.8\times10^3 \text{g } H_2O$$

L.10 (a) $H_2C_2O_4(aq) + 2\ NaOH(aq) \rightarrow Na_2C_2O_4(aq) + 2\ H_2O(l)$

$$(25.67 \text{ mL NaOH(aq)})\left(\frac{0.327 \text{ mol NaOH}}{1000 \text{ mL NaOH(aq)}}\right)$$

$= 8.39\times10^{-3}$ mol NaOH used

$$(8.39\times10^{-3} \text{ mol NaOH})\left(\frac{1 \text{ mol } H_2C_2O_4}{2 \text{ mol NaOH}}\right)$$

$= 4.20\times10^{-3}$ mol $H_2C_2O_4$ neutralized

$$\left(\frac{4.20\times10^{-3} \text{mol } H_2C_2O_4}{0.035\ 25 \text{ L}}\right) = 0.119 \text{ mol}\cdot\text{L}^{-1}\ H_2C_2O_4$$

(b) $$(0.035\ 25 \text{ L})\left(\frac{0.119 \text{ mol}}{1 \text{ L}}\right)\left(\frac{90.04 \text{ g } H_2C_2O_4}{1 \text{ mol } H_2C_2O_4}\right) = 0.328 \text{ g } H_2C_2O_4$$

L.12 (a) $H_3PO_4(aq) + 3\ KOH(aq) \rightarrow K_3PO_4(aq) + 3\ H_2O(l)$

moles of KOH = $(0.100 \text{ L})(3.0 \text{ mol}\cdot\text{L}^{-1}) = 0.030$ mol

$$\text{concentration of diluted KOH} = \frac{0.030 \text{ mole}}{0.250 \text{ L}} = 0.12 \text{ mol}\cdot\text{L}^{-1}$$

moles of OH^- required $= (0.0385 \text{ L})(0.12 \text{ mol}\cdot\text{L}^{-1}) = 4.6\times10^{-3}$ mol

$$(0.0046 \text{ mol KOH})\left(\frac{1 \text{ mol } H_3PO_4}{3 \text{ mol KOH}}\right) = 0.0015 \text{ mol } H_3PO_4 \text{ neutralized}$$

$$\text{molarity of } H_3PO_4 = \frac{0.0015 \text{ mol}}{0.010 \text{ L}} = 0.15 \text{ mol}\cdot\text{L}^{-1}$$

(b) mass of

$H_3PO_4 = (0.0015 \text{ mol } H_3PO_4)(97.99 \text{ g}\cdot\text{mol}^{-1}) = 0.15 \text{ g } H_3PO_4$

L.14 $HX(aq) + NaOH(aq) \rightarrow NaX(aq) + H_2O(l)$

$(0.014\ 56\ \text{L})\ (0.115\ \text{mol} \cdot \text{L}^{-1}) = 0.001\ 67\ \text{mol NaOH and HX}$

$$\frac{0.2037\ \text{g}}{0.001\ 67\ \text{mol}} = 122\ \text{g} \cdot \text{mol}^{-1}$$

L.16 $\text{mol}\ I_3^- = (0.0101\ \text{L})(0.0521\ \text{mol} \cdot \text{L}^{-1}) = 5.26 \times 10^{-4}\ \text{mol}$

$$\left(\frac{5.26 \times 10^{-4}\ \text{mol}\ I_3^- \cdot}{10.0\ \text{mL solution}}\right)\left(\frac{1\ \text{mol vitamin C}}{1\ \text{mol}\ I_3^-}\right)\left(\frac{176\ \text{g vitamin C}}{1\ \text{mol vitamin C}}\right)$$

$$\times\left(\frac{100.00\ \text{mL solution}}{1\ \text{tablet}}\right) = 0.926\ \text{g vitamin C per tablet}$$

No, the manufacturer was not truthful.

L.18 Let x = number of moles of $CaCl_2 \cdot 2\ H_2O$ in the mixture. Then $2x$ moles of water lost upon heating = 2.543 g – 2.312 g = 0.231 g, or

$$2x = (0.231\ \text{g}\ H_2O)\left(\frac{1\ \text{mol}\ H_2O}{18.02\ \text{g}\ H_2O}\right)$$

$x = 6.41 \times 10^{-3}$ mols

number of grams of $CaCl_2$ in the dry

$$\text{mixture} = (6.41 \times 10^{-3}\ \text{mols})\left(\frac{110.98\ \text{g}\ CaCl_2}{1\ \text{mol}\ CaCl_2}\right) = 0.711\ \text{g}\ CaCl_2$$

Therefore, the original sample consisted of (0.711 + 0.231) g = 0.942 g $CaCl_2 \cdot 2\ H_2O$ and

(2.543 – 0.942)g = 1.601 g KCl.

$$\left(\frac{0.942\ \text{g}\ CaCl_2 \cdot 2\ H_2O}{2.543\ \text{g original mixture}}\right) \times 100 = 37.1\%\ CaCl_2 \cdot 2\ H_2O$$

$$\left(\frac{1.601\ \text{g KCl}}{2.543\ \text{g original mixture}}\right) \times 100 = 62.9\%\ \text{KCl}$$

L.20 $K_2CrO_4(aq) + 2\ AgNO_3(aq) \rightarrow Ag_2CrO_4(s) + 2\ KNO_3(aq)$

$$25.0\ \text{mL}\left(\frac{0.5\ \text{mol}\ K_2CrO_4}{1000\ \text{mL solution}}\right) = 0.0125\ \text{mol}\ K_2CrO_4$$

Since the reaction requires 2 moles of $AgNO_3$ for every one mole of K_2CrO_4, the concentration of the $AgNO_3$ solution is

$$\frac{(0.0125 \text{ mol } K_2CrO_4)\left(\frac{2 \text{ mol } AgNO_3}{1 \text{ mol } K_2CrO_4}\right)}{15.0 \text{ mL } AgNO_3(aq)}\left(\frac{1000 \text{ mL}}{1 \text{ L}}\right) = 1.67 \text{ M}$$

$$NaCl(aq) + AgNO_3(aq) \rightarrow AgCl(s) + NaNO_3(aq)$$

$$? \text{ g NaCl} = 35.0 \text{ mL } AgNO_3(aq)\left(\frac{1.67 \text{ mol } AgNO_3}{1000 \text{ mL } AgNO_3(aq)}\right)$$

$$\times\left(\frac{1 \text{ mol NaCl}}{1 \text{ mol } AgNO_3}\right)\left(\frac{58.44 \text{ g NaCl}}{1 \text{ mol NaCl}}\right)$$

$$= 3.42 \text{ g NaCl}$$

L.22 $CuSO_4 \cdot xH_2O \rightarrow CuSO_4 \cdot yH_2O + H_2O$

15.00 g 10.67 g 4.33 g

$CuSO_4 \cdot yH_2O \rightarrow CuSO_4 + yH_2O$

10.67 g 9.59 g 1.08 g

$$\text{total moles of water} = \frac{5.41 \text{ g}}{18.02 \text{ g} \cdot \text{mol}^{-1}} = 0.300 \text{ mols}$$

$$\text{total moles of } CuSO_4 = \frac{9.59 \text{ g}}{159.62 \text{ g} \cdot \text{mol}^{-1}} = 0.0601 \text{ mols}$$

$$\frac{\text{mols } H_2O}{\text{mols } CuSO_4} = \frac{0.300}{0.0601} = \frac{5}{1}, \text{ so } x = 5 \text{ and } y = 4$$

$$CuSO_4 \cdot 5\, H_2O(s) \rightarrow CuSO_4(s) + 5\, H_2O(l)$$

M.2 $P_4(s) + 6\, Cl_2(g) \rightarrow 4\, PCl_3(g)$

theoretical yield:

$$(77.25 \text{ g } P_4)\left(\frac{1 \text{ mol } P_4}{123.88 \text{ g } P_4}\right)\left(\frac{4 \text{ mol } PCl_3}{1 \text{ mol } P_4}\right)\left(\frac{137.32 \text{ g } PCl_3}{1 \text{ mol } PCl_3}\right) = 342.5 \text{ g } PCl_3$$

actual yield:

$$\frac{300.5 \text{ g } PCl_3}{342.5 \text{ g } PCl_3} \times 100\% = 87.73\% \text{ yield}$$

M.4 (a) $CaO(s) + H_2O(l) \rightarrow Ca(OH)_2(s)$

$$(30.0 \text{ g CaO})\left(\frac{1 \text{ mol CaO}}{56.08 \text{ g CaO}}\right)\left(\frac{1 \text{ mol } H_2O}{1 \text{ mol CaO}}\right)\left(\frac{18.02 \text{ g } H_2O}{1 \text{ mol } H_2O}\right) = 9.64 \text{ g } H_2O$$

Because 10.0 g H_2O are present, the limiting reagent is CaO.

$$(30.0 \text{ g CaO})\left(\frac{1 \text{ mol CaO}}{56.08 \text{ g CaO}}\right)\left(\frac{1 \text{ mol Ca(OH)}_2}{1 \text{ mol CaO}}\right)\left(\frac{74.10 \text{ g Ca(OH)}_2}{1 \text{ mol Ca(OH)}_2}\right)$$

$= 39.6 \text{ g Ca(OH)}_2$

M.6 (a) $3\ FeO(s) + 2\ Al(l) \rightarrow 3\ Fe(l) + Al_2O_3(s)$

$$n_{FeO} = \frac{10.325 \text{ g FeO}}{71.85 \text{ g} \cdot \text{mol}^{-1} \text{ FeO}} = 0.1437 \text{ mol FeO}$$

$$n_{Al} = \frac{5.734 \text{ g Al}}{26.98 \text{ g} \cdot \text{mol}^{-1} \text{ Al}} = 0.2125 \text{ mol Al}$$

The reaction stoichiometry requires 2 moles of Al for 3 moles of FeO. For complete reduction of FeO

$$n_{\text{Al required to reduce all FeO}} = 0.1437 \text{ mol FeO}\left(\frac{2 \text{ mol Al}}{3 \text{ mol FeO}}\right) = 0.0958 \text{ mol Al}$$

Because there is more than 0.0958 mol Al, there is more than sufficient Al to reduce all of the FeO. FeO is, therefore, the limiting reagent.

(b) Because 1 mol Fe is produced per mol FeO, the amount of Fe produced will be 0.1437 mol.

(c) Al is present in excess. 0.2125 – 0.0958 mol = 0.1167 mol of Al will remain, or $0.1167 \text{ mol} \times 26.98 \text{ g} \cdot \text{mol}^{-1} \text{ Al} = 3.148 \text{ g Al}$.

M.8 The molar mass of ephedrine is $165.23 \text{ g} \cdot \text{mol}^{-1}$.

$$n_{\text{ephedrine}} = \frac{0.057\ 32 \text{ g}}{165.23 \text{ g} \cdot \text{mol}^{-1}.} = 3.469 \times 10^{-4} \text{ mol}$$

1 mole of ephedrine will produce 10 moles of CO_2 and 7.5 moles of H_2O.

$$m_{CO_2} = (3.469 \times 10^{-4} \text{ mol ephedrine})\left(\frac{10 \text{ mol } CO_2}{1 \text{ mol ephedrine}}\right)$$

$$\times (44.01 \text{ g} \cdot \text{mol}^{-1} \text{ } CO_2)$$

$$= 0.1527 \text{ g } CO_2$$

$$m_{H_2O} = (3.469 \times 10^{-4} \text{ mol ephedrine})\left(\frac{7.5 \text{ mol } H_2O}{1 \text{ mol ephedrine}}\right)$$

$$\times (18.02 \text{ g} \cdot \text{mol}^{-1} \text{ } H_2O)$$

$$= 0.046\ 88 \text{ g } H_2O$$

M.10 $(1.072 \text{ g } CO_2)\left(\frac{1 \text{ mol } CO_2}{44.01 \text{ g } CO_2}\right)\left(\frac{1 \text{ mol C}}{1 \text{ mol } CO_2}\right) = 0.024\ 36 \text{ mol C}$

$$(0.024\ 36 \text{ mol C})\left(\frac{12.01 \text{ g C}}{1 \text{ mol C}}\right) = 0.2926 \text{ g C}$$

$$(0.307 \text{ g } H_2O)\left(\frac{1 \text{ mol } H_2O}{18.02 \text{ g } H_2O}\right)\left(\frac{2 \text{ mol H}}{1 \text{ mol } H_2O}\right) = 0.0341 \text{ mol H}$$

$$(0.0341 \text{ mol H})(1.0079 \text{ g} \cdot \text{mol}^{-1} \text{ H}) = 0.0343 \text{ g H}$$

$$(0.068 \text{ g } N_2)\left(\frac{1 \text{ mol } N_2}{28.02 \text{ g } N_2}\right)\left(\frac{2 \text{ mol N}}{1 \text{ mol } N_2}\right) = 0.0049 \text{ mol N}$$

$$(0.0049 \text{ mol N})(14.01 \text{ g} \cdot \text{mol}^{-1} \text{ N}) = 0.068 \text{ g N}$$

Dividing each amount by 0.0049 gives C : H : N ratios = 5.0 : 7.0 : 1.0.

The empirical formula is C_5H_7N. The molecular mass is $162 \text{ g} \cdot \text{mol}^{-1}$.

Its empirical formula mass is $81 \text{ g} \cdot \text{mol}^{-1}$.

molecular formula = 2 × empirical formula = $C_{10}H_{14}N_2$

$$2\ C_{10}H_{14}N_2(s) + 27\ O_2(g) \rightarrow 20\ CO_2(g) + 14\ H_2O(l) + 2\ N_2(g)$$

M.12 $4\ HCl(aq) + MnO_2(s) \rightarrow 2\ H_2O(l) + MnCl_2(s) + Cl_2(g)$

(a) $(42.7 \text{ g } MnO_2)\left(\frac{1 \text{ mol } MnO_2}{86.94 \text{ g } MnO_2}\right)\left(\frac{1 \text{ mol } Cl_2}{1 \text{ mol } MnO_2}\right)\left(\frac{70.90 \text{ g } Cl_2}{1 \text{ mol } Cl_2}\right)$

$$= 34.8 \text{ g } Cl_2$$

(b) $(0.300\text{ L})\left(\frac{0.100\text{ mol HCl}}{1.00\text{ L}}\right)\left(\frac{1\text{ mol Cl}_2}{4\text{ mol HCl}}\right)\left(\frac{70.90\text{ g Cl}_2}{1\text{ mol Cl}_2}\right)\left(\frac{1\text{ L}}{3.17\text{ g Cl}_2}\right)$

$= 0.168\text{ L Cl}_2$

(c) $\frac{0.150\text{ L}}{0.168\text{ L}}\times 100 = 89.3\%$

M.14 First calculate the theoretical percentages of C and H for $C_{14}H_{20}O_2N$ and $C_2H_2Cl_4$.

For $C_{14}H_{20}O_2N$:

$$\%C = \frac{14(12.01\text{ g}\cdot\text{mol}^{-1})}{(234.31\text{ g}\cdot\text{mol}^{-1})}\times 100\% = 71.76\%\text{ C}$$

$$\%H = \frac{20(1.0079\text{ g}\cdot\text{mol}^{-1})}{(234.31\text{ g}\cdot\text{mol}^{-1})}\times 100\% = 8.60\%\text{ H}$$

For $C_2H_2Cl_4$:

$$\%C = \frac{2(12.01\text{ g}\cdot\text{mol}^{-1})}{(167.84\text{ g}\cdot\text{mol}^{-1})}\times 100\% = 14.31\%\text{ C}$$

$$\%H = \frac{2(1.0079\text{ g}\cdot\text{mol}^{-1})}{(167.84\text{ g}\cdot\text{mol}^{-1})}\times 100\% = 1.20\%\text{ H}$$

If x = mass of $C_{14}H_{20}O_2N$ and y = mass of $C_2H_2Cl_4$ in the sample, the total masses of C and H can be expressed as

$m_C = 0.7176x + 0.1431y$
$m_H = 0.0860x + 0.0120y$

The analysis results give these relationships for a 100 g sample:

$68.50\text{ g C} = 0.7176x + 0.1431y$
$8.18\text{ g H} = 0.0860x + 0.0120y$

Solving these two equations in two unknowns by rearranging and substitution gives

$x = 94.32$
$y = 5.69$

Therefore, the percentage purity by mass of the $C_{14}H_{20}O_2N$ is 94.32%.

M.16 (a) $C_xH_yO_z(s) + (2x + y/2 - z)\,O_2(g) \rightarrow x\,CO_2(g) + y/2\,H_2O(l)$

$$\frac{1.811\text{ g }CO_2}{44.01\text{ g}\cdot\text{mol}^{-1}} = 0.041\ 15\text{ mol }CO_2\left(\frac{12.01\text{ g C}}{1\text{ mol }CO_2}\right) = 0.4942\text{ g C}$$

$$\frac{0.3172\text{ g }H_2O}{18.02\text{ g}\cdot\text{mol}^{-1}} = 0.017\ 62\text{ mol }H_2O\left(\frac{2.0158\text{ g H}}{1\text{ mol }H_2O}\right) = 0.03552\text{ g H}$$

$$1.000\text{ g} - (0.4942 + 0.03552)\text{g} = 0.4703\text{ g O}$$

$$\frac{0.4703\text{ g O}}{16.00\text{ g}\cdot\text{mol}^{-1}} = 0.029\ 39\text{ mol O}$$

Dividing each amount of moles by 0.02939 gives a ratio of C:H:O of

1.4:1.2:1.

Multiplying through by 5 gives whole numbers as required for an empirical formula of

$C_7H_6O_5$.

(b) Since the molar mass of this empirical formula matches the given value of about 170 g/mol, the molecular formula is the same as the empirical formula.

M.18 (a)

$$3\ CuSO_4\cdot 5\ H_2O(s) + 2\ PH_3(g) \rightarrow Cu_3P_2(s) + 3\ H_2SO_4(aq) + 15\ H_2O(l)$$

(b) $CuSO_4\cdot 5\ H_2O$ copper(II) sulfate pentahydrate; PH_3 phosphine; Cu_3P_2 copper(II) phosphide; H_2SO_4 sulfuric acid; H_2O water

(c) $$\frac{4.94\text{ g}(0.85)}{33.994\text{ g}\cdot\text{mol}^{-1}} = 0.1235\text{ mol }PH_3\text{ present, therefore}$$

$$0.1235\text{ mol }PH_3\left(\frac{3\text{ mol }CuSO_4\cdot 5\ H_2O}{2\text{ mol }PH_3}\right)$$

$$= 0.1853\text{ mol }CuSO_4\cdot 5\ H_2O\text{ needed}$$

$$\frac{0.110\times 10^3\text{ g}}{249.68\text{ g}\cdot\text{mol}^{-1}} = 0.4406\text{ mol }CuSO_4\cdot 5\ H_2O\text{ present}$$

>0.1853 mol needed

PH_3 is the limiting reagent while $CuSO_4 \cdot 5\,H_2O$ is present in excess.

(d)

$$? \text{ g } Cu_3P_2 = 0.1235 \text{ mol } PH_3 \left(\frac{1 \text{ mol } Cu_3P_2}{2 \text{ mol } PH_3}\right)\left(\frac{252.56 \text{ g } Cu_3P_2}{1 \text{ mol } Cu_3P_2}\right)(0.0631)$$

$= 0.984 \text{ g } Cu_3P_2$

CHAPTER 1

ATOMS: THE QUANTUM WORLD

1.2 radio waves < infrared radiation < visible light < ultraviolet radiation

1.4 (a) $2.997\,92 \times 10^{8}\ \text{m} \cdot \text{s}^{-1} = (\lambda)(7.1 \times 10^{14}\ \text{s}^{-1})$

$$\lambda = \frac{2.997\,92 \times 10^{8}\ \text{m} \cdot \text{s}^{-1}}{7.1 \times 10^{14}\ \text{s}^{-1}}$$
$$= 4.2 \times 10^{-7}\ \text{m} = 420\ \text{nm}$$

(b) $2.997\,92 \times 10^{8}\ \text{m} \cdot \text{s}^{-1} = (\lambda)(2.0 \times 10^{18}\ \text{s}^{-1})$

$$\lambda = \frac{2.997\,92 \times 10^{8}\ \text{m} \cdot \text{s}^{-1}}{2.0 \times 10^{18}\ \text{s}^{-1}}$$
$$= 1.5 \times 10^{-10}\ \text{m} = 150\ \text{pm}$$

1.6 From Wien's law: $T\lambda_{\text{max}} = 2.88 \times 10^{-3}\ \text{K} \cdot \text{m}$.

$$(T)(715 \times 10^{-9}\ \text{m}) = 2.88 \times 10^{-3}\ \text{K} \cdot \text{m}$$
$$T \approx 4.03 \times 10^{3}\ \text{K}$$

1.8 (a) $E = h\nu$

$$= (6.626\,08 \times 10^{-34}\ \text{J} \cdot \text{s})(1.2 \times 10^{17}\ \text{s}^{-1})$$
$$= 8.0 \times 10^{-17}\ \text{J}$$

(b) The energy per mole will be 6.022×10^{23} times the energy of one atom.

$$E = (2.00\ \text{mol})(6.022 \times 10^{23}\ \text{atoms} \cdot \text{mol}^{-1})$$
$$\times (6.626\,08 \times 10^{-34}\ \text{J} \cdot \text{s})(1.2 \times 10^{17}\ \text{s}^{-1})$$
$$= 9.6 \times 10^{7}\ \text{J or } 9.6 \times 10^{4}\ \text{kJ}$$

(c) $$E = \left(\frac{2.00 \text{ g Cu}}{63.54 \text{ g} \cdot \text{mol}^{-1} \text{ Cu}}\right)$$

$$\times (6.022 \times 10^{23} \text{ atoms} \cdot \text{mol}^{-1})(8.0 \times 10^{-17} \text{ J} \cdot \text{atom}^{-1})$$

$$= 1.5 \times 10^{6} \text{ J or } 1.5 \times 10^{3} \text{ kJ}$$

1.10 From $c = \nu\lambda$ and $E = h\nu$, $E = hc\lambda^{-1}$.

$$E \text{ (for one atom)} = \frac{(6.626\,08 \times 10^{-34} \text{ J} \cdot \text{s})(2.997\,92 \times 10^{8} \text{ m} \cdot \text{s}^{-1})}{470 \times 10^{-9} \text{ m}}$$

$$= 4.23 \times 10^{-19} \text{ J} \cdot \text{atom}^{-1}$$

$$E \text{ (for 1.00 mol)} = (6.022 \times 10^{23} \text{ atoms} \cdot \text{mol}^{-1})(4.23 \times 10^{-19} \text{ J} \cdot \text{atom}^{-1})$$

$$= 2.55 \times 10^{5} \text{ J} \cdot \text{mol}^{-1} \text{ or } 255 \text{ kJ} \cdot \text{mol}^{-1}$$

1.12 (a) false. UV photons have higher energy than infrared photons. (b) false. The kinetic energy of the electron is directly proportional to the energy (and hence frequency) of the radiation in excess of the amount of energy required to eject the electron from the metal surface. (c) true.

1.14 The wavelength of radiation needed will be the sum of the energy of the work function plus the kinetic energy of the ejected electron.

$$E_{\text{work function}} = (4.37 \text{ ev})(1.6022 \times 10^{-19} \text{ J} \cdot \text{eV}^{-1}) = 7.00 \times 10^{-19} \text{ J}$$

$$E_{\text{kinetic}} = \frac{1}{2} mv^2$$

$$= \frac{1}{2}(9.10939 \times 10^{-31} \text{ kg})(1.5 \times 10^{6} \text{ m} \cdot \text{s}^{-1})^2$$

$$= 1.02 \times 10^{-18} \text{ J}$$

$$E_{\text{total}} = E_{\text{work function}} + E_{\text{kinetic}}$$

$$= 7.00 \times 10^{-19} \text{ J} + 1.02 \times 10^{-18} \text{ J}$$

$$= 1.72 \times 10^{-18} \text{ J}$$

To obtain the wavelength of radiation we use the relationships between E, frequency, wavelength, and the speed of light:

From $E = h\nu$ and $c = \nu\lambda$ we can write

$$\lambda = \frac{hc}{E}$$

$$= \frac{(6.626 \times 10^{-34}\ \text{J}\cdot\text{s})(3.00 \times 10^{8}\ \text{m}\cdot\text{s}^{-1})}{1.72 \times 10^{-18}\ \text{J}}$$

$$= 1.16 \times 10^{-7}\ \text{m or } 116\ \text{nm}$$

1.16 Use the de Broglie relationship, $\lambda = hp^{-1} = h(mv)^{-1}$.

$(200\ \text{km}\cdot\text{h}^{-1})(1000\ \text{m/km})(1\ \text{h}/3600\ \text{s}) = 55.6\ \text{m}\cdot\text{s}^{-1}$

$$\lambda = h(mv)^{-1}$$

$$= (6.626\,08 \times 10^{-34}\ \text{kg}\cdot\text{m}^2\cdot\text{s}^{-1})[(1550\ \text{kg})(55.6\ \text{m}\cdot\text{s}^{-1})]^{-1}$$

$$= 7.69 \times 10^{-39}\ \text{m}$$

1.18 The mass of one He atom is given by the molar mass of He divided by Avogadro's constant:

$$\text{mass of He atom} = \frac{4.00\ \text{g}\cdot\text{mol}^{-1}}{6.022 \times 10^{23}\ \text{atoms}\cdot\text{mol}^{-1}}$$

$$= 6.64 \times 10^{-24}\ \text{g or } 6.64 \times 10^{-27}\ \text{kg}$$

From the de Broglie relationship, $p = h\lambda^{-1}$ or $h = mv\lambda$, we can calculate wavelength.

$$\lambda = h(mv)^{-1}$$

$$= \frac{6.626\,08 \times 10^{-34}\ \text{J}\cdot\text{s}}{(6.64 \times 10^{-27}\ \text{kg})(1230\ \text{m}\cdot\text{s}^{-1})}$$

$$= \frac{6.626\,08 \times 10^{-34}\ \text{kg}\cdot\text{m}^2\cdot\text{s}^{-1}}{(6.64 \times 10^{-27}\ \text{kg})(1230\ \text{m}\cdot\text{s}^{-1})}$$

$$= 8.11 \times 10^{-11}\ \text{m}$$

1.20 Yes, degeneracies are allowed. The lowest energy states which are degenerate in energy are the $n_1 = 1, n_2 = 4$ state and the $n_1 = 2, n_2 = 2$ state.

1.22 The observed line is the third lowest energy line.

The frequency of the given line

is: $\nu = \frac{c}{\lambda} = \frac{2.9979\times 10^{8}\ \text{m s}^{-1}}{5910\times 10^{-9}\ \text{m}} = 5.073\times 10^{13}\ \text{s}^{-1}$

This frequency is closest to the frequency resulting from the $n = 9$ to $n = 6$ transition:

$$\nu = \Re\left(\frac{1}{n_2^2} - \frac{1}{n_1^2}\right) = 3.29\times 10^{15}\ \text{Hz}\left(\frac{1}{6^2} - \frac{1}{9^2}\right) = 5.08\times 10^{13}\ \text{Hz}$$

1.24 (a) The Rydberg equation gives ν when $\Re = 3.29\times 10^{15}\ \text{s}^{-1}$, from which one can calculate λ from the relationship $c = \nu\lambda$.

$$\nu = \Re\left(\frac{1}{n_2^{\ 2}} - \frac{1}{n_1^{\ 2}}\right)$$

and $c = \nu\lambda = 2.997\ 92\times 10^{8}\ \text{m}\cdot\text{s}^{-1}$

$$c = \Re\left(\frac{1}{n_2^{\ 2}} - \frac{1}{n_1^{\ 2}}\right)\lambda$$

$$2.997\ 92\times 10^{8}\ \text{m}\cdot\text{s}^{-1} = (3.29\times 10^{15}\ \text{s}^{-1})\left(\frac{1}{1} - \frac{1}{25}\right)\lambda$$

$\lambda = 9.49\times 10^{-8}\ \text{m} = 94.9\ \text{nm}$

(b) Lyman series

(c) This absorption lies in the ultraviolet region.

1.26 Here we are searching for a transition of He^{+} whose frequency matches that of the $n = 2$ to $n = 1$ transition of H. The frequency of the H transition is:

$$\nu_{\text{H}} = \Re\left(\frac{1}{1^2} - \frac{1}{2^2}\right) = \left(\frac{3}{4}\right)\Re$$

A transition of the He^{+} ion with the same frequency is the $n = 2$ to $n = 4$ transition:

$$\nu_{\text{He}^+} = (Z^2)\Re\left(\frac{1}{2^2} - \frac{1}{4^2}\right) = (2^2)\Re\left(\frac{3}{16}\right) = \left(\frac{3}{4}\right)\Re$$

1.28 (a) The highest energy photon is the one that corresponds to the ionization energy of the atom, the energy required to produce the condition in which the electron and nucleus are "infinitely" separated. This energy corresponds to the transition from the highest energy level for which $n = 1$ to the highest energy level for which $n = \infty$.

$$E = h\Re\left(\frac{1}{n_{\text{lower}}^{2}} - \frac{1}{n_{\text{upper}}^{2}}\right) = (6.626\,08 \times 10^{-34}\ \text{J}\cdot\text{s})$$

$$\times(3.29 \times 10^{15}\ \text{s}^{-1})\left(\frac{1}{1^2} - \frac{1}{\infty^2}\right)$$

$$= 2.18 \times 10^{-18}\ \text{J}$$

(b) The wavelength is obtained from

$c = \nu\lambda$ and $E = h\nu$, or $E = hc\lambda^{-1}$, or $\lambda = hcE^{-1}$.

$$\lambda = \frac{(6.626\,08 \times 10^{-34}\ \text{J}\cdot\text{s})(2.997\,92 \times 10^{8}\ \text{m}\cdot\text{s}^{-1})}{2.18 \times 10^{-18}\ \text{J}}$$

$$= 9.11 \times 10^{-8}\ \text{m} = 91.1\ \text{nm}$$

(c) ultraviolet

1.30 Because the line is in the visible part of the spectrum, it belongs to the Balmer series for which the ending n is 2. We can use the following equation to solve for the starting value of n:

$$\nu = \frac{c}{\lambda} = \frac{2.99792 \times 10^{8}\ \text{m}\cdot\text{s}^{-1}}{434 \times 10^{-9}\ \text{m}} = 6.91 \times 10^{14}\ \text{s}^{-1}$$

$$\nu = (3.29 \times 10^{15}\ \text{s}^{-1})\left(\frac{1}{n_1^{2}} - \frac{1}{n_2^{2}}\right)$$

$$6.91 \times 10^{14}\ \text{s}^{-1} = (3.29 \times 10^{15}\ \text{s}^{-1})\left(\frac{1}{2^2} - \frac{1}{n_2^{2}}\right)$$

$$0.210 = 0.250 - \frac{1}{n_2^{2}}$$

$$\frac{1}{n_2^{2}} = 0.04$$

$$n_2^2 = \frac{1}{0.04} \; .$$
$$n_2 = 5$$

This transition is from the $n = 5$ to the $n = 2$ level.

1.32 (a) The easiest way to approach this problem is to plot the radial probability distribution, $P(r) = r^2 \cdot R_{3s}^2$, ($R_{3s}$ is given in Table 1.2) and, using the plot, determine the position of the nodes. Another approach is to take the derivative of this probability distribution with respect to r and set this derivative equal to zero to find the position of the extrema of the curve. Taking the latter approach, one finds that the minima are at

$$\left(\frac{9+(3\cdot\sqrt{3})}{2}\right)a_o \text{ and } \left(\frac{9-(3\cdot\sqrt{3})}{2}\right)a_o\text{, or } 7.10\cdot a_o \text{ and } 1.90\cdot a_o\,.$$

(b) $R_{4d} = 1/96\sqrt{5}\left(6-\dfrac{Zr}{2a_o}\right)\left(\dfrac{Zr}{2a_o}\right)^2 Z^{\frac{3}{2}} e^{\frac{-Zr}{4a_o}}$ As in part (a) above, the easiest way to find the node is to plot the probability distribution function and find the position of the node by inspection. Taking the derivative of the probability distribution function and setting it equal to zero, one finds three extrema for the 4d distribution: $12a_o$, $2\left(7+\sqrt{13}\right)a_o$, and $2\left(7-\sqrt{13}\right)a_o$. The first of these three is the position of the node while the latter two are the positions of the maxima in the distribution.

1.34 The d_{xy} orbital will have its lobes pointing *between* the x and y axes, while the $d_{x^2-y^2}$ orbital will have its lobes pointing *along* the x and y axes.

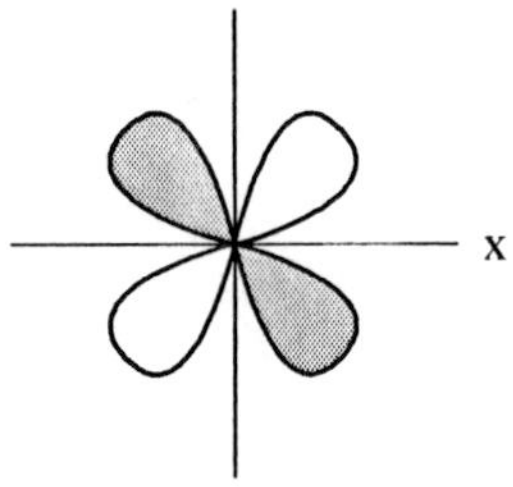

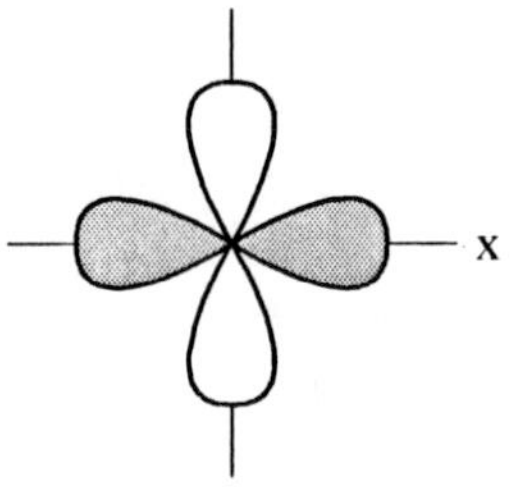

d_{xy} $d_{x^2-y^2}$

1.36 The equation demonstrated in example 1.8 can be used:

$$\frac{\psi^2\,(r = 0.65a_0, \theta, \phi)}{\psi^2(0, \theta, \phi)} = \frac{\dfrac{e^{-2(0.65a_0)/a_0}}{\pi a_0^3}}{\left(\dfrac{1}{\pi a_0^3}\right)} = 0.27$$

1.38 The radial probability distribution may be found by integrating the full wavefunction, $\psi(r,\theta,\phi)$, over all possible values of θ and ϕ. Since s-orbitals are spherically symmetric and are not a function of θ or ϕ, integration of any s-orbital over all θ and ϕ always gives the same result:

$$\int_0^{\pi}\int_0^{2\pi} \psi(r)^2 \sin\theta \; d\theta \; d\phi = \psi(r)^2 \left[\int_0^{\pi}\int_0^{2\pi} \sin\theta \; d\theta \; d\phi \right]$$

$$= \psi(r)^2 \left[\int_0^{2\pi} 2 \; d\phi \right] = \psi(r)^2 (4\pi)$$

The $\sin\theta$ term in the equation above is needed to correct for the differential volume element in spherical polar coordinates. Likewise, to integrate over all possible values of r one must evaluate the integral:

$$\int_0^{\infty} (4\pi) \; \psi(r)^2 r^2 \; dr$$

where again the r^2 term corrects for the differential volume element in spherical polar coordinates. From this expression it is clear that the probability distribution is $(4\pi) \; \psi(r)^2 r^2$.

1.40 (a) To find the maximum in the radial probability distribution function for the $3d$-orbital, one can take the derivative of the distribution with respect to r, set the result equal to zero and solve for r:

$$P = r^2 R_{3d}^2 = C^2 \cdot r^6 \cdot e^{-\frac{2r}{3a_o}} \quad \text{where } C = \frac{4}{81\cdot\sqrt{30}}\left(\frac{1}{a_o}\right)^{\frac{7}{2}}$$

$$\frac{dP}{dr} = C^2\left[r^6\left(\frac{-2}{3a_o}\right)\cdot e^{-\frac{2r}{3a_o}} + 6\,r^5\cdot e^{-\frac{2r}{3a_o}}\right] = C^2\cdot e^{-\frac{2r}{3a_o}}\left[r^6\left(\frac{-2}{3a_o}\right) + 6\,r^5\right] \equiv 0$$

The non-trivial solution to this equation is found when:

$$r^6\left(\frac{-2}{3a_o}\right) + 6\,r^5 = 0$$

Dividing both sides by r^5:

$$r\left(\frac{-2}{3a_o}\right) + 6 = 0 \text{ and, therefore, } r = 9a_o$$

The position of the maximum in the distribution for the $3d$-orbital occurs when $r = 9a_0$.

(b) The radial wavefunction for the $4s$-orbital is:

$$R_{4s} = C\left(1 - \frac{3}{4\cdot a_o}r + \frac{1}{8\cdot a_o^2}r^2 - \frac{1}{192\cdot a_o^3}r^3\right)e^{-\frac{r}{4a_o}} \quad \text{where } C = \frac{1}{4}\left(\frac{1}{a_o}\right)^{\frac{3}{2}}$$

The derivative of the radial probability distribution, $P = r^2 R_{4s}^2$, with respect to r is:

$$\frac{dP}{dr} = C^2 e^{-\frac{r}{2a_o}}\left[r^2 - \frac{3}{2\cdot a_o}r^3 + \frac{13}{16\cdot a_o^2}r^4 - \frac{19}{96\cdot a_o^3}r^5 + \frac{3}{128\cdot a_o^4}r^6 - \frac{1}{768\cdot a_o^5}r^7 + \frac{1}{36864\cdot a_o^6}r^8\right]$$

The non-trivial solutions to this equation are found when the polynomial in brackets is equal to zero. The roots are most easily found using a computer and are: $0.732a_0$, $1.872\ a_0$, $4\ a_0$, $6.611\ a_0$, $10.65\ a_0$, $15.52\ a_0$, and $24.62\ a_0$. The last of these roots, $24.62\ a_0$, is the position of the maximum of the probability distribution for the $4s$-orbital.

1.42 (a) 5: $l = 0, 1, 2, 3, 4$; (b) 5*s*, 5*p*, 5*d*, 5*f*, 5*g*; (c) $1 + 3 + 5 + 7 + 9 = 25$ orbitals

1.44 (a) 6 values: 0, 1, 2, 3, 4, 5; (b) 7 values: −3, −2, −1, 0, 1, 2, 3; (c) 1 value: 0; (d) 3 subshells: 3*s*, 3*p*, and 3*d*

1.46 (a) $n = 2$; $l = 0$; (b) $n = 6$; $l = 3$; (c) $n = 4$; $l = 2$; (d) $n = 5$; $l = 1$

1.48 (a) 0; (b) −3, −2, −1, 0, 1, 2, 3; (c) −2, −1, 0, 1, 2; (d) −1, 0, 1

1.50 (a) 2; (b) 6; (c) 10; (d) 14

1.52 (a) 3*d*, 10; (b) 5*s*, 2; (c) 7*p*, 6; (d) 4*f*, 14

1.54 (a) 6; (b) 2; (c) 2; (d) 98

1.56 (a) cannot exist; (b) cannot exist; (c) exists; (d) cannot exist

1.58 (a) The total Coulomb potential energy V(*r*) is the sum of the individual coulombic attractions and repulsions. There will be one attraction between the nucleus and each electron plus a repulsive term to represent the interaction between each pair of electrons. For beryllium, there are four protons in the nucleus and four electrons. Each attractive force will be equal to

$$\frac{(-e)(+4e)}{4\pi\varepsilon_0 r} = \frac{-4e^2}{4\pi\varepsilon_0 r} = -\frac{e^2}{\pi\varepsilon_0 r}$$

where $-e$ is the charge on the electron and $+4e$ is the charge on the nucleus, ϵ_0 is the vacuum permittivity, and r is the distance from the electron to the nucleus. The total attractive potential will thus be

$$-\frac{e^2}{\pi\varepsilon_0 r_1}-\frac{e^2}{\pi\varepsilon_0 r_2}-\frac{e^2}{\pi\varepsilon_0 r_3}-\frac{e^2}{\pi\varepsilon_0 r_4}=-\frac{e^2}{\pi\varepsilon_0}\left(\frac{1}{r_1}+\frac{1}{r_2}+\frac{1}{r_3}+\frac{1}{r_4}\right)$$

The first four terms are the attractive terms between the nucleus and each electron, and the last six terms are the repulsive interactions between all the possible combinations of electrons taken in pairs.

(b) The number of attractive terms is straightforward. There should be one term representing the attraction between the nucleus and each electron, so there should be a total of n terms representing attractions. The number of repulsive terms goes up with the number of electrons. Examining the progression, we see that

$n =$	1	2	3	4	5	6	7
# of repulsive terms =	0	1	3	6	10	15	21

Hence, the addition of an electron adds one r_{ab} term for each electron already present; so the difference in the number of repulsive terms increases by $n-1$ for each additional electron. This relation can be written as a summation to give the total number of repulsive terms:

number of repulsive terms = $\sum_{1\to n}(n-1)$

The repulsive terms will have the form

$$\frac{(-e)(-e)}{4\pi\varepsilon_0 r_{ab}}=\frac{e^2}{4\pi\varepsilon_0 r_{ab}}$$

where r_{ab} represents the distance between the two electrons a and b. The total repulsive term will thus be

$$\frac{e^2}{4\pi\varepsilon_0 r_{12}}+\frac{e^2}{4\pi\varepsilon_0 r_{13}}+\frac{e^2}{4\pi\varepsilon_0 r_{14}}+\frac{e^2}{4\pi\varepsilon_0 r_{23}}+\frac{e^2}{4\pi\varepsilon_0 r_{24}}+\frac{e^2}{4\pi\varepsilon_0 r_{34}}$$

$$\frac{e^2}{4\pi\varepsilon_0}\left(\frac{1}{r_{12}}+\frac{1}{r_{13}}+\frac{1}{r_{14}}+\frac{1}{r_{23}}+\frac{1}{r_{24}}+\frac{1}{r_{34}}\right)$$

This gives

$$V(r) = \left(\frac{-e^2}{\pi\varepsilon_0}\right)\left(\frac{1}{r_1}+\frac{1}{r_2}+\frac{1}{r_3}+\frac{1}{r_4}\right)$$

$$+\frac{e^2}{4\pi\varepsilon_0}\left(\frac{1}{r_{12}}+\frac{1}{r_{13}}+\frac{1}{r_{14}}+\frac{1}{r_{23}}+\frac{1}{r_{24}}+\frac{1}{r_{34}}\right)$$

The total number of attractive and repulsive terms will thus be equal to

$$n+\sum_{1\to n}(n-1)$$

The point of this exercise is to show that, with each added electron, we add an increasingly larger number of *e-e* repulsive terms.

1.60 (a) false. The 2*s*-electrons will be shielded by the electrons in the 1*s*-orbital and will thus experience a lower Z_{eff}. (b) false. Because the 2*p*-orbitals do not penetrate to the nucleus as the 2*s*-orbitals do, they will experience a lower Z_{eff}. (c) false. The ability of the electrons in the 2*s*-orbital to penetrate to the nucleus will make that orbital lower in energy than the 2*p*. (d) false. There are three *p*-orbitals, and the electron configuration for C will be $1s^2 2s^2 2p^2$. There will be two electrons in the *p*-orbitals, but each will go into a separate orbital and, as per quantum mechanics and Hund's rule, they will be in these orbitals with the spins parallel (i.e., the spin magnetic quantum numbers will have the same sign) for the ground-state atom. (e) false. Because the electrons are in the same orbital, they must have opposite spin quantum numbers, m_s, because the Pauli exclusion principle states that no two electrons in an atom can have the same four quantum numbers.

1.62 The atom with a $4s^2 4p^2$ valence-shell configuration is germanium, Ge. The ground-state configuration is given by (d); the other configurations represent excited states.

1.64 (a) This configuration is not possible because the maximum value *l* can have is $n-1$; because $n = 2$, $l_{max} = 1$. (b) This configuration is possible.

(c) This configuration is not possible because, for $l = 4$, m_l can only be an integer from -3 to $+3$, i.e., m_l can only equal 0, ±1, ±2, or ±3.

1.66 (a) sulfur $[Ne]3s^2 3p^4$

(b) cesium $[Xe]6s^1$

(c) polonium $[Xe]4f^{14} 5d^{10} 6s^2 6p^4$

(d) palladium $[Kr]4d^{10}$

(e) rhenium $[Xe]4f^{14} 5d^5 6s^2$

(f) vanadium $[Ar]3d^3 4s^2$

1.68 (a) Ga; (b) Na; (c) Sr; (d) Eu

1.70 (a) 4*s*; (b) 3*p*; (c) 3*p*; (d) 4*s*

1.72 (a) 5; (b) 2; (c) 7; (d) 12

For (d), note that the filled 3*d*-orbitals become core electrons, which are not available. The chemistry of Zn is dominated by its +2 oxidation number; and, consequently, Zn, Cd, and Hg are often grouped with the *p*-block elements and referred to collectively as the *post transition metals.*

1.74 (a) 2; (b) 3; (c) 1; (d) 0

1.76 (a) ns^2np^5; (b) ns^2np^4; (c) $nd^3(n+1)s^2$; (d) ns^2np^2

1.78 (a) As one goes across a period, a proton and an electron are added to each new atom. The electrons, however, are not completely shielded from the nucleus by other electrons in the same subshell, so the set of electrons experience an overall greater nuclear charge. (b) The ionization energies of the Group 16 elements of O, S, and Se lie somewhat lower than those of the Group 15 elements that preceed them. This exception may be explained by observing that, as the three p-orbitals up through Group 15 are filled, each electron goes into a separate orbital. The next electron (for Group 16) goes into an orbital already containing an electron, so electron-electron repulsions are higher. This increased repulsion makes it easier to remove the additional electron from the Group 16 elements.

1.80 While Na has a smaller effective nuclear charge than K, the outermost electron occupies a 3s-orbital. In potassium atoms, the outermost electron occupies a 4s-orbital and so on average is much further from the nucleus than the electron in the 3s-orbital of sodium. Despite the larger effective nuclear charge, the electron in potassium is more easily removed because it is, on average, further from the nucleus.

1.82 (a) silicon (118 pm) > sulfur (104 pm) > chlorine (99 pm);
(b) titanium (147 pm) > chromium (129 pm) > cobalt (125 pm);
(c) mercury (155 pm) > cadmium (152 pm) > zinc (137 pm);
(d) bismuth (182 pm) > antimony (141 pm) > phosphorus (110 pm)

1.84 (a) Ba^{2+}; (b) As^{3-}; (c) Sn^{2+}

1.86 (a) Al; (b) Sb; (c) Si

1.88 From Appendix 2D, the radii (in pm) are

Ge	122	Sb	141
Ge^{2+}	90	Sb^{3+}	89

The diagonal relationship between elements can often be attributed to the fact that the most common oxidation states for these elements give rise to ions of similar size, which consequently often show similar reaction chemistry.

1.90 (a) Ga and Si and (c) As and Sn. Note: (b) Be and Al exhibit a diagonal relationship. Because diagonal relationships often exist as a result of similarities in ionic radii, they can exist across the *s* and *p* blocks.

1.92 (c) hafnium and (d) niobium

1.94 (a) metal; (b) nonmetal; (c) metalloid; (d) metalloid; (e) nonmetal; (f) metalloid

1.96 (a) $\lambda = 0.20$ nm; $E = h\nu$ or $E = hc\lambda^{-1}$

$$E = \frac{(6.626\,08 \times 10^{-34}\ \text{J}\cdot\text{s})(2.997\,92 \times 10^{8}\ \text{m}\cdot\text{s}^{-1})}{0.20 \times 10^{-9}\ \text{m}}$$

$$= 9.9 \times 10^{-16}\ \text{J}$$

This radiation is in the x-ray region of the electromagnetic spectrum. For comparison, the K_α radiation from Cu is 0.154 439 0 nm and that from Mo is 0.0709 nm. X-rays produced from these two metals are those most commonly employed for determining structures of molecules in single crystals.

(b) From the de Broglie relationship $p = h\lambda^{-1}$, we can write $h\lambda^{-1} = mv$, or $v = hm^{-1}\lambda^{-1}$. For an electron, $m_e = 9.109\,39 \times 10^{-28}$ g. (Convert units to kg and m.)

$$v = \frac{(6.626\,08 \times 10^{-34}\ \text{J}\cdot\text{s})}{(9.109\,39 \times 10^{-31}\ \text{kg})(200 \times 10^{-12}\ \text{m})}$$

$$= \frac{(6.626\,08 \times 10^{-34}\ \text{kg}\cdot\text{m}^2\cdot\text{s}^{-1})}{(9.109\,39 \times 10^{-31}\ \text{kg})(200 \times 10^{-12}\ \text{m})}$$

$$= 3.6 \times 10^{6}\ \text{m}\cdot\text{s}^{-1}$$

(c) Solve similarly to (b). For a neutron, $m_n = 1.674\,93 \times 10^{-24}$ g. (Convert units to kg and m.)

$$v = \frac{(6.626\,08 \times 10^{-34}\ \text{J} \cdot \text{s})}{(1.674\,93 \times 10^{-27}\ \text{kg})(200 \times 10^{-12}\ \text{m})}$$

$$= \frac{(6.626\,08 \times 10^{-34}\ \text{kg} \cdot \text{m}^2 \cdot \text{s}^{-1})}{(1.67493 \times 10^{-27}\ \text{kg})(200 \times 10^{-12}\ \text{m})}$$

$$= 2.0 \times 10^3\ \text{m} \cdot \text{s}^{-1}$$

1.98 A ground-state oxygen atom has four electrons in the *p*-orbitals. This configuration means that as one goes across the periodic table in Period 2, oxygen is the first element encountered in which the *p*-electrons must be paired. This added electron-electron repulsion energy causes the ionization potential to be lower.

1.100 molar volume $(\text{cm}^3 \cdot \text{mol}^{-1}) = \text{molar mass}\ (\text{g} \cdot \text{mol}^{-1})/\text{density}\ (\text{g} \cdot \text{cm}^{-3})$

Element	Molar vol.	Element	Molar vol.
Li	13	Na	24
Be	4.87	Mg	14.0
B	4.62	Al	9.99
C	5.31	Si	12.1
N	16	P	17.0
O	14.0	S	15.5
F	17.1	Cl	17.5
Ne	16.7	Ar	24.1

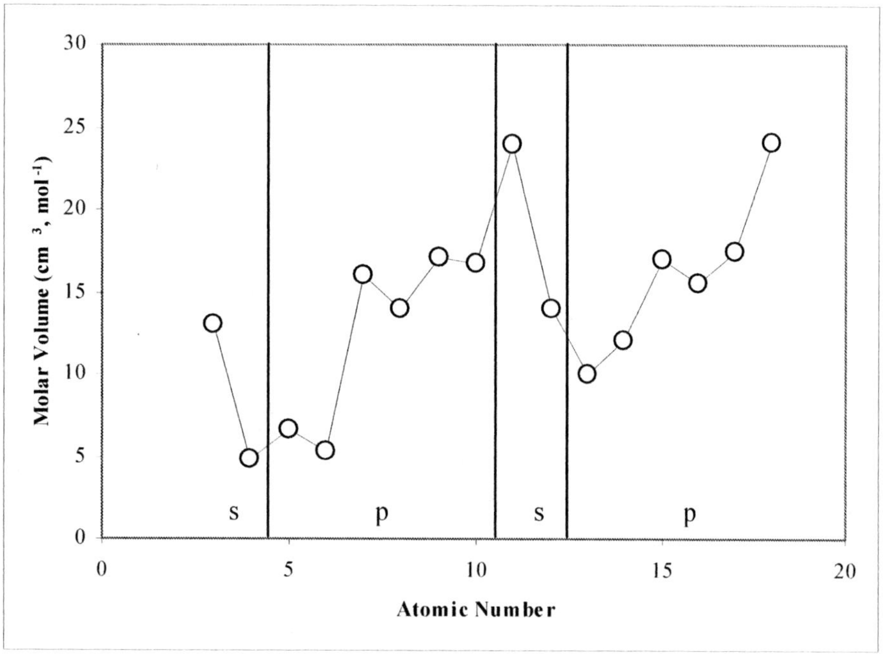

The molar volume roughly parallels atomic size (volume), which increases as the *s*-sublevel begins to fill and subsequently decreases as the *p*-sublevel fills (refer to the text discussion of periodic variation of atomic radii). In the above plot, this effect is most clearly seen in passing from Ne(10) to Na(11) and Mg(12), then to Al(13) and Si(14). Ne has a filled 2*p*-sublevel; the 3*s*-sublevel fills with Na and Mg; and the 3*p*-sublevel begins to fill with Al.

1.102 In general, as the principal quantum number increases, the energy spacing between orbitals becomes smaller. This trend indicates that it doesn't take very much change in electronic structure to cause the normal orbital energy pattern to rearrange.

1.104 (a)

$$\int_0^L \left(\sin\frac{\pi \cdot x}{L}\right)\cdot\left(\sin\frac{2\pi \cdot x}{L}\right) dx$$

$$= \frac{L}{2\pi}\left(\sin\frac{\pi \cdot x}{L}\right) - \frac{L}{6\pi}\left(\sin\frac{3\pi \cdot x}{L}\right)\Bigg|_0^L = 0$$

(b) Below is a plot of the first two wavefunctions describing the one-dimensional particle-in-a-box and the product of these two wavefunctions. Notice that the area above zero in the product exactly cancels the area below zero, making the integral of the product zero. This happens whenever a wavefunction that is unaltered by a reflection through the center of the box (wavefunctions with odd n) is multiplied by a wavefunction that changes sign everywhere when reflected through the center of the box (wavefunctions with even n).

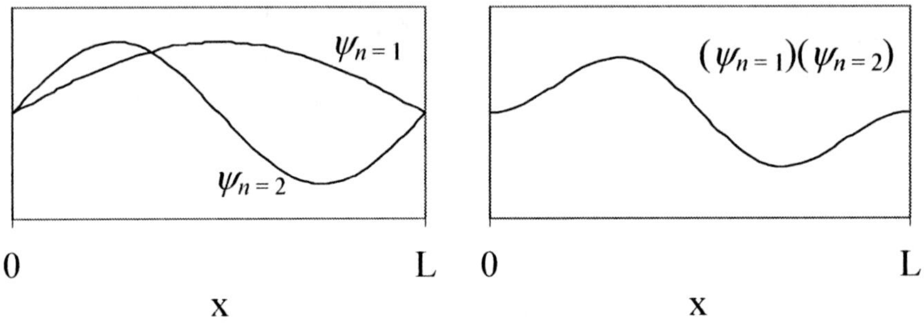

1.106 (a) $\Delta p \Delta x \geq \frac{1}{2}\hbar$

where $\hbar = \dfrac{h}{2\pi} = \dfrac{6.626\ 08 \times 10^{-34}\ \text{J} \cdot \text{s}}{2\pi} = 1.054\ 57 \times 10^{-34}\ \text{J} \cdot \text{s}$

$\Delta p \Delta x \geq \frac{1}{2}(1.054\ 57 \times 10^{-34}\ \text{J} \cdot \text{s}) = 5.272\ 85 \times 10^{-35}\ \text{J} \cdot \text{s}$

The minimum uncertainty occurs at the point where this relationship is an equality (i.e., using = rather than ≥).

The uncertainty in position will be taken as the 200 nm corresponding to the length of the box:

$\Delta p \Delta x = 5.272\ 85 \times 10^{-35}\ \text{J} \cdot \text{s}$

$$\Delta p = \frac{5.272\ 85 \times 10^{-35}\ \text{kg} \cdot \text{m}^2 \cdot \text{s}^{-1}}{200 \times 10^{-9}\ \text{m}} = 2.64 \times 10^{-28}\ \text{kg} \cdot \text{m} \cdot \text{s}^{-1}$$

$$= \Delta(mv) = m\Delta v = 2.64 \times 10^{-28}\ \text{kg} \cdot \text{m} \cdot \text{s}^{-1}$$

Because the mass of an electron is $9.109\ 39 \times 10^{-28}$ g or $9.109\ 39 \times 10^{-31}$ kg, the uncertainty in velocity will be given by

$$\Delta v = (2.63 \times 10^{-28}\ \text{kg} \cdot \text{m} \cdot \text{s}^{-1})(9.109\,39 \times 10^{-31}\ \text{kg})^{-1}$$
$$= 289\ \text{m} \cdot \text{s}^{-1}$$

The calculation is then repeated for a 1.00 mm wire where 1.00 mm is considered the uncertainty in position.

$$\Delta p = \frac{5.272\,85 \times 10^{-35}\ \text{kg} \cdot \text{m}^2 \cdot \text{s}^{-1}}{1.00 \times 10^{-3}\ \text{m}} = 5.27 \times 10^{-32}\ \text{kg} \cdot \text{m} \cdot \text{s}^{-1}$$

$$\Delta v = (5.27 \times 10^{-32}\ \text{kg} \cdot \text{m} \cdot \text{s}^{-1})(9.109\,39 \times 10^{-31}\ \text{kg})^{-1}$$
$$= 0.0578\ \text{m} \cdot \text{s}^{-1}$$

The uncertainty in velocity is much less in the macroscopic wire.

(b) The problem is solved as in (a), but now the mass is that of a Li^+ ion, which is roughly $(6.94\ \text{g} \cdot \text{mol}^{-1})/(6.022 \times 10^{23}\ \text{ions} \cdot \text{mol}^{-1})$ $= 1.15 \times 10^{-23}\ \text{g} \cdot \text{ion}^{-1}$ or $1.15 \times 10^{-26}\ \text{kg} \cdot \text{ion}^{-1}$. Technically, the mass of the Li^+ ion should be the mass of the atom less the mass of one electron, but because the mass of an electron is five orders of magnitude smaller than the mass of a neutron or a proton, the loss is insignificant in the overall mass of the lithium ion.

$$\Delta v = \frac{2.64 \times 10^{-28}\ \text{kg} \cdot \text{m}^2 \cdot \text{s}^{-1}}{1.15 \times 10^{-26}\ \text{kg}} = 0.0230\ \text{m} \cdot \text{s}^{-1}$$

(c) From the calculations, it is clear that the larger particle will have less uncertainty in the measurement of its speed and could theoretically be measured more accurately, as would be anticipated from quantum mechanics.

1.108 (a) Evaluating the integral for an $n = 1$ to $n = 3$ transition:

$$\int_0^L \sin\left(\frac{\pi x}{L}\right) \cdot x \cdot \sin\left(\frac{3\pi x}{L}\right)$$

$$= \frac{L^2}{8\pi^2}\left(\cos\left(\frac{2\pi x}{L}\right) + \frac{2\pi x}{L}\sin\frac{2\pi x}{L}\right) -$$

$$\frac{L^2}{32\pi^2}\left(\cos\left(\frac{4\pi x}{L}\right) + \frac{4\pi x}{L}\sin\frac{4\pi x}{L}\right)\Bigg|_0^L$$

$$= \frac{L^2}{8\pi^2}(1+0) - \frac{L^2}{32\pi^2}(1+0) - \left[\frac{L^2}{8\pi^2}(1+0) - \frac{L^2}{32\pi^2}(1+0)\right]$$

$$= 0$$

Because the integral is zero, one would not expect to observe a transition between the $n = 1$ and $n = 3$ states.

(b) Again, evaluating the integral:

$$\int_0^L \sin\left(\frac{\pi x}{L}\right) \cdot x \cdot \sin\left(\frac{2\pi x}{L}\right)$$

$$= \frac{L^2}{\pi^2}\left(\frac{1}{2}\left(\cos\left(\frac{\pi x}{L}\right) + \frac{\pi x}{L}\sin\left(\frac{\pi x}{L}\right)\right) - \frac{1}{18}\left(\cos\left(\frac{3\pi x}{L}\right) + \frac{3\pi x}{L}\sin\left(\frac{3\pi x}{L}\right)\right)\right)\Bigg|_0^L$$

Given the L^2 term, we see that the integral, and therefore I, will increase as the length of the box increases.

CHAPTER 2
CHEMICAL BONDS

2.2 The coulombic attraction is inversely proportional to the distance between the two oppositely charged ions (Equation 1) so the ions with the shorter radii will give the greater coulombic attraction. The answer is therefore (c) Mg^{2+}, O^{2-}.

2.4 The Br^- ion is smaller than the I^- ion (196 vs 220 pm). Because the lattice energy is related to the coulombic attraction between the ions, it will be inversely proportional to the distance between the ions (see Equation 2). Hence the larger iodide ion will have the lower lattice energy for a given cation.

2.6 (a) 7; (b) 10; (c) 7; (d) 2

2.8 (a) $[Kr]4d^{10}5s^25p^6$; (b) $[Ar]3d^8$; (c) $[Xe]4f^{14}5d^3$; (d) [Kr]

2.10 (a) [Ne]; (b) $[Kr]4d^3$; (c) [Rn]; (d) [Xe]

2.12 (a) Ca: $[Ar]4s^2$, Ti^{2+}: $[Ar]3d^2$; V^{3+}: $[Ar]3d^2$ In the d block, the energies of the n-1 d-orbitals lie below those of the ns-orbitals. Therefore, when V and Ti form ions they lose their 3s electrons before losing their 4d electrons. (b) Ca: no unpaired electrons; Ti^{2+}: two unpaired electrons; V^{3+}: two unpaired electrons. (c) Ti^{3+}: $[Ar]3d^1$ no neutral atom has this electron configuration.

2.14 (a) Au^{3+}; (b) Os^{3+}; (c) I^{3+}; (d) As^{3+}

2.16 (a) Co^{2+}; (b) Rh^{2+}; (c) Sn^{2+}; (d) Hg^{2+}

2.18 (a) 3*d*; (b) 5*s*; (c) 5*p*; (d) 4*d*

2.20 (a) +2; (b) +2 due to the inert pair effect, but 4+ is also possible; (c) +3; (d) −1; (e) −2

2.22 (a) 2; (b) 5; (c) 3; (d) 6

2.24 (a) $[Ar]3d^{10}4s^2$; no unpaired electrons; (b) $[Ar]3d^9$; one unpaired electron; (c) $[Xe]5d^{10}6s^2$; no unpaired electrons; (d) [Kr]; no unpaired electrons

2.26 (a) 3*d*; (b) 6*s*; (c) 2*p*; (d) 4*p*

2.28 (a) +6; (b) −2; (c) [Ne] for +6, [Ar] for −2; (d) Electrons are lost or added to give noble gas configuration.

2.30 (a) MnTe; (b) Ba_3As_2; (c) Si_3N_4; (d) Li_3Bi; (e) $ZrCl_4$

2.32 (a) FeS; (b) $CoCl_3$; (c) Mg_3P_2

2.34

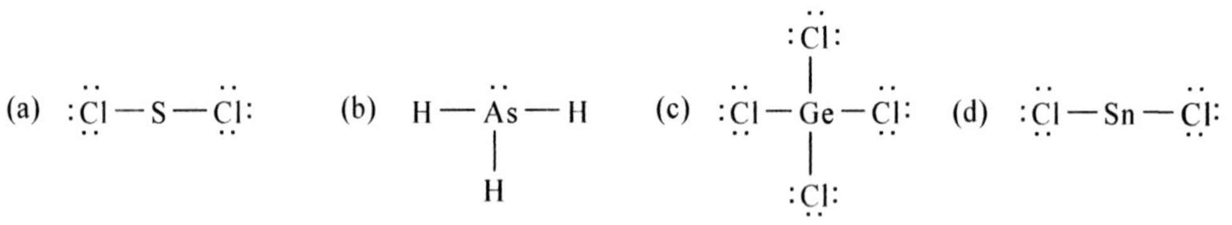

2.36

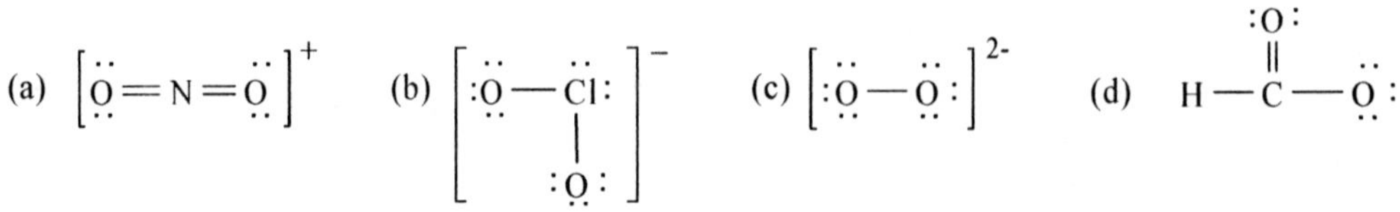

2.38 (a) 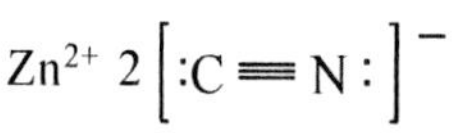 (b) 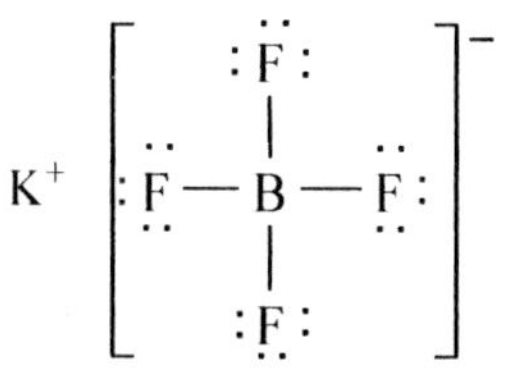 (c) 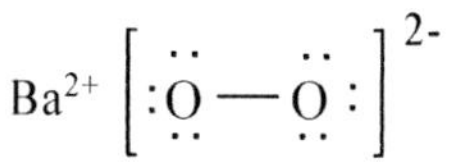

2.40 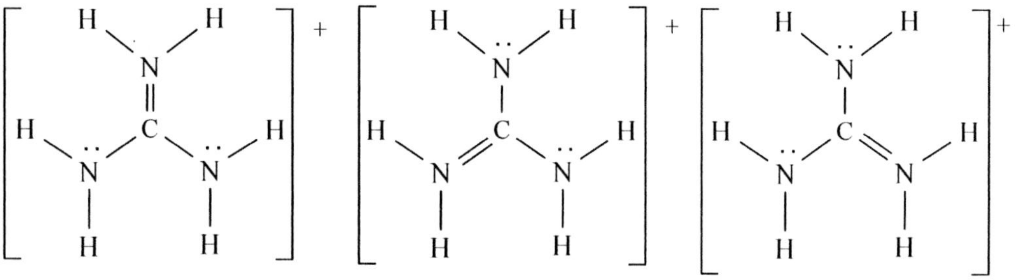

2.42 No. In order to have resonance structures, only the electrons are allowed to be rearranged. When atoms are in different relationship to each other, the result is *isomers*, not resonance forms.

2.44 (a) 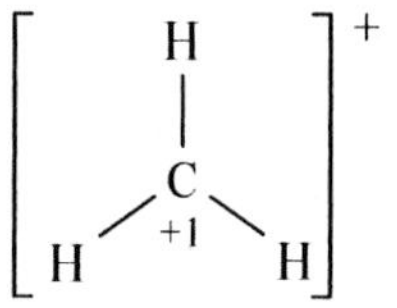

all H's are 0

(b) 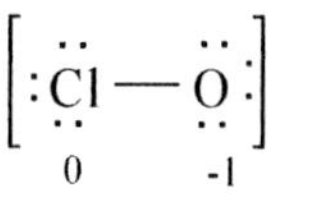 (c) 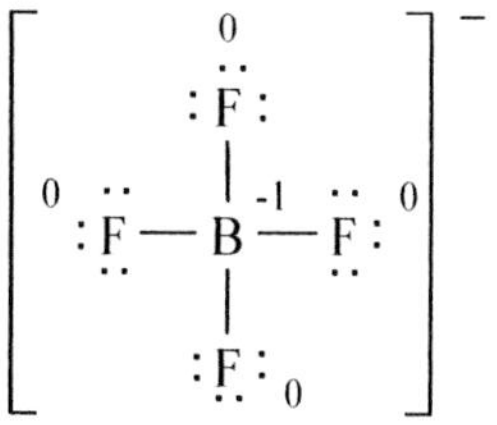

2.46 (a) 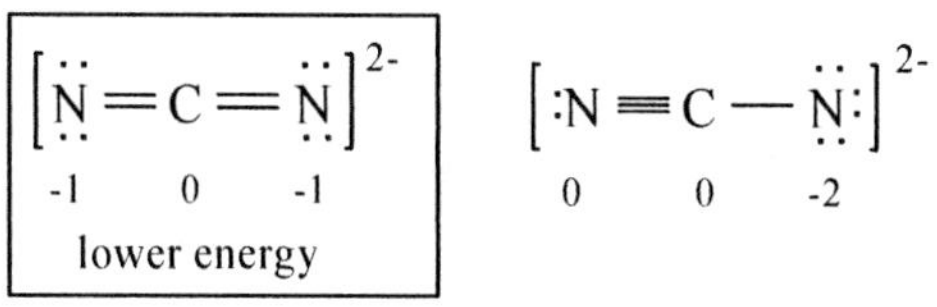

(b) 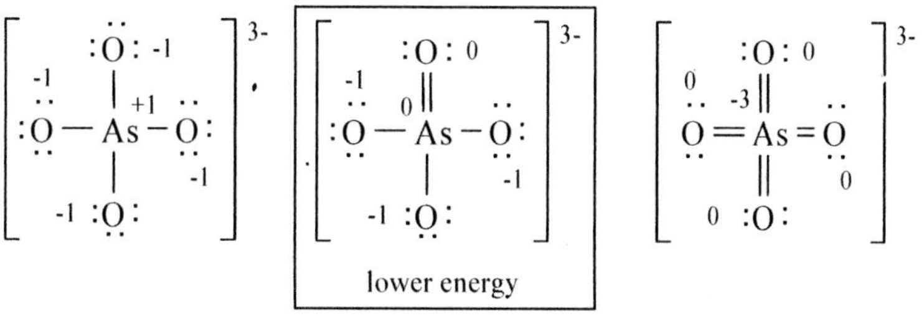

(c) 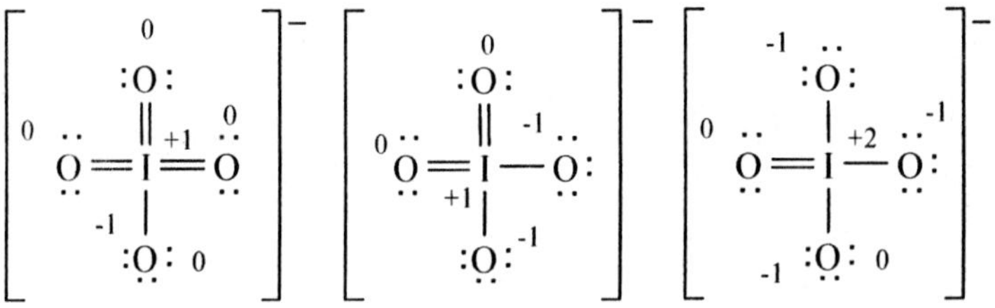

lowest

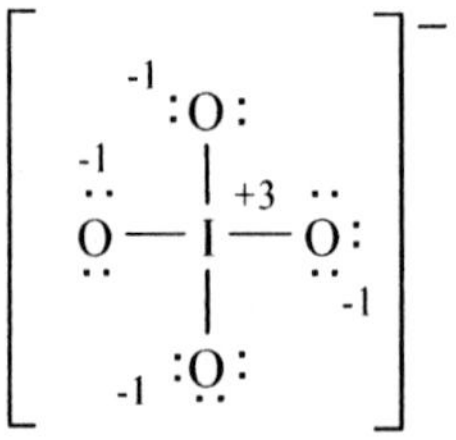

2.48 (a) The formal charge distribution is similar for both structures. In the first, the end nitrogen atom is -1, the central N atom is $+1$, and O atom is 0. For the second structure, the end N atom is 0, the central N atom is still $+1$, but the O atom is -1. The second is preferred because it places the negative formal charge on the most electronegative atom. (b) In the first structure, there are three O atoms with formal charges of -1 and one O atom with a formal charge of 0. The formal charge on the P atom is 0. In the second structure, the P atom has a formal charge of -1; there are two oxygen atoms with formal charges of -1 and two with formal charges of 0. The first structure is preferred because it places the negative formal charge at the more electronegative atom, in this case O.

2.50 (a) The dihydrogen phosphate ion has one Lewis structure that obeys the octet rule. Including one double bond to oxygen lowers the formal charge at P. There are two resonance forms that include this contribution.

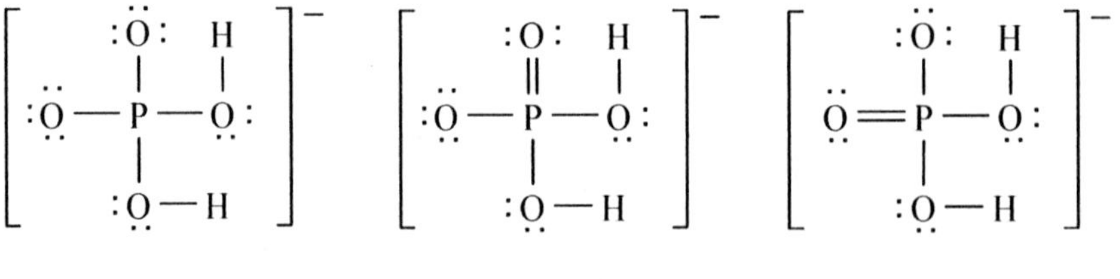

(b) There is one Lewis structure that obeys the octet rule shown below at the left. The formal charge at chlorine can be reduced to 0 by including one double bond contribution. This gives rise to two expanded octet structures.

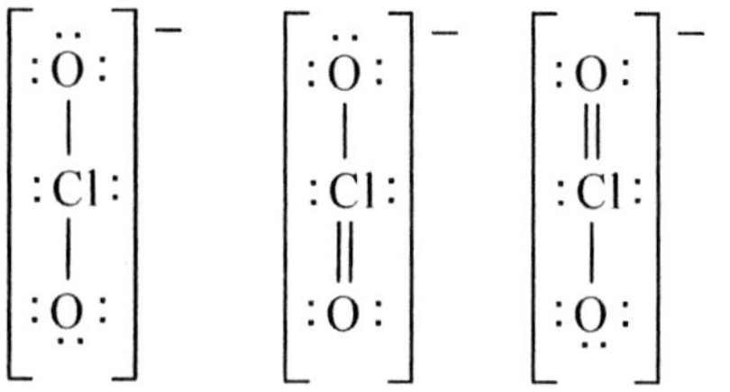

(c) As with the two preceding examples, there is one Lewis structure for the chlorate ion that obeys the octet rule. The formal charge at Cl can be reduced to 0 by including two double bond contributions, giving rise to three resonance forms.

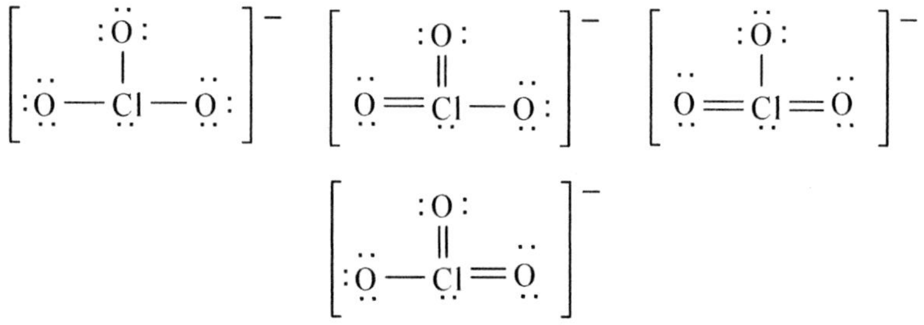

(d) For the nitrate ion, there are three resonance forms that all obey the octet rule.

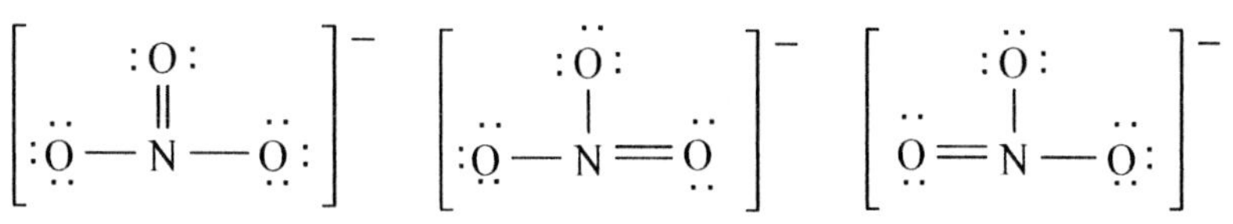

2.52 The Lewis structures are (a) 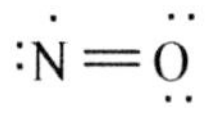 (b) 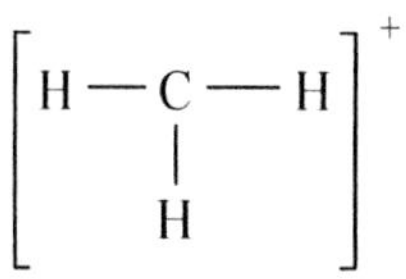

(c) 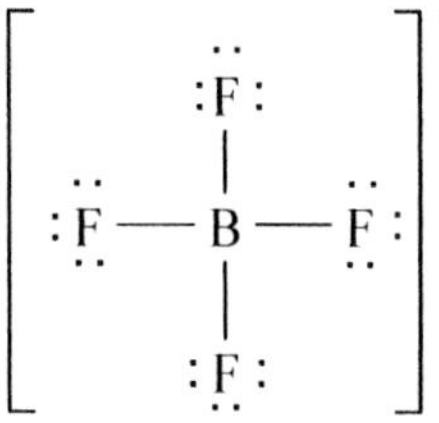 (d) 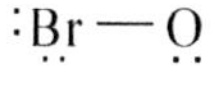

Radicals are species with an unpaired electron, therefore (a) and (d) are radicals.

2.54 (a)

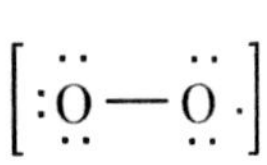

radical

(b)

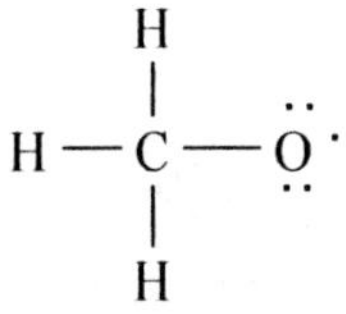

radical

(c)

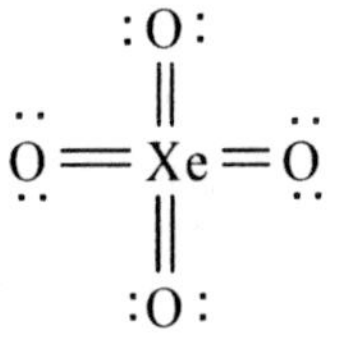

not a radical

(d)

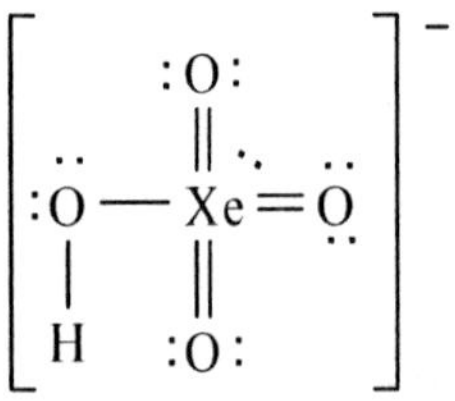

not a radical

2.56 (a)

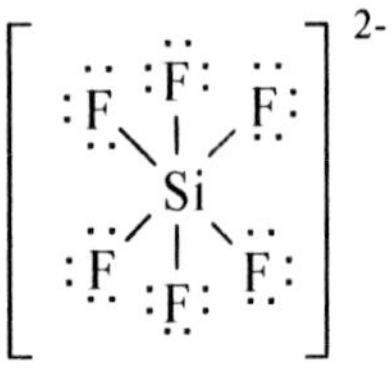

12 electrons

(b)

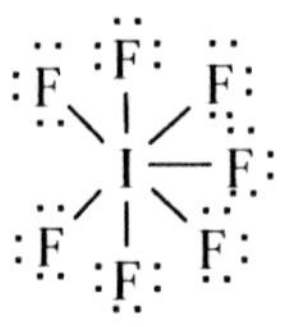

14 electrons

(c)

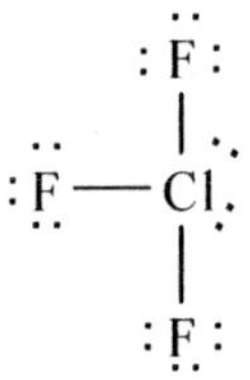

10 electrons

(d)

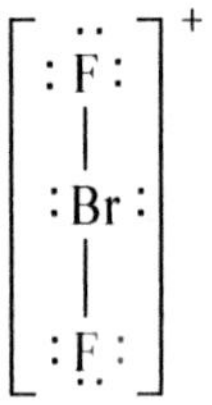

8 electrons

2.58 (a)

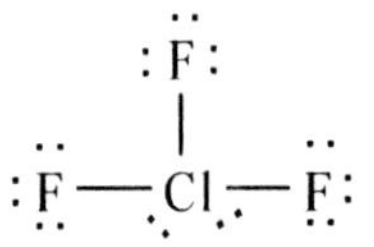

two lone pairs

(b)

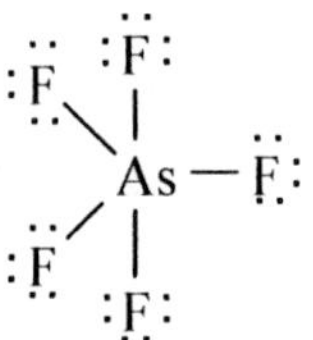

no lone pairs

(c)

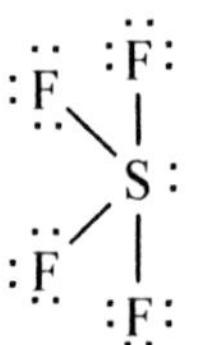

one lone pairs

2.60 Electronegativity decreases with increased mass, so the heavier alkali metals and alkaline earth metals are the most electropositive. In order of increasing *electropositive* character: Li (1.0) < Na (0.93) < K (0.82) ~ Rb (0.82) < Cs (0.79); Be (1.6) < Mg (1.3) ~ Ca (1.3) < Sr (0.95) < Ba (0.89) ~ Ra (0.9)

2.62 Si (1.9) < P (2.2) < S (2.6) ~ C (2.6) < N (3.0) < O (3.4) < F (4.0). Generally electronegativity increases as one goes from left to right across the periodic table and as one goes from heavier to lighter elements within a group.

2.64 (a) The N—H bond in NH_3 would be more ionic; the electronegativity difference between N and H (3.0 versus 2.2) is greater than between P and H (2.2 versus 2.2). (b) N and O have similar electronegativities (3.0 versus 3.4), leading us to expect that the N—O bonds in NO_2 would be fairly covalent. The electronegativity difference between S and O is greater, so S—O bonds would be expected to be more ionic (2.6 versus 3.4). (c) Difference between SF_6 and IF_5 would be small because S and I have very similar electronegativities (2.6 versus 2.7). Because I has an electronegativity closer to that of F, it may be expected that IF_5 would have more covalent bond character, but probably only slightly more, than SF_6.

2.66 $Cs^+ < K^+ < Mg^{2+} < Al^{3+}$: the smaller, more highly charged cations will be the more polarizing. The ionic radii are 170 pm, 138 pm, 72 pm, and 53 pm, respectively.

2.68 $N^{3-} < P^{3-} < I^- < At^-$: the polarizability should increase as the ion gets larger and less electronegative.

2.70 (a) $NO > NO_2 > NO_3^-$

NO_3^-. In NO the bond is a double bond, in NO_2 it is the average of a double and single bond (approximately 1.5), and in NO_3^- it is the average of three structures in which it is a single bond twice and a double bond once (approximately 1.33). The nitrate ion would, therefore, be expected to have the longest N—O bond length.

(b) $C_2H_2 > C_2H_4 > C_2H_6$

The bond would be longest in C_2H_6 in which it is a single bond. In C_2H_4 it is a double bond; in C_2H_2 it is a triple bond.

(c) $H_2CO > CH_3OH \sim CH_3OCH_3$

All the C—O bonds in CH_3OH and CH_3OCH_3 are single bonds and would be expected to be about the same length. The bond in formaldehyde H_2CO is a double bond and should be considerably shorter.

2.72 (a) The C—O bond in formaldehyde is a double bond, so the expected bond length will be 67 pm (double bond covalent radius of C) + 60 pm (double bond covalent radius of O) = 127 pm. The experimental value is 120.9 pm. (b) and (c) The C—O bonds in dimethyl ether and methanol are single bonds. The sum of the covalent single bond radii is 77 + 74 pm = 151 pm. The experimental value in methanol is 142.7 pm. (d) The C—S bond in methanethiol is a single bond. The sum of the covalent single bond radii is thus 77 + 102 pm = 179 pm. The experimental value is 181 pm.

2.74 The bond orders as determined by drawing the Lewis structures are 2 for (a), 1 for (b), and 3 for (c). Therefore, (c) HCN will have the greatest bond strength.

2.76 The Lewis structures are:

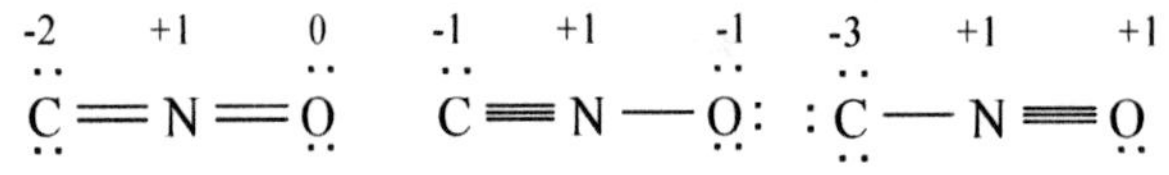

The Lewis structure in the center is probably the most important as it is the structure with the formal charges of the individual atoms closest to zero.

2.78 (a)

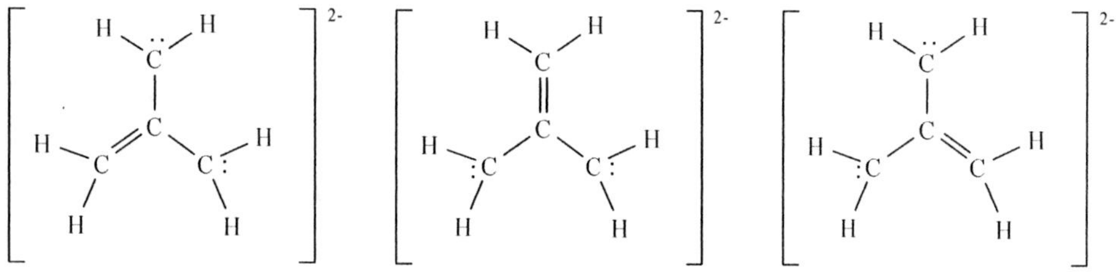

2.80

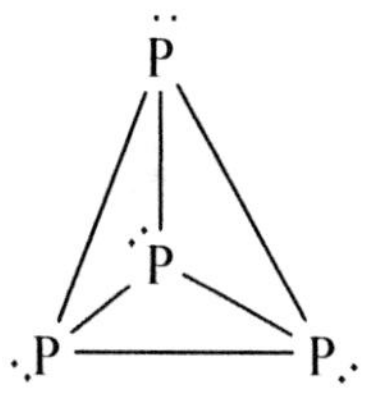

Yes, the molecule obeys the octet rule.

2.82 (a)

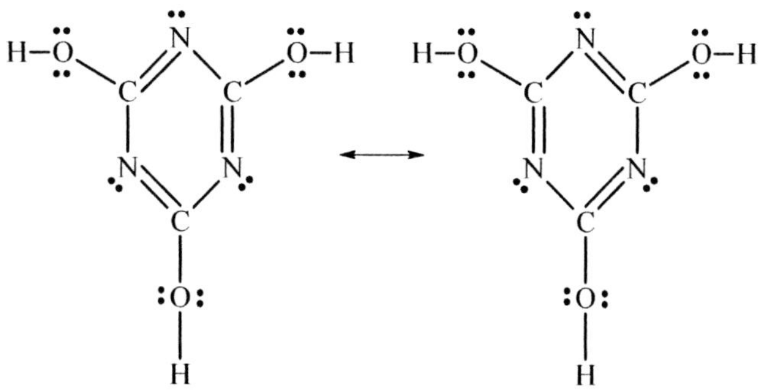

(b)

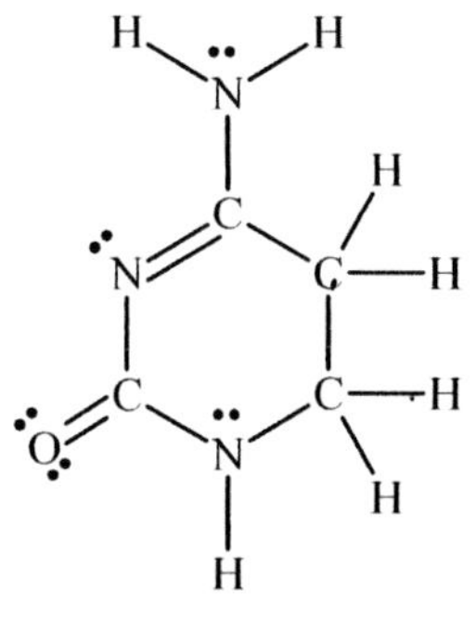

(c)

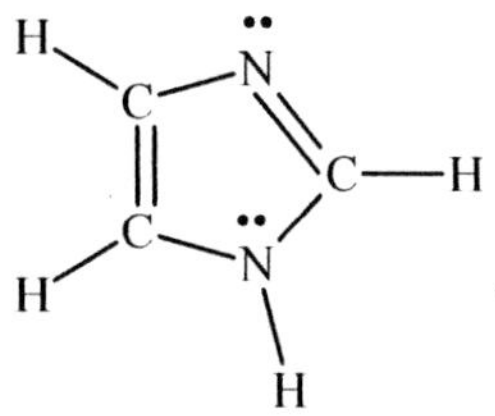

2.84 Cyclopentadiene has two C—C double bonds as shown. There are no resonance forms possible:

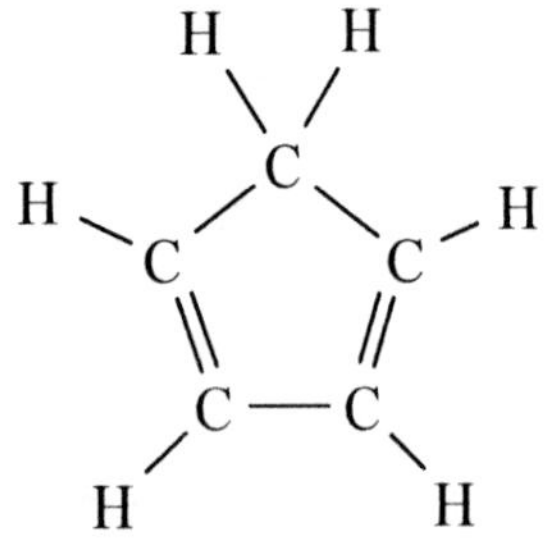

For the $[C_5H_5]^-$ ion, however, there are five resonance forms possible, as shown

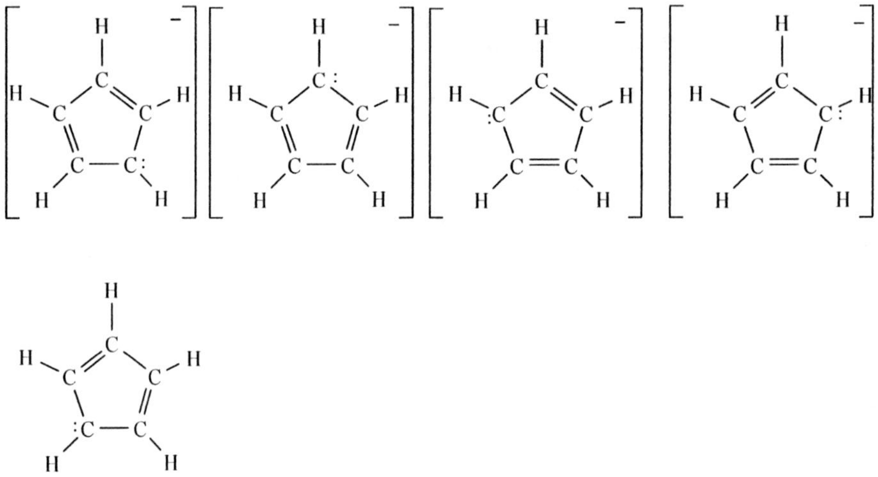

2.86

	d(K – X)	Lattice Energy, kJ/mol
Fluoride	271	826
Chloride	319	717
Iodide	358	645

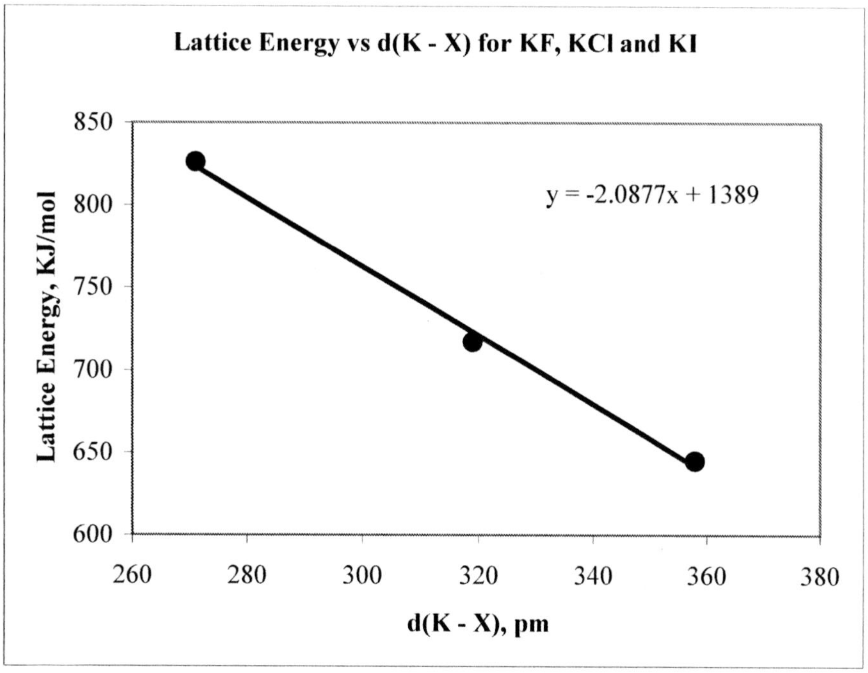

The data fit a straight line with a correlation coefficient of greater than 99%. (b) From the equation derived for the straight line relationship

Lattice Energy $= -1.984\, d_{\text{M-X}} + 1356$

and the value of $d_{\text{K-Br}} = 338$ pm, we can estimate the lattice energy of KBr to be 693 $\text{kJ} \cdot \text{mol}^{-1}$.

(c) The experimental value for the lattice energy for KBr is 689 $\text{kJ} \cdot \text{mol}^{-1}$, so the agreement is very good.

2.88 There are seven resonance structures for the tropyllium cation. All the C—C bonds will have the same bond order, which will be the average of 4 single bonds and 3 double bonds to give 1.43.

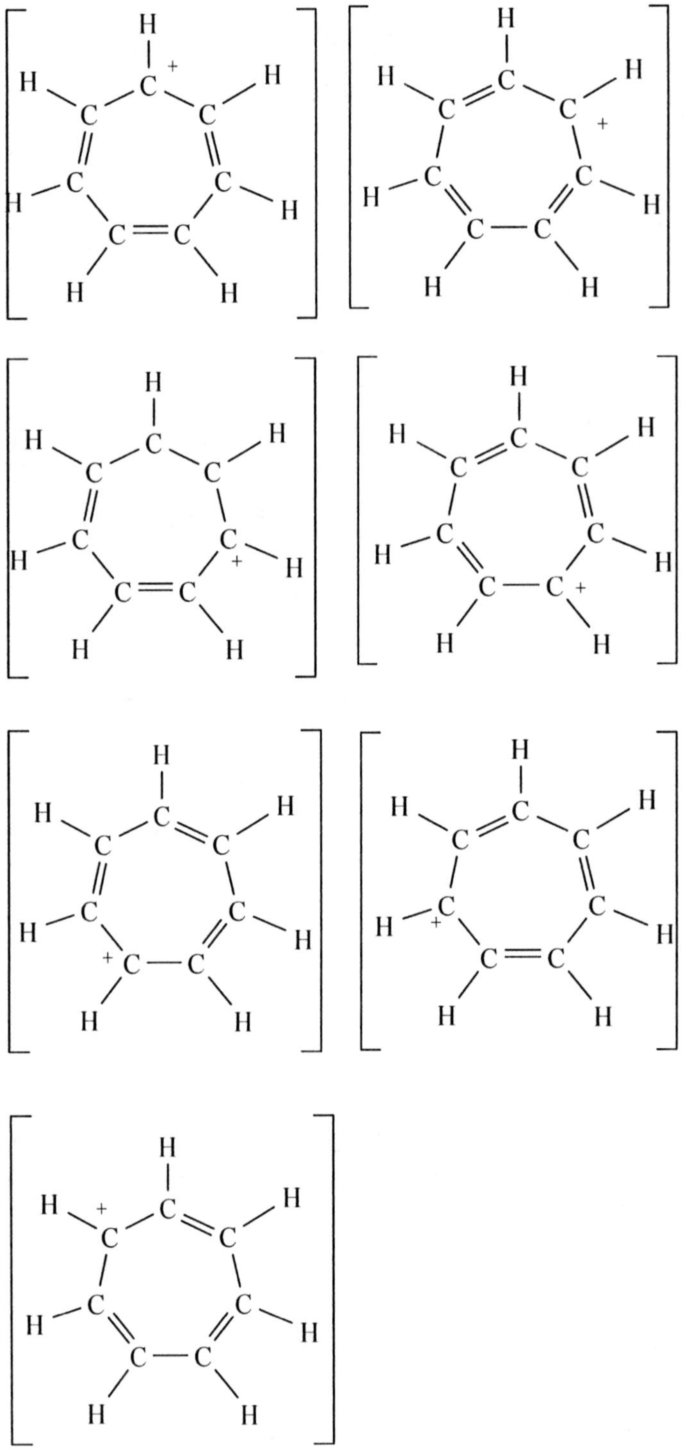

2.90 The most likely way for these to react is for the molecules to join at the atoms that possess the unpaired electrons, forming a bond that will pair the two originally unpaired electrons:

2.92

Z	Configuration	Number of unpaired e$^-$	Element	Charge	Energy state
38	[Kr]$5p^1$	1	Sr	+1	excited
45	[Kr]$4d^7$	3	Rh	+2	ground
43	[Kr]$5s^14d^5$	6	Tc	+1	ground
8	[Ne]	0	O	−2	ground
21	[Ar]$4s^13d^1$	2	Sc	+1	excited

2.94 (a) C: −1, N: 0; (b) C: −2, N: +1, O: 0; (c) terminal N's: −1, middle N: +1

2.96 (a) The structure with the lowest formal charges will be the most likely structure. All of the formal charges on the structure below are zero:

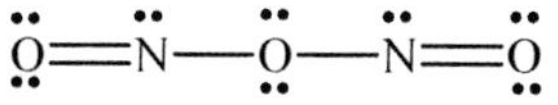

(b)

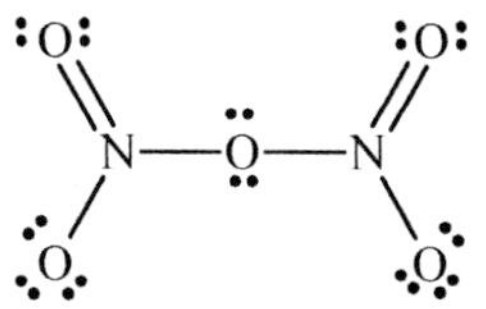

(c) $N_2O_5 + H_2O \rightarrow 2\ HNO_3$. The secondary pollutant is nitric acid.

2.98 The energy of the O−O bond is given in Table 2.3 as 157 kJ/mol. The frequency of a photon that would deliver this minimum energy needed to break the bond is given by:

$$\upsilon = \frac{E}{h} = \frac{\left(\dfrac{157\ \text{kJ/mol}}{6.02214\times 10^{23}\ \text{mol}^{-1}}\right)\left(\dfrac{1000\ \text{J}}{1\ \text{kJ}}\right)}{6.62608\times 10^{-34}\ \text{J}\cdot\text{s}} = 3.39\times 10^{14}\ \text{Hz}$$

2.100 The C—H bond, because the effective mass is lower and the C—H bond is stiffer than the C—Cl bond.

2.102 This question can be answered by examining the equation that relates the reduced mass μ to the vibrational frequency:

$$\nu = \frac{1}{2\pi}\sqrt{\frac{k}{\mu}}$$

We will assume that the force constant k is essentially the same for the Fe—H and Fe—D bonds and set up the proportionality between the frequencies of the two vibrations:

$$\frac{\nu_{\text{Fe-D}}}{\nu_{\text{Fe-H}}} = \frac{\frac{1}{2\pi}\sqrt{\frac{k}{\mu_{\text{Fe-D}}}}}{\frac{1}{2\pi}\sqrt{\frac{k}{\mu_{\text{Fe-H}}}}}$$

$$\nu_{\text{Fe-D}} = (\nu_{\text{Fe-H}})\sqrt{\frac{\mu_{\text{Fe-H}}}{\mu_{\text{Fe-D}}}}$$

$$= (1950\ \text{cm}^{-1})\sqrt{\frac{\frac{m_{\text{Fe}}m_{\text{H}}}{m_{\text{Fe}}+m_{\text{H}}}}{\frac{m_{\text{Fe}}m_{\text{D}}}{m_{\text{Fe}}+m_{\text{D}}}}}$$

$$= (1950\ \text{cm}^{-1})\sqrt{\frac{\frac{(55.85)(1.01)}{55.85+1.01}}{\frac{(55.85)(2.01)}{55.85+2.01}}}$$

$$= (1950\ \text{cm}^{-1})\sqrt{0.5113}$$

$$= 1394\ \text{cm}^{-1}$$

Note that we have also assumed the average mass for Fe to be 55.85 $\text{g}\cdot\text{mol}^{-1}$. It would be more correct to use the mass of the particular isotope of Fe bonded to the H atom. Because that is not given, the average value has been used. The change in frequency due to the use of different isotopes of iron is very small compared to the change in frequency due to the substitution of D for H, because the percentage change in much greater in the latter case. The mass essentially doubles upon replacing H with D; however, only a small percentage change is observed on going from one isotope of iron to another.

CHAPTER 3

MOLECULAR SHAPE AND STRUCTURE

3.2

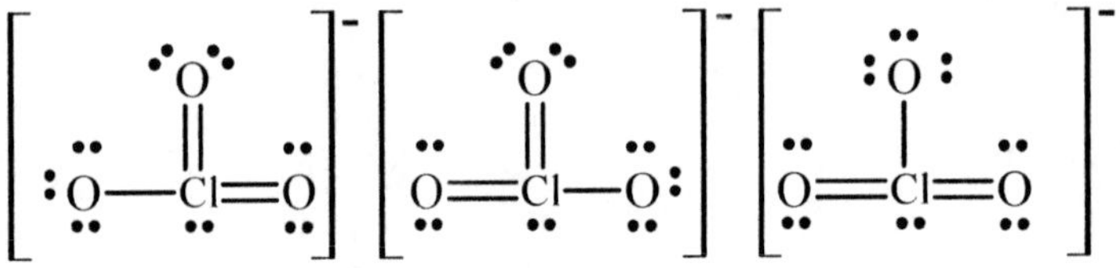

(a) The chlorate ion is trigonal pyramidal. (b) All the oxygen atoms are equivalent (resonance forms) so there should be only one O—Cl—O bond angle with a value close to 109.5°.

3.4

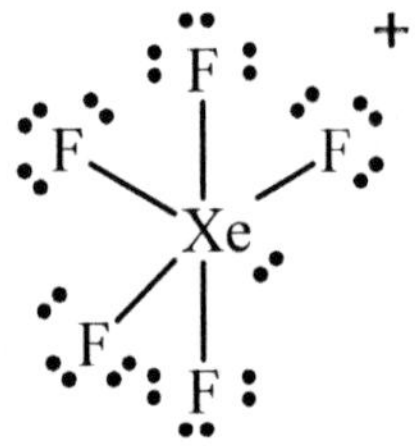

(a) The shape of the XeF_5^+ ion is square pyramidal based upon an octahedral arrangement of electrons about the Xe atom. (b) There are two types of F—Xe—Fe bond angles. (c) The F—Xe—F bond angles are approximately 90° or 180°.

3.6

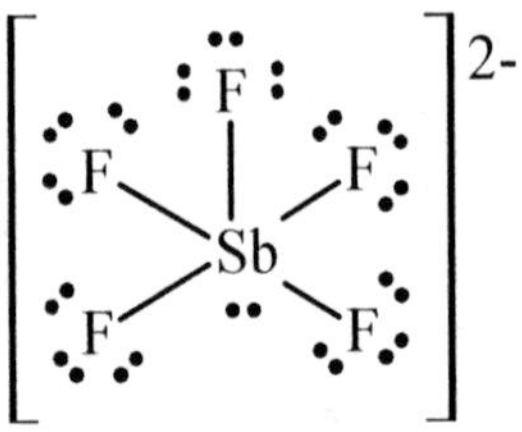

(a) Square pyramidal; (b) two; (c) 90°(adjacent), 180°(opposite)

3.8

(a) 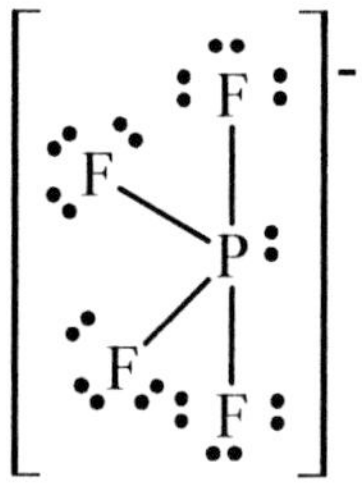

(b) 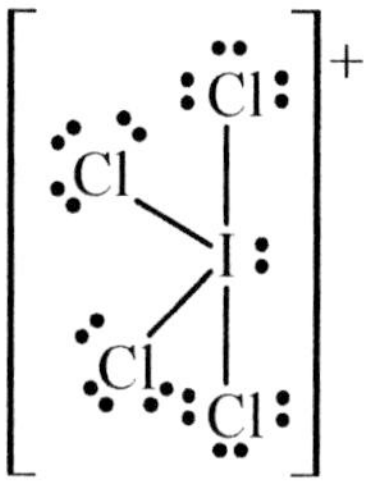

(c) 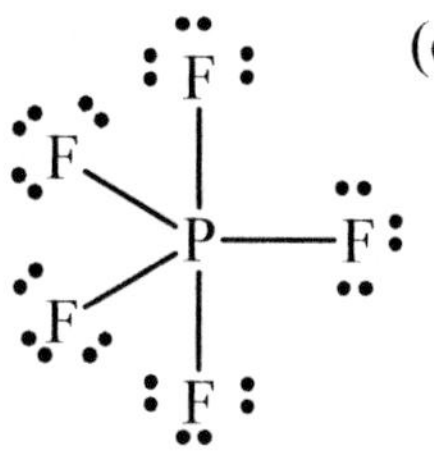

(d) 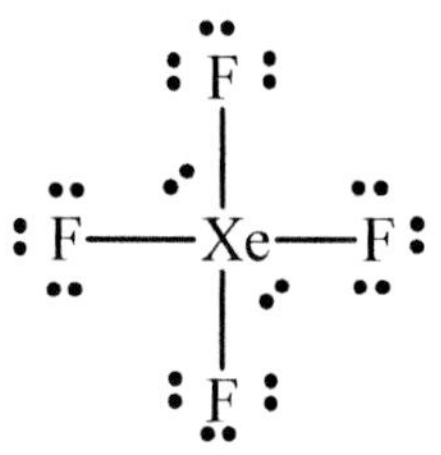

(a) The phosphorus atom in PF_4^- will have five pairs of electrons about it. There will be four bonding pairs and one lone pair. The arrangement of electron pairs will be trigonal bipyramidal with the lone pair occupying an equatorial position in order to minimize e-e repulsions. The name of the shape ignores the lone pairs, so the molecule is described best as having a seesaw structure. AX_4E; (b) The number and types of lone pairs is the same for $[ICl_4]^+$ as it is for PF_4^-. The structural arrangement of electron pairs and name are the same as in (a). AX_4E; (c) As in (a), the central P atom has five pairs of electrons about it, but this time they all are bonding pairs. The arrangement of pairs is still trigonal bipyramidal, but this time the name of the molecular shape is the same as the arrangement of electron pairs. The molecule is, therefore, a trigonal bipyramid. AX_5;

(d) Xenon tetrafluoride will have six pairs of electrons about the central atom, of which two are lone pairs and four are bonding pairs. These pairs will adopt an octahedral geometry, but because the name of the molecule ignores the lone pairs, the structure will be called square planar. The lone pairs are placed opposite each other rather than adjacent, in order to minimize e-e repulsions between the lone pairs. AX_4E_2

3.10 The Lewis structures are

(a)

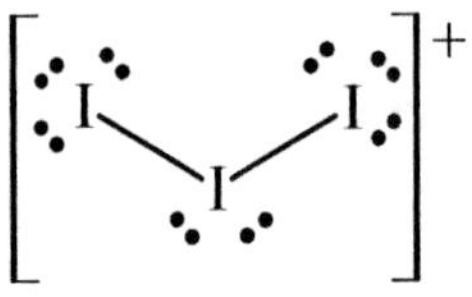

(b)

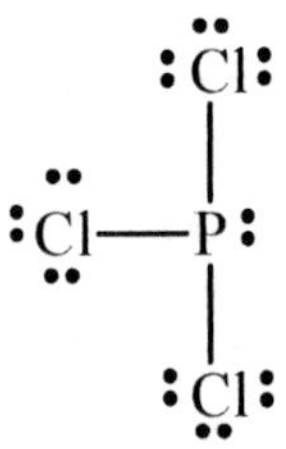

(c)

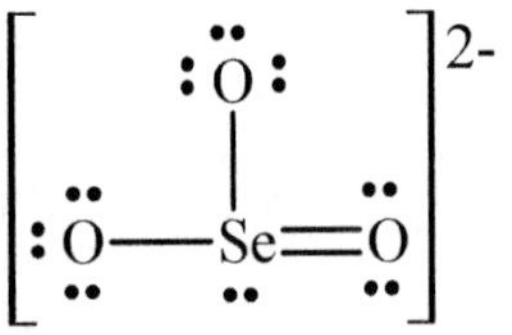

(d)

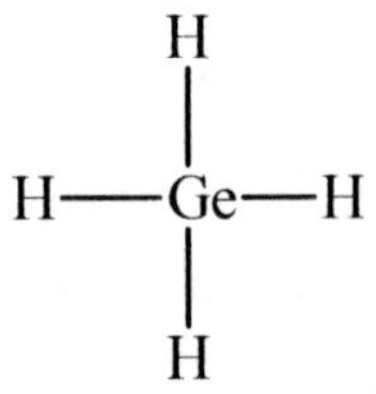

(a) I_3^+ should be bent with a bond angle of slightly less than 109.5°. AX_2E_2; (b) PCl_3 is pyramidal with Cl—P—Cl bond angles of slightly less than 109.5°. AX_3E; (c) SeO_3^{2-} is pyramidal with O—Se—O angles of slightly less than 109.5°. AX_3E; (d) GeH_4 is tetrahedral with H—Ge—H angles of 109.5°. Note that GeH_4 has the same electronic structure as CH_4 because Ge lies in the same group as C. AX_4

3.12 The Lewis structures are

(a)

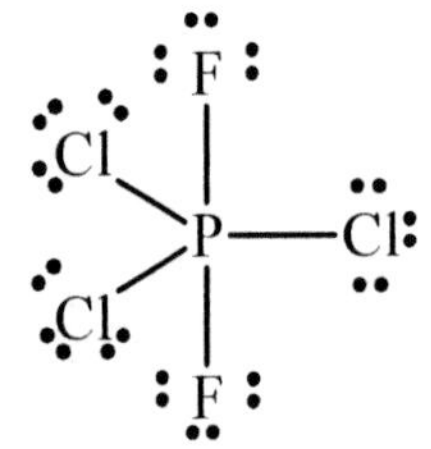

(b)

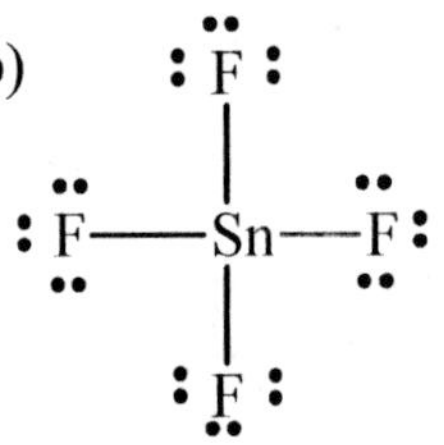

(c)

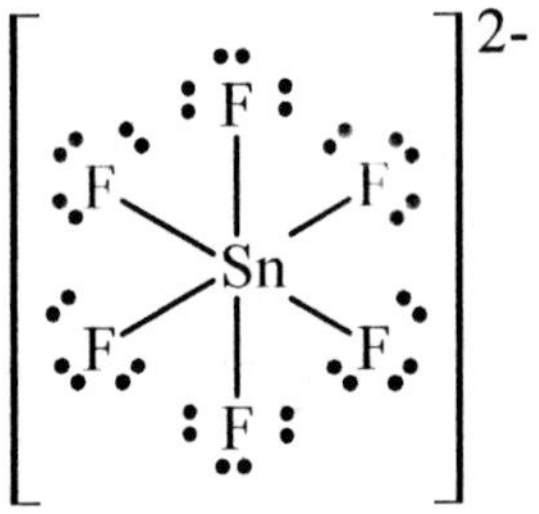

(d)

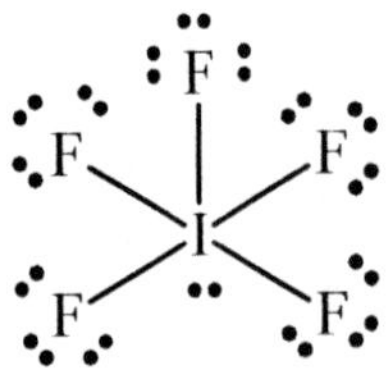

(e)

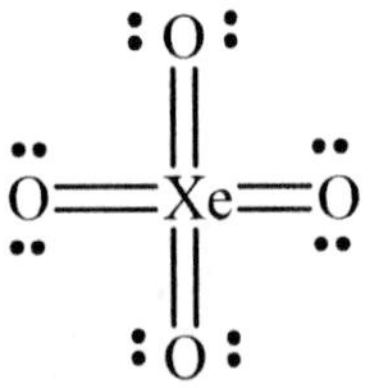

(a) PCl_3F_2 is trigonal bipyramidal with angles of 120°, 90°, and 180°. The most symmetrical structure is shown with all Cl atoms in equatorial positions. Phosphorus pentahalide compounds with more than one type of halogen atom like this have several possible geometrical arrangements of the halogen atoms. These different arrangements are known as isomers. For this type of compound, the energy differences between the different isomers are low, so that the compounds exist as mixtures of different isomers that are rapidly interconverting. AX_5 (or $AX_3X'_2$);

(b) SnF_4 is tetrahedral with F—Sn—F bond angles of 109.5°. AX_4; (c) SnF_6^{2-} is octahedral with F—Sn—F bond angles of 90° and 180°. AX_6; (d) IF_5 is square pyramidal with F—I—F angles of approximately 90° and 180°. AX_5E; (e) XeO_4 is tetrahedral with angles equal to 109.5°. AX_4

3.14 Angles a and c are expected to be approximately 120°. Angle b is expected to be around 109.5°.

3.16 (a) slightly less than 109.5°; (b) slightly less than 109.5°; (c) slightly less than 120°

3.18 The Lewis structures are

(a)

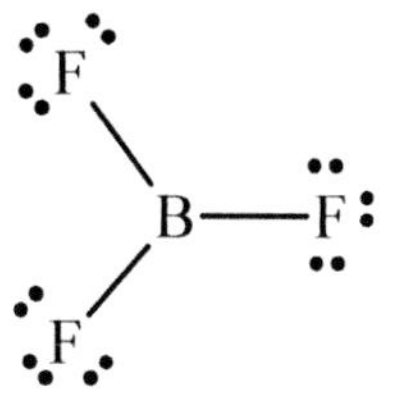

(b)

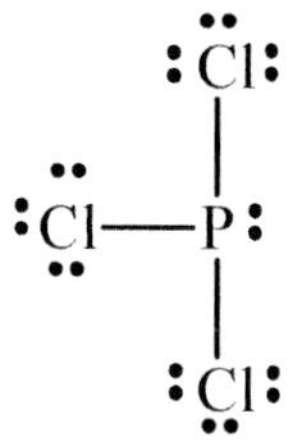

(c)

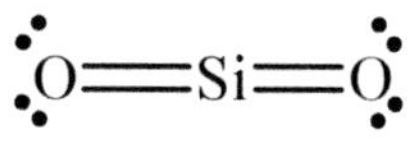

(d)

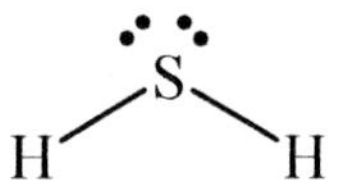

Molecules (b) and (d) will be polar; (a) and (c) will be nonpolar.

3.20 (a)

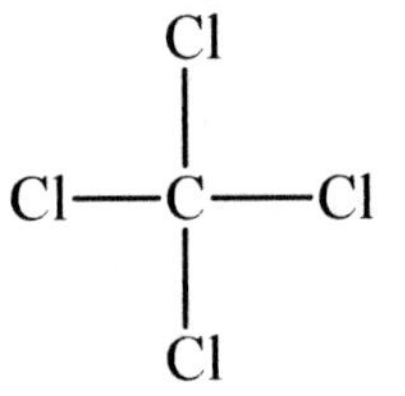

tetrachloromethane

nonpolar

(b)

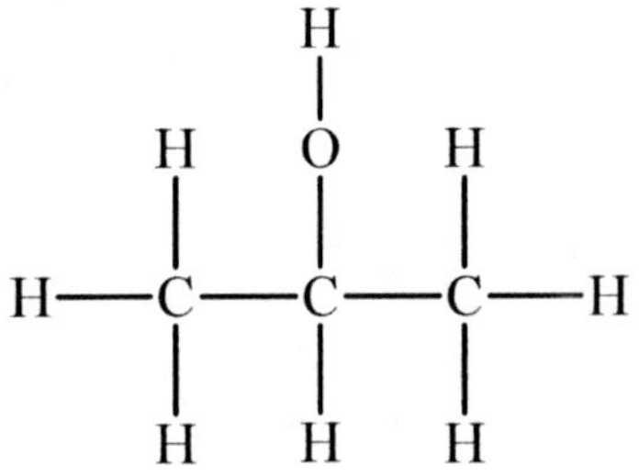

2-propanol

polar

(c)

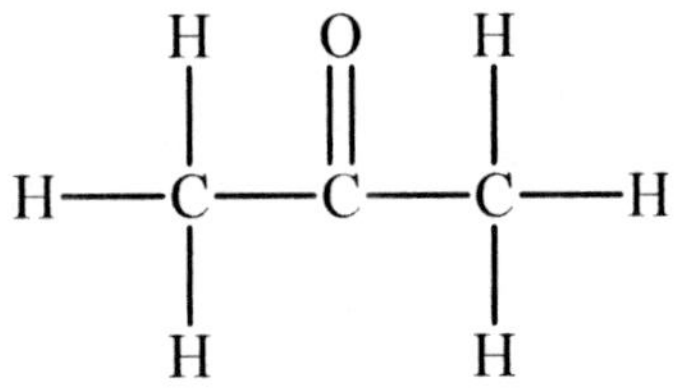

acetone

polar

3.22 (a) In **2** the C—H and C—F bond vectors oppose identical bonds on opposite ends of the molecule; the individual dipole moments will cancel so that **2** will be nonpolar. This is not true for either **1** or **3**, which will both be expected to be polar. (b) Assuming the C—F and C—H polarities are the same in molecules **1** and **3**, one can carry out a vector addition of the individual dipole moments. It is perhaps easiest to look at the resultant of the two F—C dipoles and the two C—H dipoles in each molecule. The dipoles will sum as shown:

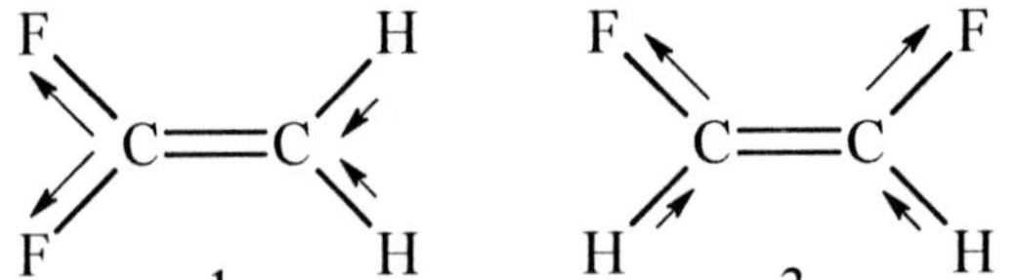

1 3

Addition of 2 C-F and 2 C-H dipoles in 1 and 3 give:

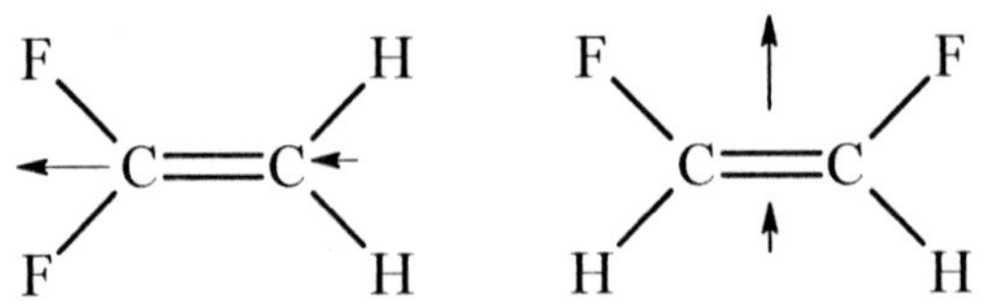

Net dipoles:

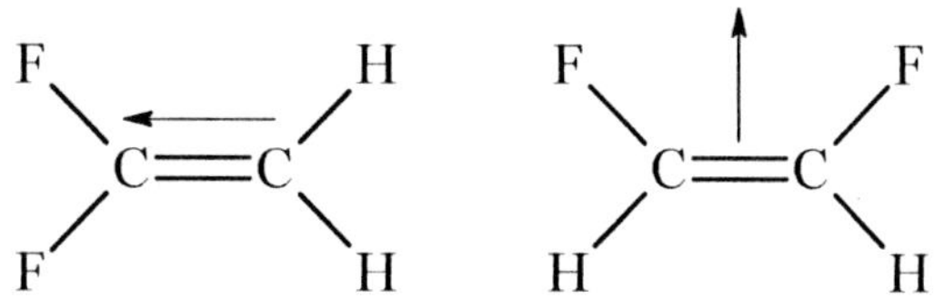

The net C—F and C—H dipoles reinforce in both these molecules, but because the F—C—F and H—C—H angles in **3** are more acute, the magnitude of the resultant will be slightly larger for **3**.

3.24

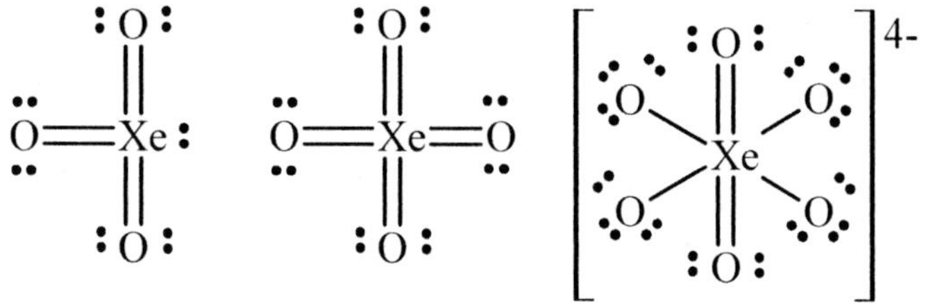

The first two molecules have four groups around the central atom, leading to tetrahedral dispositions of the bonds and lone pairs. XeO_3 is of the AX_3E type and will be pyramidal, whereas XeO_4 will be of the AX_4 type and will be tetrahedral. The XeO_6^{4-} ion will be octahedral. The hybridizations will be sp^3, sp^3, and sp^3d^2, respectively.

The Xe—O bonds should be longest in XeO_6^{2-} because each of those bonds should have a bond order of ca. $(4 \times 1 + 2 \times 2)/6 = 1.5$, whereas the bond orders in XeO_3 and XeO_4 will be about 2. This agrees with experiment: XeO_3, 174 pm; XeO_4, 176 pm; XeO_6^{2-}, 186 pm.

3.26 (a)

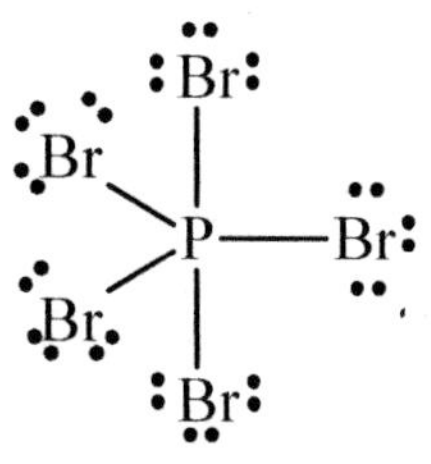

(b)

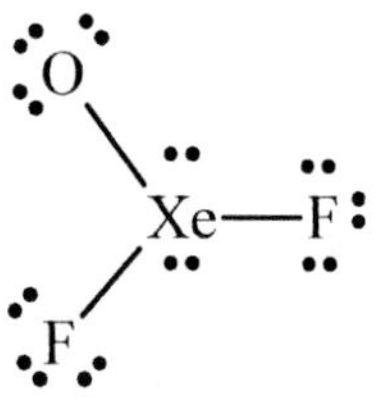

(a) no lone pairs on central atom; trigonalbipyramidyl; 90°, 120°

(b) two lone pairs on central atom; trigonal planar; 120°

(c)

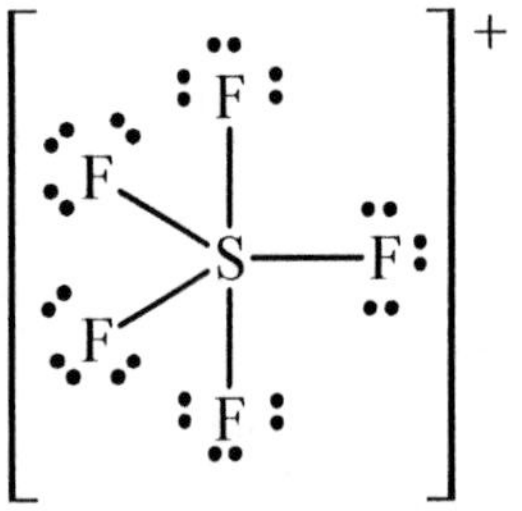

(d)

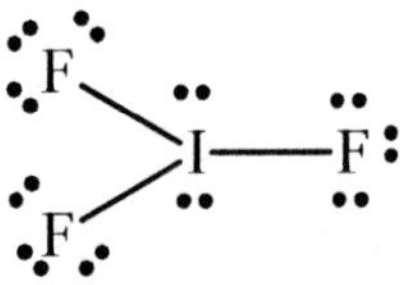

no lone pairs on central atom; tribonalbipyramidyl; 90°, 120°

two lone pairs on central atom trigonal planar; 120°

(e)

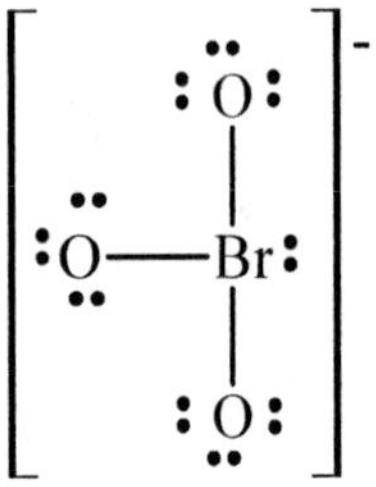

one lone pair on central atom

trigonal pyramidyl; 107°

3.28 (a)

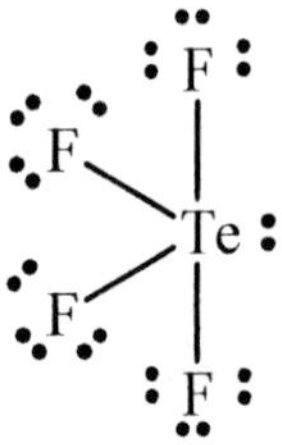

(b)

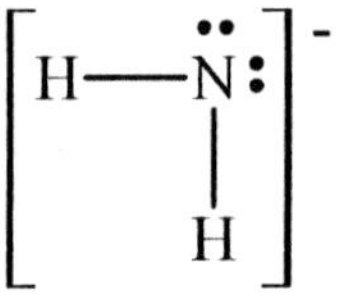

(c)

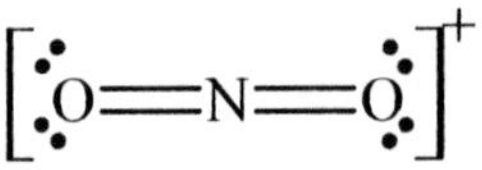

(d)

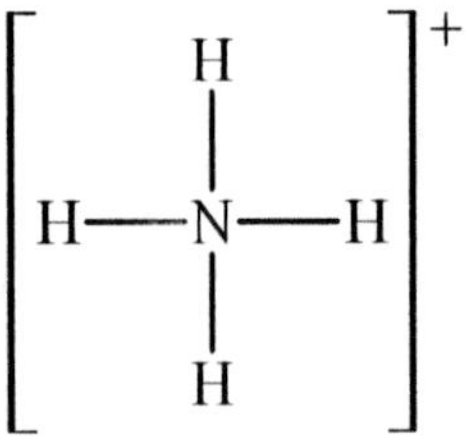

(e)

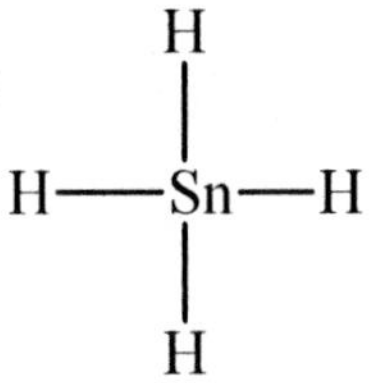

(f)

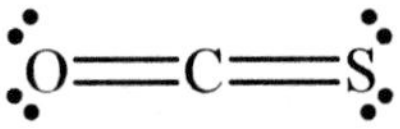

(a) seesaw; (b) angular; (c) linear; (d) tetrahedral; (e) tetrahedral; (f) linear

3.30 (a)

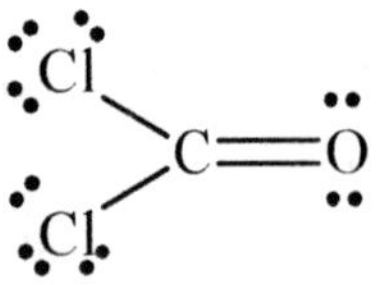

(b)

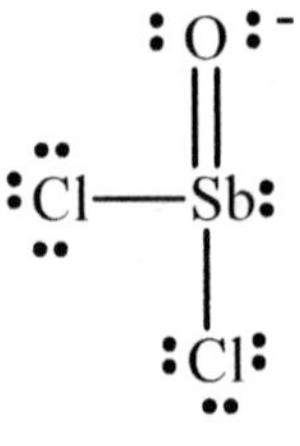

(a) trigonal planar; (b) pyramidal

3.32 (a) sp^3; (b) sp^3d; (c) sp^3d^2; (d) sp

3.34 (a) sp^3d^2; (b) sp^3; (c) sp^3; (d) sp^2

3.36 (a) sp; (b) sp^2; (c) sp^2; (d) sp^3

3.38 (a) sp; (b) sp^2; (c) sp; (d) sp^2

3.40 (a) In the NH_2^+ molecule, there is a single lone pair of electrons on the central N atom and as a result, the hybridization about the N atom is sp^2. In NH_2^-, there are two lone pairs of electrons on the central N atom and the hybridization about this atom is best described as sp^3. (b) The N2p_x orbital contributes to bonding in the NH_2^- molecule (all p-orbitals participate in the hybrid orbitals) but the N2p_x orbital does not contribute to bonding in NH_2^+. In this molecule, the N2p_x orbital lies normal to the plane containing the bonds and the lone pair, i.e. there is no overlap between the N2p_x and the orbitals responsible for bonding and the orbital containing the lone pair of electrons.

3.42 If we take the hybrid orbital to be $N \cdot h_1$, then normalization requires that:

$$\int (N \cdot h_1)^2 \, d\tau = 1$$

solving for N:

$$N^2 \cdot \int h_1^2 \, d\tau = N^2 \cdot \int (s + p_x + p_y + p_z)^2 \, d\tau = 1$$

Expanding this expression we find that the only terms that are not zero are:

$$N^2 \cdot \int s^2 + p_x^2 + p_y^2 + p_z^2 \, d\tau = N^2 \cdot 4 = 1$$

Therefore, $N = \sqrt{\frac{1}{4}}$

and the normalized wavefunction is $\sqrt{\frac{1}{4}}h_1$

3.44

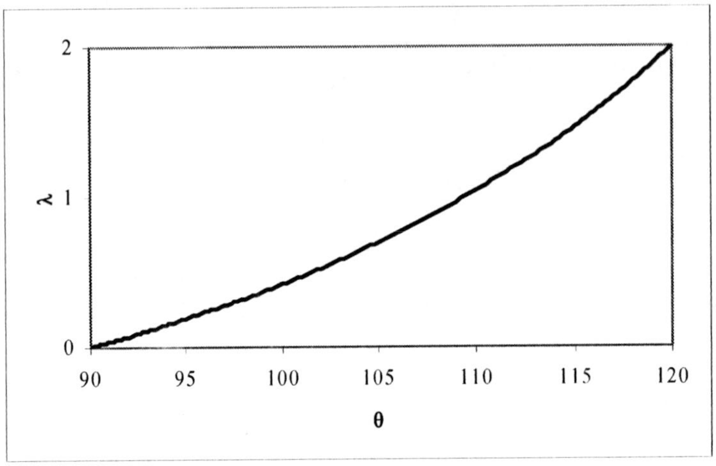

3.46 (a) B_2 BO = ½ (2 − 0) = 1 Paramagnetic, 2 unpaired electrons

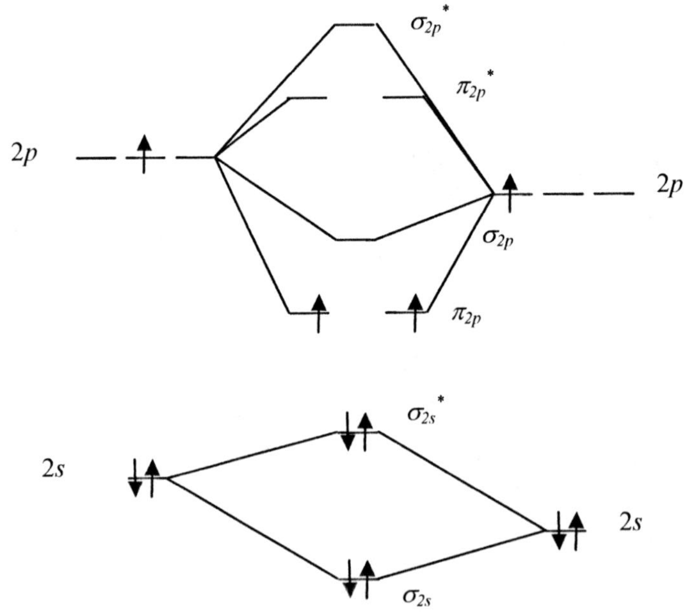

(b) B_2^- BO = ½ (2 + 3 − 2) = 3/2 Paramagnetic, one unpaired electron

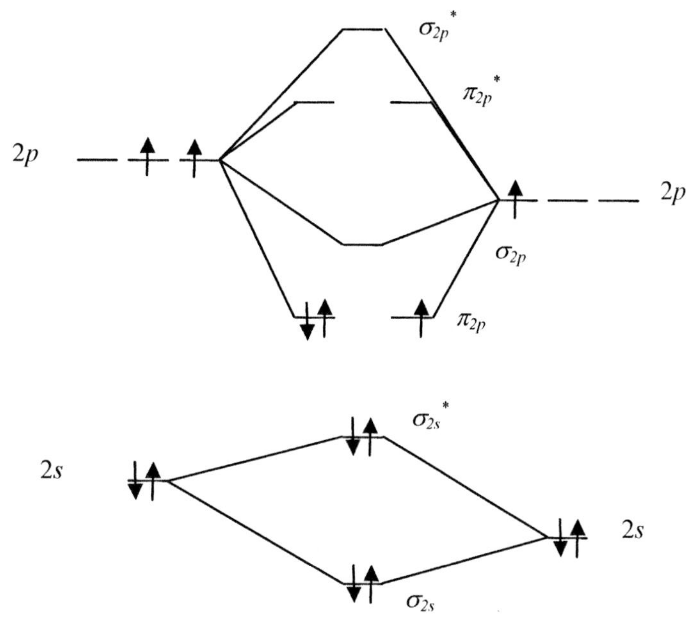

(c) B_2^+ BO = ½ (2 + 4 − 2) = 2 Paramagnetic, one unpaired electron

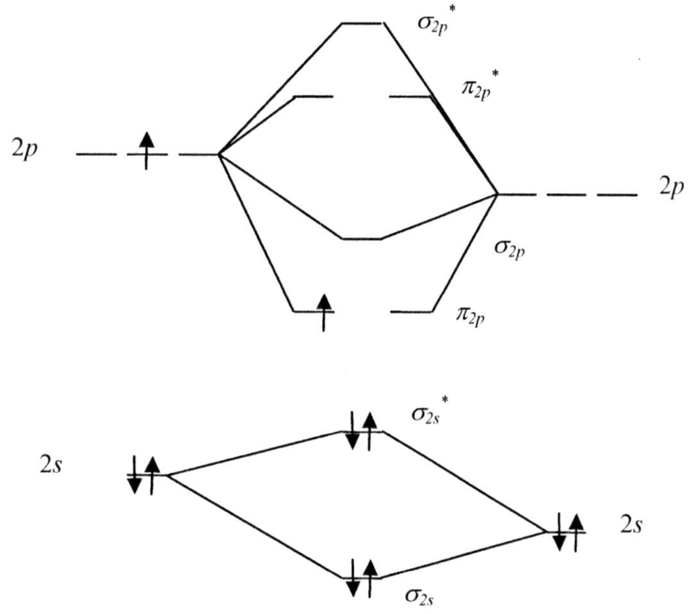

3.48 (a) (1) $(\sigma_{2s})^2(\sigma^*_{2s})^2(\pi_{2p_x})^2(\pi_{2p_y})^2(\sigma_{2p})^1$

(2) $(\sigma_{2s})^2(\sigma^*_{2s})^2(\pi_{2p_x})^2(\pi_{2p_y})^2$

(3) $(\sigma_{2s})^2(\sigma^*_{2s})^2(\pi_{2p_x})^2(\pi_{2p_y})^2(\sigma_{2p})^2(\pi^*_{2p_x})^1(\pi^*_{2p_y})^1$

(b) (1) 2.5

(2) 2

(3) 2

(c) (1) and (3) are paramagnetic

(d) (1) σ

(2) π

(3) π

3.50 (a) In molecular orbital theory, ionic and covalent bonding are extremes of the same phenomenon. According to molecular orbital theory, bonding occurs when at least one orbital on each of two atoms combines to form a bonding and antibonding set of molecular orbitals. If the orbitals on each atom that are used to make the molecular orbital are similar in energy, the resultant molecular orbitals will be composed almost equally of contributions from the two atoms. The result is a covalent bond. On the other hand, if the two orbitals that make up the molecular orbitals are quite different in energy, the bond will be highly polarized. The more electronegative atom will contribute more to the bonding orbital, and the more electropositive atom will contribute more to the antibonding orbital.
(b) The electronegativity of an atom is reflected in the energy of its atomic orbitals. The more electronegative the atom, the lower the orbitals. As the electronegativity difference becomes larger, the atomic orbitals on the two atoms being combined to form the molecular orbitals become farther apart in energy, so that the bond becomes more polarized.

3.52 (a) NO^+ (10 valence electrons): $(\sigma_{2s})^2(\sigma^*_{2s})^2(\sigma_{2p})^2(\pi_{2p_x})^2(\pi_{2p_y})^2$, bond order = 3;

(b) N_2^+ (9 valence electrons): $(\sigma_{2s})^2(\sigma^*_{2s})^2(\pi_{2p_x})^2(\pi_{2p_y})^2(\sigma_{2p})^1$, bond order = 2.5; (c) C_2^{2+}

(6 valence electrons): $(\sigma_{2s})^2(\sigma^*_{2s})^2(\pi_{2p_x})^1(\pi_{2p_y})^1$, bond order = 1.

3.54 N_2^- and F_2^+ have an odd number of electrons and must, therefore, be paramagnetic. Referring to the molecular orbital diagram for N_2 (**51**), we see that adding one electron will place it in a π^*_{2p} orbital. This will give one unpaired electron and a bond order of 2.5. The removal of an electron from F_2 to give F_2^+ will produce one unpaired electron in the π^*_{2p} orbital (see **52**). O_2^{2+} will be diamagnetic, as the removal of two electrons from O_2 will eliminate the two unpaired electrons (see **53**).

3.56 (a) C_2^+ (7 valence electrons): $(\sigma_{2s})^2(\sigma^*_{2s})^2(\pi_{2p_x})^2(\pi_{2p_y})^1$, bond order = 1.5; C_2 (8 valence electrons): $(\sigma_{2s})^2(\sigma^*_{2s})^2(\pi_{2p_x})^2(\pi_{2p_y})^2$, bond order = 2. C_2 will have the stronger bond. (b) O_2 (12 valence electrons):

$(\sigma_{2s})^2(\sigma^*_{2s})^2(\sigma_{2p})^2(\pi_{2p_x})^2(\pi_{2p_y})^2(\pi^*_{2p_x})^1(\pi^*_{2p_y})^1$,

bond order = 2; O_2^+ (11 valence electrons):

$(\sigma_{2s})^2(\sigma^*_{2s})^2(\sigma_{2p})^2(\pi_{2p_x})^2(\pi_{2p_y})^2(\pi^*_{2p_x})^1$,

bond order = 2.5. O_2^+ will have the stronger bond.

3.58 (a) and (b) The valence band is mostly filled with electrons although a small percentage of holes is present, due to electrons being promoted into the conduction band from the valence band or due to the presence of *p*-type dopants. When an electron moves from one location to a hole, the hole appears to move in the other direction, i.e., the hole is now located where the electron was. As a consequence, in semiconductor materials, holes move in the opposite direction of electrons, so if current is moving from left to right, the holes will be moving from right to left.

3.60 In order to produce an *n*-type semiconductor, we need to introduce an additional valence electron into the material. Selenium would be the element appropriate for this purpose because it has one more valence electron than arsenic. Substituting phosphorus for arsenic would not change the number of valence electrons present. Silicon would result in the formation of a *p*-type semiconductor.

3.62 Given the overlap integral $S = \int \Psi_{A1s}\Psi_{B1s}d\tau$, the bonding orbital $\Psi = \Psi_{A1s} - \Psi_{B1s}$, and the fact that the individual atomic orbitals are normalized, we are asked to find the normalization constant N which will normalize the bonding orbital Ψ such that:

$$\int N^2\Psi^2 d\tau = N^2 \int (\Psi_{A1s} - \Psi_{B1s})^2 d\tau = 1$$

$$N^2 \int (\Psi_{A1s} - \Psi_{B1s})^2 d\tau = N^2 \int (\Psi^2_{A1s} - 2\Psi_{A1s}\Psi_{B1s} + \Psi^2_{B1s})d\tau =$$

$$N^2\left(\int \Psi^2_{A1s} d\tau - 2\int \Psi_{A1s}\Psi_{B1s}d\tau + \int \Psi^2_{B1s} d\tau\right)$$

Given the definition of the overlap integral above and the fact that the individual orbitals are normalized, this expression simplifies to:

$$N^2(1 - 2S + 1) = 1$$

Therefore, $N = \sqrt{\dfrac{1}{2-2S}}$

To confirm that these two orbitals are orthogonal we evaluate the integral:

$$\int (\Psi_{A1s} - \Psi_{B1s})\cdot(\Psi_{A1s} + \Psi_{B1s})\, d\tau$$

$$= \int \Psi^2_{A1s} + \Psi_{A1s}\cdot\Psi_{B1s} - \Psi_{A1s} + \Psi_{B1s} - \Psi^2_{B1s}\, d\tau$$

$$= 1 - 1 = 0$$

The overlap integral is zero. Therefore, the two wavefunctions are orthogonal.

3.64 (a)

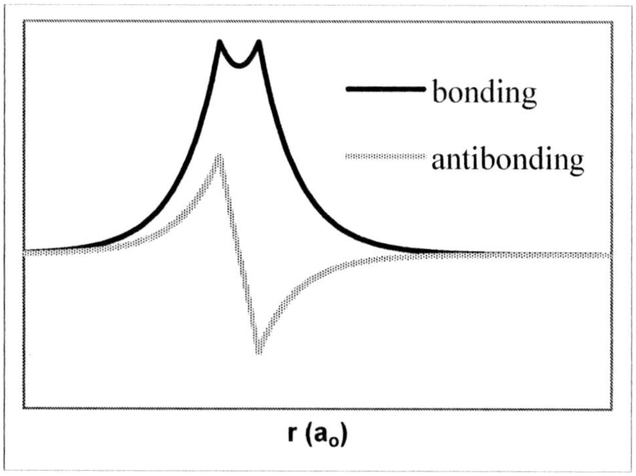
bonding
antibonding
r (a_o)

(b)

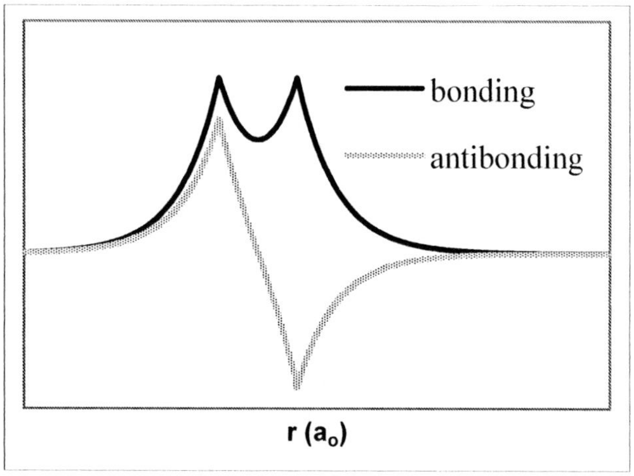
bonding
antibonding
r (a_o)

(c)

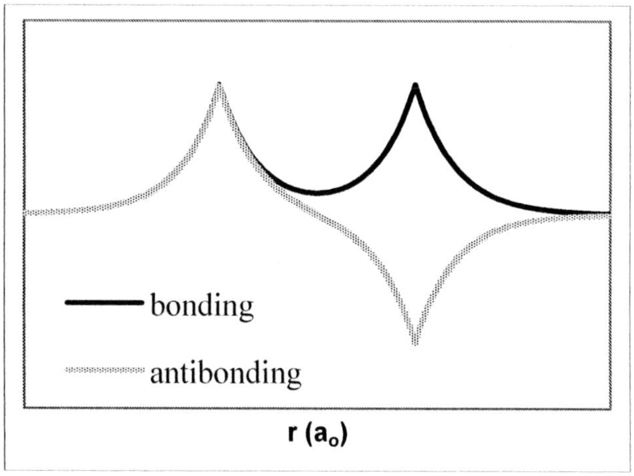
bonding
antibonding
r (a_o)

$1a_o$ is most likely closest to the real bond length as the orbital overlap integral will be greatest at this separation.

3.66 (a)

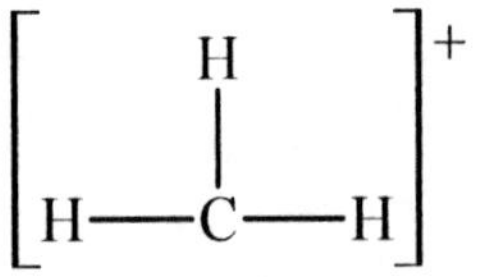

trigonal planar, sp^2, H—C—H bond angles = 120°, nonpolar

(b)

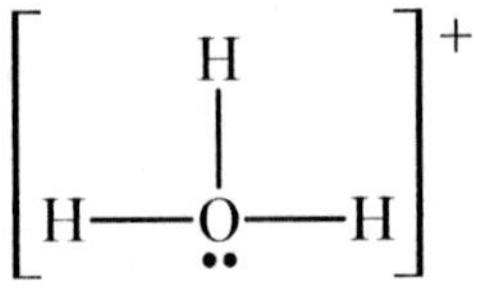

trigonal pyramidal, sp^3, H—O—H bond angles are slightly less than 109.5°, polar

(c)

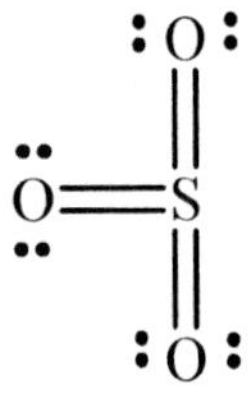

trigonal planar, sp^2, O—S—O bond angles = 120°, nonpolar

(d)

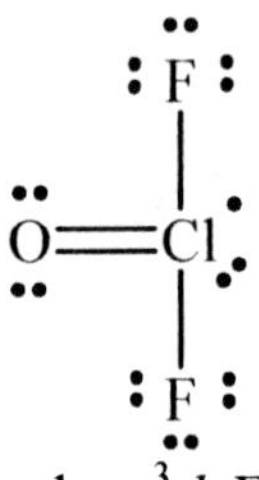

T-shaped, sp^3d, F—Cl—C, O—Cl—F bond angles are 90° and 180°, polar

(e)

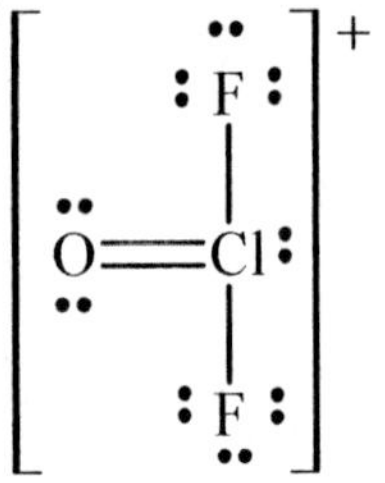

trigonal pyramidal, sp^3, bond angles are all slightly less than 109.5°, polar

3.68 (a) The XX′ molecules will be simple diatomic molecules with an X—X′ single bond. The XX'_3 molecules will have a central atom that is of the

VSEPR type AX_3E_2. The molecule will, therefore, be T-shaped. The X'—X—X' bond angles will be slightly less than 90° and 180°. The XX'_5 molecules will have central atoms of the type AX_5E, which will be a square pyramidal structure with bond angles of ca. 90° and 180°. All three types will be polar. (b) only the XX' molecules will be polar (c) The central atom is the one that is the least electronegative. A consideration of the oxidation numbers shows that the central atom is the one that is positive, and the atoms around it are negative. Thus, the central atom will be the one that retains its electrons less effectively.

3.70 (a) 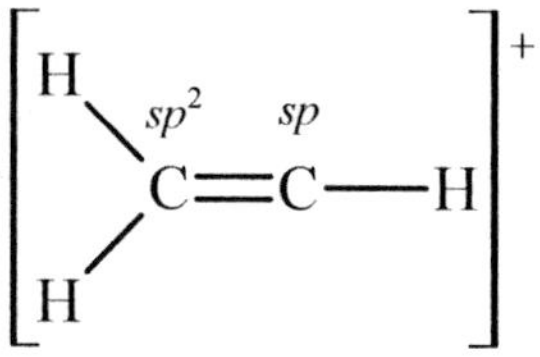

(b) 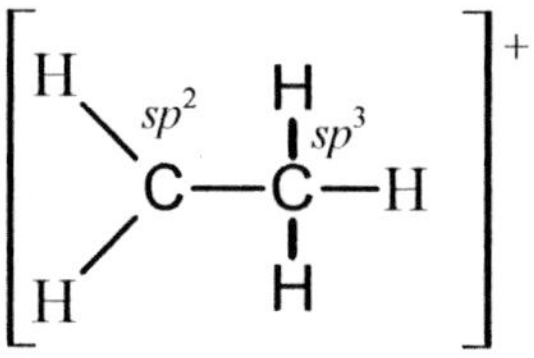

(c) 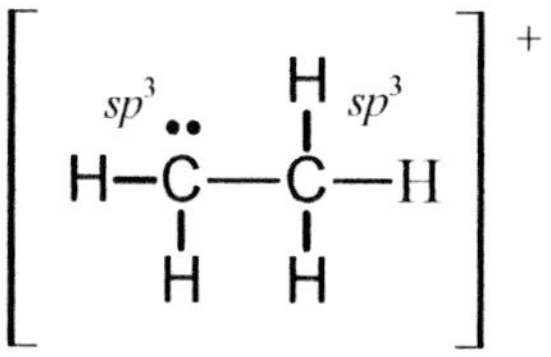

3.72 (a) The helium and hydrogen atoms have only their 1*s* orbitals available for bonding. Combination of these two will lead to a σ and a σ^* pair of orbitals. The most stable species will be one in which only the σ orbital is filled. Thus we need a species that has only two electrons. The charge on this species will be +1.

(b) The maximum bond order will be 1. (c) Adding one electron will decrease the bond order by 1/2 because an antibonding orbital will be populated. Taking an electron away will also decrease the bond order by 1/2 because it will remove one bonding electron.

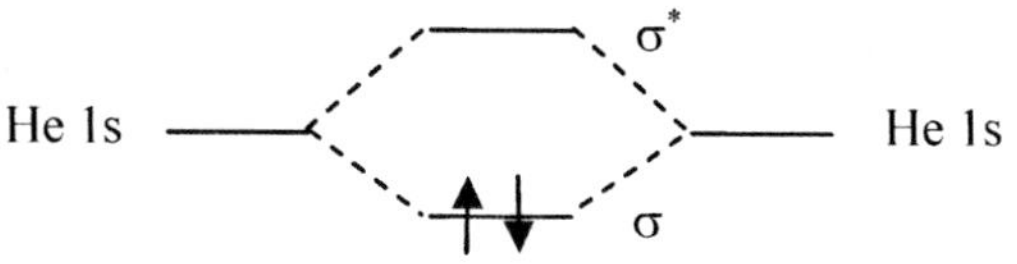

3.74 The Lewis structures that contribute to the structure of the carbamate ion are

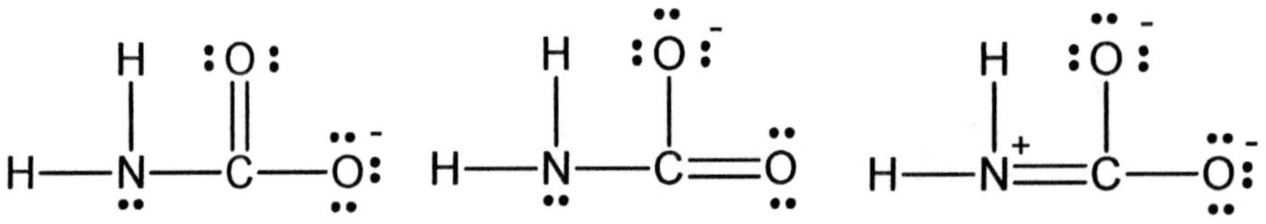

The charges show the location of the formal charges on the atoms in the Lewis structure. Note that the form that has a double bond to N is what we call a zwitterion, a structure that contains a + and a – charge in the same molecule. We might expect this structure to have a higher energy and contribute less to the overall structure than the other two forms, which have less separation of charge and which are equivalent in energy. To get insight into this question, we can compare the observed bond distances (C—O, 128 pm C—N, 136 pm) to those expected for various C—O and C—N bond orders. We can estimate these values using the data given in Table 2.3 and Figure 2.19. The following values are obtained:

Bond	Expected Bond Length, pm
C—O	151
C=O	112, 127
C—N	152
C=N	127

The C—O bond distance is very close to what we would expect for a C=O, although the average of experimental data gives a value closer to 112 pm. The second value is probably more reliable and indicates that the C—O bonds in the carbamate ion are intermediate between a double and single bond. The same is true of the C—N bond (Experimental values of C=N double bonds are close to 127 pm). It appears then that the resonance form with a C=N double bond does contribute substantially to the structure.

3.76 The three possible forms that maintain a valence of 4 at carbon are cyclopropene, allene, and propyne, which have the structures

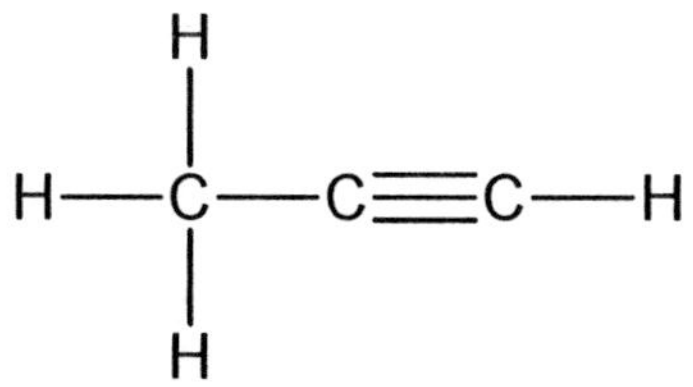

The C—C—C bond angles in cyclopropene are restricted by the cyclic structure to be approximately 60°, which is far from the ideal value of 109.47° for an sp^3 hybridized carbon atom, or 120° for an sp^2 hybridized C atom. Consequently, cyclopropene is a very strained molecule that is extremely unstable. The H—C—H bond angle at the CH_2 group would be expected to be 109.47°, but it is actually larger due to the narrow C—C—C bond angle. The carbon of the CH_2 group is sp^3 hybridized, whereas the other two carbon atoms are sp^2. Likewise the C=C—H angles, which would normally be expected to be 120° due to sp^2 hybridization at C, are somewhat larger. In allene, the middle carbon atom is *sp* hybridized; the two end carbons are sp^2 hybridized. The C—C—C bond angle is expected to be 180°; the H—C—H bond angles should be close to 120°. In propyne, one end carbon is sp^3 hybridized (109.5° angles); the other two C atoms are *sp* hybridized, with 180° bond angles. The three structures are not resonance structures of each other as they have a different spatial arrangement of atoms. For two structures to be resonance forms of each other, only the positions of the electrons may be changed. These two compounds would be known as *isomers* (see section 18.3).

3.78 (a)

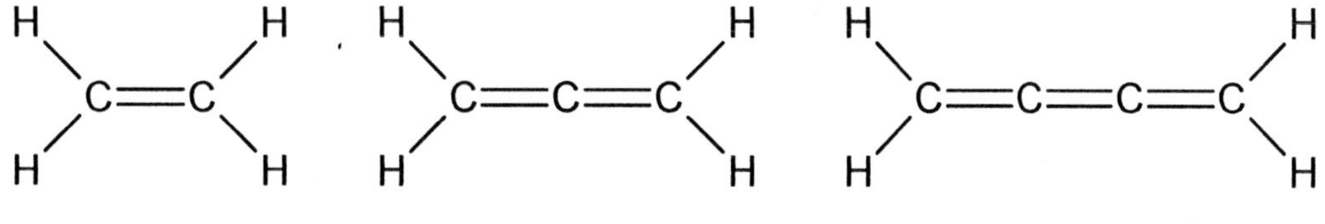

(b) The hybridization at the atoms attached to two hydrogen atoms is sp^2, whereas that at the carbon atoms attached only to two other carbon atoms is *sp*.

(c) Double bonds connect all of the carbon atoms to each other.

(d) The H—C—H and C—C—H angles should all be ca. 120°. The C—C—C angles will all be 180°.

(e) The hydrogen atoms in H_2CCH_2 and H_2CCCCH_2 lie in the same plane, whereas the planes that are defined by the two end CH_2 groups lie perpendicular to each other in H_2CCCH_2. This is because the *p* orbitals that are used by the central carbon atom to form the double bonds to the end carbon atoms are perpendicular to each other as shown.

Diagram of the interaction of the *p* orbitals used in making the C—C double bonds:

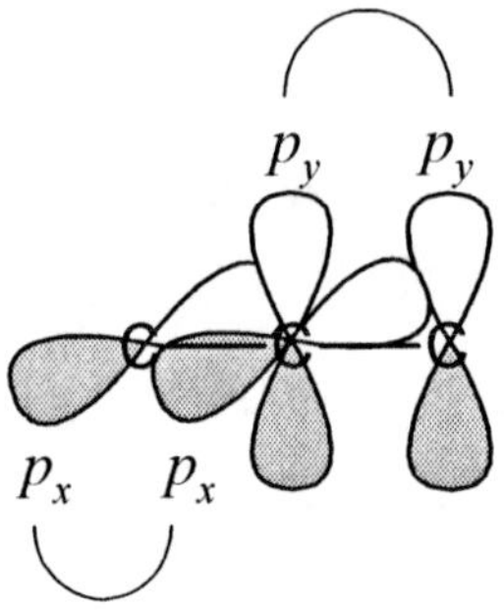

Diagram showing the orientation of the sp^2 orbitals used for bonding to the H atoms:

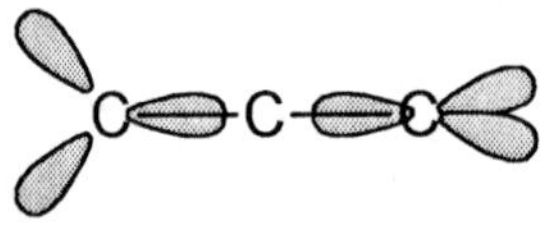

The three different *p* orbitals are commonly labeled p_x, p_y, and p_z to distinguish them. They may be thought of as pointing in the x, y, and z directions on a set of Cartesian axes that intersect at the carbon atom. The three *p* orbitals on a given carbon atom will be oriented 90° apart. Thus, if the central carbon atom is *sp* hybridized, the *sp* hybrid orbitals will be 180° apart (not shown) and will form the σ bonds to the other carbon atoms. This arrangement will use the *p* orbitals on the end carbon atoms that are oriented in the same direction (as shown here, the p_z orbitals on all three carbons have been used). The p_x and p_y orbitals on the middle carbon will be used in forming the bonds. This means that the bond to one carbon atom will be made from p_x, p_x interactions and the bond to the other will be p_y, p_y. One end carbon will be using the *s*, p_z, and p_x orbitals for forming the sp^2 hybrids and the other will be using the *s*, p_z, and p_y

orbitals. This change will result in the sp^2 orbitals on the end carbons being oriented 90° away from each other. This orientation effect will occur only if there is an odd number of carbons in the chain. If there is an even number of carbon atoms, the end C atoms will be required to use the same type of p orbitals as each other for forming the sp^2 hybrid orbitals. Thus H_2CCCCH_2 should have all the hydrogen atoms in the same plane, and in $H_2CCCCCH_2$ the planes of the end CH_2 groups will again be perpendicular to each other.

(f) If x is odd, the hydrogens will point toward the same side of the molecule. If x is even, they will point toward different sides of the molecule.

3.80 Dyes get their color from the absorption of visible light. This absorption cannot take place unless there are bonding and antibonding orbitals available that have the correct energy spacing in order to absorb this visible light. The presence of a number of multiple bonds in molecules may lead to a delocalized set of orbitals. It is these delocalized orbitals that often absorb light in the visible region, giving rise to the desired color.

3.82 To show that two π orbitals taken together are cylindrically symmetric, we assume that the z axis is the bonding axis and show that net probability distribution (the sum of the two molecular orbitals squared) is not a function of ϕ. (Refer to Figure 1.24 on phage 20 to confirm that, looking down the z axis, a cylindrically symmetric orbital will not be a function of ϕ).

$$x \cdot f(z) = f(z) \cdot C \cdot \sin\theta \cdot \cos\phi$$
$$y \cdot f(z) = f(z) \cdot C \cdot \sin\theta \cdot \sin\phi$$

where $C = \left(\frac{3}{4\pi}\right)^{\frac{1}{2}}$, and $f(z)$ is not a function of x or y

(and, therefore, $f(z)$ is not a function of ϕ). Squaring the two wavefunctions and summing them:

$$f(z)^2 C^2 \sin^2\theta \cos^2\phi + f(z)^2 C^2 \sin^2\theta \sin^2\phi$$
$$= f(z)^2 C^2 \sin^2\theta\left(\cos^2\phi + \sin^2\phi\right)$$

Using the identity $\cos^2 x + \sin^2 x = 1$ this becomes

$$= f(z)^2 C^2 \sin^2\theta$$

With two electrons in each π orbital, the electron distribution is not a function of ϕ and is, therefore, cylindrically symmetric about the bonding axis.

3.84 The overlap can be end-to-end or side-on. In the side-on overlap situation, the net overlap is zero because the areas of the wave function that are negative will cancel with the areas of the wave function that are positive. These will be equal in area and opposite in sign, giving no net overlap.

(a) 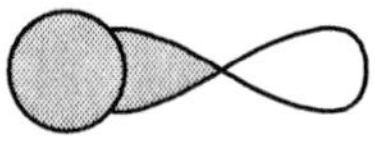 (b) 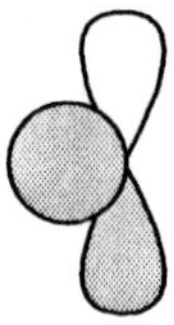

3.86

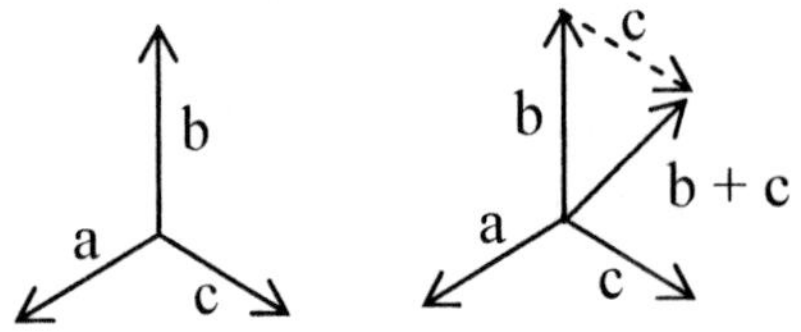

For the trigonal planar molecule, we first construct the vectors representing the individual dipole moments and then add them. This can be done most easily by combining two of the vectors first and then adding that to the third. Let's first add b and c. In order to add the vectors, we need to position them head to tail. Because the original angle where b and c come together in the AX_3 molecule is $120°$, the angle between them when the origin of c is shifted to the end of b will be $60°$. Because that angle is 60° and b is the same length as c, the resultant of b + c must also have the same length as a (and b and c). As can be seen from the diagram, the original vector a points in exactly the opposite direction from the summed vector b + c. Because the angle between b and b + c is 60° and the angle between a and b is 120°, it follows that the angle between a and b + c must be 180°. Because b + c has exactly the same magnitude as a, they will exactly cancel each other.

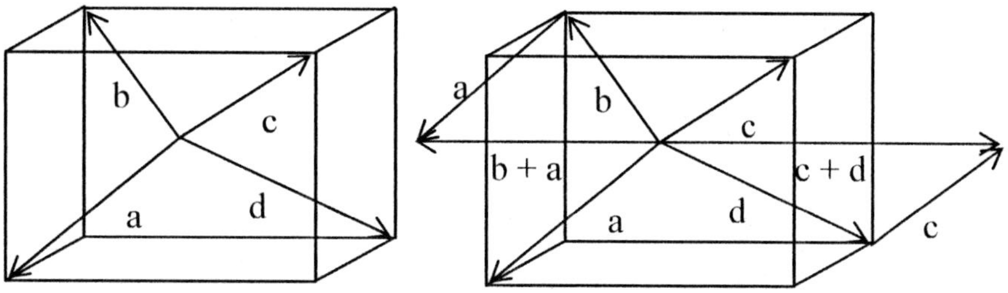

For the tetrahedron, which can be inscribed inside a cube as shown, the vectors can be added in pairs to give resultants as shown. Starting with original vectors a, b, c, and d, we can add a + b and c + d. (For an ideal tetrahedron in which all the atoms bonded to the central atom are identical, the choice of which vector is a, b, c, or d is arbitrary because they are all equivalent.) An examination of the tetrahedron will show that the vectors a and b lie in a plane that is perpendicular to the plane in which vectors c and d lie. The resultant c + d will be equal in magnitude to the resultant a + b because the vectors a, b, c, and d all have the same magnitude, and the angle between them is the same. It is easy to see that the vectors a + b and c + d lie exactly opposite each other and will add to zero.

3.88 (a) There are many, many structural reports in which the sulfate ion is the counterion. (b) The errors in terms of σ will vary, depending on the quality of the x-ray diffraction data, but generally data with three or four significant figures will be obtained. (c) Many structural reports will give results in which all the S—O distances and O—S—O angles will be the same within experimental error (within 3σ of each other). Some structural data, however, will give values that are significantly different from each other. This will be obvious if compared for a large number of compounds. Distances will be near 149 pm. (d) If the sulfate ions were well isolated from other species, as in the gas phase at low pressure, you would expect the values to be identical. However, in the solid state, there are interactions between other species in the crystal lattice and the sulfate ion. The result is that the environment around the sulfate ion is often not symmetrical, and different bond distances and angles may result. Even so, the distances will fall within a fairly close range of values.

3.90 If we assume that the size of the lanthanum atom inside a C_{60} molecule is given by the atomic radius of the lanthanum atom, then we find that the diameter of a lanthanum atom would be $2 \times 188 \text{ pm} = 376 \text{ pm}$ (from Appendix 2D). For the C_{60} molecule to encapsulate more than one lanthanum atom, it would have to have a diameter at least twice the size of a lanthanum atom $(2 \times 376 \text{ pm} = 752 \text{ pm})$. Because the diameter of C_{60} is approximately 100 pm, two lanthanum atoms could not fit inside. (c) The radius of a La^{3+} ion is 122 pm so that its diameter would be 244 pm. Two La^{3+} ions could possibly fit inside a C_{60} molecule (or $C_{60}{}^{6-}$ ion) because $2 \times 244 \text{ pm}$ is substantially smaller than the diameter of the C_{60} structure; however, placing two +3 ions in such close proximity would be electrostatically very unfavorable, and it would be unlikely to occur.

3.92 Yes. The color is due to absorption of light, which causes an electron to move from one *d* orbital to another *d* orbital (*d-d* transition).

3.94 (a) Doping silicon with P will form an n-type semiconductor. However, to make an LED one must provide a junction between an n-type seminconductor and a p-type semiconductor. Si has a completely full valence band. Electrons arriving in the conduction band of pure silicon can not fall down to the valence band, giving a photon, because the valence band in pure silicon is full.

(b) and (c) If the flow of electrons is reversed, electrons from the valence band of the p-type material would be "forced" into the conduction band of the n-type material as the valence band of the n-type material is full. One would not expect any light to be produced as electrons placed in the conduction band of the n-type material cannot fall down to the conduction band as the conduction band in an n-type semiconductor is full.

CHAPTER 4

THE PROPERTIES OF GASES

4.2 (a) 161 kPa; (b) 1.52×10^3 Torr; (c) 2.03 bar; (d) 2.00 atm

4.4 The difference in height will be 2×35.96 cmHg which will equal 71.92 cmHg. This is the same as 719.2 mmHg. (a) Because 1 mmHg is approximately equal to 1 Torr, there will be 719.2 Torr; (b) 719.2 Torr $\div$ 760 $\text{Torr} \cdot \text{atm}^{-1} = 0.9463$ atm; (c) 719.2 Torr $\times$ 133.3 $\text{Pa} \cdot \text{Torr}^{-1}$ $= 9.587 \times 10^4$ Pa; (d) 0.9587 bar

4.6 $d_1 h_1 = d_2 h_2$

$$d_1(6.14\ \text{m})(100\ \text{cm} \cdot \text{m}^{-1}) = (13.6\ \text{g} \cdot \text{cm}^{-3})\frac{(758.7\ \text{mm})}{(10\ \text{mm} \cdot \text{cm}^{-1})}$$

$$d_1 = 1.68\ \text{g} \cdot \text{cm}^{-3}$$

4.8 The pressure is inversely proportional to the density:

$$1.5\ \text{in}\ H_2O \times \left(\frac{1.0\ \text{g} \cdot \text{cm}^3}{13.5\ \text{g} \cdot \text{cm}^3}\right) = 0.11\ \text{inHg}$$

$$0.11\ \text{in Hg} \times \left(\frac{1.0\ \text{cm}}{0.3937\ \text{in}}\right)\left(\frac{10\ \text{mm}}{1\ \text{cm}}\right) = 2.8\ \text{mmHg}$$

2.8 mmHg or 2.8 Torr

4.10 (a)

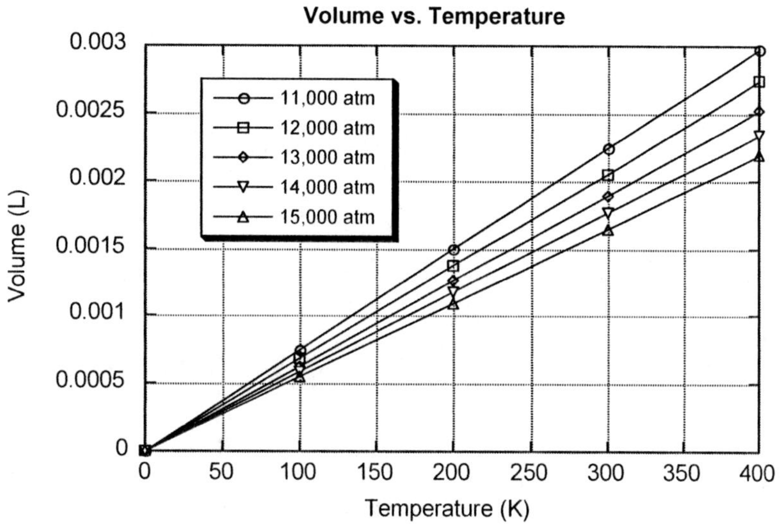

(b) The slope is equal to $\frac{nR}{P}$

Pressure, atm	$\frac{nR}{P}$, $L \cdot K^{-1}$
11 000	7.46×10^{-6}
12 000	6.83×10^{-6}
13 000	6.31×10^{-6}
14 000	5.86×10^{-6}
15 000	5.47×10^{-6}

(c) The intercept is equal to 0.00 for all the plots.

4.12 (a) From $P_1V_1 = P_2V_2$, we write

$(800.\ \text{Torr})(4.00\ \text{L}) = (P_2)(3.20\ \text{L});\ P_2 = 1.00 \times 10^3\ \text{Torr};$

(b) Similarly, from $P_1V_1 = P_2V_2$, the expression is

$(158\ \text{Pa})(936\ \text{mL}) = (P_2)(468\ \text{mL});\ P_2 = 316\ \text{Pa}.$

4.14 Because P is a constant, we can use $\frac{V_1}{T_1} = \frac{V_2}{T_2}$, which gives

$$\frac{22.5 \text{ L}}{291 \text{ K}} = \frac{V_2}{258 \text{ K}}; V_2 = 19.9 \text{ L}.$$

4.16 Because P is constant, we can use: $\frac{V_1}{T_1} = \frac{V_2}{T_2}$. Substituting, we obtain

$\frac{250 \text{ mL}}{338 \text{ K}} = \frac{V_2}{273 \text{ K}}$; $V_2 = 202$ mL. Note: It is not necessary to convert volume to liters, but the final answer will be found in the same units as the initial volume. It is, however, necessary to convert T to kelvins.

4.18 Because we want P and V to be constant, we can use the relationship $n_1T_1 = n_2T_2$. If the number of molecules is halved, the number of moles is also halved, so we can write $n_1T_1 = (2n_1)T_2$. Solving, we find $T_2 = 2T_1$. For V and P to remain constant, the temperature must double.

4.20 Because P, V, and T change, we use the relation $\frac{P_1V_1}{T_1} = \frac{P_2V_2}{T_2}$. Substituting for the appropriate values, we get

$$\frac{(1.08 \text{ atm})(355 \text{ cm}^3)}{310 \text{ K}} = \frac{(0.958 \text{ atm})(V_2)}{298 \text{ K}}; V_2 = 385 \text{ cm}^3.$$

4.22 (a) From $PV = nRT$, (1.30 atm)(0.125 L) =

$n(0.082\ 06 \text{ L}\cdot\text{atm}\cdot\text{K}^{-1}\cdot\text{mol}^{-1})(350 \text{ K})$, $n = 0.005\ 66$ mol.

(b) Substituting into the ideal gas law

$$P\ (0.120 \text{ L}) = \left(\frac{2.7\times10^{-6} \text{ g}}{32.00 \text{ g}\cdot\text{mol}^{-1}}\right)(0.082\ 06 \text{ L}\cdot\text{atm}\cdot\text{K}^{-1}\cdot\text{mol}^{-1})(290 \text{ K}).$$

$$= 1.7\times10^{-5} \text{ atm}\times\left(\frac{760 \text{ Torr}}{1 \text{ atm}}\right) = 1.3\times10^{-2} \text{ Torr}$$

(c) The number of mol N_2 present is given by $n = \frac{PV}{RT}$. The mass m will be given by $n \cdot M$ where M is the molar mass of N_2.

(d)

$$m = \frac{MPV}{RT} = \frac{(28.01\ \text{g} \cdot \text{mol}^{-1})\left(\dfrac{20\ \text{Torr}}{760\ \text{Torr} \cdot \text{atm}^{-1}}\right)(20.0\ \text{L})}{(0.082\ 06\ \text{L} \cdot \text{atm} \cdot \text{K}^{-1} \cdot \text{mol}^{-1})\ (215\ \text{K})}; m = 0.84\ \text{g}.$$

$$V = \frac{nRT}{P}; V = \frac{\left(\dfrac{16.7\ \text{g}}{83.80\ \text{g} \cdot \text{mol}^{-1}}\right)(0.082\ 06\ \text{L} \cdot \text{atm} \cdot \text{K}^{-1} \cdot \text{mol}^{-1})\ (317\ \text{K})}{\dfrac{(100 \times 10^{-3}\ \text{Torr})}{(760\ \text{Torr} \cdot \text{atm}^{-1})}}$$

$$= 3.94 \times 10^4\ \text{L}$$

(e) To calculate the number of atoms, we multiply the number of moles by the Avogadro constant. To calculate the number of moles, we use the ideal gas law:

$n = \frac{PV}{RT}$. The number of atoms of xenon, N_{Xe}, is

$$N_{\text{xe}} = \frac{(6.022 \times 10^{23}\ \text{atoms} \cdot \text{mol}^{-1})(P)(V)}{RT}$$

$$= \frac{(6.022 \times 10^{23}\ \text{atoms} \cdot \text{mol}^{-1})\left(\dfrac{2.00\ \text{Torr}}{760\ \text{Torr} \cdot \text{atm}^{-1}}\right)(2.6 \times 10^{-6}\ \text{L})}{(0.082\ 06\ \text{L} \cdot \text{atm} \cdot \text{K}^{-1} \cdot \text{mol}^{-1})\ (288\ \text{K})}$$

$$= 1.7 \times 10^{14}\ \text{atoms}$$

4.24 (a) Because n and V are constant, we simply use the relationship $\frac{P_1}{T_1} = \frac{P_2}{T_2}$:

$$\frac{1.00\ \text{atm}}{273\ \text{K}} = \frac{P_2}{373\ \text{K}}; P_2 = 1.37\ \text{atm}.$$

(b) $\frac{1.00\ \text{atm}}{273\ \text{K}} = \frac{P_2}{500\ \text{K}}; P_2 = 1.83\ \text{atm}.$

4.26 The mass of the CO_2 will be 1.04 kg – 0.74 kg = 0.30 kg.

From the ideal gas law

$$PV = nRT$$

$$P = \frac{nRT}{V}$$

$$P = \frac{\left(\dfrac{0.30\ \text{kg} \times 1000\ \text{g}\cdot\text{kg}^{-1}}{44.01\ \text{g}\cdot\text{mol}^{-1}}\right)(0.083\,145\,1\ \text{L}\cdot\text{bar}\cdot\text{K}^{-1}\cdot\text{mol}^{-1})(293\ \text{K})}{0.250\ \text{L}}$$

$$= 6.6 \times 10^2\ \text{bar}$$

4.28 Because P is constant, we can use

$$\frac{V_1}{T_1} = \frac{V_2}{T_2}$$

$$\frac{(4.60\ \text{L})}{(400.2\ \text{K})} = \frac{(0.115\ \text{L})}{T_2}$$

$$T_2 = 10.0\ \text{K or } -263.2^\circ\text{C}$$

4.30 $\dfrac{V_1}{T_1} = \dfrac{V_2}{T_2}$ or $\dfrac{V_2}{V_1} = \dfrac{T_2}{T_1}$

We want V_2 to be 1.14 times V_1:

$$1.14 = \frac{T_2}{T_1} = \frac{T_2}{340\ \text{K}}$$

$$T_2 = 388\ \text{K}$$

4.32 Assuming the balloon is spherical, the volume will be determined by $V = \frac{4}{3}\pi r^3$. Using the ideal gas law, we can calculate P:

$$P_1V_1 = P_2V_2$$

$$P_2 = \frac{P_1V_1}{V_2} = \frac{(1\ \text{atm})\left(\dfrac{4}{3}\pi(1.0\ \text{m})^3\right)}{\dfrac{4}{3}\pi(3.0\ \text{m})^3} = 0.037\ \text{atm}$$

4.34 The key to this problem is to realize that the same number of moles of gas will exert the same pressure, regardless of what the gas is (assuming the gases are all reasonably ideal). The answer will be given by

$$\left(\frac{12\text{ mg H}_2\text{S}}{34.08\text{ g}\cdot\text{mol}^{-1}\text{ H}_2\text{S}}\right)(17.03\text{ g}\cdot\text{mol}^{-1}\text{ NH}_3) = 6.0\text{ mg NH}_3$$

4.36 Density is proportional to the molar mass of the gas as seen from the ideal gas law:

$$PV = nRT$$

$$PV = \frac{m}{M}RT$$

$$\text{density} = \text{mass per unit volume} = \frac{m}{V} = \frac{MP}{RT}$$

The molar masses of the gases in question are 28.01 $\text{g}\cdot\text{mol}^{-1}$ for $N_2(g)$, 17.03 $\text{g}\cdot\text{mol}^{-1}$ for $NH_3(g)$, and 46.01 $\text{g}\cdot\text{mol}^{-1}$ for $NO_2(g)$. The most dense will be the one with the highest molar mass, which in this case is NO_2.

The order of increasing density will be $NH_3 < N_2 < NO_2$.

4.38 (a) Density is proportional to the molar mass of the gas as seen from the ideal gas law. See Section 4.9.

$$d = \frac{(34.08\text{ g}\cdot\text{mol}^{-1})(1.00\text{ atm})}{(0.082\ 06\text{ L}\cdot\text{atm}\cdot\text{K}^{-1}\cdot\text{mol}^{-1})(298\text{ K})} = 1.39\text{ g}\cdot\text{L}^{-1}$$

(b) $$d = \frac{(34.08\text{ g}\cdot\text{mol}^{-1})(0.962\text{ atm})}{(0.082\ 06\text{ L}\cdot\text{ atm}\cdot\text{K}^{-1}\cdot\text{mol}^{-1})(308.2\text{ K})} = 1.30\text{ g}\cdot\text{L}^{-1}$$

4.40 (a) $$0.943\text{ g}\cdot\text{L}^{-1} = \frac{MP}{RT} = \frac{M\left(\dfrac{727\text{ Torr}}{760\text{ Torr}\cdot\text{atm}^{-1}}\right)}{(0.082\ 06\text{ L}\cdot\text{atm}\cdot\text{K}^{-1}\cdot\text{mol}^{-1})(420\text{ K})}$$

$$M = 34.0\text{ g}\cdot\text{mol}^{-1}$$

(b) $$d = \frac{(34.0\text{ g}\cdot\text{mol}^{-1})(1.00\text{ atm})}{(0.082\ 06\text{ L}\cdot\text{atm}\cdot\text{K}^{-1}\cdot\text{mol}^{-1})(298\text{ K})} = 1.39\text{ g}\cdot\text{L}^{-1}$$

4.42 The empirical formula derived from the elemental analyses is $(CH_2)_n$.

The problem may be solved using the ideal gas law:

$$PV = nRT$$

$$PV = \frac{m}{M}RT$$

$$M = \frac{mRT}{PV}$$

$$M = \frac{(1.77\ \text{g})(0.082\ 06\ \text{L}\cdot\text{atm}\cdot\text{K}^{-1}\cdot\text{mol}^{-1})(290\ \text{K})}{\left(\dfrac{508\ \text{Torr}}{760\ \text{Torr}\cdot\text{atm}^{-1}}\right)(1.500\ \text{L})} = 42.0\ \text{g}\cdot\text{mol}^{-1}$$

The empirical formula mass is 14.01 $\text{g}\cdot\text{mol}^{-1}$. The value of n in the formula $(CH_2)_n$ is, therefore, equal to 3.

4.44 $PV = nRT$

$$PV = \frac{m}{M}RT$$

$$M = \frac{mRT}{PV} = \frac{(115\times10^{-3}\ \text{g})(0.082\ 06\ \text{L}\cdot\text{atm}\cdot\text{K}^{-1}\cdot\text{mol}^{-1})(553.2\ \text{K})}{\left(\dfrac{48.3\ \text{Torr}}{760\ \text{Torr}\cdot\text{atm}^{-1}}\right)(0.5000\ \text{L})}$$

$$= 164\ \text{g}\cdot\text{mol}^{-1}$$

From the combustion analyses, we can calculate the following:

50.0 mg CO_2 = 0.001 14 mol CO_2; contains 0.0136 g C

12.4 mg H_2O = 0.0006 88 mol H_2O; contains 0.001 39 g H

This accounts for 15.0 mg (13.6 + 1.39 mg) of the original 18.8 mg burned. The remaining 3.8 mg must be due to O. 3.8 mg O corresponds to 0.000 24 mol O. The molar ratio of C:H:O can be calculated and is found to be 4.8:5.7:1. If these numbers are rounded to the nearest whole numbers, the ratio will be 5:6:1. The formula C_5H_6O has a mass of 82.10 $\text{g}\cdot\text{mol}^{-1}$, which is almost exactly half the molecular weight found above. The molecular formula is $C_{10}H_{12}O_2$.

4.46 The molar mass of nitroglycerin is 227.1 $g \cdot mol^{-1}$. All products are gases under these conditions, so the total amount of gas produced is the sum of the pressures of the individual gases; the pressure is independent of the type of gas. From the equation we can see that 29 mol of gas is produced for 4 mol of nitroglycerin detonated. We can use this information and the ideal gas equation to calculate the volume of gases produced (nit. = nitroglycerin):

$$V = \frac{n_{\text{gas, total}} RT}{P}$$

$$= \frac{\left(\frac{29\text{ mol gas}}{4\text{ mol nit.}}\right)\left(\frac{454\text{ g nit.}}{227.1\text{ g}\cdot\text{mol}^{-1}\text{ nit.}}\right)}{215\text{ kPa}} \times (8.314\,51\text{ L}\cdot\text{kPa}\cdot\text{K}^{-1}\cdot\text{mol}^{-1})(548\text{ K})$$

$$= 307\text{ L}$$

4.48 To answer this, we need to know the number of moles of $C_2H_4(g)$ present in each case. Because the combustion reaction is the same in both cases, as are the temperature and pressure, the larger number of moles of $C_2H_4(g)$ produces the larger volume of CO_2. We use the ideal gas equation to solve for n in the first case:

$$n = \frac{PV}{RT} = \frac{(2.00\text{ atm})(1.00\text{ L})}{(0.082\,06\text{ L}\cdot\text{atm}\cdot\text{K}^{-1}\cdot\text{mol}^{-1})(318\text{ K})} = 0.0766\text{ mol CH}_4$$

1.20 g of $C_2H_4(g)$ equals $\frac{1.20\text{ g}}{28.05\text{ g}\cdot\text{mol}^{-1}} = 0.0428\text{ mol}$

The first case has the greatest number of moles of CH_4 and produces the largest amount of $CO_2(g)$.

4.50 The molar mass of urea is 60.06 $g \cdot mol^{-1}$. From the stoichiometry, it is readily seen that the same number of moles of CO_2 is needed as urea produced, and we need twice as many moles of NH_3. From the information given, we can readily calculate the number of moles of urea and use the ideal gas equation to compute the volume that those gases would occupy.

$$V_{CO_2} = \frac{\dfrac{(2.50\text{ kg urea})(1000\text{ g}\cdot\text{kg}^{-1})}{(60.06\text{ g}\cdot\text{mol}^{-1}\text{ urea})}\left(\dfrac{1\text{ mol CO}}{1\text{ mol urea}}\right)}{200\text{ atm}} \times$$

$$(0.082\,06\text{ L}\cdot\text{atm}\cdot\text{K}^{-1}\cdot\text{mol}^{-1})(723\text{ K})$$

$$= 12.3\text{ L}$$

$$V_{NH_3} = \frac{\dfrac{(2.50\text{ kg urea})(1000\text{ g}\cdot\text{kg}^{-1})}{(60.06\text{ g}\cdot\text{mol}^{-1}\text{ urea})}\left(\dfrac{2\text{ mol NH}_3}{1\text{ mol urea}}\right)}{200\text{ atm}} \times$$

$$(0.082\,06\text{ L}\cdot\text{atm}\cdot\text{K}^{-1}\cdot\text{mol}^{-1})(723\text{ K})$$

$$= 24.7\text{ L}$$

4.52 First, balance the equation:

$$C_2H_4(g) + 3\,O_2(g) \rightarrow 2\,CO_2(g) + 2\,H_2O(l)$$

Because the reaction remains at 1.00 atm pressure and 298 K throughout, we can use the volumes directly to solve the problem. The volume is directly proportional to the number of moles of each species. 1.00 L of ethene requires 3.00 L of O_2 to react in order to go to completion. C_2H_4 is thus the limiting reactant. All the C_2H_4 will react with 3.00 L of O_2. This leaves 1.00 L of unreacted O_2 and produces 2.00 L CO_2.

The total volume is the sum of the volume of O_2 remaining plus the volume of CO_2 produced.

V = 1.00 L + 2.00 L = 3.00 L

4.54 (a) The total volume after opening the stopcock is 14.0 L. Because the temperature is constant, we need only be concerned with the effect that increasing the volume has on the pressure of each gas. The pressure of N_2 and Ar will decrease as shown:

$$P_{N_2} = 803\text{ kPa}\left(\frac{4.0\text{ L}}{14.0\text{ L}}\right) = 230\text{ kPa};\ P_{Ar} = 47.2\text{ kPa}\left(\frac{10.0\text{ L}}{14.0\text{ L}}\right) = 33.7\text{ kPa}$$

(b) The total pressure is simply the sum of these two partial pressures:

230 kPa + 33.7 kPa = 260 kPa

4.56 $n_{CH_4} = \dfrac{0.320\text{ g}}{16.04\text{ g}\cdot\text{mol}^{-1}} = 0.0200\text{ mol}$

$n_{Ar} = \dfrac{0.175\text{ g}}{39.95\text{ g}\cdot\text{mol}^{-1}} = 0.004\,38\text{ mol}$

$n_{N_2} = \dfrac{0.225\text{ g}}{28.01\text{ g}\cdot\text{mol}^{-1}} = 0.008\,03\text{ mol}$

$n_{total} = 0.0200\text{ mol} + 0.004\,38\text{ mol} + 0.008\,03\text{ mol} = 0.0324\text{ mol}$

(a) $P_{N_2} = 15.2\text{ kPa}$

$$P_{Ar} = \frac{n_{Ar}}{n_{N_2}} \times P_{N_2} = \frac{0.004\,38\text{ mol}}{0.008\,03\text{ mol}} \times 15.2\text{ kPa} = 8.29\text{ kPa Ar}$$

$$P_{CH_4} = \frac{n_{CH_4}}{n_{N_2}} \times P_{N_2} = \frac{0.0200\text{ mol}}{0.008\,03\text{ mol}} \times 15.2\text{ kPa} = 37.9\text{ kPa CH}_4$$

$P_{total} = 15.2\text{ kPa} + 8.29\text{ kPa} + 37.9\text{ kPa} = 61.4\text{ kPa}$

(b) $PV = nRT$

$$\left(\frac{61.4\text{ kPa}}{101.325\text{ kPa}\cdot\text{atm}^{-1}}\right)V$$
$$= (0.0324\text{ mol})(0.082\,06\text{ L}\cdot\text{atm}\cdot\text{K}^{-1}\cdot\text{mol}^{-1})(300\text{ K})$$
$$V = 1.32\text{ L}$$

4.58 At 21°C the vapor pressure of water is 18.65 Torr, so that of the total pressure of 755 Torr in the apparatus, only 755 Torr – 18.65 Torr = 736 Torr is due to N_2O(g). We can now use $P_1V_1 = P_2V_2$ to determine the volume of an equal amount of dry N_2O(g) measured at 755 Torr:

(736 Torr) (126 mL) = (755 Torr) (V_2); V_2 = 123 mL

4.60 Graham's law of effusion states that the rate of effusion of a gas is inversely proportional to the square root of its molar mass:

$$\text{rate of effusion} = \frac{1}{\sqrt{M}}$$

If we have two different gases whose rates of effusion are measured under identical conditions, we can take the ratio:

$$\frac{\text{rate}_1}{\text{rate}_2} = \frac{\dfrac{1}{\sqrt{M_1}}}{\dfrac{1}{\sqrt{M_2}}} = \sqrt{\frac{M_2}{M_1}}$$

If a compound takes 2.7 times as long to effuse as XeF_2, the rate of effusion of XeF_2 is 2.7 times that of the compound. We can now plug into the expression to calculate the molar mass:

$$\frac{2.7}{1} = \sqrt{\frac{M_2}{169.3\ \text{g} \cdot \text{mol}^{-1}}}$$

$$M_2 = 1.2 \times 10^3\ \text{g} \cdot \text{mol}^{-1}$$

4.62 The number of collisions a molecule makes is proportional to the average speed of the molecules. Because the kinetic energy of any ideal gas sample is the same at a given temperature and is independent of the mass of the gas, it follows that the gas molecules with more mass travel slower on average than lighter gas molecules (see Figure 4.27). The result is that lighter gases have more collisions per unit time than heavier gases. Because neon is lighter than CO_2 ($20.18\ \text{g} \cdot \text{mol}^{-1}$ versus $44.01\ \text{g} \cdot \text{mol}^{-1}$), the sample of neon will have more collisions per second.

4.64 The elemental analyses give an empirical formula of $(C_2H_3Cl)_n$. This formula mass is $62.49\ \text{g} \cdot \text{mol}^{-1}$. Because the volume of the gas is proportional to the number of moles and hence molecules of gas that effuse, we can define the rates of effusion as follows:

$$\text{rate}_{\text{unknown}} \propto \frac{(V_{\text{unknown}})}{\text{time}} = \frac{(V_{\text{unknown}})}{7.73\ \text{min}}$$

$$\text{rate}_{\text{Ar}} \propto \frac{(V_{\text{Ar}})}{\text{time}} = \frac{(V_{\text{Ar}})}{6.18\ \text{min}}$$

The rate of effusion is inversely proportional to the square root of the molar mass of the substance that is effusing:

$$\text{rate of effusion} \propto \frac{1}{\sqrt{M}}$$

If we have two different gases whose effusion is measured under identical conditions, we can take the ratio of their effusion rates. This is convenient because all of the constants that relate V to number of molecules will be the same for both gases and will cancel from the expression.

$$\frac{\text{rate}_1}{\text{rate}_2} = \frac{\dfrac{1}{\sqrt{M_1}}}{\dfrac{1}{\sqrt{M_2}}} = \sqrt{\frac{M_2}{M_1}}$$

$$\frac{\dfrac{(V_{\text{unknown}})}{7.73\text{ min}}}{\dfrac{(V_{\text{Ar}})}{6.18\text{ min}}} = \sqrt{\frac{39.95\text{ g}\cdot\text{mol}^{-1}}{M_{\text{unknown}}}}$$

This expression simplifies because the volume of the unknown and the volume of Ar effusing are the same.

$$\frac{\dfrac{1}{7.73\text{ min}}}{\dfrac{1}{6.18\text{ min}}} = \sqrt{\frac{39.95\text{ g}\cdot\text{mol}^{-1}}{M_{\text{unknown}}}}$$

$$\left(\frac{6.18}{7.73}\right)^2 = \frac{39.95\text{ g}\cdot\text{mol}^{-1}}{M_{\text{unknown}}}$$

$$M_{\text{unknown}} = (39.95\text{ g}\cdot\text{mol}^{-1})\left(\frac{7.73}{6.18}\right)^2 = 62.5\text{ g}\cdot\text{mol}^{-1}$$

This mass corresponds to the mass of the empirical formula; the molecular formula is C_2H_3Cl.

4.66 (a) The average kinetic energy is obtained from the expression:

$$\text{average kinetic energy} = \frac{3}{2}RT$$

The value is independent of the nature of the monatomic ideal gas. The numerical values are:

(a) $3718\ \text{J}\cdot\text{mol}^{-1}$; (b) $3731\ \text{J}\cdot\text{mol}^{-1}$;

(c) $3731\ \text{J}\cdot\text{mol}^{-1} - 3718\ \text{J}\cdot\text{mol}^{-1} = 13\ \text{J}\cdot\text{mol}^{-1}$

4.68 The root mean square speed is calculated from the following equation:

$$c = \sqrt{\frac{3\,RT}{M}}$$

(a) Fluorine gas, F_2, $M = 38.00\ \text{g}\cdot\text{mol}^{-1}$

$$c = \sqrt{\frac{3(8.314\ \text{kg}\cdot\text{m}^2\cdot\text{s}^{-2}\cdot\text{K}^{-1}\cdot\text{mol}^{-1})(623\ \text{K})}{3.800\times10^{-2}\ \text{kg}\cdot\text{mol}^{-1}}}$$

$$= 639\ \text{m}\cdot\text{s}^{-1}$$

(b) Chlorine gas, Cl_2, $M = 70.90\ \text{g}\cdot\text{mol}^{-1}$

$$c = \sqrt{\frac{3(8.314\ \text{kg}\cdot\text{m}^2\cdot\text{s}^{-2}\cdot\text{K}^{-1}\cdot\text{mol}^{-1})(623\ \text{K})}{7.090\times10^{-2}\ \text{kg}\cdot\text{mol}^{-1}}}$$

$$= 468\ \text{m}\cdot\text{s}^{-1}$$

(c) Bromine gas, Br_2, $M = 159.82\ \text{g}\cdot\text{mol}^{-1}$

$$c = \sqrt{\frac{3(8.314\ \text{kg}\cdot\text{m}^2\cdot\text{s}^{-2}\cdot\text{K}^{-1}\cdot\text{mol}^{-1})(623\ \text{K})}{1.5982\times10^{-1}\ \text{kg}\cdot\text{mol}^{-1}}}$$

$$= 312\ \text{m}\cdot\text{s}^{-1}$$

4.70 Note that Equation 29, the Maxwell distribution of speeds, should read $f(v) = 4\pi\left(\frac{M}{2\pi RT}\right)^{3/2} v^2 e^{\frac{-Mv^2}{2RT}}$. (Earlier printings of the textbook might have a misprint of 1/2 for the 3/2 exponent.)

The first derivative with respect to speed is

$$\frac{d\,f(v)}{dv}=\left(\frac{M}{2\pi RT}\right)^{3/2}e^{\frac{-Mv^2}{2RT}}\left[8\pi v+4\pi v^2\left(\frac{-Mv}{RT}\right)\right]=0$$

Solving for v gives the most probable speed in terms of T.

$$v_{mp}=\left(\frac{2RT}{M}\right)^{1/2}$$

Substitution into $f(v)$ gives

$$f(v_{mp})=4\pi\left(\frac{M}{2\pi RT}\right)^{3/2}\left(\frac{2RT}{M}\right)\cdot e^{-1}=4\pi\left(\frac{M}{2\pi RT}\right)^{1/2}e^{-1}$$

The problem asks for the temperature that satisfies this relationship:

$$f(v_{mp})_T=\frac{1}{2}f(v_{mp})_{300}$$

$$4\pi\left(\frac{M}{2\pi RT}\right)^{1/2}e^{-1}=\left(\frac{1}{2}\right)4\pi\left(\frac{M}{2\pi R(300.\text{ K})}\right)^{1/2}e^{-1}$$

$$\left(\frac{1}{T}\right)^{1/2}=\left(\frac{1}{2}\right)\left(\frac{1}{300.\text{ K}}\right)^{1/2}$$

$$\left(\frac{1}{T}\right)=\left(\frac{1}{4}\right)\left(\frac{1}{300.\text{ K}}\right)$$

$$T=4\times 300.\text{ K}=1.20\times 10^3\text{ K}$$

4.72 (a)

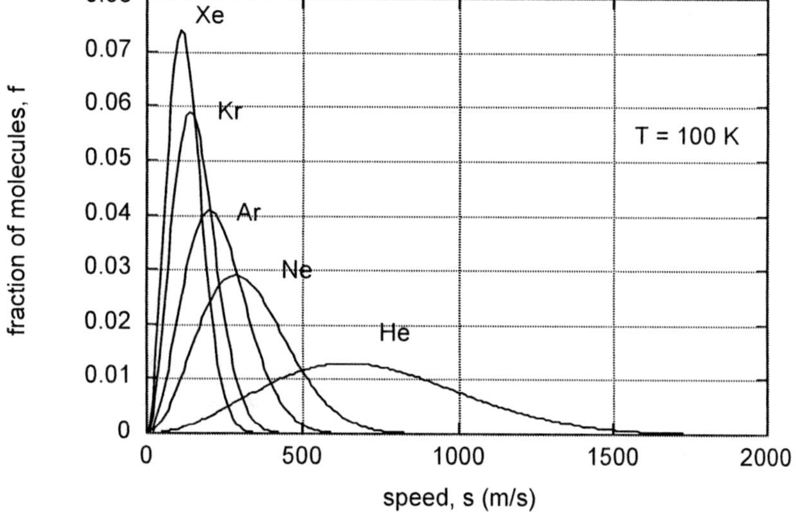

(b) As the molar mass increases, the distribution shifts toward lower velocities. The distribution becomes narrower and the fraction of atoms with the most probable speed becomes higher as the molar mass becomes larger. (Note that all distributions are normalized to the total number of atoms. Since total kinetic energy is constant for a given T, there are one-tenth as many Ar atoms as He atoms because its mass is ten times as large.)

(c)

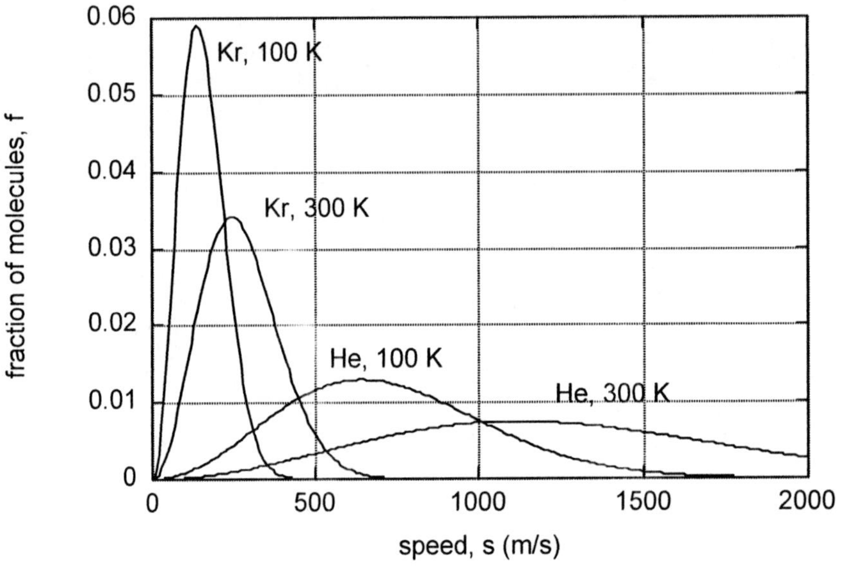

(d) The two distributions for He are clearly much broader and displaced toward higher velocities than that distribution for Kr. As T increases, the curve maximum shifts toward higher speed for each type of atom.

4.74 (a) Real gases are more compressible than ideal gases when the intermolecular forces of attraction dominate. (b) They are less compressible when the intermolecular repulsions dominate.

4.76 The pressures are calculated from the ideal gas law:

$$P = \frac{nRT}{V} = \frac{(1.00\ \text{mol})(0.082\ 06\ \text{L}\cdot\text{atm}\cdot\text{K}^{-1}\cdot\text{mol}^{-1})\ (298\text{K})}{V}$$

(a) Calculating for the specific volumes requested, we obtain $P =$ (a) 0.815 atm; (b) 24.5 atm; (c) 489 atm.

The calculations can now be repeated using the van der Waals equation:

$$\left(P + \frac{an^2}{V^2}\right)(V - nb) = nRT$$

We can rearrange this to solve for P:

$$P = \left(\frac{nRT}{V - nb}\right) - \left(\frac{an^2}{V^2}\right)$$

$$= \left(\frac{(1.00\text{ mol})(0.082\,06\text{ L}\cdot\text{atm}\cdot\text{K}^{-1}\cdot\text{mol}^{-1})(298\text{ K})}{V - (1.00\text{ mol})(0.026\,61\text{ L}\cdot\text{mol}^{-1})}\right)$$

$$- \left(\frac{(0.2476\text{ L}^2\cdot\text{atm}\cdot\text{mol}^{-2})(1.00)^2}{V^2}\right)$$

(a) Using the three values for V, we calculate $P =$ (a) 0.816 atm; (b) 24.9 atm; (c) 946 atm. Note that at larger volume (lower pressure), the ideal gas law gives essentially the same values as the van der Waals equation but at smaller volume (higher pressure) there is a very significant difference. Since the pressure values calculated from the van der Waals equation are higher than those from the ideal gas law, we would conclude that the repulsive forces represented by b are dominant for hydrogen under these conditions of temperature and volume.

4.78 (a) (1) ideal gas:

$PV = nRT$

$$P = \frac{nRT}{V} = \frac{(1.00\text{ mol})(0.082\,06\text{ L}\cdot\text{atm}\cdot\text{K}^{-1}\cdot\text{mol}^{-1})(273.15\text{ K})}{22.414\text{ L}}$$

$= 1.00$ atm

(2) van der Waal's gas:

$$\left(P + a\left(\frac{n}{V}\right)^2\right)(V - nb) = nRT$$

$$P = \frac{nRT}{V - nb} - a\left(\frac{n}{V}\right)^2$$

$$P = \frac{(1.00\text{ mol})(0.082\,06\text{ L}\cdot\text{atm}\cdot\text{K}^{-1}\cdot\text{mol}^{-1})(273.15\text{ K})}{22.414\text{ L} - (1.00\text{ mol})(0.0638\text{ L}\cdot\text{mol}^{-1})}$$

$$- 4.562\text{ L}^2\cdot\text{atm}\cdot\text{mol}^{-2}\left(\frac{1.00\text{ mol}}{22.414\text{ L}}\right)^2 = 0.994\text{ atm}$$

(b) (1) ideal gas:

$$P = \frac{(1.00\text{ mol})(0.082\,06\text{ L}\cdot\text{atm}\cdot\text{K}^{-1}\cdot\text{mol}^{-1})(1000\text{ K})}{0.100\text{ L}} = 821\text{ atm}$$

(2) van der Waal's gas:

$$P = \frac{(1.00\text{ mol})(0.082\,06\text{ L}\cdot\text{atm}\cdot\text{K}^{-1}\cdot\text{mol}^{-1})(1000\text{ K})}{0.100\text{ L} - (1.00\text{ mol})(0.0638\text{ L}\cdot\text{mol}^{-1})}$$

$$- 4.562\text{ L}^2\cdot\text{atm}\cdot\text{mol}^{-2}\left(\frac{1.00\text{ mol}}{0.100\text{ L}}\right)^2 = 1.81\times10^3\text{ atm}$$

4.80

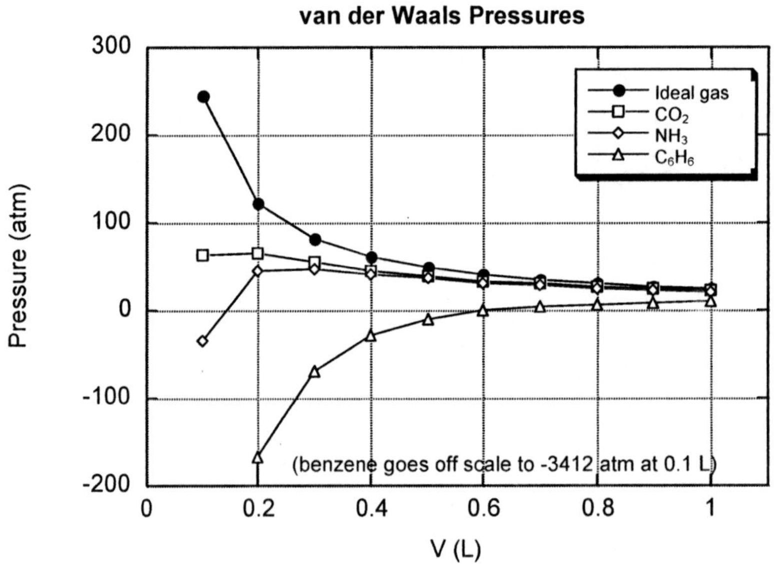

Pressures for the real gases can be calculated from Eq. 33 and the appropriate constants.

$$P = \frac{nRT}{V - nb} - a\frac{n^2}{V^2}$$

v.d.W. Constants	carbon dioxide	ammonia	benzene
a, $L^2 \cdot atm \cdot mol^{-1}$	3.64	4.225	18.24
b, $L \cdot mol^{-1}$	0.042 67	0.037 07	0.1154

The different shapes of these calculated curves can be attributed to deviations from ideality caused by intermolecular interactions. The constant a corrects for attractions while the constant b corrects for repulsions. As a becomes larger going from carbon dioxide to ammonia to benzene, the effect of the second term in the van der Waals equation is increased proportionally such that it reduces the pressure at any volume. Carbon dioxide deviates least from ideality while benzene deviates most. This effect is more pronounced at small volumes since pressure is reduced by the inverse square of volume multiplied by a. Attractions dominate the non-ideal behavior for all three real gases across the whole volume range where $P_{vdW} < P_{ideal}$. We would expect the real gases to liquefy at their equilibrium vapor pressures, or near the volume that corresponds to zero pressure for each gas (~0.65 L for benzene, ~0.15 L for ammonia).

4.82 $(CH_3)_2N_2H_2(s) + 2\ N_2O_4(l) \rightarrow 2\ CO_2(g) + 3\ N_2(g) + 4\ H_2O(l)$

$P_{tot} = P_{CO_2} + P_{N_2} + P_{H_2O} = 2.50\text{ atm}$

Since the mole ratio of the gaseous products is 2:3:4 and all three gases are present in the same vessel at the same temperature, the partial pressures will exhibit this ratio as well. Therefore,

$$P_{tot} = 2x + 3x + 4x = 9x = 2.50\text{ atm}$$
$$x = 0.278\text{ atm}$$

$$P_{CO_2} = 2x = 0.556\text{ atm}$$
$$P_{N_2} = 3x = 0.834\text{ atm}$$
$$P_{H_2O} = 4x = 1.11\text{ atm}$$

4.84 First, balance the equation:

$$C_2H_4(g) + 3\,O_2(g) \rightarrow 2\,CO_2(g) + 2\,H_2O(l)$$

Because the reaction remains at 1.00 atm pressure and 298 K throughout, we can use the volumes directly to solve the problem. The volume is directly proportional to the number of moles of each species. 2.00 L of ethene requires 6.00 L of O_2 to react in order to go to completion. O_2 is thus the limiting reactant. All the O_2 will react with 0.667 L of ethene (2.00 L ÷ 3 mol O_2 per mol ethene). This leaves 1.33 L of unreacted ethene and produces 1.33 L CO_2:

$$\text{L of CO}_2\text{ produced} = (2.0\text{ L O}_2\text{ consumed})\left(\frac{2\text{ L CO}_2\text{ produced}}{3\text{ L O}_2\text{ consumed}}\right)$$

$$= 1.33\text{ L CO}_2\text{ produced}$$

The total volume is the sum of the volume of ethene remaining plus the volume of CO_2 produced.

$V = 1.33\text{ L} + 1.33\text{ L} = 2.66\text{ L}$

4.86 (a) Because the gas was collected over water, it contains water vapor. Of the 737.7 Torr of pressure in the vessel, only 737.7 Torr − 9.21 Torr = 728.5 Torr is due to $H_2(g)$. The amount of dry H_2 can be calculated from the ideal gas law:

$$\frac{P_1V_1}{T_1} = \frac{P_2V_2}{T_2}$$

$$\frac{\left(\dfrac{728.5\text{ Torr}}{760\text{ Torr}\cdot\text{atm}^{-1}}\right)(0.127\text{ L})}{283\text{ K}} = \frac{(1.00\text{ atm})(V)}{298\text{ K}}$$

$V = 0.128\text{ L}$

(b)

$$\left(\frac{728.5\text{ Torr}}{760\text{ Torr}\cdot\text{atm}^{-1}}\right)(0.128\text{ L}) = n\,(0.082\,06\text{ L}\cdot\text{atm}\cdot\text{K}^{-1}\cdot\text{mol}^{-1})(283\text{ K})$$

$n = 5.28 \times 10^{-3}\text{ mol}$

(c) To determine this, we must first write the balanced equation for the reaction of Zn metal with HCl:

$$Zn(s) + 2\ HCl(aq) \rightarrow ZnCl_2(aq)\ + H_2(g)$$

There should be 1 mol $H_2(g)$ produced per 1 mol Zn(s) consumed. If the zinc metal were 100% pure, the number of moles of zinc would be $0.40\ g \div 65.37\ g \cdot mol^{-1} = 6.12 \times 10^{-3}$ mol and the same number of moles of H_2 should have been produced. Because not enough H_2 was produced for all of the original sample to be Zn, the sample was clearly impure. We can calculate the mass of Zn present in the original sample from the number of moles of H_2 produced, which is also equal to the number of moles of Zn in the initial sample:

$5.24 \times 10^{-3}\ mol \times 65.37\ g \cdot mol^{-1}\ Zn = 0.34\ g$. The percent purity is given by $\dfrac{0.34\ g}{0.40\ g} \times 100 = 85\%$

4.88 The information in the first sentence allows us to calculate the empirical formula of the compound. If 2.36 g of the mass of the phosphorus chloride is phosphorus, then 8.14 g must be Cl. Upon calculation, this gives an empirical formula of PCl_3, which has an empirical formula mass of $137.32\ g \cdot mol^{-1}$. Because the phosphorus chloride took 1.77 times as long to effuse, the rate of effusion of CO_2 must be 1.77 times that of the phosphorus compound. We can then write:

$$\frac{rate_1}{rate_2} = \frac{\dfrac{1}{\sqrt{M_1}}}{\dfrac{1}{\sqrt{M_2}}} = \sqrt{\frac{M_2}{M_1}}$$

$$1.77 = \sqrt{\frac{M_2}{44.01\ g \cdot mol^{-1}}}$$

$$M_2 = 138\ g \cdot mol^{-1}$$

The molar mass determined is close to that of the formula mass, and so the empirical formula must also correspond to the molecular formula.

4.90 (a) The number of moles of nitrogen required to fill the air bag can be found from the ideal gas law.

$$?\text{ mol N}_2 = \frac{pV}{RT} = \frac{(1.37\text{ atm})(57.0\text{ L N}_2)}{(0.0821\text{ L}\cdot\text{atm}\cdot\text{K}^{-1}\cdot\text{mol}^{-1})(298\text{ K})} = 3.19\text{ mol N}_2$$

The mass of sodium azide required to produce this amount of nitrogen depends upon the balanced chemical reaction.

$$2\text{ NaN}_3(\text{s}) \rightarrow 3\text{ N}_2(\text{g}) + 2\text{ Na(s)}$$

$$?\text{ g NaN}_3 = 3.19\text{ mol N}_2\left(\frac{2\text{ mol NaN}_3}{3\text{ mol N}_2}\right)\left(\frac{65.02\text{ g NaN}_3}{1\text{ mol NaN}_3}\right) = 138\text{ g NaN}_3$$

(b) $$v_{rms} = \sqrt{\frac{3RT}{M}}$$

$$= \left(\frac{3(8.314\text{ J}\cdot\text{K}^{-1}\cdot\text{mol}^{-1})(298\text{ K})}{(28.0\text{ g}\cdot\text{mol}^{-1})}\left(\frac{\text{kg}\cdot\text{m}^{-2}\cdot\text{s}^{-2}}{\text{J}}\right)\left(\frac{1000\text{ g}}{\text{kg}}\right)\right)^{1/2}$$

$$= 515\text{ m}\cdot\text{s}^{-1}$$

4.92 To answer this question, we need to find which gases are lighter than the average density of the atmosphere in which the balloons are supposed to float. The density will depend on the molecular weight of the sample:

$$PV = nRT$$

$$PV = \frac{m}{M}RT$$

$$d = \frac{m}{V} = \frac{PM}{RT}$$

The problem then reduces to finding which of these gases have a molar mass less than the average molar mass of the atmosphere. The average molar mass of the atmosphere will be the weighted average of the molar masses of the components. Because Ar and O_2 make up the greatest part of the atmospheric composition, we can neglect the other minor components that are present. Average molar mass $= 0.79\,(39.95\text{ g}\cdot\text{mol}^{-1}) + 0.20\,(32.00\text{ g}\cdot\text{mol}^{-1}) = 38\text{ g}\cdot\text{mol}^{-1}$. The molar masses of Kr, CO_2, CH_4, C_3H_8, N_2, CO are 83.80, 44.01, 16.04, 44.09, 28.02, and 28.01 $g \cdot mol^{-1}$, respectively. The gases that would produce balloons that would float include methane, nitrogen and carbon monoxide. Because methane is flammable and carbon monoxide is toxic, the best choice would be nitrogen, which is chemically inert.

4.94 We can use the ideal gas law to determine the number of moles of hydrocarbon equivalent to 2.55 g in order to get the molar mass of the compound.

$$? \text{ mol} = \frac{PV}{RT} = \frac{(0.950 \text{ atm})(3.00 \text{ L})}{(0.0821 \text{ L} \cdot \text{atm} \cdot \text{K}^{-1} \cdot \text{mol}^{-1})(355 \text{ K})} = 0.0978 \text{ mol}$$

$$\text{molar mass} = \frac{2.55 \text{ g}}{0.0978 \text{ mol}} = 26.1 \text{ g} \cdot \text{mol}^{-1}$$

This molar mass fits a molecular formula of C_2H_2 so the hydrocarbon is acetylene.

$H{-}C{\equiv}C{-}H$

4.96 (a) $4 \text{ FeS}_2(s) + 11 \text{ O}_2(g) \rightarrow 2 \text{ Fe}_2\text{O}_3(s) + 8 \text{ SO}_2(g)$

(b) First we calculate the number of moles of O_2 present at the conditions given, using the ideal gas law:

$$PV = nRT$$

$$(2.33 \text{ atm})(75.0 \text{ L}) = n\,(0.082\,06 \text{ L} \cdot \text{atm} \cdot \text{K}^{-1} \cdot \text{mol}^{-1})(423 \text{ K})$$

$$n = 5.03 \text{ mol}$$

Then, using the stoichiometry of the reaction and the molar mass of Fe_2O_3, we can calculate the mass of Fe_2O_3 produced:

$$\text{mass of Fe}_2\text{O}_3 = (159.70 \text{ g Fe}_2\text{O}_3 \cdot \text{mol}^{-1} \text{ Fe}_2\text{O}_3)(5.03 \text{ mol O}_2) \times \left(\frac{2 \text{ mol Fe}_2\text{O}_3}{11 \text{ mol O}_2}\right) = 146 \text{ g Fe}_2\text{O}_3$$

(c) The number of moles of SO_2 produced can be obtained from the stoichiometry:

$$n_{SO_2} = (5.03\ n_{O_2})\left(\frac{8\ n_{SO_2}}{11\ n_{O_2}}\right) = 3.66\text{ mol}$$

3.66 mol of SO_2 will dissolve in 5.00 L of water to form a solution that is (3.66 mol ÷ 5.00 L) = 0.732 M in H_2SO_3, according to the equation $SO_2(g) + H_2O(l) \rightarrow H_2SO_3(aq)$.

(d) $$\text{mass of } SO_2 = 1\times10^3\text{ kg coal}\left(\frac{5\times10^3\text{ g FeS}_2}{100\text{ kg coal}}\right)\left(\frac{1\text{ mol FeS}_2}{119.99\text{ g FeS}_2}\right)$$
$$\times\left(\frac{8\text{ mol SO}_2}{4\text{ mol FeS}_2}\right)\left(\frac{64.07\text{ g SO}_2}{1\text{ mol SO}_2}\right)$$
$$= 5\times10^4\text{ g SO}_2 = 50\text{ kg SO}_2$$

(e) $$n_{SO_2} = 5\times10^4\text{ g SO}_2\left(\frac{1\text{ mol SO}_2}{64.07\text{ g SO}_2}\right) = 800\text{ mol SO}_2$$

$$V = \frac{nRT}{P} = \frac{(800\text{ mol})(0.082\,06\text{ L}\cdot\text{atm}\cdot\text{K}^{-1}\cdot\text{mol}^{-1})(298\text{ K})}{(1.00\text{ atm})}$$
$$= 2\times10^4\text{ L}$$

4.98 (a), (b) and (c)

The values of the three quantities are:

$$v = \sqrt{\frac{3\ RT}{M}}$$

$$v_{\text{average}} = \sqrt{\frac{8\ RT}{\pi M}}$$

$$v_{\text{most probable}} = \sqrt{\frac{2\ RT}{M}}$$

The quantities can be rewritten to emphasize their relationship because all contain the same $\sqrt{\frac{RT}{M}}$ term:

$$v = \sqrt{3} \times \sqrt{\frac{RT}{M}} = 1.732\sqrt{\frac{RT}{M}}$$

$$v_{\text{average}} = \sqrt{\frac{8}{\pi}} \times \sqrt{\frac{RT}{M}} = 1.596\sqrt{\frac{RT}{M}}$$

$$v_{\text{most probable}} = \sqrt{2} \times \sqrt{\frac{RT}{M}} \doteq 1.414\sqrt{\frac{RT}{M}}$$

It can be easily seen that $v_{\text{most probable}}$ is the smallest followed by v_{average} and then v. The numerical ratio between them will be 1.414 : 1.596 : 1.732 or 1 : 1.13 : 1.22.

4.100 (a) $\frac{\text{rate}_1}{\text{rate}_2} = \sqrt{\frac{M_2}{M_1}}$

Because the rate is proportional to the number of moles of compound, we can write

$$\frac{\dfrac{0.9560\text{ g}}{16.01\text{ g}\cdot\text{mol}^{-1}}}{\dfrac{2.292\text{ g}}{M_2}} = \sqrt{\frac{M_2}{16.01\text{ g}\cdot\text{mol}^{-1}}}$$

$$\left[\left(\frac{0.9560\text{ g}}{16.01\text{ g}\cdot\text{mol}^{-1}}\right)\left(\frac{M_2}{2.292\text{ g}}\right)\right]^2 = \frac{M_2}{16.01\text{ g}\cdot\text{mol}^{-1}}$$

$$\left(\frac{1}{16.01\text{ g}\cdot\text{mol}^{-1}}\right)\left(\frac{0.9560\text{ g}}{2.292\text{ g}}\right)^2 (M_2)^2 - M_2 = 0$$

$$M_2\left[\left(\frac{1}{16.01\text{ g}\cdot\text{mol}^{-1}}\right)\left(\frac{0.9560\text{ g}}{2.292\text{ g}}\right)^2 M_2 - 1\right] = 0$$

$$M_2 = 92.02\text{ g}\cdot\text{mol}^{-1}$$

(b) C_7H_8. We could consider molecules with other carbon counts; however, these prove to be impossible. Molecules with six carbon atoms or fewer would have too many hydrogen atoms. The formula C_6H_{20} would be required to have a molar mass of 92, but if there are 20 C—H bonds and each C atom can at most make four bonds to other atoms, then there are only $(6 \times 4) - 20 = 4$ bonds remaining, and that is too few to bond together six carbon atoms. If we had eight carbon atoms, the mass would be 96 from carbon alone and that exceeds the observed mass.

(c) There are many structures possible. The formula C_7H_8, however, can provide clues about the types of molecules that can be drawn. The saturated hydrocarbon based on seven carbon atoms has a formula of C_7H_{16}. There are 6 C—C and 16 C—H bonds in that molecule. For every two H atoms we remove, we create a site of unsaturation. That is, we introduce either a double bond or a ring structure. A triple bond counts as two sites of unsaturation. Thus C_7H_8 has eight fewer hydrogen atoms and should contain four sites of unsaturation. Examples include:

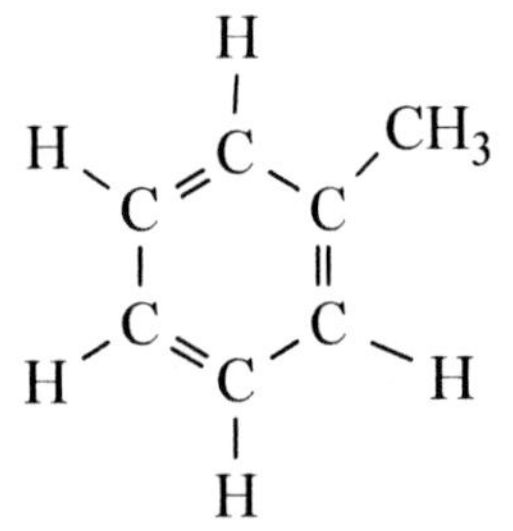

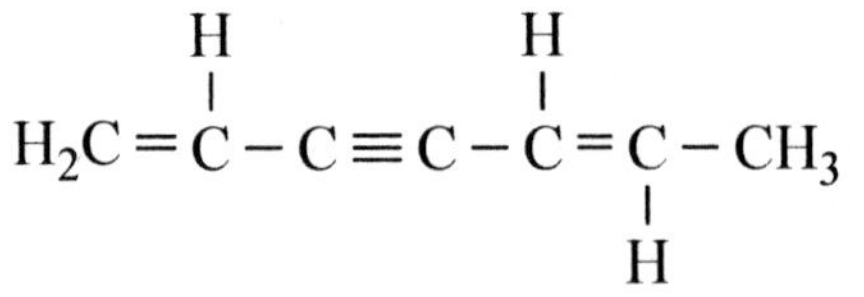

4.102 First we can use the ideal gas law to determine the number of moles of arsine present initially in the flask.

$$? \text{ mol} = \frac{PV}{RT} = \frac{(300.\text{ Torr})(0.500\ 0\text{ L})}{(0.08206\text{ L}\cdot\text{atm}\cdot\text{K}^{-1}\cdot\text{mol}^{-1})(223\text{ K})}\left(\frac{1\text{ atm}}{760.\text{ Torr}}\right)$$
$$= 0.0108\text{ mol AsH}_3$$

If this amount of arsine were to decompose completely according to this balanced reaction

$$\text{AsH}_3(\text{g}) \rightarrow \text{As(s)} + \tfrac{3}{2}\text{H}_2(\text{g})$$

then (1.5)(0.0108 mol) = 0.0162 mol of hydrogen gas would be produced for 100% decomposition. If less than 100% decomposes, the observed pressure in the flask will be due to both unreacted arsine and product hydrogen. The actual number of moles of gas in the flask for the final conditions is

$$? \text{ mol} = \frac{PV}{RT} = \frac{(508.\text{ Torr})(0.500\ 0\text{ L})}{(0.08206\text{ L}\cdot\text{atm}\cdot\text{K}^{-1}\cdot\text{mol}^{-1})(273\text{ K})}\left(\frac{1\text{ atm}}{760.\text{ Torr}}\right)$$

$= 0.0149$ mol gas

Setting x = number of moles of arsine that decomposed also requires that $1.5x$ moles of hydrogen were formed. Then

$$n_{\text{total}} = 0.0149\text{ mol} = (0.0108\text{ mol} - x) + 1.5x$$

$$0.0041\text{ mol} = 0.5x$$

$$x = 0.0082\text{ mol}$$

Then (1.5)(0.0082 moles) = 0.0123 moles of hydrogen were formed so

$\frac{0.0123\text{ mol}}{0.0162\text{ mol}} \times 100 = 75.9\%$ of the arsine decomposed.

(Equivalently, $\frac{0.0082\text{ mol}}{0.0108\text{ mol}} \times 100 = 75.9\%$.)

4.104 Since the van der Waals a parameter represents the role of attractions, it will become larger as the strength of the intermolecular attractions increases. Acetonitrile is polar and would be expected to have the strongest intermolecular forces and the highest value of a. The three non-polar species can be ranked in order of the polarizability of their electron clouds. Carbon dioxide has the largest, most polarizable electron cloud, so it should have the strongest London dispersion forces of these three species and the highest value of a after acetonitrile. Methane would be next, while neon would exhibit the weakest London dispersion forces and the lowest value of a. We can check this reasoning by first noting that acetonitrile is the only one of these four substances that is a liquid at room temperature. In fact, the normal boiling points of these chemical substances, as given in the third column of the table, fall in the same order as their van der Waals a parameters.

a ($L^2 \cdot atm \cdot mol^{-2}$)	substance	normal b. pt. (°C)
17.58	CH_3CN	81.6
3.392	CO_2	−78.5 (subl.)
2.253	CH_4	−164.76
0.2107	Ne	−246.048

CHAPTER 5
LIQUIDS AND SOLIDS

5.2 (a) London forces, dipole-dipole, hydrogen bonding; (b) London forces, dipole-dipole; (c) London forces, dipole-dipole; (d) London forces

5.4 (b) and (d), O_2 and CO_2 do not have dipole moments.

5.6 Ionic solids are high melting because the energy to get the ions to move past each other is very strong, on the order of 250 $kJ \cdot mol^{-1}$. Dissolution in water, however, compensates for the energy required to separate the ions by hydrating those ions. The energy produced when the ions are hydrated offsets the energy required to separate the ions. In network solids the bonds are covalent. Because the atoms there are not in ionic form, there is no hydration energy (or solvation energy if other solvents are used) to offset the breaking of the covalent bonds. Also, covalent bonds such as C—C bonds are very strong, with an average value of 318 $kJ \cdot mol^{-1}$ for a single bond. Comparisons between ionic solids and network solids are complicated by the fact that there is more than one interaction—i.e., an ion will bond to a number of ions in the solid state just as the atoms in a network solid form covalent bonds to a number of other atoms.

5.8 Only molecules with H attached to the electronegative atoms N, O, and F can hydrogen bond. Of the molecules given, only (b) CH_3COOH and

(c) CH_3CH_2OH have hydrogen attached to oxygen, so these are the only ones that can undergo hydrogen bonding.

5.10 Arrangement III should possess the strongest intermolecular attractions because the molecules are oriented with hydrogen atoms directed toward the lone pairs on nitrogen atoms of neighboring molecules. This geometry allows hydrogen bonding to occur more readily than in arrangement I or II.

5.12 (a) H_2O (100°C vs. −60.3°C) because hydrogen bonding is important for H_2O but not for H_2S; (b) NH_3 (−33°C vs. −78°C) because hydrogen bonding is important in ammonia but not phosphine; (c) KBr (1435°C vs. 3.6°C) because it is an ionic compound as opposed to a molecular compound; (d) SiH_4 (−112°C vs. −164°C) because it has more electrons with which to form stronger London forces.

5.14 (a) BF_3 is a trigonal planar molecule whereas ClF_3 is T-shaped. The latter will be polar and should have the higher boiling point. BF_3 boils at −99.9°C; ClF_3 boils at 11.3°C. (b) SF_4 is seesaw shaped whereas CF_4 is tetrahedral. The former should be polar and have the higher boiling point. The greater number of electrons of SF_4 also contribute to the higher boiling point. SF_4 boils at −40°C; CF_4 boils at −129°C. (c) Both molecules are planar but the *cis* form will have a dipole moment and the *trans* form will not. This will give the *cis* form dipole-dipole interactions not present in the other molecule, giving it the higher boiling point. The *cis* compound boils at 60.3°C whereas the *trans* compound boils at 47.5°C.

5.16 The ionic radius of Ca^{2+} is 100 pm and that of In^{3+} is 72 pm. The ratio of energies will be given by

$$E_p \propto \frac{-|z|\mu}{d^2}$$

$$E_{pCa^{2+}} \propto \frac{-|z| \times \mu}{d^2} = \frac{-|2|\mu}{(100)^2}$$

$$E_{pIn^{3+}} \propto \frac{-|z|\mu}{d^2} = \frac{-|3|\mu}{(72)^2}$$

The electric dipole moment of the water molecule (μ) will cancel:

$$\text{ratio}\left(\frac{E_{pCa^{2+}}}{E_{pIn^{3+}}}\right) = \frac{-|2|\mu/(100)^2}{-|3|\mu/(72)^2} = \frac{2(72)^2}{3(100)^2} = 0.35$$

The attraction of the Ca^{2+} ion will be less than that of the In^{3+} ion because it has both a larger radius and a lower charge.

5.18 The ionic radius of Na^+ is 102 pm and that of K^+ is 138 pm. The ratio of energies will be given by

$$E_p \propto \frac{-|z|\mu}{d^2}$$

$$E_{pNa^+} \propto \frac{-|1|\mu}{(102 \text{ pm})^2}$$

$$E_{pK^+} \propto \frac{-|1|\mu}{(138 \text{ pm})^2}$$

The electric dipole moment of the water molecule (μ) will cancel:

$$\text{ratio}\left(\frac{E_{pNa^+}}{E_{pK^+}}\right) = \frac{\dfrac{-|1|\mu}{(102 \text{ pm})^2}}{\dfrac{-|1|\mu}{(138 \text{ pm})^2}} = \frac{(138 \text{ pm})^2}{(102 \text{ pm})^2} = 1.83$$

The water molecule will be more strongly attracted to the Na^+ ion because of its smaller radius.

5.20 (a) network; (b) ionic; (c) molecular; (d) molecular; (e) network

5.22 Yes, we would expect the energy of interaction between two rotating molecules to be temperature dependent. As the temperature is raised, the volume will increase, the molecules will be farther apart, *r* in the denominator of Eq. 4 will increase, and the potential energy of interaction will be reduced in magnitude. In addition, the rate of rotation will increase with the temperature so the time when the dipoles are aligned favorably is likely to be reduced.

5.24 (a) Ethanol should have a greater viscosity than dimethyl ether because ethanol can hydrogen bond while dimethyl ether cannot. (b) Propanone (a.k.a. acetone) should be more viscous than butane because it is more polar than butane.

5.26 $CH_3CH_2CH_3 < C_6H_6 < H_2O < CH_3CH_2OH < CH_2OHCH_2OH < CH_2OHCHOHCH_2OH$

In order to predict relative viscosities we need to consider both the strength of intermolecular forces and the tendency of the molecules to get tangled like spaghetti (see Figure 5.12). Since glycerol, $CH_2OHCHOHCH_2OH$, can form several hydrogen bonds per molecule and it has a long chain structure, it has both strong intermolecular forces and the tendency to get tangled. Ethylene glycol, CH_2OHCH_2OH, is just one –CH-OH- unit smaller than glycerol, so we would expect it to be less viscous than glycerol but more viscous than ethanol, CH_3CH_2OH. The relative viscosities of ethanol, water, and benzene are given in Figure 5.11. Finally, since propane, $CH_3CH_2CH_3$, has the weakest intermolecular forces it would be expected to be the least viscous. Note that it is the only substance in the group that is a gas under normal ambient conditions.

5.28 Molecules with stronger intermolecular forces tend to have higher surface tension. Using this principle along with the data given in Table 5.3 we can match the values with the substances as follows: H_2O 72.75,

$CH_3(CH_2)_4CH_3$ 18.43, C_6H_6 28.85, CH_3CH_2OH 22.75, CH_3COOH 27.80. Note that units for surface tension are mN/m.

5.30 Just as molecules have polarity, surfaces of solids have polarity. By carrying out the silylation reaction, the surface of the glass will become less polar because the $Si(CH_3)_3$ groups are nonpolar. Therefore, polar liquids will not adhere to the surface quite so readily.

5.32 Assuming that we can ignore any elliptical deformation of the cross-sectional area occupied by the liquid in the tube as it is tilted, and that the surface tension is a constant, all of the factors in the expression seem to be independent of the tip angle, θ. However, we should note that when the tube is verticle, i.e., $\theta = 0°$,

$$h = l\cos\theta = \frac{2\gamma}{gdr}$$

where l is the length of the tube occupied by liquid and $\cos\left(0^\circ\right) = 1$. As θ gets larger, $\cos\theta$ gets smaller. Since all of the factors on the right-hand side of the equation are constant, it appears that we can expect the height of the meniscus above the zero level, h, to remain the same while the length of the liquid in the tube, l, grows in counterbalance to the decrease in $\cos\theta$.

[Note that some physics textbooks represent the height to which a liquid will rise in a capillary tube in terms of a contact angle, θ, between the liquid and the side of the tube. This angle reflects the difference between liquid-solid and liquid-vapor surface tensions. It is not the same as the tip angle called θ in this problem.]

5.34 For a picture of the simple, a.k.a. primitive, cubic unit cell, see Figures 5.28 and 5.33a. (a) There are eight atoms at the eight corners of the unit cell. One-eighth of each of these atoms will lie in the unit cell for a total of one atom per unit cell. (b) Each atom will be bonded to six atoms that form an octahedron. (c) The edge length of the unit cell for a primitive cubic cell will be twice the atomic radius of the atom or 334 pm in this case.

5.36 (a) a = length of side for a unit cell; for an fcc unit cell,

$a = \sqrt{8}\, r$ or $2\sqrt{2}\, r = 354$ pm

$V = a^3 = (354 \text{ pm} \times 10^{-12} \text{ m} \cdot \text{pm}^{-1})^3 = 4.42 \times 10^{-29} \text{ m}^3 = 4.42 \times 10^{-23} \text{ cm}^3$

For an fcc unit cell there are 4 atoms per unit cell; therefore we have

$$\text{mass (g)} = 4 \text{ Ni atoms} \times \frac{1 \text{ mol Ni atoms}}{6.022 \times 10^{23} \text{ atoms} \cdot \text{mol}^{-1}} \times \frac{58.71 \text{ g}}{\text{mol Ni atoms}}$$

$$= 3.90 \times 10^{-22} \text{ g}$$

$$d = \frac{3.90 \times 10^{-22} \text{ g}}{4.42 \times 10^{-23} \text{ cm}^3} = 8.82 \text{ g} \cdot \text{cm}^3$$

(b) $a = \dfrac{4r}{\sqrt{3}} = \dfrac{4 \times 250 \text{ pm}}{\sqrt{3}} = 577 \text{ pm}$

$V = (577 \times 10^{-12} \text{ m})^3 = 1.92 \times 10^{-28} \text{ m}^3 = 1.92 \times 10^{-22} \text{ cm}^3$

Given 2 atoms per bcc unit cell:

$$\text{mass (g)} = 2 \text{ Rb atoms} \times \frac{1 \text{ mol Rb atoms}}{6.022 \times 10^{23} \text{ atoms} \cdot \text{mol}^{-1}} \times \frac{85.47 \text{ g}}{\text{mol Rb atoms}}$$

$$= 2.84 \times 10^{-22} \text{ g}$$

$$d = \frac{2.84 \times 10^{-22} \text{ g}}{1.92 \times 10^{-22} \text{ cm}^3} = 1.5 \text{ g} \cdot \text{cm}^3$$

5.38 a = length of unit cell edge

$$V = \frac{\text{mass of unit cell}}{d}$$

(a) $V = a^3 = \dfrac{(1 \text{ unit cell})\left(\dfrac{107.87 \text{ g Ag}}{\text{mol Ag}}\right)}{10.5\ 00 \text{ g} \cdot \text{cm}^3} \times$

$$\dfrac{\left(\dfrac{1 \text{ mol Ag}}{6.02\ 21 \times 10^{23} \text{ atoms Ag}}\right)\left(\dfrac{4 \text{ atoms}}{1 \text{ unit cell}}\right)}{10.5\ 00 \text{ g} \cdot \text{cm}^3}$$

$a = 4.08\ 64 \times 10^{-8}$ cm

For an fcc cell, $a = \sqrt{8}\, r, r = \dfrac{\sqrt{2}\text{ a}}{4} = \dfrac{\sqrt{2}\,(4.08\ 64 \times 10^{-8} \text{ cm})}{4}$

$= 1.44\ 48 \times 10^{-8}$ cm or 144.48 pm

(b) $V = a^3 = \dfrac{(1 \text{ unit cell})\left(\dfrac{52.0\ 0 \text{ g Cr}}{1 \text{ mol Cr}}\right)}{7.19\ 0 \text{ g} \cdot \text{cm}^3}$

$$\dfrac{\left(\dfrac{1 \text{ mol Cr}}{6.02\ 2 \times 10^{23} \text{ atoms Cr}}\right)\left(\dfrac{2 \text{ atoms}}{1 \text{ unit cell}}\right)}{7.19\ 0 \text{ g} \cdot \text{cm}^3}$$

$a = 2.88\ 5 \times 10^{-8}$ cm

$r = \dfrac{\sqrt{3}\, a}{4} = \dfrac{\sqrt{3}\,(2.88\ 5 \times 10^{-8} \text{ cm})}{4} = 1.24\ 9 \times 10^{-8} \text{ cm} = 124.9 \text{ pm}$

5.40 (a) In the fcc geometry, the unit cell will contain four atoms of Kr. The density will be given by

mass in unit cell $= \dfrac{4 \text{ atoms}}{\text{unit cell}} \times \dfrac{83.80 \text{ g Kr}}{\text{mol Kr}} \times \dfrac{1 \text{ mol Kr}}{6.022 \times 10^{23} \text{ atoms}}$

$= 5.566 \times 10^{-22} \text{ g} \cdot \text{unit cell}^{-1}$

volume of unit cell $= \left(559 \text{ pm} \times \dfrac{10^{-12} \text{ m}}{\text{pm}} \times \dfrac{100 \text{ cm}}{\text{m}}\right)^3$

$= 1.75 \times 10^{-22} \text{ cm}^3 \cdot \text{unit cell}^{-1}$

density $= \dfrac{\text{mass in unit cell}}{\text{volume of unit cell}} = \dfrac{5.566 \times 10^{-22} \text{ g} \cdot \text{unit cell}^{-1}}{1.75 \times 10^{-22} \text{ cm}^3 \cdot \text{unit cell}^{-1}} = 3.18 \text{ g} \cdot \text{cm}^3$

(b) The face diagonal of the unit cell will equal four times the radius of the Kr atom:

$\sqrt{2}\, a = 4\, r$

$r = \dfrac{\sqrt{2}\, a}{4} = \dfrac{\sqrt{2}\ (559\ \text{pm})}{4} = 198\ \text{pm}$

(c) $V = \dfrac{4}{3}\pi\, r^3 = \dfrac{4}{3}\pi\ (198\ \text{pm})^3 = 3.25 \times 10^7\ \text{pm}^3$

(d) Volume occupied by the four Kr atoms in one unit cell $= 4(3.25 \times 10^7\ \text{pm}^3) = 1.30 \times 10^8\ \text{pm}^3$. The volume of the unit will be $(559\ \text{pm})^3 = 1.75 \times 10^8\ \text{pm}^3$. The percent of occupied space will be given by

$\dfrac{1.30 \times 10^8\ \text{pm}^3}{1.75 \times 10^8\ \text{pm}^3} \times 100 = 74.3\%$

5.42 Recall from plane geometry that in an equilateral triangle, the altitudes, perpendicular bisectors of the sides, and angle bisectors coincide and divide each other in the ratio 2:1.

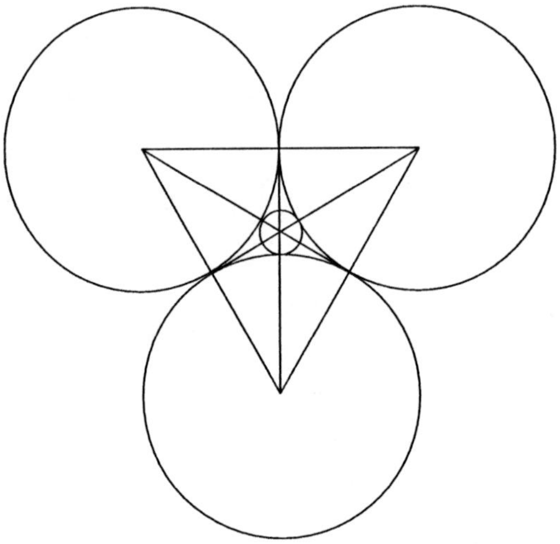

Since the vertical line in the sketch above is the altitude, a, of the equilateral triangle that connects the centers of the large circles, we can set $a = 3x$. Then two-thirds of its length is below the center of the small circle. From the sketch, we can see that $2x = r_{\text{large}} + r_{\text{small}}$.

We can use the Pythagorean Theorem to get x in terms of r_{large}.

$$(2r_{\text{large}})^2 = r_{\text{large}}^2 + a^2$$

$$a^2 = 3r_{\text{large}}^2$$

$$a = \sqrt{3}r_{\text{large}} = 3x$$

$$x = \frac{r_{\text{large}}}{\sqrt{3}}$$

Recall $2x = r_{\text{large}} + r_{\text{small}}$, so

$$2\left(\frac{r_{\text{large}}}{\sqrt{3}}\right) = r_{\text{large}} + r_{\text{small}}$$

$$r_{\text{small}} = 1.1547r_{\text{large}} - r_{\text{large}} = 0.1547r_{\text{large}}$$

In other words, the radius of the cavity is 15.5% as large as the radius of the large disks. A CD has a radius of 61 +/−1 mm with a measured cavity radius between three close-packed disks of 9 +/−1 mm. This experimental result is in good agreement with our geometrical treatment since $0.155 \times 61 = 9.5$.

5.44 (a) anions: $8 \text{ corners} \times \frac{1}{8} \text{ atom} \cdot \text{corner}^{-1} + 6 \text{ faces} \times \frac{1}{2} \text{ atom} \cdot \text{face}^{-1} = 4$ atoms; cations: $12 \text{ edges} \times \frac{1}{4} \text{ atom} \cdot \text{edge}^{-1} + 1 \text{ atom in center} = 4$ atoms; the cation to anion ratio is thus 4 : 4 or 1 : 1.

(b) calcium ions at corners of cell and at face centers:

$8 \text{ corners} \times \frac{1}{8} \text{ atom} \cdot \text{corner}^{-1} + 6 \text{ faces} \times \frac{1}{2} \text{ atom} \cdot \text{face}^{-1} = 4$ atoms; fluoride ions:

8 atoms in tetrahedral sites within the face-centered cubic lattice of Ca^{2+} ions; these atoms lie completely within the unit cell.

ratio of Ca : F = 4 : 8 or 1 : 2, giving an empirical formula of CaF_2

5.46 Ti atoms at corners: $8 \text{ corners} \times \frac{1}{8} \text{ atom} \cdot \text{corner}^{-1} = 1$ Ti

Ca atoms: 1 atom at center = 1 Ca

O atoms: 12 atoms, one on each edge gives $12 \text{ edges} \times \frac{1}{4} \text{ atom} \cdot \text{edge}^{-1}$ or 3 atoms:

3 O

empirical formula = $CaTiO_3$

5.48 (a) ratio $= \dfrac{138\text{ pm}}{196\text{ pm}} = 0.704$, predict rock-salt structure with (6,6) coordination;

(b) ratio $= \dfrac{58\text{ pm}}{196\text{ pm}} = 0.29$, predict zinc-blende structure with (4,4) coordination; however, LiBr actually adopts the rock-salt structure.

(c) ratio $= \dfrac{136\text{ pm}}{140\text{ pm}} = 0.971$, predict cesium-chloride structure with (8,8) coordination; however, BaO actually adopts the rock-salt structure.

5.50 (a) In the rock-salt structure, the unit cell edge length is equal to two times the distance between centers of the oppositely charged ions. Thus for NaI, a = 644 pm. The volume of the unit cell will be given by converting to cm^3:

$$V = \left(644\text{ pm} \times \frac{10^{-12}\text{ m}}{\text{pm}} \times \frac{100\text{ cm}}{\text{m}}\right)^3 = 2.67 \times 10^{-22}\text{ cm}^3$$

There are four formula units in the unit cell (see Exercise 5.44) so the mass in the unit cell will be given by

$$\text{mass in unit cell} = \frac{\left(\dfrac{4\text{ formula units}}{1\text{ unit cell}}\right) \times \left(\dfrac{149.89\text{ g NaI}}{1\text{ mol NaI}}\right)}{6.022 \times 10^{23}\text{ molecules} \cdot \text{mol}^{-1}} = 9.96 \times 10^{-22}\text{ g}$$

The density will be given by the mass in the unit cell divided by the volume of the unit cell:

$$d = \frac{9.96 \times 10^{-22}\text{ g}}{2.67 \times 10^{-22}\text{ cm}^3} = 3.73\text{ g} \cdot \text{cm}^3$$

(b) For the cesium-chloride structure, it is the body diagonal that represents two times the distance between the cation and anion centers. Thus the body diagonal is equal to 712 pm.

For a cubic cell, the body diagonal = $\sqrt{3}\ a = 712$ pm

$$a = 411 \text{ pm}$$

$$a^3 = V = \left(411 \text{ pm} \times \frac{10^{-12} \text{ m}}{\text{pm}} \times \frac{100 \text{ cm}}{\text{m}}\right)^3 = 6.94 \times 10^{-23} \text{ cm}^3$$

There is one formula unit of CsCl in the unit cell, so the mass in the unit cell will be given by

$$\text{mass in unit cell} = \frac{\left(\dfrac{1 \text{ formula unit}}{1 \text{ unit cell}}\right)\left(\dfrac{168.36 \text{ g CsCl}}{1 \text{ mol CsCl}}\right)}{6.022 \times 10^{23} \text{ molecules} \cdot \text{mol}^{-1}} = 2.796 \times 10^{-22} \text{ g}$$

$$d = \frac{2.796 \times 10^{-22} \text{ g}}{6.94 \times 10^{-23} \text{ cm}^3} = 4.03 \text{ g} \cdot \text{cm}^3$$

5.52 (a) Solid methane is held together only by London forces, whereas chloromethane is held in solid form by both London forces and dipole-dipole forces. Methanol will exhibit both London forces and dipole-dipole interactions, in addition to hydrogen bonding. (b) The melting points should increase in the order

CH_4 (−183°C) < CH_3Cl (−97°C) < CH_3COOH (16°C).

5.54 boron, carbon, phosphorus, silicon

5.56 (a) The alloy is substitutional—the phosphorus atoms are similar in size to the silicon atoms and can replace them in the crystal lattice.

(b) Because phosphorus has one more valence electron than silicon, the added electron will be forced into the conduction band (see Chapter 3) making it easier for this electron to move through the solid. The conductivity of the doped silicon is thus higher than the pure material.

5.58 The formula of zinc oxide is ZnO with one zinc atom per oxygen atom. After heating in a vacuum, some of the oxygen atoms are able to leave the lattice as $O_2(g)$. Because the oxidation state of oxygen in zinc oxide is -2 and in O_2 it is 0, some of the zinc atoms must have been reduced from Zn^{2+} to Zn^0. Because this process is reversible, the lattice must remain mostly intact. The result is that the ZnO_x ($x < 1$) will have more electrons that can move into the conduction band, enhancing the electrical conductivity of the semiconductor. Regaining these electrons reverses this process and lowers the electrical conductivity to the starting value.

5.60 (a) The residual composition of Wood's metal is Bi; this constitutes 51% of the alloy by mass. The number of moles per 100 g of alloy is: 0.105 mol Sn, 0.111 mol Cd, 0.11 mol Pb, and 0.24 mol Bi. The atom ratio is 2.3 Bi : 1.1 Cd : 1.1 Pb : 1 Sn.

(b) Of 100 g of steel, there are 1.75 g C and 98.25 g iron.

$$\text{mol C} = \frac{1.75\text{ g}}{12.01\text{ g}\cdot\text{mol}^{-1}}$$

$$\text{mol C} = 0.146\text{ mol}$$

$$\text{mol Fe} = \frac{98.25\text{ g}}{55.85\text{ g}\cdot\text{mol}^{-1}}$$

$$\text{mol Fe} = 1.759\text{ mol}$$

Ratio Fe : C = 1.759 mol $\div$ 0.146 mol = 12.0 : 1

5.62 (a) The N atoms fit the pattern for a trigonal planar arrangement of three objects—two bonds and one lone pair for one nitrogen atom, and three bonds for the other. This corresponds to sp^2 hybridization. The bond angles are all close to 120°.

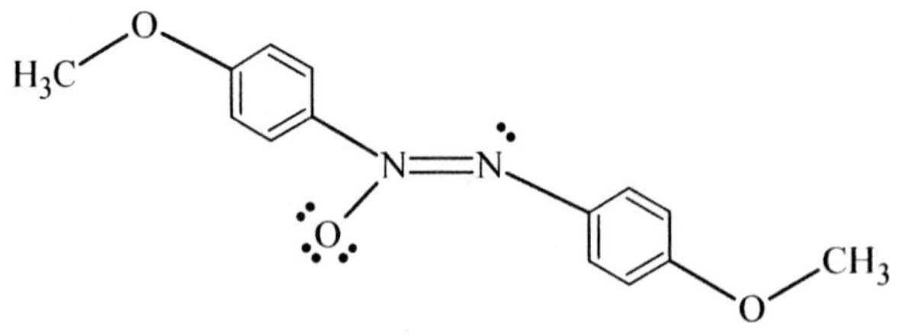

(b) The multiple bond structure of this molecule gives rise to the rigid rod-like nature of the molecule. The alternating double bonds that allow electrons to be delocalized extend from one ring across the bridging N atoms to the other ring. To maintain the delocalization, the phenyl rings do not rotate about the O—N axis. Thus, the entire molecule is like a stiff, flat rod. Only the CH_3 groups rotate. Another feature that enhances the tendency to form liquid crystalline materials is the inclusion of the aromatic rings. The π-bonds in aromatic rings show a strong tendency to stack one upon another; this helps orient the molecules in the liquid crystalline array. (c) There is an infinite number of answers here. Basically, any molecule that has a fairly rigid backbone structure without too many dangling appendages to interfere with the packing will be a possible liquid crystalline material.

5.64 The strategy employed here is simply to modify the attached groups with a view toward disrupting some of the London interactions that the molecule will have with its neighbors, thus allowing the molecules more freedom to move with respect to each other—if the intermolecular forces are too strong, the material will be held in the solid state longer. One must be careful, because too much disruption of these forces will also destroy the liquid crystal order as well. This is still a matter of trial and error in many cases. For example, the molecule

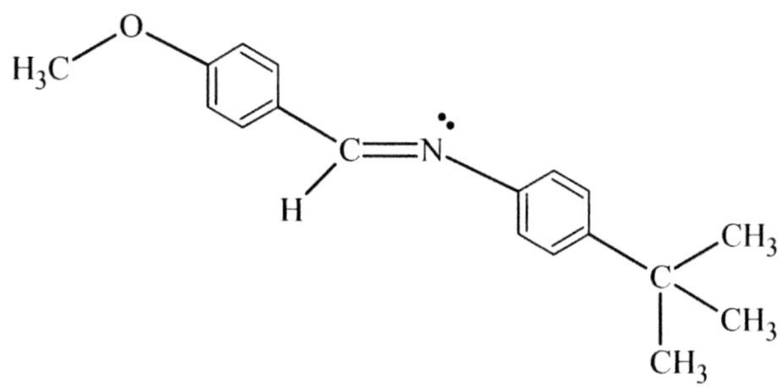

is very similar to *p*-azoxyanisole but has a liquid crystal range of 21°C–47°C. It would be expected that the bulky *t*-butyl group ($-C(CH_3)_3$) would help to disrupt the orderly packing of the molecule.

5.66 (a) Note that each kink in the long chains represents one $-CH_2-$ group while the terminus of each chain is a $-CH_3$ group.

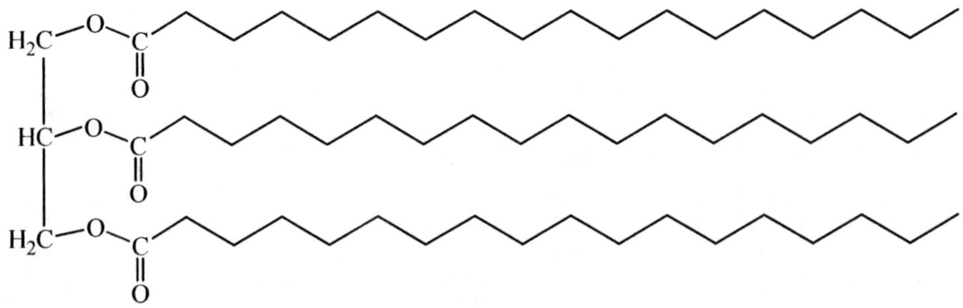

(b) London forces are very important here. Notice that the long chains are aligned parallel to each other, as may be expected because they each have a large surface area that can interact with the other chains.

(c) Although this molecule does have some oxygen atoms with lone pairs, the long hydrocarbon chains are very nonpolar. Given that stearic acid itself is only slightly soluble in water, one would expect this molecule to be insoluble, which is indeed the case.

5.68 (a)

(i)

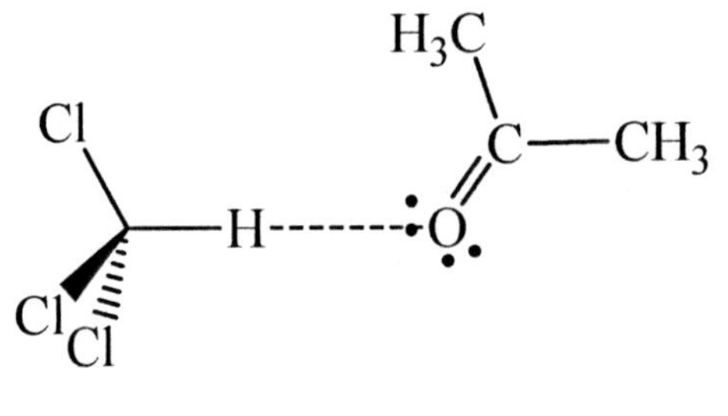

(ii)

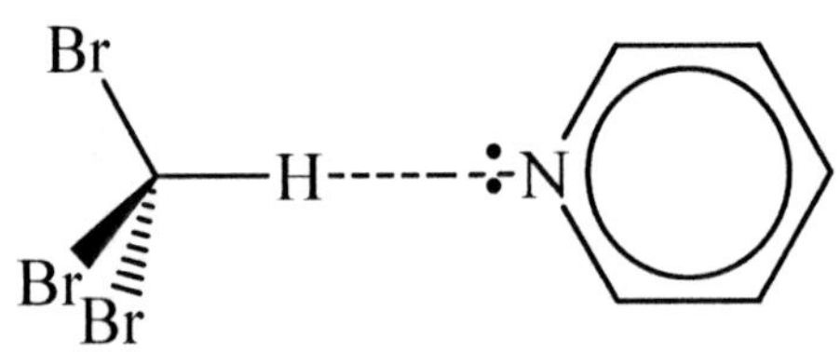

(b) The hydrogen atom that has low electron density in both cases is attached to carbon on the trihalomethane molecule. The hydrogen bonds formed by these pairs of molecules are referred to as weak hydrogen bonds. The hydrogen atoms are not as electropositive as they would be if they were bound directly to O, N, or F. But electronegative atoms attached to a neighboring carbon atom draw enough electron density away from the H through the bonds to allow it to participate in hydrogen bonding with a lone pair on an electronegative atom in another molecule.

5.70 (a) A body-centered cubic lattice has two atoms per unit cell. For this cell, the relation between the radius of the atom r and the unit cell edge length a is derived from the body diagonal of the cell, which is equal to four times the radius of the atom. The body diagonal is found from the Pythagorean theorem to be equal to

$\sqrt{3}\,a$.

$$4\,r = \sqrt{3}\,a$$

$$\text{a} = \frac{4\,r}{\sqrt{3}}$$

The volume of the unit cell is given by

$$V = a^3 = \left(\frac{4\,r}{\sqrt{3}}\right)^3$$

If r is given in pm, then a conversion factor to cm is required:

$$V = a^3 = \left(\frac{4\,r}{\sqrt{3}} \times \frac{10^{-12}\ \text{m}}{\text{pm}} \times \frac{10\ \text{cm}}{\text{m}}\right)^3$$

Because there are two atoms per bcc unit cell, the mass in the unit cell will be given by

$$\text{mass} = \left(\frac{2\ \text{atoms}}{\text{unit cell}}\right)\left(\frac{M}{6.022 \times 10^{23}\ \text{atoms}\cdot\text{mol}^{-1}}\right)$$

The density will be given by

$$d = \frac{\text{mass of unit cell}}{\text{volume of unit cell}} = \frac{\left(\dfrac{2\ \text{atoms}}{\text{unit cell}}\right)\left(\dfrac{M}{6.022 \times 10^{23}\ \text{atoms} \cdot \text{mol}^{-1}}\right)}{\left(\dfrac{4\,r}{\sqrt{3}} \times \dfrac{10^{-12}\ \text{m}}{\text{pm}} \times \dfrac{100\ \text{cm}}{\text{m}}\right)^3}$$

$$= \frac{\left(\dfrac{2\ \text{atoms}}{\text{unit cell}}\right)\left(\dfrac{M}{6.022 \times 10^{23}\ \text{atoms} \cdot \text{mol}^{-1}}\right)}{\left(2.309 \times 10^{-10}\ r\right)^3}$$

$$= \frac{\left(2.698 \times 10^{5}\right) M}{r^3}$$

or

$$r = \sqrt[3]{\frac{\left(2.698 \times 10^{5}\right) M}{\text{density}}}$$

where M is the molar mass in $\text{g} \cdot \text{mol}^{-1}$ and r is the radius in pm.

For the fcc unit cell, the relation between the radius of the atom r and the unit cell edge length a is

$$4\,r = \sqrt{2}\,a$$

$$a = \frac{4\,r}{\sqrt{2}}$$

The volume of the unit cell is given by

$$V = a^3 = \left(\frac{4\,r}{\sqrt{2}}\right)^3$$

If r is given in pm, then a conversion factor to cm is required:

$$V = a^3 = \left(\frac{4\,r}{\sqrt{2}} \times \frac{10^{-12}\ \text{m}}{\text{pm}} \times \frac{100\ \text{cm}}{\text{m}}\right)^3$$

Because there are four atoms per fcc unit cell, the mass in the unit cell will be given by

$$\text{mass} = \left(\frac{4\ \text{atoms}}{\text{unit cell}}\right)\left(\frac{M}{6.022 \times 10^{23}\ \text{atoms} \cdot \text{mol}^{-1}}\right)$$

The density will be given by

$$d = \frac{\text{mass of unit cell}}{\text{volume of unit cell}} = \frac{\left(\dfrac{4 \text{ atoms}}{\text{unit cell}}\right)\left(\dfrac{M}{6.022 \times 10^{23} \text{ atoms} \cdot \text{mol}^{-1}}\right)}{\left(\dfrac{4\,r}{\sqrt{2}} \times \dfrac{10^{-12} \text{ m}}{\text{pm}} \times \dfrac{100 \text{ cm}}{\text{m}}\right)^3}$$

$$= \frac{(2.936 \times 10^5)M}{r^3}$$

or

$$r = \sqrt[3]{\frac{(2.936 \times 10^5)M}{d}}$$

where M is the molar mass in $\text{g} \cdot \text{mol}^{-1}$ and r is the radius in pm.

Setting these bcc and fcc equations equal to each other (because both are equal to r) and cubing both sides, we obtain

$$\frac{(2.936 \times 10^5)M}{d_{\text{fcc}}} = \frac{(2.698 \times 10^5)M}{d_{\text{bcc}}}$$

The molar mass M is the same and will cancel from the equation.

$$\frac{(2.936 \times 10^5)}{d_{\text{fcc}}} = \frac{(2.698 \times 10^5)}{d_{\text{bcc}}}$$

Rearranging, we get

$$d_{\text{fcc}} = \frac{(2.936 \times 10^5)}{(2.698 \times 10^5)} d_{\text{bcc}}$$

$$= 1.088\ d_{\text{bcc}}$$

$$= 1.088 \times 19.6 \text{ g} \cdot \text{cm}^3$$

$$= 21.0 \text{ g} \cdot \text{cm}^3$$

(c) For the different alkali metals, we calculate the results given in the following table:

Gas	Density bcc ($g \cdot cm^3$)	Density fcc ($g \cdot cm^3$)	Molar mass ($g \cdot mol^{-1}$)	Radius (pm)
Li	0.53	0.58	6.94	152
Na	0.97	1.0	22.99	186
K	0.86	0.94	39.10	231
Rb	1.53	1.66	85.47	247
Cs	1.87	2.03	132.91	268

(d) Li, Na, and K all have densities less than that of water and should float, but not for long, as they all react violently with water to form MOH(aq) and $H_2(g)$.

5.72 (a) The ion-ion potential energy ratio will be given by

$$V \propto \frac{z_1 z_2}{d}$$

$$V_{LiCl} \propto \frac{z_1 z_2}{d_{Li\text{-}Cl}}$$

$$V_{KCl} \propto \frac{z_1 z_2}{d_{K\text{-}Cl}}$$

$$\text{ratio}\left(\frac{V_{Li^+}}{V_{K^+}}\right) = \frac{\dfrac{z_1 z_2}{d_{Li\text{-}Cl}}}{\dfrac{z_1 z_2}{d_{K\text{-}Cl}}} = \frac{\dfrac{1}{d_{Li\text{-}Cl}}}{\dfrac{1}{d_{K\text{-}Cl}}} = \frac{d_{K\text{-}Cl}}{d_{Li\text{-}Cl}}$$

From the ionic radii in Figure 1.41 we can calculate the LiCl and KCl distances to be

$$d_{Li\text{-}Cl} = 58 \text{ pm} + 181 \text{ pm} = 239 \text{ pm}$$

$$d_{K\text{-}Cl} = 138 \text{ pm} + 181 \text{ pm} = 319 \text{ pm}$$

$$\text{ratio} = \frac{319 \text{ pm}}{239 \text{ pm}} = 1.33$$

(b) The ratio for the ion-dipole interaction is derived as follows:

$$V \propto \frac{-|z|\mu}{d^2}$$

$$V_{Li^+} \propto \frac{-|z|\mu}{d^2} = \frac{-|1|\mu}{(58)^2}$$

$$V_{K^+} \propto \frac{-|z|\mu}{d^2} = \frac{-|1|\mu}{(138)^2}$$

The electric dipole moment of the water molecule (μ) will cancel:

$$\text{ratio}\left(\frac{V_{Li^+}}{V_{K^+}}\right) = \frac{-|1|\mu/(58)^2}{-|1|\mu/(138)^2} = \frac{(138)^2}{(58)^2} = 5.66$$

(c) Because the ion-dipole interactions are proportional to d^2 and the ion-ion interactions directly proportional to *d,* the relative importance of hydration will be much larger for the smaller lithium ion.

5.74 The number of oxide ions is equal to 12 edges × $\frac{1}{4}$ oxide ion in the unit cell per edge, for a total of three. For the niobium atoms, there will be six faces with $\frac{1}{2}$ of the niobium atom per face inside the cell, also for a total of three. The empirical formula will be NbO, with three formula units in the unit cell.

5.76 (a) The oxidation state on uranium must balance the charge due to the oxide ions. If 2.17 oxide ions are present with a charge of −2 per oxide ion, then the uranium must have an average oxidation state of +4.34.

(b) This is most easily solved by setting up two equations in two unknowns. We know that the total charge on the uranium atoms present must equal +4.34, so if we multiply the charge on each type of uranium by the fraction of uranium present in that oxidation state and sum the values, we should get 4.34.

Let x = fraction of U^{4+}, y = fraction of U^{5+}, then

4x + 5y = 4.34

Also, because we are assuming that all the uranium is either +4 or +5, the fractions of each present must add up to 1:

x + y = 1

Solving these two equations simultaneously, we obtain y = 0.34, x = 0.66.

5.78 (a) False. In order for the unit cell to be considered body centered, the atom at the center must be identical to the atoms at the corners of the unit cell. (b) True. The properties of the unit cell in general must match the properties of the bulk material. (c) True. This is the basis for Bragg's law. (d) False. The angles that define the values of the unit cell can have any value, the only restriction being that opposing faces of the unit cell must be parallel.

5.80 There are several ways to draw unit cells that will repeat to generate the entire lattice. Some examples are shown below where T represents a tree, D a dog, and L a leaf. The choice of unit cell is determined by conventions that are beyond the scope of this text (the smallest unit cell that indicates all of the symmetry present in the lattice is typically the one of choice).

(a)

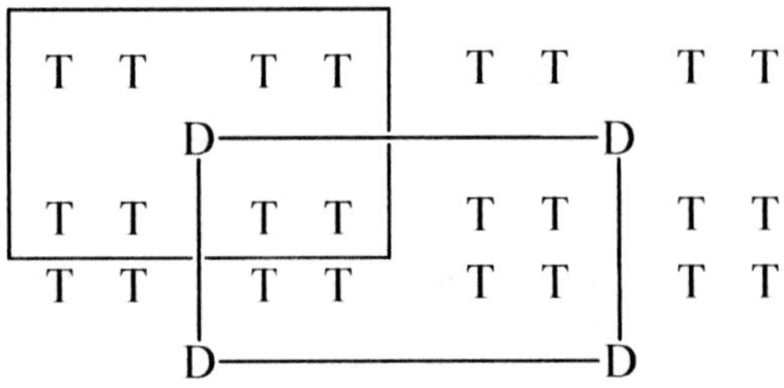

(b)

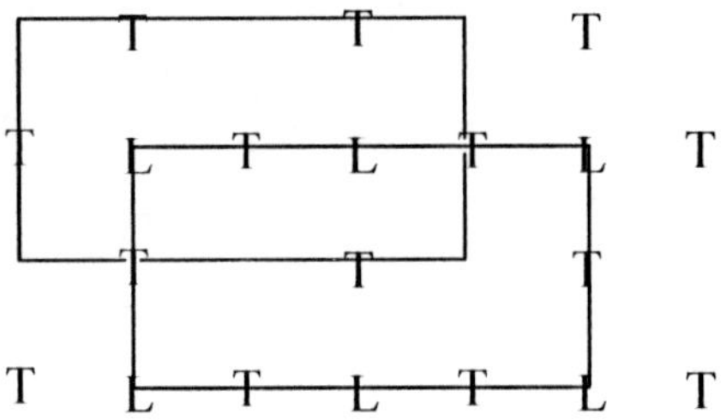

5.82 (a)

Molecule	Molar mass $(g \cdot mol^{-1})$	Boiling point (°C)
(1) CH_4	16.04	−164
(2) C_2H_6	30.07	−88.6
(3) $CH_3CH_2CH_3$	44.11	−42.1
(4) $CH_3CH_2CH_2CH_3$	58.13	−0.5
(5) $CH_3CH_2CH_2CH_2CH_3$	72.15	36.1
(6) $(CH_3)_3CH$	58.13	−11.7
(7) $CH_3CH_2CH(CH_3)_2$	72.15	27.8
(8) $C(CH_3)_4$	72.15	9.5

(b) The easiest comparison to make is among those compounds that are all straight chain compounds (those which do not have carbon atoms attached to more than two other carbon atoms). This includes [1] methane, [2] ethane, [3] propane, [4] butane, and [5] pentane. From the data it is clear that the boiling point increases regularly with an increase in molar mass. The increase in boiling point arises because the bigger molecules have more surface area and more electrons with which to interact with other molecules (i.e., the effect of the London forces will be greater if there is more surface area to interact via instantaneous dipoles).

(c) The molecular shape issue is best examined by comparing molecules of the same molar mass. The best comparisons are, therefore, between butane and 2-methylpropane (isobutane [6]), and between pentane, 2-methylbutane (isopentane [7]) and 2,2-dimethylpropane (neopentane [8]). What is clear is that the boiling point decreases as the number of side groups increases. If one constructs models of these compounds, one can see that the more branched a molecule is, the more spherical and compact it becomes. More spherical molecules have less surface area (which is exposed to other molecules for the formation of London forces).

5.84 Hydrogen ions can "migrate" through a solution much faster than other ions because they essentially do not really have to move at all. The hydrogen bonding network readily allows the shifting of hydrogen bonds, so that a proton can become available in solution almost instantaneously anywhere without any specific individual proton actually having to move any great distance. It amounts to the migration of a unit of positive charge through the network of hydrogen-bonded water molecules. Any other molecule or ion would have to migrate through the solution normally, as an intact unit.

5.86 (a) The gases with the larger van der Waals' *a* parameter are: (i) Ne; (ii) O_2; (iii) CO_2; (iv) H_2CO; (v) $CH_3(CH_2)_{10}CH_3$. (b) The gases with the larger van der Waals' *b* parameter are: (i) Br_2; (ii) F_2; (iii) $CH_3CH_2CH_3$; (iv) Kr; (v) SO_2.

5.88 (a) The octahedral hole is considerably bigger than the tetrahedral hole and can accommodate an ion with a radius about 3.7 times that of the tetrahedral hole.

(b) If the anions are close packed in a cubic close-packed array, the unit cell will be a fcc unit cell. The octahedral sites will lie at the body center of the cell and at the center of each edge. If the unit cell is divided into octants, then there will be a tetrahedral site at the center of each octant. In the cubic close-packed geometry, the face diagonal of the unit cell will represent four times the radius of the anion:

$$4\, r_{\text{Anion}} = a\sqrt{2}$$

If a cation occupies the octahedral site along the unit cell edge (a cation at the very center of the cell will be identical), then the maximum size it can have if the anions are close-packed will be when $2\, r_{\text{Anion}} + 2\, r_{\text{Cation}} = a$.

Combining these two relationships, we see that the radius of the cation will be

$$2\,r_{\text{Cation, octahedral}} = a - 2\,r_{\text{Anion}} = a - 2\left(\frac{a\sqrt{2}}{4}\right) = a - \frac{a\sqrt{2}}{2} = a\left(\frac{2-\sqrt{2}}{2}\right)$$

The distance between the cation and the anion in a tetrahedral site will be given by $\frac{1}{4}$ the body diagonal of the cell, which will correspond to $r_{\text{Cation, tetrahedral}} + r_{\text{Anion}}$.

$$r_{\text{Cation, tetrahedral}} + r_{\text{Anion}} = \frac{a\sqrt{3}}{4}$$

$$r_{\text{Cation, tetrahedral}} = \frac{a\sqrt{3}}{4} - r_{\text{Anion}} = \frac{a\sqrt{3}}{4} - \frac{a\sqrt{2}}{4} = \frac{a(\sqrt{3}-\sqrt{2})}{4}$$

The ratio of $r_{\text{Cation, octahedral}}$ to $r_{\text{Cation, tetrahedral}}$ will thus be given by

$$\text{ratio} = \frac{a\left(\dfrac{2-\sqrt{2}}{2}\right)}{a(\sqrt{3}-\sqrt{2})/4} = \frac{\left(\dfrac{2-\sqrt{2}}{2}\right)}{(\sqrt{3}-\sqrt{2})/4} = 3.7$$

Notice that the value of a will cancel in the calculation, so that the ratio is independent of the actual edge length.

(c) If half the tetrahedral holes are filled, there will be four cations in the unit cell. The fcc cell will have a total of four anions from contributions of atoms at the corners and face centers, so the empirical formula will be MA.

5.90 The calcium fluoride lattice is based upon a face-centered cubic unit cell with the Ca^{2+} ions occupying the corners and face centers. The F^- ions are located in the centers of the eight tetrahedral cavities found within the face-centered cell. In the unit cell there are four formula units of CaF_2. Using this information and the density given in Ex. 5.79, we can calculate the volume of the unit cell.

$$V = \frac{\left(\dfrac{78.08\ \text{g}\cdot\text{mol}^{-1}}{6.022\times 10^{23}\ \text{formula units}\cdot\text{mol}^{-1}}\right)(4\ \text{formula units}\cdot\text{unit cell}^{-1})}{3.180\ \text{g}\cdot\text{cm}^{-3}}$$

$$V = 1.631\times 10^{-22}\ \text{cm}^3 = 1.631\times 10^{8}\ \text{pm}^3$$

The occupied volume is calculated from the size of the ions using their ionic radii [$r(\text{Ca}^{2+}) = 100$ pm; $r(\text{F}^-) = 133$ pm].

$$V_{\text{Ca}^{2+}} = \frac{4}{3}\pi\, r^3 = \frac{4}{3}\pi(100\text{ pm})^3 = 4.19 \times 10^6\ \text{pm}^3$$

$$V_{\text{F}^-} = \frac{4}{3}\pi\, r^3 = \frac{4}{3}\pi(133\text{ pm})^3 = 9.85 \times 10^6\ \text{pm}^3$$

The total volume occupied is

$4(4.19 \times 10^6\ \text{pm}^3) + 8(9.85 \times 10^6\ \text{pm}^3) = 9.56 \times 10^7\ \text{pm}^3$. The percent of empty space will be given by

$$\frac{(1.631 \times 10^8\ \text{pm}^3 - 9.56 \times 10^8\ \text{pm}^3)}{1.631 \times 10^8\ \text{pm}^3} \times 100 = 41.4\%$$

41.4% of the space is empty.

5.92 The difficulty with this problem is choosing a proper unit cell for the hexagonal lattice. This cell is represented below:

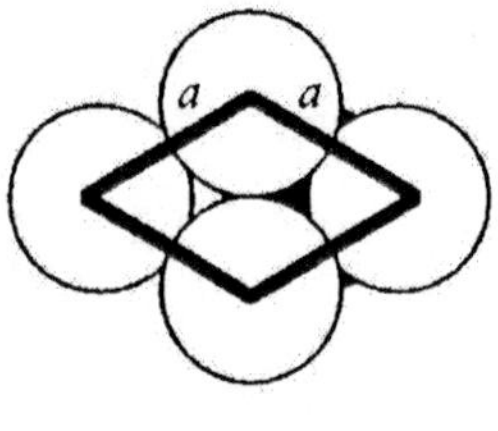

top view

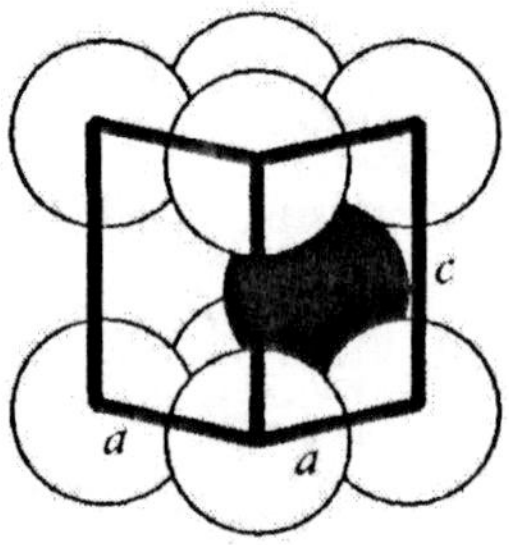

side view

The unit cell has the hexagonal shape where there are two edges with the same length *a* and a third that is different from *a*, which is labeled *c*. In this unit cell there are a total of two atoms. One is present from the sum of the parts of the atoms on the corners; the second is the atom that is completely enclosed (see side view). All the atoms are equivalent, but the one inside the unit cell is shaded for easy reference.

The distance *a* is equal to twice the radius of the atom, but *c* must be calculated using geometry and trigonometry. It too is related to the radius of the atom for a close-packed lattice of a single element. Upon examining the geometry in order to set up the calculation, what we note is that the totally enclosed atom forms a regular tetrahedron with three atoms in the face. The center of this tetrahedron lies exactly in the face of the unit cell. The distance from this point to the center of the enclosed atom is equal to $\frac{1}{2}c$, so if we can calculate that distance, we will be able easily to find the lattice parameter c. This is illustrated in the diagrams below.

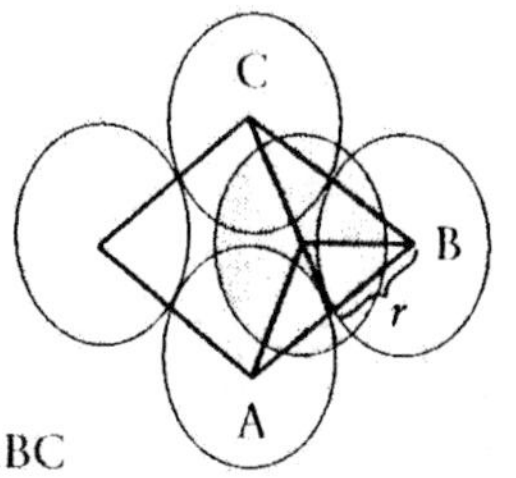

point D is at the midpoint of triangle ABC

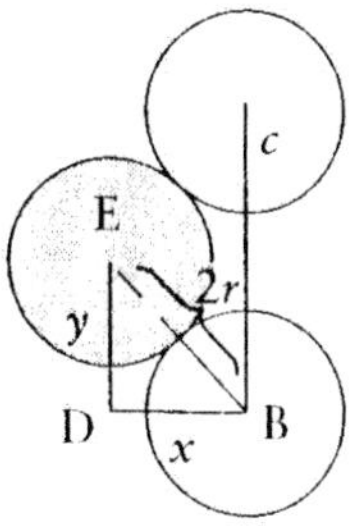

The distance labeled y is $\frac{1}{2}c$. The distance labeled x in the figure on the right is the same as the line segment BD in the figure on the left. The triangle ABC is an equilateral triangle with all sides equal to a or $2r$. A line drawn from C perpendicular to line segment AB will bisect that segment. The angle ABC is $60°$ so that the angle ABD is equal to $30°$. We can then write

$$\cos 30° = \frac{r}{x}$$

$$x = \frac{r}{\cos 30°}$$

$$x = \frac{2}{\sqrt{3}}r$$

To calculate y, we use the Pythagorean theorem:

$$y^2 + x^2 = (2r)^2$$

$$y^2 = 4r^2 - \left(\frac{2}{\sqrt{3}}r\right)^2$$

$$y^2 = 4r^2 - \frac{4}{3}r^2$$

$$y^2 = \frac{8}{3}r^2$$

$$y = \frac{2\sqrt{2}}{\sqrt{3}}r$$

$$c = 2y = \frac{4\sqrt{2}}{\sqrt{3}}r$$

The volume of the unit cell is determined by multiplying c by the area of the unit cell face. This area is found by noting that the face is a parallelogram whose area will equal base × height $(b \times h)$, where $b = 2r$. We can obtain h from

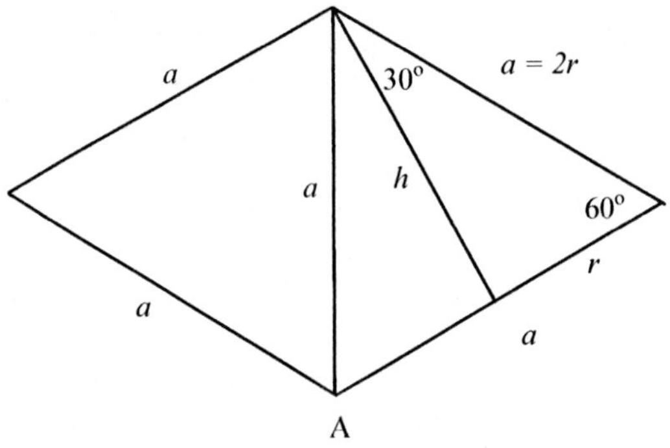

$$\sin 60° = \frac{h}{2r}$$

$$h = 2r \sin 60°$$

$$h = \frac{2\sqrt{3}}{2}r = r\sqrt{3}$$

The area of the face is then

$$\text{face area} = 2r \times r\sqrt{3} = 2\sqrt{3}\,r^2$$

The volume of the unit cell is then

$$V = 2\sqrt{3}\, r^2 \times \frac{4\sqrt{2}}{\sqrt{3}} r$$

$$= 8\sqrt{2}\, r^3$$

The volume of an individual atom will be given by

$$V_{\text{atom}} = \frac{4}{3}\pi r^3$$

The total volume occupied by the atoms in the unit cell will, therefore, be

$$V_{\text{total}} = \frac{8}{3}\pi r^3$$

The fraction of occupied space will be given by

$$\frac{\left(\dfrac{8\,\pi r^3}{3}\right)}{(8\sqrt{2}\, r^3)} = \frac{\pi}{3\sqrt{2}} = 0.74$$

We can see that this is exactly equal to the fraction of occupied space for the face-centered cubic lattice as derived in the book (see section 5.9).

5.94 To solve this problem, we refer to Bragg's law: $\lambda = 2\, d \sin\theta$ where λ is the wavelength of radiation, d is the interplanar spacing, and θ is the angle of incidence of the x-ray beam.

154 pm = $2\, d \sin 7.42°$

d = 596 pm

5.96 1 eV = $1e^-$ through a potential of 1 V

$$?\ \text{eV} = \frac{1}{2} m_e v^2 \text{ when } \lambda = 100 \text{ pm} = \frac{h}{mv}$$

$$= \frac{1}{2}\left(\frac{h}{\lambda}\right)\left(\frac{h}{m\lambda}\right) = \frac{1}{2} \cdot \frac{(6.626 \times 10^{-34}\ \text{J} \cdot \text{s})^2}{(9.109 \times 10^{-31}\ \text{kg})(1 \times 10^{-10} \text{m})^2}$$

$$= 2.41 \times 10^{-17}\ \text{J}\left(\frac{1\ \text{eV}}{1.6022 \times 10^{-19}\ \text{J}}\right) = 150\ \text{eV}$$

Therefore, electrons must be accelerated through a potential of 150 V in order for them to have a wavelength of 100 pm. Under these conditions, electrons can be used in diffraction studies.

5.98 The Coulomb potential energy of interaction for the point charge with the dipole can be represented by the sum of the repulsive and attractive terms as given by equation 1, page 162, $E_p = \frac{q_1 q_2}{4\pi\varepsilon_o r}$. To avoid confusion, we will call the point charge z since each end of the dipole is q or $-q$. The distance to one end of the dipole is $r + \frac{l}{2}$ and the distance to the oppositely charged end is $r - \frac{l}{2}$. So the potential energy at the point charge is

$$E_p = |z|\left(\frac{kq}{r+\frac{l}{2}} - \frac{kq}{r-\frac{l}{2}}\right) = |z|\frac{kq}{r}\left(\frac{1}{1+\frac{l}{2r}} - \frac{1}{1-\frac{l}{2r}}\right)$$

where $k = \frac{1}{4\pi\varepsilon_o}$ and z is the magnitude of the point charge.

Since $l << r$, it is also true that $\frac{l}{2r} << 1$. Using the expansion given in the problem, $\frac{1}{(1+x)} = 1 - x + ...$, with $x = \frac{l}{2r}$, gives

$$\frac{1}{1+\frac{l}{2r}} \approx 1 - \frac{l}{2r} \text{ and } \frac{1}{1-\frac{l}{2r}} \approx 1 + \frac{l}{2r}.$$

Then the expression for the potential energy becomes

$$E_p = |z|\frac{kq}{r}\left(1 - \frac{l}{2r} - 1 - \frac{l}{2r}\right) = |z|\frac{kq}{r}\left(-\frac{2l}{2r}\right) = |z|\frac{-kql}{r^2}$$

Substituting $k = \frac{1}{4\pi\varepsilon_o}$ and $\mu = ql$

$$E_p = -\frac{|z|\mu}{4\pi\varepsilon_o r^2}$$

Note that this expression agrees with equation 2, page 163.

CHAPTER 6

THERMODYNAMICS: THE FIRST LAW

6.2 (a) The internal energy of an open system could be increased by (1) adding matter to the system, (2) doing work on the system, and (3) adding heat to the system. (b) Matter cannot be added to a closed system, so only (2) doing work on the system and (3) adding heat to the system could be used in increase the internal energy. (c) The internal energy of an isolated system cannot be changed.

6.4 (a) Given four cylinders operating at 60 cycles per minute and displacing 2.5 L each under a constant pressure of 1400 bar, the work done is given by:

$$(4\text{ cylinders})\cdot\left(60\ \frac{\text{cycles}}{\text{min}}\right)\cdot\left(-1400\text{ bar}\cdot 10^5\ \frac{\text{Pa}}{\text{bar}}\right)\cdot\left(2.5\text{ L}\cdot\frac{0.001\text{ m}^3}{\text{L}}\right)$$

$$= -8.4\times 10^7\text{ J}$$

(b) The expanding gas in the cylinder does work on the pistons (and to the load connected to the pistons by the crankshaft of the engine). Therefore, work done by the gas is negative.

6.6 (a) The change in internal energy ΔU is given simply by summing the two energy terms involved in this process. We must be careful, however, that the signs on the energy changes are appropriate. In this case, internal energy will be added to the gas sample by heating, but the expansion will remove internal energy from the sample so it will be a negative number: $\Delta U = 325\text{ kJ} - 515\text{ kJ} = -190\text{ kJ}$

(b) If the heat added had exactly matched the amount of energy lost due to the work of the gas (i.e., if q had been 235 kJ), then ΔU would have been 0 and the temperature of the gas would not have changed. Because less heat was added, however, and the gas was allowed to expand further, the temperature of the gas would have had to decrease, and consequently the pressure of the gas would be lower at the end.

6.8 To calculate this, we use the relationship $\Delta U = q + w$, which arises from the first law of thermodynamics. Because the system releases heat, q will be a negative number as will ΔU, because it is a decrease in internal energy:

$$-125 \text{ kJ} = -346 \text{ kJ} + w$$

$$w = +221 \text{ kJ}$$

Because work is positive, the surroundings will do work on the system.

6.10 If the heater operates as rated, then the total amount of heat transferred to the cylinder will be

$$(100 \text{ J} \cdot \text{s}^{-1})(10 \text{ min})(60 \text{ s} \cdot \text{min}^{-1}) = 6.0 \times 10^4 \text{ J or } 60 \text{ kJ}$$

Work will be given by $w = -P_{\text{ext}}\Delta V$ in this case because it is an expansion against a constant opposing pressure:

$$w = -(0.975 \text{ atm})(10.00 \text{ L} - 2.00 \text{ L}) = -7.80 \text{ L} \cdot \text{atm}$$

In order to combine the two terms to get the internal energy change, we must first convert the units on the energy terms to the same values. A handy trick to get the conversion factor for $\text{L} \cdot \text{atm}$ to J or *vice versa* is to make use of the equivalence of the ideal gas constant R in terms of $\text{L} \cdot \text{atm}$ and J:

$$w = -7.80 \text{ L} \cdot \text{atm}\left(\frac{8.314 \text{ J} \cdot \text{K}^{-1} \cdot \text{mol}^{-1}}{0.082\,06 \text{ L} \cdot \text{atm} \cdot \text{K}^{-1} \cdot \text{mol}^{-1}}\right) = -790 \text{ J or } -0.790 \text{ kJ}$$

The internal energy change is then the sum of these two numbers:

$$\Delta U = q + w = 60 \text{ kJ} + (-0.790 \text{ kJ}) \cong 59 \text{ kJ}$$

The energy change due to the work term turns out to be negligible in this problem.

6.12 This is calculated from $\Delta U = q + w$ where we know $\Delta U = -892.4\ \text{kJ}$ and $w = -492\ \text{kJ}$:

$$-892.4\ \text{kJ} = q - 492\ \text{kJ}$$
$$q = -400\ \text{kJ}$$

4.00×10^2 kJ of heat are lost from the system.

6.14 (a) true only if $q = 0$; (b) always true, volume is fixed and no work can be done; (c) never true; (d) always true; (e) true if $q = 0$

6.16 (a) The heat change will be made up of two terms: one term to raise the temperature of the stainless steel and the other to raise the temperature of the water:

$$q = (450.0\ \text{g})(4.18\ \text{J}\cdot(°\text{C})^{-1}\cdot\text{g}^{-1})(100.0°\text{C} - 25.0°\text{C})$$
$$+ (500.0\ \text{g})(0.51\ \text{J}\cdot(°\text{C})^{-1}\cdot\text{g}^{-1})(100°\text{C} - 25°\text{C})$$
$$q = 141\ \text{kJ} + 19\ \text{kJ} = 1.60\times10^2\ \text{kJ}$$

(b) The percentage of heat attributable to raising the temperature of the water will be

$$\left(\frac{141\ \text{kJ}}{160\ \text{kJ}}\right)(100) = 88.1\%$$

(c) The use of the copper kettle is more efficient, as a larger percentage of the heat goes into heating the water and not the container holding it.

6.18 heat lost by metal = $-$heat gained by water

$$(20.0\ \text{g})(25.7°\text{C} - 100.0°\text{C})(C_s)$$
$$= -(50.7\ \text{g})(4.18\ \text{J}\cdot(°\text{C})^{-1}\cdot\text{g}^{-1})(25.7°\text{C} - 22.0°\text{C})$$
$$C_s = 0.53\ \text{J}\cdot(°\text{C})^{-1}\cdot\text{g}^{-1}$$

6.20 $$C_{\text{cal}} = \frac{-\left(\dfrac{1.236 \text{ g benzoic acid}}{122.12 \text{ g}\cdot\text{mol}^{-1} \text{ benzoic acid}}\right)(-3227 \text{ kJ}\cdot\text{mol}^{-1} \text{ benzoic acid})}{2.345°\text{C}}$$

$$= 13.93 \text{ kJ}\cdot(°\text{C})$$

6.22 (a) Because the process is isothermal, $\Delta U = 0$ and $q = -w$. For a reversible process,

$$w = -nRT \ln\frac{V_2}{V_1}$$

n is obtained from the ideal gas law:

$$n = \frac{PV}{RT} = \frac{(2.57 \text{ atm})(3.42 \text{ L})}{(0.082\ 06 \text{ L}\cdot\text{atm}\cdot\text{K}^{-1}\cdot\text{mol}^{-1})(298 \text{ K})} = 0.359 \text{ mol}$$

$$w = -(0.359 \text{ mol})(8.314 \text{ J}\cdot\text{K}^{-1}\cdot\text{mol}^{-1})(298 \text{ K}) \ln\frac{7.39}{3.42} = -685 \text{ J}$$

$$q = +\ 685 \text{ J}$$

(b) For step 1, because the volume is constant, $w = 0$ and $\Delta U = q$.

In step 2, there is an irreversible expansion against a constant opposing pressure, which is calculated from $w = -P_{\text{ex}}\Delta V$

The constant opposing pressure is given, and ΔV can be obtained from

$$V_{\text{final}} - V_{\text{initial}} = 7.39 \text{ L} - 3.42 \text{ L}$$

$$w = -(1.19 \text{ atm})(7.39 \text{ L} - 3.42 \text{ L})$$

$$= -4.72 \text{ L}\cdot\text{atm} = (-4.72 \text{ L}\cdot\text{atm})\left(\frac{101.325 \text{ J}}{1 \text{ L}\cdot\text{atm}}\right) = -479 \text{ J}$$

The total work for part B is $0 \text{ J} + (-479 \text{ J}) = -479 \text{ J}$

6.24 Molecules have a higher heat capacity than monatomic gases because they are able to store energy in the form of rotations and vibrations, in addition to simple translational kinetic energy, which is the only energy storage mode available to monatomic gases. The heat capacity of C_2H_6 is larger than that of CH_4 because it is a more complicated molecule that has more possibilities for the molecule to rotate and for more bonds to vibrate than does CH_4.

6.26 (a) The molar heat capacity of a monatomic ideal gas at constant pressure is $C_{P,m} = \frac{5}{2}R$. The heat released will be given by

$$q = \left(\frac{10.35 \text{ g}}{20.18 \text{ g} \cdot \text{mol}^{-1}}\right)(50°\text{C} - 25°\text{C})(20.8 \text{ J} \cdot \text{mol}^{-1} \cdot (°\text{C})^{-1}) = 2.7 \times 10^2 \text{ J}$$

(b) Similarly, the molar heat capacity of a monatomic ideal gas at constant volume is $C_{V,m} = \frac{3}{2}$. The heat released will be given by

$$q = \left(\frac{10.35 \text{ g}}{20.18 \text{ g} \cdot \text{mol}^{-1}}\right)(50°\text{C} - 25°\text{C})(12.5 \text{ J} \cdot \text{mol}^{-1} \cdot (°\text{C})^{-1}) = 1.6 \times 10^2 \text{ J}$$

6.28 (a) BF_3 is a polyatomic, nonlinear molecule. The contribution from molecular motions will be 3R.

(b) N_2O is a polyatomic, linear molecule. The contribution from molecular motions will be 5/2 R.

(c) HCl is a linear molecule. The contribution from molecular motions will be 5/2 R.

(d) SO_2 is a polyatomic, nonlinear molecule. The contribution from molecular motions will be 3R.

6.30 No, the gasses do not have the same final temperature. The CH_4 molecules have more internal vibrations in which to store energy than do N_2 molecules. As a result, it will require more energy to increase the temperature of CH_4(g), and if the same amount of energy is supplied to both gasses, the temperature of N_2 will be higher than that of CH_4.

6.32 Not counting internal vibrations, a nonlinear molecule will have a heat capacity of $C_{V,m} = 3R$. The temperatures at which the three vibrations of SO_2 become available are 1968 K, 1680 K, and 768 K. If a vibration is accessible, it will contribute a factor of R to the heat capacity. Therefore, (a) $C_{V,m} = 6R$; (b) $C_{V,m} = 4R$; (c) $C_{V,m} = 3R$

6.34 (a) $\Delta H_{\text{freezing}} = -\Delta H_{\text{fus}}$

$$\Delta H_{\text{fus}} = -\frac{(-4.01\ \text{kJ})}{\left(\dfrac{25.23\ \text{g}}{32.04\ \text{g}\cdot\text{mol}^{-1}}\right)} = 5.09\,\text{kJ}\cdot\text{mol}^{-1}$$

(b) $$\Delta H_{\text{vap}} = \frac{37.5\ \text{kJ}}{\left(\dfrac{95\ \text{g}}{78.11\ \text{g}\cdot\text{mol}^{-1}}\right)} = 31\ \text{kJ}\cdot\text{mol}^{-1}$$

6.36 This process is composed of two steps: raising the temperature of the liquid water from 30°C to 100°C and then converting the liquid water to steam at 100°C.

Step 1: $\Delta H = (155\ \text{g})(4.18\ \text{J}\cdot(°\text{C})^{-1}\cdot\text{g}^{-1})(100°\text{C} - 30°\text{C}) = 45.4\ \text{kJ}$

Step 2: $$\Delta H = \left(\frac{155\ \text{g}}{18.02\ \text{g}\cdot\text{mol}^{-1}}\right)(40.7\ \text{kJ}\cdot\text{mol}^{-1}) = 350\ \text{kJ}$$

Total heat required = 45.4 kJ + 350 kJ = 395 kJ

6.38 The heat lost by the metal must equal the heat gained by the water. Also, the final temperature of the metal must be the same as that of the water. We can set up the following relationship and solve for the specific heat capacity of the metal:

$\Delta H_{\text{metal}} = -\Delta H_{\text{water}}$

$$(25.0\ \text{g})(\text{Specific heat})_{\text{metal}}(29.8°\text{C} - 90.0°\text{C})$$
$$= -(50.0\ \text{g})(4.184\ \text{J}\cdot(°\text{C})^{-1}\cdot\text{g}^{-1})(29.8°\text{C} - 25.0°\text{C})$$

$(\text{Specific heat})_{\text{metal}} = 0.667\ \text{J}\cdot(°\text{C})^{-1}\cdot\text{g}^{-1}$

6.40 (a) $$\text{heat absorbed} = (1.55\ \text{mol NO})\left(\frac{180.6\ \text{kJ}}{2\ \text{mol NO}}\right) = 140\ \text{kJ}$$

(b) $n_{N_2} = \dfrac{PV}{RT} = \dfrac{1.00\text{ atm} \times 5.45\text{ L}}{0.082\,06\text{ L}\cdot\text{atm}\cdot\text{K}^{-1}\cdot\text{mol}^{-1} \times 273\text{ K}} = 0.243\text{ mol}$

$$(0.243\text{ mol N}_2)\left(\frac{180.6\text{ kJ}}{\text{mol N}_2}\right) = 43.9\text{ kJ}$$

(c) $n_{N_2} = \dfrac{0.492\text{ kJ}}{180.6\text{ kJ}\cdot\text{mol}^{-1}} = 0.002\,72\text{ mol}$

$$n_{N_2} = (0.002\,72\text{ mol})(28.02\text{ g}\cdot\text{mol}^{-1}) = 0.0762\text{ g}$$

6.42 (a) $\text{heat} = \dfrac{(7.0\text{ cm} \times 6.0\text{ cm} \times 5.0\text{ cm})(1.5\text{ g}\cdot\text{cm}^{-3}\text{ C})}{12.01\text{ g}\cdot\text{mol}^{-1}\text{ C}}(-394\text{ kJ}\cdot\text{mol}^{-1}\text{ C})$

$$= -1.0\times10^4\text{ kJ}$$

(b) $\text{mass of water} = \dfrac{(1.0\times10^7\text{ J})}{(4.18\text{ J}\cdot(°\text{C})^{-1}\cdot\text{g}^{-1})(100°\text{C} - 25°\text{C})} = 3.2\times10^4\text{ g}$

6.44 $n_{SO_2} = \dfrac{PV}{RT} = \dfrac{(1.00\text{ atm})(13.4\text{ L})}{(0.082\,06\text{ L}\cdot\text{atm}\cdot\text{K}^{-1}\cdot\text{mol}^{-1})(273\text{ K})} = 0.598\text{ mol}$

$$n_{O_2} = \frac{15.0\text{ g O}_2}{32.0\text{ g}\cdot\text{mol}^{-1}\text{ O}_2} = 0.469\text{ mol}$$

The SO_2 is the limiting reactant and can react with 0.299 mol of O_2:

$$\text{heat evolved} = (0.598\text{ mol SO}_2)\left(\frac{-198\text{ kJ}}{2\text{ mol SO}_2}\right) = -59.2\text{ kJ}$$

6.46 $\Delta U = -65$ kJ and $w = -48$ kJ. At constant pressure, $\Delta U = \Delta H + w$.

Therefore, $\Delta H = \Delta U - w = -65\text{ kJ} - (-48\text{ kJ}) = -17\text{ kJ}$

6.48 Given the composition of "synthesis gas" it is easiest to first calculate the enthalpy of combustion of 1.000 L of the gas. In 1.000 L of synthesis gas there are 0.40 L of CO. The moles of CO present is given by:

$$n_{CO} = \frac{P\cdot V}{R\cdot T} = \frac{(1\text{ atm})\cdot(0.40\text{ L})}{(0.0820574\ \frac{\text{L}\cdot\text{atm}}{\text{K}\cdot\text{mol}})\cdot(298\text{ K})} = 0.0164\text{ mol of CO}$$

The enthalpy of combustion of CO is calculated, using enthalpies of formation and the balanced reaction $2CO(g) + O_2(g) \rightarrow 2CO_2(g)$, to be −566.0 kJ/mol. Therefore, if 1.000L of synthesis gas were to burn,

$(0.0164 \text{ mol})\cdot(-566.0 \text{ kJ/mol}) = -9.26 \text{ kJ}$ of energy would be released due to the combustion of CO. The energies released due to the combustion of $H_2(g)$ and $CH_4(g)$ when 1.000 L of synthesis gas burns is calculated in a similar way:

$$n_{H_2} = \frac{P\cdot V}{R\cdot T} = \frac{(1 \text{ atm})\cdot(0.25 \text{ L})}{(0.0820574 \frac{\text{L}\cdot\text{atm}}{\text{K}\cdot\text{mol}})\cdot(298 \text{ K})} = 0.0102 \text{ mol of } H_2$$

$$(0.0102 \text{ mol of } H_2)\cdot(-571.66 \text{ kJ}\cdot\text{mol}^{-1}) = -5.84 \text{ kJ}$$

$$n_{CH_4} = \frac{P\cdot V}{R\cdot T} = \frac{(1 \text{ atm})\cdot(0.25 \text{ L})}{(0.0820574 \frac{\text{L}\cdot\text{atm}}{\text{K}\cdot\text{mol}})\cdot(298 \text{ K})} = 0.0102 \text{ mol of } CH_4$$

$$(0.0102 \text{ mol of } H_2)\cdot(-890 \text{ kJ}\cdot\text{mol}^{-1}) = -9.10 \text{ kJ}$$

All together, when 1 L of synthesis gas burns, −24.2 kJ of energy is released.

The amount of energy needed to heat 5.5 L of $H_2O(l)$ by 5°C is given by:

$$5.5 \text{ L}\cdot\left(\frac{1000 \text{ g}}{\text{L}}\right)\cdot(4.184 \text{ J g}^{-1}\ {}^{\circ}\text{C}^{-1})\cdot(5^{\circ}\text{C}) = 115 \text{ kJ}$$

Therefore, the volume of synthesis gas that must be burned to provide this much energy is:

$$\frac{115 \text{ kJ}}{24.2 \text{ kJ}\cdot\text{L}^{-1}} = 4.75 \text{ L}$$

6.50 (a) and (b) The balanced chemical equations and enthalpies of combustion for methane, octane, and glucose are:

$CH_4(g) + 2\ O_2(g) \rightarrow CO_2(g) + 2\ H_2O(l)$ $\Delta H = -890$ kJ/mol

$C_8H_{18} + 12\frac{1}{2}\ O_2(g) \rightarrow 8\ CO_2(g) + 9\ H_2O(l)$ $\Delta H = -5470$ kJ/mol

$C_6H_{12}O_6(s) + 6\ O_2(g) \rightarrow 6\ CO_2(g) + 6\ H_2O(l)$ $\Delta H = -890$ kJ/mol

Dividing these combustion enthalpies by the moles of $CO_2(g)$ produced we find:

For combustion of $CH_4(g)$: $\Delta H = -890$ kJ/mol $CO_2(g)$

For combustion of C_8H_{18}: $\Delta H = -684$ kJ/mol $CO_2(g)$

For combustion of $C_6H_{12}O_6(s)$: $\Delta H = -468$ kJ/mol $CO_2(g)$

(c) combustion of glucose produces more CO_2 than does generating the same amount of energy using octane.

6.52 The combustion of the monoclinic sulfur is reversed and added to the combustion reaction of rhombohedral sulfur:

$$\text{S(rhombic)} + O_2(g) \longrightarrow SO_2(g) \quad \Delta H° = -296.83 \text{ kJ}$$

$$SO_2(g) \longrightarrow \text{S(monoclinic)} + O_2(g) \quad \Delta H° = +297.16 \text{ kJ}$$

$$\text{S(rhombic)} \longrightarrow \text{S(monoclinic)} \quad \Delta H° = +0.33 \text{ kJ}$$

6.54 The first reaction is reversed and added to the second:

$$2\ NO(g) \longrightarrow N_2(g) + O_2(g) \quad \Delta H° = -180.5 \text{ kJ}$$

$$N_2(g) + 2\ O_2(g) \longrightarrow 2\ NO_2(g) \quad \Delta H° = +66.4 \text{ kJ}$$

$$2\ NO(g) + O_2(g) \longrightarrow 2\ NO_2(g)$$

$$\Delta H° = -180.5 \text{ kJ} + 66.4 \text{ kJ} = -114.1 \text{ kJ}$$

6.56 First, write the balanced equations for the known reactions:

$$CH_4(g) + 2\ O_2(g) \longrightarrow CO_2(g) + 2\ H_2O(l) \quad \Delta H° = -890 \text{ kJ}$$

$$CO(g) + \tfrac{1}{2}\ O_2(g) \longrightarrow CO_2(g) \quad \Delta H° = -283.0 \text{ kJ}$$

The desired reaction can be obtained by multiplying the first reaction by 2 and adding it to the reverse of the second reaction, also multiplied by 2.

$$2\ CH_4(g) + 4\ O_2(g) \longrightarrow 2\ CO_2(g) + 4\ H_2O(l) \quad \Delta H° = -1780 \text{ kJ}$$

$$2\ CO_2(g) \longrightarrow 2\ CO(g) + O_2(g) \quad \Delta H° = +566.0 \text{ kJ}$$

$$2\ CH_4(g) + 3\ O_2(g) \longrightarrow 2\ CO(g) + 4\ H_2O(l)$$

$$\Delta H° = -1780 \text{ kJ} + 566.0 \text{ kJ} = -1214 \text{ kJ}$$

6.58 $\Delta H^\circ = \Delta H^\circ_f(CaCl_2, aq) + \Delta H^\circ_f(H_2O, l) + \Delta H^\circ_f(CO_2, g)$

$- [\Delta H^\circ_f(CaCO_3, s) + 2(\Delta H^\circ_f(HCl, aq)]$

$= -877.1\ kJ \cdot mol^{-1} + (-285.83\ kJ \cdot mol^{-1}) + (-393.51\ kJ \cdot mol^{-1})$

$- [(-1206.9\ kJ \cdot mol^{-1}) + 2(-167.16\ kJ \cdot mol^{-1})]$

$= -15.2\ kJ \cdot mol^{-1}$

6.60 The desired reaction is obtained adding reaction 1 to 6 times reaction 2, 3 times reaction 3, and 2 times the reverse of reaction 4:

$2\ Al(s) + 6\ HCl(aq) \longrightarrow 2\ AlCl_3(aq) + 3\ H_2(g)$ $\quad \Delta H^\circ = -1049\ kJ$

$6[HCl(g) \longrightarrow HCl(aq)]$ $\quad 6[\Delta H^\circ = -74.8\ kJ]$

$3[H_2(g) + Cl_2(g) \longrightarrow 2\ HCl(g)]$ $\quad 3[\Delta H^\circ = -185\ kJ]$

$2[AlCl_3(aq) \longrightarrow AlCl_3(s)]$ $\quad 2[\Delta H^\circ = +323\ kJ]$

$2\ Al(s) + 3\ Cl_2(g) \longrightarrow 2\ AlCl_3(s)$

$\Delta H^\circ = -1049\ kJ + 6(-74.8\ kJ) + 3(-185\ kJ) + 2(+323\ kJ) = -1407\ kJ$

6.62 From Appendix 2A, $\Delta H^\circ_f(O_3) = +142.7\ kJ \cdot mol^{-1}$

$NO_2(g) + O_2(g) \longrightarrow NO(g) + O_3(g)$ $\quad \Delta H^\circ = +200\ kJ$

$O_3(g) \longrightarrow \frac{3}{2} O_2(g)$ $\quad \Delta H^\circ = -142.7\ kJ$

$\frac{1}{2} O_2(g) \longrightarrow O(g)$ $\quad \Delta H^\circ = +249.2\ kJ$

$NO_2(g) \longrightarrow NO(g) + O(g)$ $\quad \Delta H^\circ = +306\ kJ$

6.64 (a) $3\ NO_2(g) + H_2O(l) \longrightarrow 2\ HNO_3(aq) + NO(g)$

$\Delta H^\circ_r = 2\ \Delta H^\circ_f(HNO_3, aq)$

$+ \Delta H^\circ_f(NO, g - [3\ \Delta H^\circ_f(NO_2, g) + \Delta H^\circ_f(H_2O, l)]$

$= 2(-207.36\ kJ \cdot mol^{-1}) + 90.25\ kJ \cdot mol^{-1} - [3(+33.18\ kJ \cdot mol^{-1})$

$+ (-285.83\ kJ \cdot mol^{-1})]$

$= -138.18\ kJ$

(b) $B_2O_3(s) + 3\ CaF_2(s) \longrightarrow 2\ BF_3(g) + 3\ CaO(s)$

$$\Delta H°_r = 2\,\Delta H°_f(BF_3, g) + 3\,\Delta H°_f(CaO, s)$$
$$- [\Delta H°_f(B_2O_3, s) + \Delta H°_f(CaF_2, s)]$$
$$= 2(-1137.0\ kJ \cdot mol^{-1}) + 3(-635.09\ kJ \cdot mol^{-1})$$
$$- [-1272.8\ kJ \cdot mol^{-1} + 3(-1219.6\ J \cdot mol^{-1})]$$
$$= +752.3\ kJ$$

(c) $H_2S(aq) + 2\ KOH(aq) \longrightarrow K_2S(aq) + 2\ H_2O(l)$

$$\Delta H°_r = \Delta H°_f(K_2S, aq) + 2\,\Delta H°_f(H_2O, l)$$
$$- [\Delta H°_f(H_2S, aq) + 2\,\Delta H°_f(KOH, aq)]$$
$$= -417.5\ kJ \cdot mol^{-1} + 2(-285.83\ kJ \cdot mol^{-1})$$
$$- [-39.7\ kJ \cdot mol^{-1} + 2\,(-482.37\ kJ \cdot mol^{-1})]$$
$$= -38.7\ kJ$$

6.66 (a) The reaction of interest is: $\frac{1}{2}N_2(g) + 1\frac{1}{2}H_2(g) \rightarrow NH_3(l)$. The reaction enthalpy may be calculated using individual bond enthalpies and the enthalpy of vaporization of NH_3. During the reaction, $\frac{1}{2}$ of a mole of N_2 bonds are broken as are $1\frac{1}{2}$ moles of H-H bonds while three N-H bonds are formed. In addition, −23.4 kJ will be released due to condensation of the product. Using the suggested data tables, the enthalpy of formation is estimated to be:

$$\tfrac{1}{2}\left(+944\ kJ \cdot mol^{-1}\right) + 1\tfrac{1}{2}\left(436\ kJ \cdot mol^{-1}\right)$$
$$+ 3\left(-388\ kJ \cdot mol^{-1}\right) + \left(-23.4\ kJ \cdot mol^{-1}\right)$$
$$= -61.4\ kJ \cdot mol^{-1}$$

(b) The reaction of interest is $2\ C(gr) + 3\ H_2(g) + \frac{1}{2}O_2(g) \rightarrow C_2H_6O(l)$. As in part (a) above, the reaction enthalpy may be estimated using the bond enthalpies, the enthalpy of sublimation of C(gr) and the enthalpy of vaporization of the product. In this reaction, 3 moles of H-H bonds and 0.5 moles of O=O bonds, are broken while 1 mole of C-C bonds, 5 moles of C-H bonds, 1 mole of O-H bonds, and 1 mole of C-O bonds are formed. Furthermore, 717 kJ mol^{-1} are needed to vaporize the graphite and −43.5

kJ mol^{-1} are released when the product condenses into a liquid. Using the suggested data tables, the enthalpy of formation is estimated to be:

$$3(+436\text{ kJ}\cdot\text{mol}^{-1})+\tfrac{1}{2}(496\text{ kJ}\cdot\text{mol}^{-1})+2(717\text{ kJ}\cdot\text{mol}^{-1})$$
$$+(-348\text{ kJ}\cdot\text{mol}^{-1})+5(-412\text{ kJ}\cdot\text{mol}^{-1})+(-463\text{ kJ}\cdot\text{mol}^{-1})$$
$$+2(-743\text{ kJ}\cdot\text{mol}^{-1})+(-43.5\text{ kJ}\cdot\text{mol}^{-1})$$
$$=-668\text{ kJ}\cdot\text{mol}^{-1}$$

(c) The reaction of interest is $3C(gr) + 3H_2(g) + \frac{1}{2}O_2(g) \rightarrow C_3H_6O(l)$. As in parts (a) and (b) above, 3 moles of H-H bonds and 0.5 moles of O-O bonds are broken and three moles of C(gr) is vaporized while 2 moles of C-C bonds, 1 mole of C=O bonds and 6 moles of C-H bonds are formed and −21.9 kJ mol^{-1} are released when one mole of acetone condenses from a gas to a liquid. Using the suggested data tables, the enthalpy of formation is estimated to be:

$$3(+436\text{ kJ}\cdot\text{mol}^{-1})+\tfrac{1}{2}(496\text{ kJ}\cdot\text{mol}^{-1})+3(717\text{ kJ}\cdot\text{mol}^{-1})$$
$$+2(-348\text{ kJ}\cdot\text{mol}^{-1})+6(-412\text{ kJ}\cdot\text{mol}^{-1})+(-743\text{ kJ}\cdot\text{mol}^{-1})$$
$$+(-29.1\text{ kJ}\cdot\text{mol}^{-1})$$
$$=-233\text{ kJ}\cdot\text{mol}^{-1}$$

6.68 For the reaction $AlCl_3(s) \longrightarrow Al^{3+}(g) + 3\,Cl^-(g)$

$$\Delta H_L = \Delta H^\circ_f(\text{Al, g}) + 3\,\Delta H^\circ_f(\text{Cl, g}) + I_1(\text{Al}) + I_2(\text{Al}) + I_3(\text{Al}) - E_{ea}(\text{Cl}) - \Delta H_f(\text{AlCl}_3, \text{s})$$

$$\Delta H_L = 326\text{ kJ}\cdot\text{mol}^{-1} + 3(121\text{ kJ}\cdot\text{mol}^{-1}) + 577\text{ kJ}\cdot\text{mol}^{-1} + 1817\text{ kJ}\cdot\text{mol}^{-1} + 2744\text{ kJ}\cdot\text{mol}^{-1} - 349\text{ kJ}\cdot\text{mol}^{-1} + 704.2\text{ kJ}\cdot\text{mol}^{-1}$$

$$\Delta H_L = 6182\text{ kJ}\cdot\text{mol}^{-1}$$

6.70 The process can be broken down into the following steps:

	$\Delta H°$, $kJ \cdot mol^{-1}$
$Na^+(g) + 2\,Cl^-(g) \longrightarrow NaCl_2(s)$	-2524*
$Na(s) \longrightarrow Na(g)$	$2(+107.32)$
$Na(g) \longrightarrow Na^+(g) + e^-$	$+494$
$Na^+(g) \longrightarrow Na^{2+}(g) + e^-$	$+4560$
$Cl_2(g) \longrightarrow 2\,Cl(g)$	$+242$
$2\,(Cl(g) + e^- \longrightarrow Cl^-, g)$	$2(-349)$
$Na(s) + 2\,Cl_2(g) \longrightarrow NaCl_2(s)$	$\Delta H°_f = +2289$

For comparison, the enthalpy of formation of

$NaCl(s) = -411.15\ kJ \cdot mol^{-1}$. Because the enthalpy of formation of

$NaCl_2$ is a large positive number, $NaCl_2$ would be considerably unstable.

The disproportionation reaction

$2\,NaCl_2(s) \longrightarrow 2\,NaCl(s) + Cl_2(g)$

would have an energy of

$2(NaCl_2(s) \longrightarrow Na(s) + 2\,Cl_2, g)\ 2(-2289\ kJ \cdot mol^{-1})$

$2(Na(s) + \frac{1}{2}\,Cl_2(g) \longrightarrow NaCl(s)\ 2(-411.15\ kJ \cdot mol^{-1})$

$2NaCl_2(s) \longrightarrow 2\,NaCl(s) + Cl_2(g)\ -5400\ kJ \cdot mol^{-1}$

This would be very likely to take place, so $NaCl_2$ would not be isolated.

6.72 (a)

break:	1 mol H—Cl bonds	$431\ kJ \cdot mol^{-1}$
	1 mol F—F bonds	$158\ kJ \cdot mol^{-1}$
form:	1 mol H—F bonds	$-565\ kJ \cdot mol^{-1}$
	1 mol Cl—F bonds	$-256\ kJ \cdot mol^{-1}$
	Total	$-232\ kJ \cdot mol^{-1}$

*From the assumption that ΔH_L is the same as that of $MgCl_2$.

(b) break:	1 mol H—Cl bonds	431 $kJ \cdot mol^{-1}$
	1 mol C=C bonds	612 $kJ \cdot mol^{-1}$
form:	1 mol C—C bonds	−348 $kJ \cdot mol^{-1}$
	1 mol C—H bonds	−412 $kJ \cdot mol^{-1}$
	1 mol C—Cl bonds	−338 $kJ \cdot mol^{-1}$
	Total	−55 $kJ \cdot mol^{-1}$
(c) break:	1 mol H—H bonds	436 $kJ \cdot mol^{-1}$
	1 mol C=C bonds	612 $kJ \cdot mol^{-1}$
form:	1 mol C—C bonds	−348 $kJ \cdot mol^{-1}$
	2 mol C—H bonds	2(−412) $kJ \cdot mol^{-1}$
	Total	−124 $kJ \cdot mol^{-1}$

6.74 (a) The Lewis structures for NO and NO_2 show that the bond order in NO is a double bond and that in NO_2 is 1.5 on average.

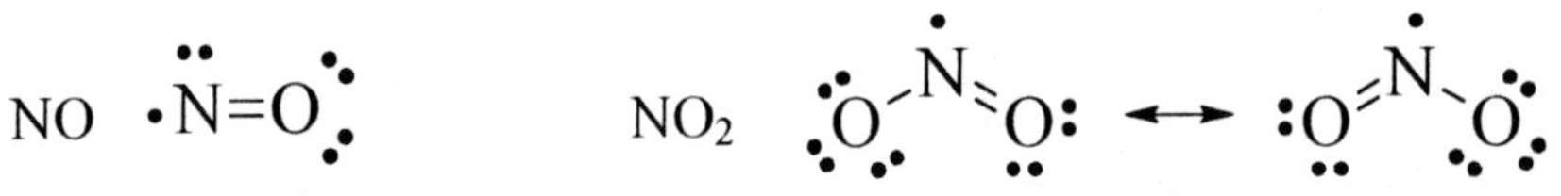

The bond enthalpy of NO at 632 $kJ \cdot mol^{-1}$ is close to the N=O value listed in the table, whereas the N—O bond enthalpy in NO_2 of 469 $kJ \cdot mol^{-1}$ is slightly greater than the average of a N—O single and N=O double bond:

$\frac{1}{2}(630 \text{ kJ} + 210 \text{ kJ}) = 420 \text{ kJ}$

The extra stability is due to resonance stabilization.

(b) The bond energies are the same for the two bonds, because the bonds are equivalent due to resonance.

6.76

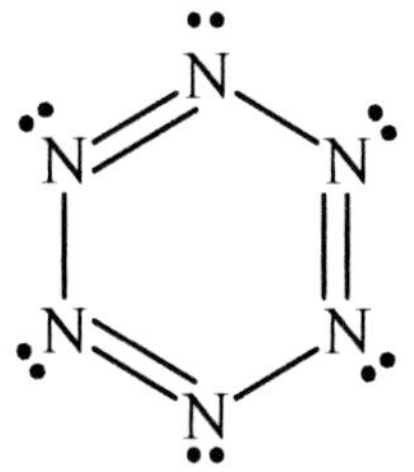

The reaction in question is $N_6 \longrightarrow 3\ N_2(g)$

For this to occur, we will need to break 3 N—N and 3 NRN bonds and form 3 N—N triple bonds:

$3(163\ kJ \cdot mol^{-1})$

$3(409\ kJ \cdot mol^{-1})$

$-3(944\ kJ \cdot mol^{-1})$

$-1116\ kJ \cdot mol^{-1}$

The reaction to form $N_2(g)$ is very exothermic and so the reaction is likely to occur. Although N_6 might be resonance-stabilized like benzene, it is unlikely that resonance stabilization would overcome the tendency to form the very strong N≡N triple bond; thus we do not expect N_6 to be a stable molecule.

6.78 (a) The enthalpy of vaporization is the heat required for the conversion $CH_3OH(l) \longrightarrow CH_3OH(g)$ at constant pressure. The value at 298.2 K will be given by

$$\begin{aligned}\Delta H^\circ_{\text{vap at 298 K}} &= \Delta H^\circ_f(CH_3OH,\ g) - \Delta H^\circ_f(CH_3OH,\ 1)\\ &= -200.66\ kJ \cdot mol^{-1} - (-238.86\ kJ \cdot mol^{-1})\\ &= 38.20,kJ \cdot mol^{-1}\end{aligned}$$

(b) In order to take into account the difference in temperature, we need to use the heat capacities of the reactants and products in order to raise the temperature of the system to 64.5°C. We can rewrite the reactions as follows, to emphasize temperature:

$$CH_3OH(l)_{\text{at 298 K}} \longrightarrow CH_3OH(g)_{\text{at 298 K}} \qquad \Delta H^\circ = 38.20\ kJ$$

$$CH_3OH(l)_{\text{at 298 K}} \longrightarrow CH_3OH(l)_{\text{at 337.6 K}} \quad \Delta H° = (1\text{ mol}) \times (337.8\text{ K} - 298.2\text{ K}) \times (81.6\text{ J}\cdot\text{mol}^{-1}\cdot\text{K}^{-1}) = 3.23\text{ kJ}$$

$$CH_3OH(g)_{\text{at 298 K}} \longrightarrow CH_3OH(g)_{\text{at 337.6 K}} \quad \Delta H° = (1\text{ mol}) \times (337.8\text{ K} - 298.2\text{ K}) \times (43.89\text{ J}\cdot\text{mol}^{-1}\cdot\text{K}^{-1}) = 1.74\text{ kJ}$$

To add these together to get the overall equation at 337.6 K, we must reverse the second equation:

$$CH_3OH(l)_{\text{at 298 K}} \longrightarrow CH_3OH(g)_{\text{at 298 K}} \quad \Delta H° = 38.20\text{ kJ}$$

$$CH_3OH(l)_{\text{at 337.8 K}} \longrightarrow CH_3OH(l)_{\text{at 298 K}} \quad \Delta H° = -3.23\text{ kJ}$$

$$CH_3OH(g)_{\text{at 298 K}} \longrightarrow CH_3OH(g)_{\text{at 337.8 K}} \quad \Delta H° = 1.74\text{ kJ}$$

$$CH_3OH(1)_{\text{at 337.8 K}} \longrightarrow CH_3OH(g)_{\text{at 337.8 K}} \quad \Delta H^{\circ} = 34.71\text{ kJ}$$

(c) The value in the table is 35.3 $\text{kJ}\cdot\text{mol}^{-1}$ for the enthalpy of vaporization of methanol. The value is close to that calculated as corrected by heat capacities. At least part of the error can be attributed to the fact that heat capacities are not strictly independent of temperature.

6.80 The change in the reaction enthalpy is given by:

$\Delta H_r(T_2) - \Delta H_r(T_1) = (T_2 - T_1)\Delta C_p$ where

$$\Delta C_p = \sum n \cdot C_p(\text{prod.}) - \sum n \cdot C_p(\text{prod.})$$

given N_2 and H_2 are linear molecules, their heat capacities are both approximately $\frac{7}{2}R$ while that of NH_3 is $4R$. Therefore, for this reaction the change in the reaction enthalpy is:

$$\Delta H_r(T_2) - \Delta H_r(T_1) = (500\text{ K} - 300\text{ K})\left[2(4R) - 3(\tfrac{7}{2}R) - \tfrac{7}{2}R\right]$$
$$= -9.98\text{ kJ}\cdot\text{mol}^{-1}$$

6.82 (a) The amount of heat lost by the piece of stainless steel must equal the amount of heat absorbed by the water. The specific heat of stainless steel is found in Table 6.1. This problem is complicated because the water may undergo one or more phase changes during this process. In order to answer this question, we will need to analyze each step of the reaction to determine if the steel can transfer enough heat to the water to cause the given change.

For stainless steel, the enthalpy change will be given by

$155.7\ \text{g} \times 0.51\ \text{J} \cdot (°\text{C})^{-1} \cdot \text{g}^{-1} \times \Delta T = 79\ \text{J} \cdot (°\text{C})^{-1} \times \Delta T$

For water, there are five separate processes that may be involved: (1) heating solid water to 0°C, (2) converting the solid water to liquid at 0°C, (3) raising the temperature of the water from 0°C to 100°C, (4) converting the liquid water to vapor at 100°C, (5) heating the vapor above 100°C.

To heat the solid water from $-24°\text{C}$ to 0°C:

$25.34\ \text{g} \times 2.03\ \text{J} \cdot (°\text{C})^{-1} \cdot \text{g}^{-1} \times 24°\text{C} = 1.23\ \text{kJ}$

For this process the corresponding decrease in temperature of the steel would be

$1.23\ \text{kJ} \div 0.079\ \text{kJ} \cdot (°\text{C})^{-1} = 16°\text{C}$

Temperature of the steel after heating the solid water to 0°C will be $475°\text{C} - 16°\text{C} = 459°\text{C}$.

Because the temperature of the steel is still above that of the water, there is adequate heat for the next transformation to occur.

To convert all the solid water to liquid water at 0°C, we would require

$$\left(\frac{25.34\ \text{g H}_2\text{O}}{18.01\ \text{g} \cdot \text{mol}^{-1}\ \text{H}_2\text{O}}\right)(6.01\ \text{kJ} \cdot \text{mol}^{-1}) = 8.46\ \text{kJ}$$

This change would require the temperature of the steel to decrease further by

$8.46\ \text{kJ} \div 0.079\ \text{kJ} \cdot (°\text{C})^{-1} = 107°\text{C}$

The temperature of the steel after melting all of the solid water to 0°C will be $459°C - 107°C = 352°C$. The temperature of the steel is still above that of the water, so more heat can be transferred to heat the liquid water.

To heat the liquid water from 0°C to 100°C:

$$25.34\ g \times 4.184\ J \cdot (°C)^{-1} \cdot g^{-1} \times 100°C = 10.6\ kJ$$

This change would require the temperature of the steel to decrease further by

$$10.6\ kJ \div 0.079\ kJ \cdot (°C)^{-1} = 134°C$$

The steel is still sufficiently hot enough to cause this transformation, so all of the water will be heated to 100°C. The temperature of the steel after melting all of the solid water to 0°C will be $352°C - 134°C = 218°C$. The steel still has enough heat to convert at least some of the water from liquid to vapor.

$$\left(\frac{25.34\ g\ H_2O}{18.01\ g \cdot mol^{-1}\ H_2O}\right)(40.7\ kJ \cdot mol^{-1}) = 57.3\ kJ$$

$$57.3\ kJ \div 0.079\ kJ \cdot (°C)^{-1} = 725°C$$

The temperature of the steel is not sufficient to convert all the water to vapor. The final temperature will then be 100°C, with some of the water converted to vapor and some remaining in the liquid phase.

(b) To determine the amount of liquid and gaseous water present, we must determine the amount of heat transferred to the water from the steel at 100°C. The steel should still be at 218°C when all the liquid water has reached 100°C. The final temperature of the steel must also be 100°C, so its change in temperature will be $-118°C$. To go from 218°C to 100°C, the steel will lose

$$0.079\ kJ \cdot (°C)^{-1} \times 118°C = 9.3\ kJ$$

We can then calculate the amount of water that can be converted to vapor at 100°C by 9.3 kJ.

$$\left(\frac{9.3\ \text{kJ}}{40.7\ \text{kJ}\cdot\text{mol}^{-1}}\right)(18.02\ \text{g}\cdot\text{mol}^{-1}) = 4.1\ \text{g}$$

4.1 g of water will be converted to vapor, leaving

$25.34\ \text{g}\ -\ 4.1\ \text{g} = 21.2\ \text{g}$ of water in the liquid phase.

6.84 (a) First, we calculate the amount of heat needed to raise the temperature of the water. Converting the temperatures from °F to °C:

$°\text{C} = \frac{5}{9}(°\text{F} - 32°)$, so 65°F corresponds to 18°C and 108°F corresponds to 42°C.

$$\begin{aligned}\Delta H &= (100\ \text{gal})(3.785\ \text{L}\cdot\text{gal}^{-1})(1000\ \text{cm}^3\cdot\text{L}^{-1})(1.00\ \text{g}\cdot\text{cm}^{-3})\\ &\quad\times(4.18\ \text{J}\cdot(°\text{C})^{-1}\cdot\text{g}^{-1})(42°\text{C} - 18°\text{C})\\ &= 38\ \text{MJ}\end{aligned}$$

The enthalpy of combustion of methane is $-890\ \text{kJ}\cdot\text{mol}$ (from Table 6.4). The mass of methane required will be calculated as follows:

$$\left(\frac{38\times10^3\ \text{kJ}}{890\ \text{kJ}\cdot\text{mol}^{-1}}\right)(16.04\ \text{g}\cdot\text{mol}^{-1}\ \text{CH}_4) = 6.8\times10^2\ \text{g CH}_4$$

(b) This quantity can be obtained from the ideal gas law. 6.8×10^2 g of CH_4 is 42 mol of CH_4:

$$V = \frac{nRT}{P} = \frac{(42\ \text{mol})(0.082\ 06\ \text{L}\cdot\text{atm}\cdot\text{K}^{-1}\cdot\text{mol}^{-1})(298\ \text{K})}{1.00\ \text{atm}} = 1030\ \text{L}$$

6.86 (a) $w = -(1.00\ \text{mol})(8.314\ \text{J}\cdot\text{K}^{-1}\cdot\text{mol}^{-1})(298.2\ \text{K})\ln\frac{15.50\ \text{L}}{7.00\ \text{L}}$

$= -1.97\ \text{kJ}$

(b) $w = -P_{\text{ext}}\Delta V = -\left(\frac{743\ \text{Torr}}{760\ \text{Torr}\cdot\text{atm}^{-1}}\right)(15.5\ \text{L} - 7.00\ \text{L})\left(\frac{101.325\ \text{J}}{1\ \text{L}\cdot\text{atm}}\right)$

$= -0.842\ \text{kJ}$

(c) The fact that the expansion occurs adiabatically—which means that the system is isolated from its surroundings so that no heat is transferred—is important. It tells us that $q = 0$, and therefore $\Delta U = w = -0.842$ kJ. But

ΔU will also be equal to $= nCv\Delta T$ (see equation 19) because U is a state function. The heat capacity of an ideal gas is 12.5 $\text{J}\cdot\text{K}^{-1}\cdot\text{mol}^{-1}$ (we will assume it is constant over this temperature range). Therefore, $-0.842\ \text{J} =$ $(1.00\ \text{mol})(12.5\ \text{J}\cdot\text{K}^{-1}\cdot\text{mol}^{-1})(\Delta T)$. $\Delta T = -67.4\ \text{K}$. The final temperature will be $298.2\text{K} - 67.4\text{K} = 230.6\text{K}$.

6.88 In order to solve this problem, we must be able to add reactions (a), (b), and (c) to give the reaction desired. The most logical starting point for solving this problem is to find the reactant or product that appears only once in (a), (b), or (c), as this will determine how that equation must be used. Because O_2 only appears in (a), equation (a) must be multipled by $\frac{3}{4}$ and left in its original direction in order to obtain the correct amount of $O_2(g)$.

$\frac{3}{4}\,CH_4(g) + \frac{3}{2}\,O_2(g) \longrightarrow \frac{3}{4}\,CO_2(g) + \frac{3}{2}\,H_2O(g) \qquad \frac{3}{4}(-802\ \text{kJ})$

We now consider how to combine this equation with (b) and (c) to give the desired overall reaction. Because $H_2O(g)$ appears in (c) and not in (b), we must add equation (c) to the transformed (a) so as to obtain a total of two $H_2O(g)$. To do this, we must reverse (c) and multiply it by $\frac{1}{2}$:

$\frac{1}{2}\,CO(g) + \frac{3}{2}\,H_2(g) \longrightarrow \frac{1}{2}\,CH_4(g) + \frac{1}{2}\,H_2O(g) \qquad \frac{1}{2}(-247\ \text{kJ})$

Equation (b) must be multiplied by $\frac{3}{4}$ and left in its original direction in order that the $CO_2(g)$ will be cancelled from the overall equation:

$\frac{3}{4}\,CH_4(g) + \frac{3}{4}\,CO_2(g) \longrightarrow \frac{3}{2}\,CO(g) + \frac{3}{2}\,H_2(g) \qquad \frac{3}{4}(+206\ \text{kJ})$

The net result is

$\frac{3}{4}\,CH_4(g) + \frac{3}{2}\,O_2(g) \longrightarrow \frac{3}{4}\,CO_2(g) + \frac{3}{2}\,H_2O(g)\ \frac{3}{4}(-802\ \text{kJ})$

$\frac{1}{2}\,CO(g) + \frac{3}{2}\,H_2(g) \longrightarrow \frac{1}{2}\,CH_4(g) + \frac{1}{2}\,H_2O(g)\ \frac{1}{2}(-247\ \text{kJ})$

$\frac{3}{4}\,CH_4(g) + \frac{3}{4}\,CO_2(g) \longrightarrow \frac{3}{2}\,CO(g) + \frac{3}{2}\,H_2(g)\ \frac{3}{4}(+206\ \text{kJ})$

$CH_4(g) + \frac{3}{2}\,O_2(g) \longrightarrow CO(g) + 2\,H_2O(g)\ \Delta H^\circ = -570\ \text{kJ}$

6.90 The formation reaction is

$$\tfrac{1}{2}\,N_2(g) + \tfrac{7}{2}\,H_2(g) + 2\,C(s) + O_2(g) \longrightarrow NH_4CH_3CO_2(s)$$

The combustion reaction that we want is:

$$NH_4CH_3CO_2(s) + O_2(g) \longrightarrow 2\,CO_2(g) + \tfrac{7}{2}\,H_2O(l) + \tfrac{1}{2}\,N_2(g)$$

Using Hess' Law

$$\begin{aligned}\Delta H^\circ_{\text{combustion}} &= 2\,\Delta H^\circ_f(CO_2, g) + \tfrac{7}{2}\,\Delta H^\circ_f(H_2O, l) \\ &\quad - \Delta H^\circ_f(NH_4CH_3CO_2, s) \\ &= 2\,(-393.15\ \text{kJ}\cdot\text{mol}^{-1}) + \tfrac{7}{2}(-285.83\ \text{kJ}\cdot\text{mol}^{-1}) \\ &\quad - (-616.14\ \text{kJ}\cdot\text{mol}^{-1}) \\ &= -117.28\ \text{kJ}\cdot\text{mol}^{-1}\end{aligned}$$

6.92 (a) The heat output of each sample is found using the heat capacity of the calorimeter and the observed change in temperature of the calorimeter. For Brand X, ΔT was 8.8°C and the energy content of 1.00 g is: $(8.8^\circ\text{C})\cdot(600\ \text{J}\cdot\text{K}^{-1}) = 5300\ \text{J}\cdot\text{g}^{-1}$. For the ABC product, ΔT was 8.5°C and the energy content of 1.00 g is: $(8.5^\circ\text{C})\cdot(600\ \text{J}\cdot\text{K}^{-1}) = 5100\ \text{J}\cdot\text{g}^{-1}$.

(b) 30 g of Brand X cereal contains

$(30\ \text{g})(5.3\ \text{kJ}\cdot\text{g}^{-1})\left(\dfrac{1\ \text{Calorie}}{4.184\ \text{kJ}}\right) = 38$ Calories while 30 g of ABC product contains $(30\ \text{g})(5.1\ \text{kJ}\cdot\text{g}^{-1})\left(\dfrac{1\ \text{Calorie}}{4.184\ \text{kJ}}\right) = 36$ Calories

6.94 (a) and (b)

compound	M $(\text{g}\cdot\text{mol}^{-1})$	ΔH_c $(\text{kJ}\cdot\text{mol}^{-1})$	$\text{kJ}\cdot\text{g}^{-1}$	density $(\text{g}\cdot\text{cm}^{-3})$	$\text{L}\cdot\text{mol}^{-1}$	$\text{MJ}\cdot\text{L}^{-1}$
benzene	78.11	−3268	−41.84	0.8786	0.088 90	−36.76
ethanol	46.07	−1368	−29.69	0.7893	0.058 37	−23.44
hexane	86.18	−4163	−48.30	0.6603	0.1305	−31.89
octane	114.23	−5471	−47.89	0.7025	0.1626	−33.64

(c) On a per-liter basis, the chemicals increase in cost in the following order: ethanol < benzene ~ hexane < octane; (d) The choice of fuel is not a simple matter. Although it would appear that benzene is the best as far as heat production per liter is concerned, the cost of ethanol is roughly two-thirds that of the cost of benzene, making the heat produced per dollar about the same. Other issues such as cleanliness of burning, ecological impact of the production of the fuel, capacity of fuel tanks (and how far a car can drive on a tank of gas) then become important issues.

6.96 In step 1, $\Delta U = q_V = 50\text{ J}$ and $w = 0$. In step 2, $q = -5\text{ J}$ and $\Delta U = -50\text{ J}$; therefore, $-50\text{ J} = -5\text{ J} + w$; $w = -45\text{ J}$. Because $w = -P\Delta V$ and $P = 1.00$ atm

$$\Delta V = \left(\frac{+45\text{ J}}{1.00\text{ atm}}\right)\left(\frac{0.082\,06\text{ L}\cdot\text{atm}\cdot\text{K}^{-1}\cdot\text{mol}^{-1}}{8.314\text{ J}\cdot\text{K}^{-1}\cdot\text{mol}^{-1}}\right) = +0.44\text{ L}$$

The gas constants are used to convert the units of J to $\text{L}\cdot\text{atm}$. Because the value ΔV is positive, this change will represent an expansion.

6.98 The heat given off by the reaction that was absorbed by the calorimeter is given by

$\Delta H = -(488.1\text{ J}\cdot(°\text{C})^{-1})(21.34°\text{C} - 20.00°\text{C}) = -654\text{ J}$

This, however, is not all the heat produced, as the 50.0 mL of solution resulting from mixing also absorbed some heat. If we assume that the volume of NaOH and HCl are negligible compared to the volume of water present and that the density of the solution is $1.00\text{ g}\cdot\text{mL}^{-1}$, then the change in enthalpy of the solution is given by

$\Delta H = -(21.34°\text{C} - 20.00°\text{C})(4.18\text{ J}\cdot(°\text{C})^{-1}\cdot\text{g}^{-1})(50.0\text{ g}) = -280\text{ J}$

The total heat given off will be $-654\text{ J} + (-280\text{ J}) = -934\text{ J}$

This heat is for the reaction of (0.700 M)(0.0250 L) = 0.0175 mol HCl, so the amount of heat produced per mole of HCl will be given by

$$\frac{-934\text{ J}}{0.0175\text{ mol}} = -53.4\text{ kJ}\cdot\text{mol}^{-1}\text{ HCl}$$

6.100 From Appendix 2A we can find the enthalpies of combustion of the gases involved:

Compound	Enthalpy of combustion $(\text{kJ}\cdot\text{mol}^{-1})$
CH_4(g), methane	−890
C_2H_6(g), ethane	−1560
C_3H_8(g), propane	−2220
C_4H_{10}(g), butane	−2878

(a) To calculate the mass of CO_2 produced, we simply need to realize that there will be one mole of CO_2 produced per mole of carbon present in the starting compound. So there will be one mole of CO_2 produced per mole of CH_4, two per mole of C_2H_6, three per mole of C_3H_8, and four per mole of C_4H_{10}.

Total moles of CO_2 produced per minute

$$= \frac{1\text{ mol CO}_2}{1\text{ mol CH}_4} \times 9.3\text{ mol CH}_4$$

$$+ \frac{2\text{ mol CO}_2}{1\text{ mol C}_2\text{H}_6} \times 3.1\text{ mol C}_2\text{H}_6$$

$$+ \frac{3\text{ mol CO}_2}{1\text{ mol C}_3\text{H}_8} \times 0.40\text{ mol C}_3\text{H}_8$$

$$+ \frac{4\text{ mol CO}_2}{1\text{ mol C}_4\text{H}_{10}} \times 0.20\text{ mol C}_4\text{H}_{10}$$

$$= 17.5\text{ mol}\cdot\text{min}^{-1}$$

The mass of CO_2 produced per minute will be

$$17.5\text{ mol}\cdot\text{min}^{-1} \times 44.01\text{ g}\cdot\text{mol}^{-1} = 7.70\times10^{2}\text{ g}\cdot\text{min}^{-1}$$

(b) The heat released per minute will be given by the enthalpies of combustion multiplied by the number of moles of each gas combusted in that time period:

heat released $= (9.3\text{ mol CH}_4)(-890\text{ kJ}\cdot\text{mol}^{-1}\text{ CH}_4)$

$$+(3.1 \text{ mol } C_2H_6)(-1560 \text{ kJ}\cdot\text{mol}^{-1} \ C_2H_6)$$
$$+(0.40 \text{ mol } C_3H_8)(-2220 \text{ kJ}\cdot\text{mol}^{-1})$$
$$+(0.20 \text{ mol } C_4H_{10})(-2878 \text{ kJ}\cdot\text{mol}^{-1} \ C_4H_{10})$$
$$= -1.5\times 10^4 \text{ kJ}\cdot\text{min}^{-1}$$

6.102 The total amount of heat released by the reaction will be given by

$-5.270 \text{ kJ}\cdot(°\text{C})^{-1} \times 1.140°\text{C} = -6.008 \text{ kJ}$

In this case, the heats of reaction for the two different outcomes will be given simply by the $\Delta H°_f$ values for SO_2 ($-296.83 \text{ kJ}\cdot\text{mol}^{-1}$) and SO_3 ($-395.72 \text{ kJ}\cdot\text{mol}^{-1}$). The number of moles of sulfur is $0.6192 \text{ g S(s)} \div 32.06 \text{ kJ}\cdot\text{mol}^{-1} = 0.01931 \text{ mol S(s)}$. We do not know the relative amounts of SO_2 versus SO_3 formed, but these can be determined using the following two relationships, which are required by the stoichiometries of the reactions.

Let x = number of moles of SO_2 formed

Let y = number of moles of SO_3 formed

Then $x + y = 0.01931$ mol

$(-296.83 \text{ kJ}\cdot\text{mol}^{-1})x + (-395.72 \text{ kJ}\cdot\text{mol}^{-1})y = -6.008 \text{ kJ}$

$x = 0.01652$ mol

$y = 0.00279$ mol

The ratio of SO_2 to SO_3 will be $0.01652 \text{ mol} \div 0.00279 \text{ mol} = 5.92 : 1$.

6.104 In order of increasing enthalpy of vaporization: N_2, CH_4, C_6H_6, H_2O, NaCl. A large enthalpy of vaporization indicates strong intermolecular interactions in the liquid phase. The stronger these interactions the larger the amount of energy needed to free a molecule into the gas phase. Nitrogen, being a homonuclear diatomic molecule interacts through other N_2 molecules in the condensed phase through very weak London forces. CH_4, being a nonpolar molecule also interacts with other CH_4 molecules through weak London forces. However, the electron distribution along the

C-H bonds is not uniform leading to stronger interactions than in N_2. C_6H_6 has no net dipole moment but, due to p bonding, above and below the ring of carbons tends to be negative compared to the hydrogen atoms along the edge of the ring giving rise to stronger bonding interactions than in CH_4 and N_2. H_2O is a highly polar molecule that can bind to other water molecules through relatively strong hydrogen bonding interactions, making its enthalpy of vaporization higher than that of the previous molecules. Finally, NaCl forms an ionic solid with strong ionic bonds extending throughout a three-dimensional crystalline lattice, requiring a large amount of energy to free a NaCl molecule from the condensed phase.

6.106 The energy released by a falling body is given by: $E = mgh$ where m is the mass of the object, g is the acceleration due to free fall, and h is the height though which the object falls. The energy deposited into the water was therefore, $E = (15\ \text{kg})(20\ \text{m})(9.8067\ \text{m}\cdot\text{s}^{-2}) = 2940\ \text{J}$. If all of this energy is deposited into the water upon impact, the change in the temperature of the water would be:

$(2500\ \text{g})(4.184\ \text{J}\cdot\text{g}^{-1}\cdot{}^{\circ}\text{C}^{-1})\Delta T = 2940\ \text{J}$

$\Delta T = 0.28\ {}^{\circ}\text{C}$

If the falling body does 900.0 J of work before impacting the water, only 2040 J will be deposited into the water, resulting in a temperature change of:

$(2500\ \text{g})(4.184\ \text{J}\cdot\text{g}^{-1}\cdot{}^{\circ}\text{C}^{-1})\Delta T = 2040\ \text{J}$

$\Delta T = 0.20\ {}^{\circ}\text{C}$

6.108 The amount of calcium chloride in the pipes is

$(45.6\ \text{kJ}\cdot\text{pipe}^{-1})(60\ \text{pipes}) = 2736\ \text{kg}$ or 1.25×10^4 moles of $CaCl_2{\cdot}6H_2O$. Given the enthalpy of fusion of $CaCl_2{\cdot}6H_2O$, the amount of energy released as the Ca recrystalizes is:

$(27\ \text{kJ} \cdot \text{mol}^{-1})(1.25 \times 10^4\ \text{mol}) = 3.37 \times 10^5\ \text{kJ}$. The mass of water that may be heated by 10°C (from 15 °C to 25 °C) is:

$$m(4.184\ \text{J} \cdot \text{g}^{-1} \cdot {}^\circ\text{C}^{-1})(10\,{}^\circ\text{C}) = 3.37 \times 10^5\ \text{kJ}$$

$$m = 8100\ \text{g} \quad \text{or} \quad 8.1\ \text{kg}$$

CHAPTER 7

THERMODYNAMICS: THE SECOND AND THIRD LAWS

7.2 (a) rate of entropy generation $= \dfrac{\Delta S_{\text{surroundings}}}{\text{time}} = -\dfrac{q_{\text{rev}}}{\text{time} \cdot T}$

$$= -\frac{\text{rate of heat generation}}{T} = \frac{-(1.8 \times 10^3\ \text{J} \cdot \text{s}^{-1})}{298\ \text{K}} = -\,6.0\ \text{J} \cdot \text{K}^{-1} \cdot \text{s}^{-1}$$

(b) $\Delta S_{\text{day}} = (7\ \text{J} \cdot \text{K}^{-1} \cdot \text{s}^{-1})(60\ \text{sec} \cdot \text{min}^{-1})(60\ \text{min} \cdot \text{hr}^{-1})(24\ \text{hr} \cdot \text{day}^{-1})$

$= 5.2 \times 10^5\ \text{kJ} \cdot \text{K}^{-1} \cdot \text{day}^{-1}$

(c) More, because in the equation $\Delta S = \dfrac{-\Delta H}{T}$, ΔS is larger if T is smaller.

7.4 (a) $\Delta S = \dfrac{\Delta H}{T} = \dfrac{235}{273\ \text{K}} = 0.861\ \text{J} \cdot \text{K}^{-1}$

(b) $\Delta S = \dfrac{235\ \text{J}}{372\ \text{K}} = 0.632\ \text{J} \cdot \text{K}^{-1}$

(c) The entropy change is smaller at higher temperatures because the matter is already more chaotic. The same amount of heat has a greater effect on entropy changes when transferred at lower temperatures.

7.6 (a) The relationship to use is $\text{d}S = \dfrac{\text{d}q}{T}$. At constant pressure, we can substitute

$\text{d}q = nC_{\text{p}}\ \text{d}T$:

$$\text{d}S = \frac{nC_{\text{p}}\ \text{d}T}{T}$$

Upon integration, this gives $\Delta S = nCp \ln \frac{T_2}{T_1}$. The answer is calculated by simply plugging in the known quantities. Remember that for an ideal monatomic gas $C_P = \frac{5}{2}R$:

$$\Delta S = (2.92 \text{ mol})(\tfrac{5}{2} \times 8.314 \text{ J}\cdot\text{K}^{-1}\cdot\text{mol}^{-1}) \ln \frac{220.76 \text{ K}}{380.50 \text{ K}} = -33.0 \text{ J}\cdot\text{K}^{-1}$$

(b) A similar analysis using C_v gives $\Delta S = nC_v \ln \frac{T_2}{T_1}$, where C_v for a monatomic ideal gas is $\frac{3}{2}R$:

$$\Delta S = (2.92 \text{ mol})(\tfrac{3}{2} \times 8.314 \text{ J}\cdot\text{K}^{-1}\cdot\text{mol}^{-1}) \ln \frac{220.76 \text{ K}}{380.50 \text{ K}} = -19.8 \text{ J}\cdot\text{K}^{-1}$$

7.8 Because the process is isothermal and reversible, the relationship $dS = \frac{dq}{T}$ can be used. Because the process is isothermal, $\Delta U = 0$ and hence $q = -w$, where $w = -PdV$. Making this substitution, we obtain

$$dS = \frac{PdV}{T} = \frac{nRT}{TV}dV = \frac{nR}{V}dV$$

$$\therefore \Delta S = nR \ln \frac{V_2}{V_1}$$

but because $P_1V_1 = P_2V_2$, we can also write $\frac{P_1}{P_2} = \frac{V_2}{V_1}$

$$\Delta S = nR \ln \frac{P_1}{P_2}$$

Substituting the known quantities, we obtain

$$\Delta S = (6.32 \text{ mol})(8.314 \text{ J}\cdot\text{mol}^{-1}\cdot\text{K}^{-1}) \ln \frac{6.72 \text{ atm}}{13.44 \text{ atm}}$$

$$= -36.4 \text{ J}\cdot\text{K}^{-1}$$

7.10 In order for this process to occur spontaneously, entropy cannot decrease. However, to get as much work out of the transfer of energy as possible, we do not want to waste energy increasing entropy unnecessarily. Therefore, the most work will be done when ΔS for the process is infinitesimally greater than zero, just enough to make the process spontaneous. This will ensure as little energy is lost to entropy as possible, but enough energy is lost to make the process spontaneous. The change in entropy due to the extraction of 100 J of energy from the hot energy source is:

$\Delta S_{hot} = \frac{q}{T} = \frac{-100\ \text{J}}{573\ \text{K}} = -0.1745\ \text{J} \cdot \text{K}^{-1}$. In order for the net change in entropy to be greater than zero, a slightly greater increase in entropy must be realized at the cold energy sink. The minimum energy needed to affect an offsetting increase in entropy is given by:

$$\Delta S_{cold} = \frac{q}{T} = \frac{q}{293\ \text{K}} = +0.1745\ \text{J} \cdot \text{K}^{-1}$$

$$q = 51\ \text{J}$$

Therefore, of the 100 J extracted from the heat source, slightly more than 51 J must be deposited into the cold energy sink to make the ΔS positive and the transfer spontaneous. Slightly less than 49 J is therefore available to do work.

7.12 (a) Using Table 6.2, the entropy for the vaporization of 1.00 mol of water at the boiling point under a constant pressure of 1.00 atm is found by:

$$\Delta S = \frac{q}{T} = \frac{\Delta H}{T} = \frac{40700\ \text{J} \cdot \text{mol}^{-1}}{373.2\ \text{K}} = 109\ \text{J} \cdot \text{mol}^{-1} \cdot \text{K}^{-1}$$

(b) As in part (a) the molar change in entropy is:

$$\Delta S = \frac{q}{T} = \frac{\Delta H}{T} = \frac{-4600\ \text{J} \cdot \text{mol}^{-1}}{158.7\ \text{K}} = -109\ \text{J} \cdot \text{mol}^{-1} \cdot \text{K}^{-1}$$

the change in entropy for freezing 3.33g of C_2H_5OH is:

$$\left(-109\ \text{J} \cdot \text{mol}^{-1} \cdot \text{K}^{-1}\right)\left(\frac{3.33\ \text{g}}{46.069\ \text{g} \cdot \text{mol}^{-1}}\right) = 7.88\ \text{J} \cdot \text{K}^{-1}$$

7.14 (a) The boiling point of a liquid may be obtained from the relationship $\Delta S_{\text{vap}} = \dfrac{\Delta H_{\text{vap}}}{T_{\text{B}}}$, or $T_{\text{B}} = \dfrac{\Delta H_{\text{vap}}}{\Delta S_{\text{vap}}}$. This relationship should be rigorously true if we have the actual enthalpy and entropy of vaporization. The data in the Appendix, however, are for 298 K. Thus, calculation of $\Delta H°_{\text{vap}}$ or $\Delta S°_{\text{vap}}$, using the enthalpy and entropy differences between the gas and liquid forms at 298 K, will give a good approximation of these quantities but the values will not be exact. For $Br_2(l) \longrightarrow Br_2(g)$, the data in the appendix give

$$\Delta H_{\text{vap}} \cong 30.91\ \text{kJ}\cdot\text{mol}^{-1}$$

$$\Delta S_{\text{vap}} \cong 245.46\ \text{J}\cdot\text{K}^{-1}\cdot\text{mol}^{-1} - 152.23\ \text{J}\cdot\text{K}^{-1}\cdot\text{mol}^{-1} = 93.23\ \text{J}\cdot\text{K}^{-1}\cdot\text{mol}^{-1}$$

$$T_{\text{B}} = \frac{30.91\times 10^{3}\ \text{J}\cdot\text{mol}^{-1}}{93.23\ \text{J}\cdot\text{K}^{-1}\cdot\text{mol}^{-1}} = 331.5\ \text{K}$$

(b) The boiling point of bromine is 58.4°C or 332 K.

(c) These numbers are in excellent agreement.

7.16 (a) Trouton's rule indicates that the entropy of vaporization for a number of organic liquids is approximately $85\ \text{J}\cdot\text{K}^{-1}\cdot\text{mol}^{-1}$. From this information we can calculate T_{B}:

$T_{\text{B}} = \dfrac{\Delta H°_{\text{vap}}}{\Delta S°_{\text{vap}}} = \dfrac{25.76\times 10^{3}\ \text{J}\cdot\text{mol}^{-1}}{85\ \text{J}\cdot\text{K}^{-1}\cdot\text{mol}^{-1}} = 303\ \text{K}$. (b) The experimental boiling point of ethanal, 294 K, agrees reasonably although not perfectly with the calculated value.

7.18 (a) The value can be estimated from:

$$\begin{aligned}\Delta H°_{\text{vap}} &= T\Delta S°_{\text{vap}}\\ &= (329.4\ \text{K})(85\ \text{J}\cdot\text{mol}^{-1}\cdot\text{K}^{-1})\\ &= 28.0\ \text{kJ}\cdot\text{mol}^{-1}\end{aligned}$$

(b) $\Delta S°_{\text{surr}} = -\dfrac{\Delta H°_{\text{system}}}{T}$

$$\Delta S^{\circ}_{\text{surr}} = -\left(\frac{10\ \text{g}}{58.08\ \text{g}\cdot\text{mol}^{-1}}\right)\times 85\ \text{J}\cdot\text{K}^{-1}\cdot\text{mol}^{-1} = -15\ \text{J}\cdot\text{K}^{-1}$$

7.20 We would expect NO and N_2O to be the most likely to have a residual entropy at 0 K. This is because the structures are set up so that the O and N atoms, which are of similar size, could be oriented in one of two ways without perturbing the lattice of the solid, as shown below. Because CO_2 and Cl_2 are symmetrical, switching ends of the molecule does not result in increased disorder.

N=O	N=O	N=O	N≡N—O	N≡N—O	N≡N—O
N=O	N=O	N=O	N≡N—O	N≡N—O	O—N≡N
N=O	O=N	N=O	N≡N—O	N≡N—O	N≡N—O
N=O	N=O	N=O			

7.22 There are three orientations that PH_2F could adopt in the solid state.

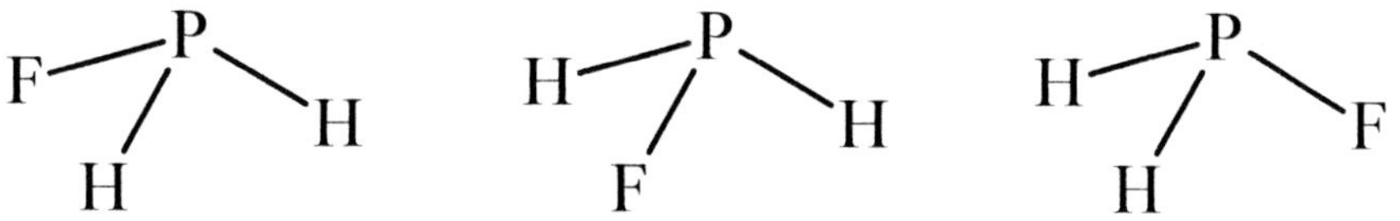

The Boltzmann entropy calculation then becomes:

$$S = k\ \ln 3^{6.02\times 10^{23}} = (1.38\times 10^{-23}\ \text{J}\cdot\text{K}^{-1})\ln 3^{6.02\times 10^{23}}$$

$$S = 9.13\ \text{J}\cdot\text{K}^{-1}$$

7.24 (a) $C_2H_6(g)$, because it has a greater molecular complexity; (b) KCl(aq) because ions distributed in solution will be arranged more randomly than ions localized in a crystal lattice; (c) Kr(g), because Kr is more massive and has more elementary particles; (d) $O_2(g)$ at 450 K, because of increased molecular randomness (1 mol of O_2 at 1.00 atm pressure will occupy a larger volume at 450 K than at 273 K).

7.26 Gases will have a higher entropy than liquids so we expect $H_2O(l)$ to have the lowest molar entropy. The gases will increase in entropy in the order

Ne(g) < Ar(g) < CO_2(g). Ne and Ar are both atoms so they should have less entropy than a molecular substance, which will have more complexity. Ar will have a higher entropy than Ne because it has a larger mass and more fundamental particles. The final order is H_2O(l) < Ne(g) < Ar(g) < CO_2(g).

7.28 Benzene is a rigid molecule in which the carbon and hydrogen atoms are all constrained to lie in a plane. Benzene has some added entropy due to the delocalization of electrons caused by resonance. However, this effect does not outweigh the added flexibility in cyclohexane, in which the C—C bonds can twist and turn. The additional H atoms also add to the molecular complexity, giving more possibilities for bond vibrations.

7.30 (a) increases. Generally, entropy increases when a solid is dissolved because the molecules or ions are dispersed in the solvent, giving them more locations over which to arrange themselves. (b) decreases. In this example, twelve moles of reactants form seven moles of products, so we might expect immediately that entropy would decrease; however, we must also consider the states of the various reactants and products. There are six moles of gas on each side of the equation so this might roughly balance; however, we still expect a decrease in entropy for the gases because CO_2(g) (a triatomic molecule) is more complex than O_2 (a diatomic molecule). The remainder of the reaction is six moles of liquid going to one mole of solid, which again would predict a decrease in overall entropy. (c) In the evaporation of water, a liquid is converted to a gas, which should be accompanied by an increase in entropy.

7.32 The total change in entropy is given by:

$$\Delta S = \Delta S_{\mathrm{Ne}} + \Delta S_{\mathrm{F_2}} = n_{\mathrm{Ne}} \cdot C_{V,\mathrm{Ne}} \cdot \ln\frac{T_2}{T_1} + n_{\mathrm{F_2}} \cdot C_{V,\mathrm{F_2}} \cdot \ln\frac{T_2}{T_1}$$

where the heat capacities of the monatomic gas Ne and diatomic gas F_2 are 3/2 R and 5/2 R, respectively. This equation can be reduced to one unknown by the fact that the sum $n_{Ne} + n_{F_2}$ is equal to the total number of moles of gas present, n_{tot} calculated using the ideal gas law:

$$n_{tot} = \frac{P \cdot V}{R \cdot T} = \frac{3.32 \text{ atm} \cdot 2.5 \text{ L}}{\left(0.0820574 \text{ L} \cdot \text{atm} \cdot \text{K}^{-1} \cdot \text{mol}^{-1}\right) \cdot 273.2 \text{ K}} = 0.370 \text{ mol}.$$

Therefore,

$$\Delta S = 0.345 \text{ J} \cdot \text{K}^{-1} = n_{Ne} \cdot C_{V,Ne} \cdot \ln\frac{T_2}{T_1} + \left(0.370 \text{ mol} - n_{Ne}\right) \cdot C_{V,F_2} \cdot \ln\frac{T_2}{T_1}$$

$$0.345 \text{ J} \cdot \text{K}^{-1} = n_{Ne} \cdot \frac{3}{2} R \cdot \ln\frac{288.15 \text{ K}}{273.15 \text{ K}}$$

$$+\left(0.370 \text{ mol} - n_{Ne}\right)\frac{5}{2} R \cdot \ln\frac{288.15 \text{ K}}{273.15 \text{ K}}$$

$$0.345 = n_{Ne} \cdot 0.667 \text{ J} \cdot \text{K}^{-1} \cdot \text{mol}^{-1}$$

$$+\left(0.370 \text{ mol} - n_{Ne}\right) 1.111 \text{ J} \cdot \text{K}^{-1} \cdot \text{mol}^{-1}$$

$$n_{Ne} = 0.150 \text{ mol}$$

and

$$n_{F_2} = 0.370 \text{ mol} - 0.15 \text{ mol} = 0.22 \text{ mol}$$

7.34 (a) $\Delta S°_r = S°_m(CS_2, l) + 2S°_m(H_2S, g)$

$$- [S°_m(CH_4, g) + 4S°_m(S(s), \text{rhombic})]$$

$$= 151.34 \text{ J} \cdot \text{K}^{-1} \cdot \text{mol}^{-1} + 2(205.79 \text{ J} \cdot \text{K}^{-1} \cdot \text{mol}^{-1})$$

$$- [186.26 \text{ J} \cdot \text{K}^{-1} \cdot \text{mol}^{-1} + 4(31.80 \text{ J} \cdot \text{K}^{-1} \cdot \text{mol}^{-1})]$$

$$= +249.46 \text{ J} \cdot \text{K}^{-1} \cdot \text{mol}^{-1}$$

The entropy change is positive because there are two more moles of gas in the products than in the reactants. In addition, the molecular structure of the products indicates more disorder.

(b) $\Delta S°_r = S°_m(Ca(OH)_2, s) + S°_m(C_2H_2, g)$

$$- [S°_m(CaC_2, s) + 2S°_m(H_2O, l)]$$

$$= 83.39 \text{ J} \cdot \text{K}^{-1} \cdot \text{mol}^{-1} + 200.94 \text{ J} \cdot \text{K}^{-1} \cdot \text{mol}^{-1}$$

$$- [69.96 \text{ J} \cdot \text{K}^{-1} \cdot \text{mol}^{-1} + 2(69.91 \text{ J} \cdot \text{K}^{-1} \cdot \text{mol}^{-1})]$$

$$= 74.55 \text{ J} \cdot \text{K}^{-1} \cdot \text{mol}^{-1}$$

The entropy change is positive because the reaction produces one mole of gas. Originally there were no gases.

(c)

$$\begin{aligned}\Delta S^\circ_\mathrm{r} &= 4S^\circ_\mathrm{m}(\mathrm{NO, g}) + 6S^\circ_\mathrm{m}(\mathrm{H_2O, l}) - [4S^\circ_\mathrm{m}(\mathrm{NH_3, g}) + 5S^\circ_\mathrm{m}(\mathrm{O_2, g})] \\ &= 4(210.76\ \mathrm{J\cdot K^{-1}\cdot mol^{-1}}) + 6(69.91\ \mathrm{J\cdot K^{-1}\cdot mol^{-1}}) \\ &\quad -[4(192.45\ \mathrm{J\cdot K^{-1}\cdot mol^{-1}}) + 5(205.14\ \mathrm{J\cdot K^{-1}\cdot mol^{-1}})] \\ &= -533.00\ \mathrm{J\cdot K^{-1}\cdot mol^{-1}}\end{aligned}$$

The entropy change is large and negative because the number of moles of gas has decreased by five.

(d)

$$\begin{aligned}\Delta S^\circ_\mathrm{r} &= S^\circ_\mathrm{m}(\mathrm{CO(NH_2)_2, s}) + S^\circ_\mathrm{m}(\mathrm{H_2O, l}) - [S^\circ_\mathrm{m}(\mathrm{CO_2(g)} + 2S^\circ_\mathrm{m}\,\mathrm{NH_3(g)}] \\ &= 104.60\ \mathrm{J\cdot K^{-1}\cdot mol^{-1}}) + 69.91\ \mathrm{J\cdot K^{-1}\cdot mol^{-1}} \\ &\quad -[213.74\ \mathrm{J\cdot K^{-1}\cdot mol^{-1}}) + 2(192.45\ \mathrm{J\cdot K^{-1}\cdot mol^{-1}})] \\ &= -424.13\ \mathrm{J\cdot K^{-1}}\end{aligned}$$

The change in entropy is negative because there are no moles of gas in the products, but there are three moles of gas in the reactants.

7.36 The standard molar entropy of a substance at temperature T_i may be found by plotting C_p/T vs. T for that substance and finding the area under the curve from $T = 0$ to $T = T_i$. If at low temperatures $C_p \propto T^3$, the standard molar entropy of the substance is:

$$S(T) = \int_{0\,\mathrm{K}}^{T} \frac{C_p}{T}\,dT = \int_{0\,\mathrm{K}}^{T} \frac{T^3}{T}\,dT = \frac{1}{3}T^3$$

7.38 For C_2H_5O, the molar enthalpy of vaporization is found by dividing the amount of energy supplied to cause the phase transition by the number of moles of C_2H_5O that underwent the transition:

$$\Delta H_{vap} = \frac{(500\ \mathrm{J\cdot s^{-1}})(4.0\ \mathrm{min})(60\ \mathrm{s\cdot min^{-1}})}{\left(\dfrac{400.15\ \mathrm{g} - 271.15\ \mathrm{g}}{46.688\ \mathrm{g\cdot mol^{-1}}}\right)} = 43\ \mathrm{kJ\cdot mol^{-1}}$$

The associated change in entropy is then found using:

$$\Delta S_{vap} = \frac{\Delta H_{vap}}{T} = \frac{43 \text{ kJ} \cdot \text{mol}^{-1}}{351.5 \text{ K}} = 120 \text{ J} \cdot \text{mol}^{-1} \cdot \text{K}^{-1}$$

For C_4H_{10} the enthalpy and entropy of vaporization are:

$$\Delta H_{vap} = \frac{\left(500 \text{ J} \cdot \text{s}^{-1}\right)(4.0 \text{ min})\left(60 \text{ s} \cdot \text{min}^{-1}\right)}{\left(\frac{398.05 \text{ g} - 74.95 \text{ g}}{58.123 \text{ g} \cdot \text{mol}^{-1}}\right)} = 22 \text{ kJ} \cdot \text{mol}^{-1}$$

$$\Delta S_{vap} = \frac{\Delta H_{vap}}{T} = \frac{22 \text{ kJ} \cdot \text{mol}^{-1}}{273.2 \text{ K}} = 81 \text{ J} \cdot \text{mol}^{-1} \cdot \text{K}^{-1}$$

For CH_4O the enthalpy and entropy of vaporization are:

$$\Delta H_{vap} = \frac{\left(500 \text{ J} \cdot \text{s}^{-1}\right)(4.0 \text{ min})\left(60 \text{ s} \cdot \text{min}^{-1}\right)}{\left(\frac{395.15 \text{ g} - 294.25 \text{ g}}{32.042 \text{ g} \cdot \text{mol}^{-1}}\right)} = 38 \text{ kJ} \cdot \text{mol}^{-1}$$

$$\Delta S_{vap} = \frac{\Delta H_{vap}}{T} = \frac{38 \text{ kJ} \cdot \text{mol}^{-1}}{337.7 \text{ K}} = 110 \text{ J} \cdot \text{mol}^{-1} \cdot \text{K}^{-1}$$

7.40 For this problem we can treat the two different samples of ethanol as different systems, and find the total change in entropy by addition. The change in entropy for both samples of ethanol is found using:

$\Delta S = C \ln \frac{T_2}{T_1}$ so our first job is to determine the final equilibrium temperature after the two samples mix:

$$T_F = \frac{m_1 \cdot T_1 + m_2 \cdot T_2}{m_1 + m_2} = \frac{250.0 \text{ g} \cdot 329.2 \text{ K} + 150.0 \text{ g} \cdot 295.15 \text{ K}}{250.0 \text{ g} + 150.0 \text{ g}}$$
$$= 316.4 \text{ K}$$

The change in enthalpy for the ethanol initially at 22°C is:

$$\Delta S_{sys} = (150.0 \text{ g}) \cdot \left(2.42 \text{ J} \cdot \text{g}^{-1} \cdot \text{K}^{-1}\right) \ln \frac{316.4 \text{ K}}{295.2 \text{ K}} = 25.2 \text{ J} \cdot \text{K}^{-1}$$

and the change in enthalpy for the ethanol initially at 56°C is:

$$\Delta S_{sys} = (250.0 \text{ g}) \cdot \left(2.42 \text{ J} \cdot \text{g}^{-1} \cdot \text{K}^{-1}\right) \ln \frac{316.4 \text{ K}}{329.2 \text{ K}} = -23.9 \text{ J} \cdot \text{K}^{-1}.$$

The total change in entropy is that of the entire system (there is no change in entropy of the surroundings as the vessel isolates the system from the surroundings):

$$\Delta S_{tot} = 25.2\ \text{J}\cdot\text{K}^{-1} - 23.9\ \text{J}\cdot\text{K}^{-1} = +1.3\ \text{J}\cdot\text{K}^{-1}$$

7.42 (a) The change in entropy will be given by

$$\Delta S_{\text{surr}} = \frac{-\Delta H_{\text{system}}}{T} = \frac{1.00\ \text{mol} \times -5.65 \times 10^3\ \text{J}\cdot\text{mol}^{-1}}{195.4\ \text{K}} = -28.9\ \text{J}\cdot\text{K}^{-1}$$

$$\Delta S_{\text{system}} = \frac{\Delta H_{\text{system}}}{T} = \frac{1.00\ \text{mol} \times 5.65 \times 10^3\ \text{J}\cdot\text{mol}^{-1}}{195.4\ \text{K}} = +28.9\ \text{J}\cdot\text{K}^{-1}$$

(b)

$$\Delta S_{\text{surr}} = \frac{-\Delta H_{\text{system}}}{T} = \frac{-(1.00\ \text{mol} \times -3.16 \times 10^3\ \text{J}\cdot\text{mol}^{-1})}{175.2\ \text{K}} = +18.0\ \text{J}\cdot\text{K}^{-1}$$

$$\Delta S_{\text{system}} = \frac{\Delta H_{\text{system}}}{T} = \frac{1.00\ \text{mol} \times -3.16 \times 10^3\ \text{J}\cdot\text{mol}^{-1}}{175.2\ \text{K}} = -18.0\ \text{J}\cdot\text{K}^{-1}$$

(c)

$$\Delta S_{\text{surr}} = \frac{-\Delta H_{\text{system}}}{T} = \frac{-(1.00\ \text{mol} \times 40.7 \times 10^3\ \text{J}\cdot\text{mol}^{-1})}{373.2\ \text{K}} = -109\ \text{J}\cdot\text{K}^{-1}$$

$$\Delta S_{\text{system}} = \frac{\Delta H_{\text{system}}}{T} = \frac{1.00\ \text{mol} \times 40.7 \times 10^3\ \text{J}\cdot\text{mol}^{-1}}{373.2\ \text{K}} = +109\ \text{J}\cdot\text{K}^{-1}$$

7.44 (a) The total entropy change is given by $\Delta S_{\text{tot}} = \Delta S_{\text{surr}} + \Delta S$. ΔS for an isothermal, reversible process is calculated from

$\Delta S = \dfrac{q_{\text{rev}}}{T} = \dfrac{-w_{\text{rev}}}{T} = nR \ln \dfrac{V_2}{V_1}$. To do the calculation

we need the value of n, which is obtained by use of the ideal gas law: $(0.6789\ \text{atm})(12.62\ \text{L}) = n(0.082\,06\ \text{L}\cdot\text{atm}\cdot\text{K}^{-1}\cdot\text{mol}^{-1})(412\ \text{K})$; $n =$ 0.253 mol.

$$\Delta S = (0.253\ \text{mol})(8.314\ \text{J}\cdot\text{K}^{-1}\cdot\text{mol}^{-1}) \ln \frac{19.44\ \text{L}}{12.62\ \text{L}} = +0.909\ \text{J}\cdot\text{K}^{-1}.$$

Because the process is reversible,

$\Delta S_{tot} = 0$, so $\Delta S_{surr} = -\Delta S = -0.909\ J\cdot K^{-1}$. (b) For the irreversible process, ΔS also $= +0.909\ J\cdot K^{-1}$. No work is done in free expansion (see Section 6.6), so $w = 0$. Because $\Delta U = 0$, it follows that $q = 0$. Therefore, no heat is transferred into the surroundings, and their entropy is unchanged:

$\Delta S_{surr} = 0$. The total change in entropy is therefore $\Delta S_{tot} = +0.909\ J\cdot K^{-1}$.

7.46 Under constant temperature and pressure conditions, it is the sign of ΔG_r that determines whether or not a process is spontaneous. If $\Delta G_r < 0$, (ΔG_r is negative); the process is spontaneous. ΔG_r is related by the equation $\Delta G_r = \Delta H_r - T\Delta S_r$ to the enthalpy and entropy changes in a reaction. If a reaction is endothermic ΔH_r is positive), then the reaction will be spontaneous only if $-T\Delta S_r$ is larger than ΔH_r. So, effectively, a reaction that is endothermic can be spontaneous only if the entropy change in the system outweighs the enthalpy change's effect on the entropy of the surroundings.

7.48 (a)

$$\Delta H^\circ_r = -2279.7\ kJ\cdot mol^{-1} - [-771.36\ kJ\cdot mol^{-1} + 5(-285.83\ kJ\cdot mol^{-1})]$$
$$= -79.19\ kJ\cdot mol^{-1}$$

$$\Delta S^\circ_r = 300.4\ J\cdot K^{-1}\cdot mol^{-1} - [109\ J\cdot K^{-1}\cdot mol^{-1} + 5(69.91\ J\cdot K^{-1}\cdot mol^{-1})]$$
$$= -158\ J\cdot K^{-1}\cdot mol^{-1}$$

$$\Delta G^\circ_r = -1879.7\ kJ\cdot mol^{-1} - [-661.8\ kJ\cdot mol^{-1} + 5(-237.13\ kJ\cdot mol^{-1})]$$
$$= -32.3\ kJ\cdot mol^{-1}$$

ΔG°_r may also be calculated from ΔH°_r and ΔS°_r (the numbers calculated differ slightly from the two methods due to rounding differences):

$$\Delta G°_r = \Delta H°_r - T\Delta S°_r$$
$$= -79.19 \text{ kJ}\cdot\text{mol}^{-1} - (298 \text{ K})(-158 \text{ J}\cdot\text{K}^{-1}\cdot\text{mol}^{-1})/(1000 \text{ J}\cdot\text{kJ}^{-1})$$
$$= -32.1 \text{ kJ}\cdot\text{mol}^{-1}$$

(b) The major process occurring upon dissolution of H_2SO_4 in water is

$$H_2SO_4(l) \longrightarrow H^+(aq) + HSO_4^-(aq)$$

This ionization is complete.

The second ionization occurs to a much smaller extent:

$$HSO_4^-(aq) \longrightarrow H^+(aq) + SO_4^{2-}(aq)$$

Using the first equation:

$$\Delta H°_r = -887.34 \text{ kJ}\cdot\text{mol}^{-1} - [-813.99 \text{ kJ}\cdot\text{mol}^{-1})$$
$$= -74.35 \text{ kJ}\cdot\text{mol}^{-1}$$
$$\Delta S°_r = +131.8 \text{ J}\cdot\text{K}^{-1}\cdot\text{mol}^{-1} - (156.90 \text{ J}\cdot\text{K}^{-1}\cdot\text{mol}^{-1})$$
$$= -25.1 \text{ J}\cdot\text{K}^{-1}\cdot\text{mol}^{-1}$$
$$\Delta G°_r = -755.91 \text{ kJ}\cdot\text{mol}^{-1} - (-690.00 \text{ kJ}\cdot\text{mol}^{-1})$$
$$= -65.91 \text{ kJ}\cdot\text{mol}^{-1}$$

or

$$\Delta G°_r = -74.35 \text{ kJ}\cdot\text{mol}^{-1} - \frac{(298 \text{ K})(-25.1 \text{ J}\cdot\text{K}^{-1}\cdot\text{mol}^{-1})}{1000 \text{ J}\cdot\text{kJ}^{-1}}$$
$$= -66.87 \text{ kJ}\cdot\text{mol}^{-1}$$

(c)

$$\Delta H°_r = -986.09 \text{ kJ}\cdot\text{mol}^{-1} - [(-635.09 \text{ kJ}\cdot\text{mol}^{-1}) + (-285.83 \text{ kJ}\cdot\text{mol}^{-1})]$$
$$= -65.17 \text{ kJ}\cdot\text{mol}^{-1}$$
$$\Delta S°_r = 83.39 \text{ J}\cdot\text{K}^{-1}\cdot\text{mol}^{-1} - [39.75 \text{ J}\cdot\text{K}^{-1}\cdot\text{mol}^{-1} + 69.91 \text{ J}\cdot\text{K}^{-1}\cdot\text{mol}^{-1}]$$
$$= -26.27 \text{ J}\cdot\text{K}^{-1}\cdot\text{mol}^{-1}$$
$$\Delta G°_r = -898.49 \text{ kJ}\cdot\text{mol}^{-1} - [(-604.03 \text{ kJ}\cdot\text{mol}^{-1}) + (-237.13 \text{ kJ}\cdot\text{mol}^{-1})]$$
$$= -57.33 \text{ kJ}\cdot\text{mol}^{-1}$$

or

$$\Delta G°_r = \Delta H°_r - T\Delta S°_r$$
$$= -65.17 \text{ kJ}\cdot\text{mol}^{-1} - (298 \text{ K})(-26.27 \text{ J}\cdot\text{K}^{-1}\cdot\text{mol}^{-1})/(1000 \text{ J}\cdot\text{kJ}^{-1})$$
$$= -57.34 \text{ kJ}\cdot\text{mol}^{-1}$$

(d) In order to rank the ability of these compounds to remove water, we can examine the free energies for the reactions involved. The one with the greatest driving force (most negative $\Delta G°_r$) is the hydration of calcium oxide, followed closely by that of sulfuric acid, with the hydration of copper sulfate falling somewhat further behind. In practice, both sulfuric acid and calcium oxide are used as water-scavenging agents; copper sulfate is ineffective.

7.50 (a) $\frac{1}{2}H_2(g) + \frac{1}{2}Cl_2(g) \longrightarrow HCl(g)$

$$\Delta H°_r = \Delta H°_f(HCl, g) = -92.31\ kJ\cdot mol^{-1}$$

$$\Delta S°_r = S°_m(HCl, g) - [\tfrac{1}{2}S°_m(H_2, g) + \tfrac{1}{2}S°_m(Cl_2, g)]$$

$$= 186.91\ J\cdot K^{-1}\cdot mol^{-1} - [\tfrac{1}{2}(130.68\ J\cdot K^{-1}\cdot mol^{-1}) + \tfrac{1}{2}(223.07\ J\cdot K^{-1}\cdot mol^{-1})]$$

$$= 10.04\ J\cdot K^{-1}\cdot mol^{-1}$$

$$\Delta G°_r = -92.31\ kJ\cdot mol^{-1} - (298\ K)(10.04\ J\cdot K^{-1}\cdot mol^{-1})/(1000\ J\cdot kJ^{-1})$$

$$= -95\ kJ\cdot mol^{-1}$$

(b) $6\ C(s), \text{graphite} + 3\ H_2(g) \longrightarrow C_6H_6(l)$

$$\Delta H°_r = \Delta H°_f(C_6H_6, l) = 49.0\ kJ\cdot mol^{-1}$$

$$\Delta S°_r = S°_m(C_6H_6, l) - [6S°_m(C, s) + 3S°_m(H_2, g)]$$

$$= 173.3\ J\cdot K^{-1}\cdot mol^{-1} - [6(5.740\ J\cdot K^{-1}\cdot mol^{-1}) + 3(130.68\ J\cdot K^{-1}\cdot mol^{-1})]$$

$$= -253.2\ J\cdot K^{-1}\cdot mol^{-1}$$

$$\Delta G°_r = 49.0\ kJ\cdot mol^{-1} - (298\ K)(-253.2\ J\cdot K^{-1}\cdot mol^{-1})/(1000\ J\cdot kJ^{-1})$$

$$= 124.5\ kJ\cdot mol^{-1}$$

$$S°_m(C_6H_6, l) = 173.3\ J\cdot K^{-1}\cdot mol^{-1}$$

(c) $Cu(s) + S(s) + 9/2\ O_2(g) + 5\ H_2(g) \longrightarrow CuSO_4\cdot 5\ H_2O$

$$\Delta H^\circ_r = \Delta H^\circ_f(\text{Cu}_6\text{SO}_4 \cdot 5\,\text{H}_2\text{O, s}) = -2279.7\ \text{kJ}\cdot\text{mol}^{-1}$$

$$\begin{aligned}\Delta S^\circ_r &= S^\circ_m(\text{CuSO}_4 \cdot 5\,\text{H}_2\text{O, s}) - [S^\circ_m(\text{Cu, s}) + S^\circ_m(\text{S, s}) \\ &\quad + \tfrac{9}{2} S^\circ_m(\text{O}_2, \text{g}) + 5\, S^\circ_m(\text{H}_2, \text{g})] \\ &= 300.4\ \text{J}\cdot\text{K}^{-1}\cdot\text{mol}^{-1} - [(33.15\ \text{J}\cdot\text{K}^{-1}\cdot\text{mol}^{-1}) \\ &\quad + (31.80\ \text{J}\cdot\text{K}^{-1}\cdot\text{mol}^{-1}) + \tfrac{9}{2}(205.14\ \text{J}\cdot\text{K}^{-1}\cdot\text{mol}^{-1}) \\ &\quad + 5\,(130.68\ \text{J}\cdot\text{K}^{-1}\cdot\text{mol}^{-1})] \\ &= -1341.1\ \text{J}\cdot\text{K}^{-1}\cdot\text{mol}^{-1}\end{aligned}$$

$$\begin{aligned}\Delta G^\circ_r &= -2279.7\ \text{kJ}\cdot\text{mol}^{-1} \\ &\quad - (298\ \text{K})(-1341.1\ \text{J}\cdot\text{K}^{-1}\cdot\text{mol}^{-1})/(1000\ \text{J}\cdot\text{kJ}^{-1}) \\ &= -1880.0\ \text{kJ}\cdot\text{mol}^{-1}\end{aligned}$$

$$S^\circ_m(\text{C}_2\text{H}_5\text{OH, l}) = 160.7\ \text{J}\cdot\text{K}^{-1}\cdot\text{mol}^{-1}$$

(d) $\text{Ca(s)} + \text{C(s), graphite} + \tfrac{3}{2}\,\text{O}_2(\text{g}) \longrightarrow \text{CaCO}_3(\text{s})$

$$\Delta H^\circ_r = \Delta H^\circ_f(\text{CaCO}_3, \text{s}) = -1206.9\ \text{kJ}\cdot\text{mol}^{-1}$$

$$\begin{aligned}\Delta S^\circ_r &= S^\circ_m(\text{CaCO}_3, \text{s}) - [S^\circ_m(\text{Ca(s)}) + S^\circ_m(\text{C, s}) + \tfrac{3}{2} S^\circ_m(\text{O}_2, \text{g})] \\ &= 92.9\ \text{J}\cdot\text{K}^{-1}\cdot\text{mol}^{-1} - [41.42\ \text{J}\cdot\text{K}^{-1}\cdot\text{mol}^{-1} + 5.740\ \text{J}\cdot\text{K}^{-1}\cdot\text{mol}^{-1} \\ &\quad + \tfrac{3}{2}(205.14\ \text{J}\cdot\text{K}^{-1}\cdot\text{mol}^{-1})] \\ &= -262.0\ \text{J}\cdot\text{K}^{-1}\cdot\text{mol}^{-1}\end{aligned}$$

$$\begin{aligned}\Delta G^\circ_r &= -1206.9\ \text{kJ}\cdot\text{mol}^{-1} \\ &\quad - (298\ \text{K})(-262.0\ \text{J}\cdot\text{K}^{-1}\cdot\text{mol}^{-1})/(1000\ \text{J}\cdot\text{kJ}^{-1}) \\ &= -1.1288\times 10^3\ \text{kJ}\cdot\text{mol}^{-1}\end{aligned}$$

$$S^\circ_m(\text{CaCO}_3, \text{s}) = 92.9\ \text{J}\cdot\text{K}^{-1}\cdot\text{mol}^{-1}$$

The ΔS°_f value is negative because of the reduction of the number of moles of gas during the reaction. For all of these, the important point to gain is that the S°_m value of a compound is not the same as the ΔS°_f for the formation of that compound. ΔS°_f is often negative because one is bringing together a number of elements to form that compound.

7.52 Use the relationship $\Delta G^\circ_r = \Sigma\, \Delta G^\circ_f(\text{products}) - \Sigma\, \Delta G^\circ_r(\text{reactants})$:

(a) $\Delta G^\circ_r = 2\Delta G_f^\circ(NH_3, g) + \Delta G^\circ_f(HCl, g) - [\Delta G^\circ_f(NH_4Cl, s)]$

$$= (-16.45\ kJ\cdot mol^{-1}) + (-95.30\ kJ\cdot mol^{-1}) - [-202.87\ kJ\cdot mol^{-1}]$$
$$= +91.12\ kJ\cdot mol^{-1}$$

The reaction is not spontaneous.

(b) $\Delta G^\circ_r = \Delta G^\circ_f(H_2O, l) - [\Delta G^\circ_f(D_2O, l)]$

$$= (-237.13\ kJ\cdot mol^{-1}) - [-243.44\ kJ\cdot mol^{-1}]$$
$$= +6.31\ kJ\cdot mol^{-1}$$

The reaction is not spontaneous.

(c) $\Delta G^\circ_r = \Delta G^\circ_f(N_2O, g) + \Delta G^\circ_f(NO, g) - [\Delta G^\circ_f(NO_2, g)]$

$$= 104.20\ kJ\cdot mol^{-1} + 86.55\ kJ\cdot mol^{-1} - 51.31\ kJ\cdot mol^{-1}$$
$$= +139.44\ kJ\cdot mol^{-1}$$

The reaction is not spontaneous.

(d) $\Delta G^\circ_r = 2\Delta G^\circ_f(CO_2, g) + 4\Delta G^\circ_f(H_2O, l) - [2\Delta G^\circ_f(CH_3OH, g)]$

$$= 2(-394.36\ kJ\cdot mol^{-1}) + 4(-237.13\ kJ\cdot mol^{-1})$$
$$- [2(-161.96\ kJ\cdot mol^{-1})]$$
$$= -1413.32\ kJ\cdot mol^{-1}$$

7.54 To answer this question, we examine the standard free energies of formation of the compounds. These values from the Appendix are:

(a) $CuO(s)$, $-129.7\ kJ\cdot mol^{-1}$;

(b) $C_6H_{12}(l)$, cyclohexane, $26.7\ kJ\cdot mol^{-1}$;

(c) $PCl_3(g)$, $-267.8\ kJ\cdot mol^{-1}$;

(d) $N_2H_4(l)$, $149.34\ kJ\cdot mol^{-1}$. Those compounds with a negative free energy of formation are stable, whereas those with a positive free energy of formation are unstable with respect to the elements that compose them. Accordingly, compounds (b) and (d) are thermodynamically unstable, whereas (a) and (c) are thermodynamically stable.

7.56 To understand what happens to ΔG°_r as the temperature is raised, we use the relationship $\Delta G^\circ_{\mathrm{r}} = \Delta H^\circ_{\mathrm{r}} - T\Delta S^\circ_{\mathrm{r}}$. From this it is clear that the free energy of the reaction becomes less favorable (more positive) as temperature increases only if ΔS°_r is a negative number. Therefore, we only have to find out whether the standard entropy of formation of the compound is a negative number. This is calculated for each compound as follows:

(a) $\mathrm{Cu(s)} + \frac{1}{2}\,\mathrm{O_2(g)} \longrightarrow \mathrm{CuO(s)}$

$$\begin{aligned}\Delta S^\circ_{\mathrm{r}} &= S^\circ_{\mathrm{m}}(\mathrm{CuO, s}) - [S^\circ_{\mathrm{m}}(\mathrm{Cu, s}) + \tfrac{1}{2}S^\circ_{\mathrm{m}}(\mathrm{O_2, g})]\\ &= 42.63\ \mathrm{J\cdot K^{-1}\cdot mol^{-1}} - [33.15\ \mathrm{J\cdot K^{-1}\cdot mol^{-1}}\\ &\qquad + \tfrac{1}{2}(205.14\ \mathrm{J\cdot K^{-1}\cdot mol^{-1}})]\\ &= -93.09\ \mathrm{J\cdot K^{-1}\cdot mol^{-1}}\end{aligned}$$

The compound is less stable at higher temperatures.

(b) 6 C(s), graphite + 6 H_2(g) – C_6H_{12}(l)

$$\begin{aligned}\Delta S^\circ_{\mathrm{r}} &= S^\circ_{\mathrm{m}}(\mathrm{C_6H_{12}, l}) - [6S^\circ_{\mathrm{m}}(\mathrm{C, s}) + 6S^\circ_{\mathrm{m}}(\mathrm{H_2, g})]\\ &= 204.4\ \mathrm{J\cdot K^{-1}\cdot mol^{-1}} - [6(5.740\ \mathrm{J\cdot K^{-1}\cdot mol^{-1}})\\ &\qquad + 6(130.68\ \mathrm{J\cdot K^{-1}\cdot mol^{-1}})]\\ &= -614.1\ \mathrm{J\cdot K^{-1}\cdot mol^{-1}}\end{aligned}$$

The compound is less stable at higher temperatures.

(c) $\mathrm{P(s)} + \frac{3}{2}\,\mathrm{Cl_2(g)} \longrightarrow \mathrm{PCl_3(g)}$

$$\begin{aligned}\Delta S^\circ_{\mathrm{r}} &= S^\circ_{\mathrm{m}}(\mathrm{PCl_3, g}) - [S^\circ_{\mathrm{m}}(\mathrm{P, s}) + \tfrac{3}{2}S^\circ_{\mathrm{m}}(\mathrm{Cl_2, g})]\\ &= 311.78\ \mathrm{J\cdot K^{-1}\cdot mol^{-1}} - [41.09\ \mathrm{J\cdot K^{-1}\cdot mol^{-1}}\\ &\qquad + \tfrac{3}{2}(223.07\ \mathrm{J\cdot K^{-1}\cdot mol^{-1}})]\\ &= -63.92\ \mathrm{J\cdot K^{-1}\cdot mol^{-1}}\end{aligned}$$

The compound is less stable at higher temperatures.

(d) N_2(g) + 2 H_2(g) $\longrightarrow$ N_2H_4(l)

$$\begin{aligned}\Delta S^\circ_{\mathrm{r}} &= S^\circ_{\mathrm{m}}(\mathrm{N_2H_4, l}) - [S^\circ_{\mathrm{m}}(\mathrm{N_2, g}) + 2S^\circ_{\mathrm{m}}(\mathrm{H_2, g})]\\ &= 121.21\ \mathrm{J\cdot K^{-1}\cdot mol^{-1}} - [191.61\ \mathrm{J\cdot K^{-1}\cdot mol^{-1}}\\ &\qquad + 2(130.68\ \mathrm{J\cdot K^{-1}\cdot mol^{-1}})]\\ &= -331.76\ \mathrm{J\cdot K^{-1}\cdot mol^{-1}}\end{aligned}$$

The compound is less stable at higher temperatures.

7.58 (a) $CH_4(g) + H_2O(g) \longrightarrow CO(g) + 3\ H_2(g)$

$$\Delta S^\circ_r = S^\circ_m(CO, g) + 2S^\circ_m(H_2, g) - [S^\circ_m(CH_4, g) + S^\circ_m(H_2O, g)]$$
$$= 197.67\ J\cdot K^{-1}\cdot mol^{-1} + 3(130.68\ J\cdot K^{-1}\cdot mol^{-1})$$
$$- [186.26\ J\cdot K^{-1}\cdot mol^{-1} + 188.83\ J\cdot K^{-1}\cdot mol^{-1}]$$
$$= +\ 214.62\ J\cdot K^{-1}\cdot mol^{-1}$$

$$\Delta H^\circ_r = \Delta H^\circ_f(CO, g) - [\Delta H^\circ_f(CH_4, g) + \Delta H^\circ_f(H_2O, g)]$$
$$= (-110.53\ kJ\cdot mol^{-1}) - [(-74.81\ kJ\cdot mol^{-1}) + (-241.82\ kJ\cdot mol^{-1})]$$
$$= +206.10\ kJ\cdot mol^{-1}$$

$$\Delta G^\circ_r = \Delta G^\circ_f(CO, g) - [\Delta G^\circ_f(CH_4, g) + \Delta G^\circ_f(H_2O, g)]$$
$$= (-137.17\ kJ\cdot mol^{-1}) - [(-50.72\ kJ\cdot mol^{-1})$$
$$+ (-228.57\ kJ\cdot mol^{-1})]$$
$$= +142.12\ kJ\cdot mol^{-1}$$

ΔG°_r can also be calculated from ΔS°_r and ΔH°_r using the relationship:

$$\Delta G^\circ_r = \Delta H^\circ_r - T\Delta S^\circ_r = +206.1\ kJ\cdot mol^{-1}$$
$$- (298\ K)(+214.62\ J\cdot K^{-1}\cdot mol^{-1})$$
$$= +142.1\ kJ\cdot mol^{-1}$$

(b) $NH_4NO_3(s) \longrightarrow N_2O(g) + 2\ H_2O(g)$

$$\Delta S^\circ_r = S^\circ_m(N_2O, g) + 2S^\circ_m(H_2O, g) - S^\circ_m(NH_4NO_3, s)$$
$$= 219.85\ J\cdot K^{-1}\cdot mol^{-1} + 2(188.83\ J\cdot K^{-1}\cdot mol^{-1})$$
$$- 151.08\ J\cdot K^{-1}\cdot mol^{-1}$$
$$= +446.43\ J\cdot K^{-1}\cdot mol^{-1}$$

$$\Delta H^\circ_r = \Delta H^\circ_f(N_2O, g) + 2\Delta H^\circ_f(H_2O, g) - [\Delta H^\circ_f(NH_4NO_3, s)]$$
$$= 82.05\ kJ\cdot mol^{-1} + 2(-241.82\ kJ\cdot mol^{-1}) - [-365.56\ kJ\cdot mol^{-1}]$$
$$= -36.03\ kJ\cdot mol^{-1}$$

$$\Delta G^\circ_r = \Delta G^\circ_f(N_2O, g) + 2\Delta G^\circ_f(H_2O, g) - [\Delta G^\circ_f(NH_4NO_3, s)]$$
$$= 104.20\ kJ\cdot mol^{-1} + 2(-228.57\ kJ\cdot mol^{-1}) - [-183.87\ kJ\cdot mol^{-1}]$$
$$= -169.07\ kJ\cdot mol^{-1}$$

ΔG°_r can also be calculated from ΔS°_r and ΔH°_r using the relationship:

$$\Delta G°_r = \Delta H°_r - T\Delta S°_r$$
$$= -36.03\ \text{kJ}\cdot\text{mol}^{-1} - (298\ \text{K})(446.43\ \text{J}\cdot\text{K}^{-1}\cdot\text{mol}^{-1})/(1000\ \text{J}\cdot\text{kJ}^{-1})$$
$$= -169.07\ \text{kJ}\cdot\text{mol}$$

7.60 In order to find $\Delta G°_r$ at a temperature other than 298 K, we must first calculate $\Delta H°_r$ and $\Delta S°_r$ and then use the relationship $\Delta G°_r = \Delta H°_r + T\Delta S°_r$ to calculate $\Delta G°_r$.

(a) $$\Delta H°_r = \Delta H°_f(CH_3NH_2, g) - [\Delta H°_f(HCN, g)$$
$$= -22.97\ \text{kJ}\cdot\text{mol}^{-1} - [135.1\ \text{kJ}\cdot\text{mol}^{-1}]$$
$$= -158.1\ \text{kJ}\cdot\text{mol}^{-1}$$

$$\Delta S°_r = S°_m(CH_3NH_2, g) - [S°_m(HCN, g) + 2S°_m(H_2, g)]$$
$$= 243.41\ \text{J}\cdot\text{K}^{-1}\cdot\text{mol}^{-1}$$
$$- [201.78\ \text{J}\cdot\text{K}^{-1}\cdot\text{mol}^{-1} + 2(130.68\ \text{J}\cdot\text{K}^{-1}\cdot\text{mol}^{-1})]$$
$$= -219.73\ \text{J}\cdot\text{K}^{-1}\cdot\text{mol}^{-1}$$

$$\Delta G°_r = -158.1\ \text{kJ}\cdot\text{mol}^{-1}$$
$$- (223\ \text{K})(-219.73\ \text{J}\cdot\text{K}^{-1}\cdot\text{mol}^{-1})/(1000\ \text{J}\cdot\text{kJ}^{-1})$$
$$= -109.1\ \text{kJ}\cdot\text{mol}^{-1}$$

In order to determine over what range the reaction will be spontaneous, we consider the relative signs of $\Delta H°_r$ and $\Delta S°_r$ and their effect on $\Delta G°_r$. Because $\Delta H°_r$ is negative and $\Delta S°_r$ is also negative, we expect the reaction to be spontaneous at low temperatures, where the term $T\Delta S°_r$ will be less than $\Delta H°_r$. To find the temperature of the cutoff, we calculate the temperature at which $\Delta G°_r = 0$. For this reaction, this temperature is

$$\Delta G°_r = 0 = -158.1\ \text{kJ}\cdot\text{mol}^{-1} - (T)(-219.73\ \text{J}\cdot\text{K}^{-1}\cdot\text{mol}^{-1})/(1000\ \text{J}\cdot\text{kJ}^{-1})$$
$$T = 719.4\ \text{K}.$$

The reaction should be spontaneous below 719.4 K. (Although the number of significant figures here is calculated according to the numbers given, the actual number of figures is less, because the values of reaction enthalpy and entropy are not exactly constant with temperature, especially over a larger temperature range such as this.)

(b) $\Delta H^\circ_r = \Delta H^\circ_f(Cu^{2+}, aq) - [2\Delta H^\circ_f(Cu^+, aq)]$

$= 64.77\ kJ \cdot mol^{-1} - [2(71.67\ kJ \cdot mol^{-1})]$

$= -78.57\ kJ \cdot mol^{-1}$

$\Delta S^\circ_r = S^\circ_m(Cu, s) + S^\circ_m(Cu^{2+}, aq) - [2S^\circ_m(Cu^+, aq)]$

$= 33.15\ J \cdot K^{-1} \cdot mol^{-1} + (-99.6\ J \cdot K^{-1} \cdot mol^{-1})$

$- [2(40.6\ J \cdot K^{-1} \cdot mol^{-1})]$

$= -147.6\ J \cdot K^{-1} \cdot mol^{-1}$

$\Delta G^\circ_r = -78.57\ kJ \cdot mol^{-1} - (223\ K)(-146.6\ J \cdot K^{-1} \cdot mol^{-1})/(1000\ J \cdot kJ^{-1})$

$= -45.9\ kJ \cdot mol^{-1}$

Because ΔH°_r is negative and ΔS°_r is negative, the reaction will be spontaneous at lower temperatures as in (a).

$\Delta G^\circ_r = 0 = -78.57\ kJ \cdot mol^{-1} - (T)(-147.6\ J \cdot K^{-1} \cdot mol^{-1})/(1000\ J \cdot kJ^{-1})$

$T = 532\ K$

The reaction is spontaneous at temperatures below 532 K.

(c) $\Delta H^\circ_r = \Delta H^\circ_f(CaF_2, s) + 2\Delta H^\circ_f(HCl, g)$

$- [\Delta H^\circ_f(CaCl_2, s) + 2\Delta H^\circ_f(HF, g)]$

$= (-1219.6\ kJ \cdot mol) + 2(-92.31\ kJ \cdot mol^{-1})$

$- [(-795.8\ kJ \cdot mol^{-1}) + 2(-271.1\ kJ \cdot mol^{-1})]$

$= -66.2\ kJ \cdot mol^{-1}$

$\Delta S^\circ_r = S^\circ_m(CaF_2, s) + 2S^\circ_m(HCl, g) - [S^\circ_m(CaCl_2, s) + 2S^\circ_m(HF, g)]$

$= 68.87\ J \cdot K^{-1} \cdot mol^{-1} + 2(186.91\ J \cdot K^{-1} \cdot mol^{-1})$

$- [104.6\ J \cdot K^{-1} \cdot mol^{-1} + 2(173.78\ J \cdot K^{-1} \cdot mol^{-1})]$

$= -9.5\ J \cdot K^{-1} \cdot mol^{-1}$

$\Delta G^\circ_r = -66.2\ kJ \cdot mol^{-1} - (223\ K)(-9.5\ J \cdot K^{-1} \cdot mol^{-1})/(1000\ J \cdot kJ^{-1})$

$= -64.1\ kJ \cdot mol^{-1}$

Because ΔH°_r is negative and ΔS°_r is negative, the reaction will be spontaneous at lower temperatures as in (a).

$\Delta G^\circ_r = 0 = -66.2\ kJ \cdot mol^{-1} - (T)(-9.5\ J \cdot K^{-1} \cdot mol^{-1})/(1000\ J \cdot kJ^{-1})$

$T = 6.97 \times 10^3\ K$

The reaction is spontaneous at temperatures below 6.97×10^3 K, which essentially covers the entire accessible temperature range.

7.62 The balanced chemical reaction is:

$$2AlCl_3(g) + \tfrac{9}{2}O_2(g) \longrightarrow Al_2O_3(s) + 6ClO(g)$$

and ΔG°_r for the reaction at 2000 K is:

$$\Delta G^{\circ}_r = -1034 \text{ kJ} \cdot \text{mol}^{-1} + 6(75 \text{ kJ} \cdot \text{mol}^{-1}) - 2(-467 \text{ kJ} \cdot \text{mol}^{-1})$$
$$\Delta G^{\circ}_r = 350 \text{ kJ} \cdot \text{mol}^{-1}$$

Given the sign of ΔG°_r, it does not appear that the reaction will be spontaneous and, therefore, this reaction does not appear to be a good candidate for rocket propulsion.

7.64 (a) and (b) Normally one would expect dissolution of a substance in another to increase the entropy of the substance. The ability of acetic acid to hydrogen bond to water molecules makes the solution more ordered than in the pure liquid in spite of the mixing.

7.66 1,2-difluorobenzene has six possible orientations:

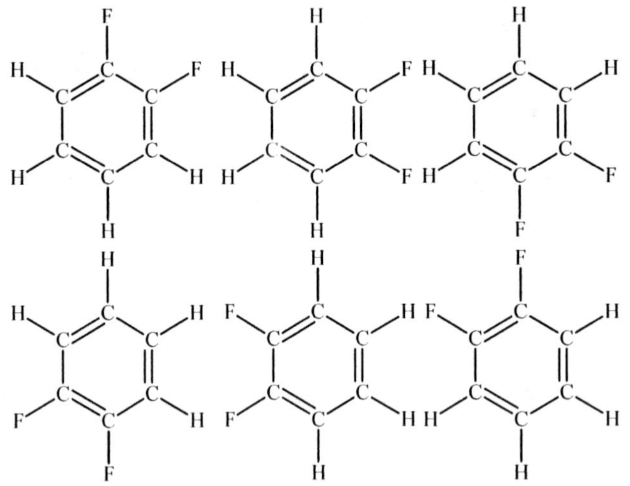

1,3-difluorobenzene also has six possible orientations:

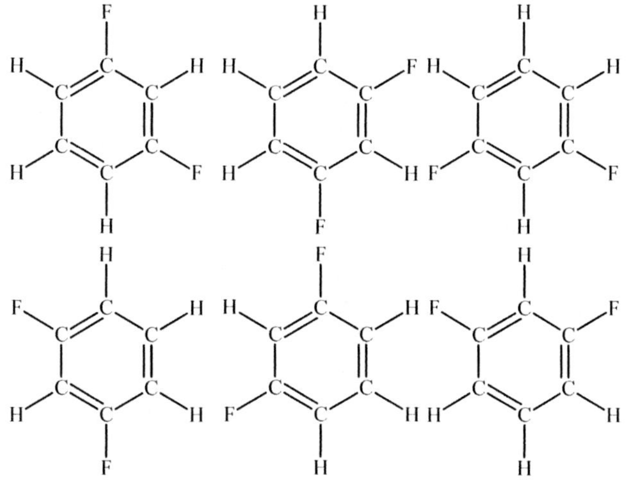

1,4-difluorobenzene has only three possible orientations:

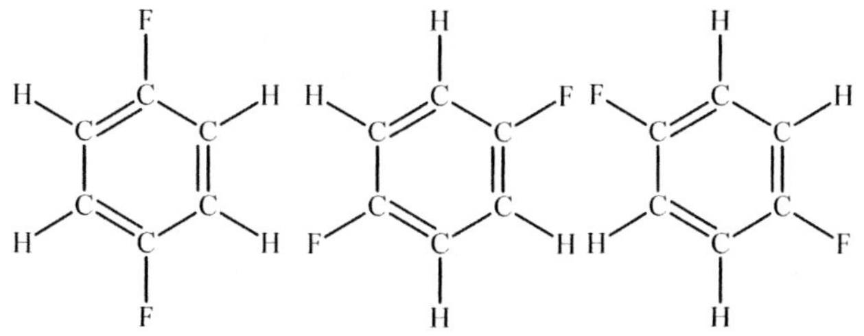

The least residual molar entropy will be exhibited for the compound that has the fewest possible orientations, in this case 1,4-difluorobenzene. The 1,2- and 1,3- isomers should have a higher residual entropy (assuming that all three are disordered) and the residual entropies should be about the same for these two compounds.

7.68 First, calculate the ΔG°_r for both reactions:

(a) $\Delta G^\circ_r = 2\Delta G^\circ_f(\text{CO, g}) - [\Delta G^\circ_r(\text{TiO}_2\text{, s})]$
$= 2(-200\ \text{kJ}\cdot\text{mol}^{-1}) - [(-762\ \text{kJ}\cdot\text{mol}^{-1})]$
$= +\ 362\ \text{kJ}\cdot\text{mol}^{-1}$

(b) $\Delta G^\circ_r = \Delta G^\circ_f(\text{CO}_2\text{, g}) - [\Delta G^\circ_r(\text{TiO}_2\text{, s})]$
$= (-396\ \text{kJ}\cdot\text{mol}^{-1}) - [-762\ \text{kJ}\cdot\text{mol}^{-1}]$
$= +\ 362\ \text{kJ}\cdot\text{mol}^{-1}$

Neither reaction is spontaneous, so TiO_2 cannot be reduced by carbon at 1000 K.

7.70 (a) $\Delta H°_r = \Delta H°_f(C_2H_6, g) - [\Delta H°_f(C_2H_2, g)]$

$$= (-84.68\ kJ\cdot mol^{-1}) - [226.73\ kJ\cdot mol^{-1}]$$
$$= -311.41\ kJ\cdot mol^{-1}$$

$$\Delta S°_r = S°_m(C_2H_6, g) - [S°_m(C_2H_2, g) + 2\,S°_m(H_2, g)]$$
$$= 229.60\ J\cdot K^{-1}\cdot mol^{-1}$$
$$- [200.94\ J\cdot K^{-1}\cdot mol^{-1} + 2(130.68\ J\cdot K^{-1}\cdot mol^{-1})]$$
$$= -232.70\ J\cdot K^{-1}\cdot mol^{-1}$$

(b) $\Delta G°_r = \Delta H°_r - T\Delta S°_r$

$$= -311.41\ kJ\cdot mol^{-1}$$
$$- (298\ K)(-232.7\ J\cdot K^{-1}\cdot mol^{-1})/(1000\ J\cdot kJ^{-1})$$
$$= -242.07\ kJ\cdot mol^{-1}$$

(c) The entropy change of the reaction is negative, because there are fewer moles of gas on the reactant side of the equation than on the product side. The reaction enthalpy is exothermic, because one obtains more energy in the formation of four C—H bonds than one gives up to break 2 H—H bonds.

(d) $\Delta G°_r = \Delta H°_r - T\Delta S°_r$

$$= -311.41\ kJ\cdot mol^{-1} - (T)(-232.7\ J\cdot K^{-1}\cdot mol^{-1})/(1000\ J\cdot kJ^{-1}) = 0$$

T = 1338 K

(e) This is the temperature at which the reaction switches from being spontaneous to becoming nonspontaneous.

7.72 (a) First, balance the equation for the combustion of 8.50 mol of $C_6H_6(l)$ to give carbon dioxide and water vapor:

$$8.5\ C_6H_6(l) + 63.75\ O_2(g) \longrightarrow 51\ CO_2(g) + 25.5\ H_2O(g)$$

The work term will be dominated by the change in the number of moles of gas, which in this case is $76.5 \text{ moles} - 63.75 \text{ moles} = 12.75 \text{ moles}$. Work will be given by

$$
\begin{aligned}
w &= -P\Delta V \\
\Delta V &= \frac{RT}{P}\Delta n \\
w &= -P\frac{RT}{P}\Delta n \\
&= -RT\Delta n \\
&= -(8.314 \text{ J}\cdot\text{K}^{-1}\cdot\text{mol}^{-1})(298 \text{ K})(12.76 \text{ mol}) \\
&= -31.6 \text{ kJ}
\end{aligned}
$$

(b) $$
\begin{aligned}
\Delta H^\circ &= 51(-393.51 \text{ kJ}\cdot\text{mol}^{-1}) + 25.5(-241.82 \text{ kJ}\cdot\text{mol}^{-1}) \\
&\quad - 8.5(49.0 \text{ kJ}\cdot\text{mol}^{-1}) \\
&= -26651 \text{ kJ}
\end{aligned}
$$

(c) $\Delta U^\circ = \Delta H^\circ + w = -26650 \text{ kJ} - 31.6 \text{ kJ} = -26680 \text{ kJ}$

7.74 (a) Reactions with negative reaction free energies are thermodynamically favored, but the thermodynamics will not tell us how fast a process takes place. For example, some reactions with large negative free energies do not happen unless initiated, as in the case of the reaction of hydrogen gas with oxygen gas to produce water.

(b) This statement is false because the sample of the element must be in its standard state. Not all forms of the element at a given temperature have the same energy; the one chosen as the standard state will have the lowest energy.

(c) False. For this process, ΔH°_r is a negative number, ΔS°_r will be positive because the number of moles of gas increases. According to the relationship $\Delta G^\circ_r = \Delta H^\circ_r - T\Delta S^\circ_r$, if ΔH°_r is negative and ΔS°_r is positive, ΔG°_r must be negative.

7.76 According to the second law of thermodynamics, the formation of complex molecules from simpler precursors would not be spontaneous, because such processes create order from disorder. If there is an external input of energy, however, a more ordered system could be created. One of the challenges remaining to evolutionary biology is to explain how the exceedingly complex and highly organized biological structures were created from randomly occurring chemical reactions.

7.78 The reactions are

(a) $CH_3OH(l) + CO(g) \longrightarrow CH_3COOH(l)$

(b) $C_2H_5OH(l) + O_2(g) \longrightarrow CH_3COOH(l) + H_2O(l)$

(c) $CH_4(g) + CO_2(g) \longrightarrow CH_3COOH(l)$

To understand these three reactions completely, we must calculate $\Delta H°_r$, $\Delta S°_r$, and $\Delta G°_r$ for each reaction:

(a)
$$\begin{aligned}\Delta H°_r &= \Delta H°_f(CH_3COOH, l) \\ &\quad - [\Delta H°_f(CH_3OH, l) + \Delta H°_f(CO, g)] \\ &= (-484.5\ kJ\cdot mol^{-1}) - [(-238.86\ kJ\cdot mol^{-1}) \\ &\quad + (-110.53\ kJ\cdot mol^{-1})] \\ &= -135.1\ kJ\cdot mol^{-1}\end{aligned}$$

$$\begin{aligned}\Delta S°_r &= S°_m(CH_3COOH, l) - [S°_m(CH_3OH, l) + S°_m(CO, g)] \\ &= 159.8\ J\cdot K^{-1}\cdot mol^{-1} - [126.8\ J\cdot K^{-1}\cdot mol^{-1} \\ &\quad + 197.67\ J\cdot K^{-1}\cdot mol^{-1}] \\ &= -164.7\ J\cdot K^{-1}\cdot mol^{-1}\end{aligned}$$

$$\begin{aligned}\Delta G°_r &= \Delta G°_f(CH_3COOH, l) - [\Delta G°_f(CH_3OH, l) + \Delta G°_f(CO, g)] \\ &= (-389.9\ kJ\cdot mol^{-1}) - [(-166.27\ kJ\cdot mol^{-1}) \\ &\quad + (-137.17\ kJ\cdot mol^{-1})] \\ &= -86.5\ kJ\cdot mol^{-1}\end{aligned}$$

(b) $\Delta H°_r = \Delta H°_f(CH_3COOH, l)$

$$+ \Delta H°_f(H_2O, l) - [\Delta H°_f(C_2H_5OH, l)]$$
$$= (-484.5 \text{ kJ}\cdot\text{mol}^{-1}) + (-285.83 \text{ kJ}\cdot\text{mol}^{-1})$$
$$- [-277.69 \text{ kJ}\cdot\text{mol}^{-1}]$$
$$= -492.6 \text{ kJ}\cdot\text{mol}^{-1}$$

$$\Delta S°_r = S°_m(CH_3COOH, l) + S°_m(H_2O(l))$$
$$- [S°_m(C_2H_5OH, l) + S°_m(O_2, g)]$$
$$= 159.8 \text{ J}\cdot\text{K}^{-1}\cdot\text{mol}^{-1} + 69.91 \text{ J}\cdot\text{K}^{-1}\cdot\text{mol}^{-1}$$
$$- [160.7 \text{ J}\cdot\text{K}^{-1}\cdot\text{mol}^{-1} + 205.14 \text{ J}\cdot\text{K}^{-1}\cdot\text{mol}^{-1}]$$
$$= -136.1 \text{ J}\cdot\text{K}^{-1}\cdot\text{mol}^{-1}$$

$$\Delta G°_r = \Delta G°_f(CH_3COOH, l) + \Delta G°_f(H_2O, l) - [\Delta G°_f(C_2H_5OH, l)]$$
$$= (-389.9 \text{ kJ}\cdot\text{mol}^{-1}) + (-237.13 \text{ kJ}\cdot\text{mol}^{-1}) - [-174.78 \text{ kJ}\cdot\text{mol}^{-1}]$$
$$= -452.3 \text{ kJ}\cdot\text{mol}^{-1}$$

(c) $\Delta H°_r = \Delta H°_f(CH_3COOH, l)$

$$- [\Delta H°_f(CH_4, g) + \Delta H°_f(CO_2, g)]$$
$$= (-484.5 \text{ kJ}\cdot\text{mol}^{-1}) - [(-74.81 \text{ kJ}\cdot\text{mol}^{-1}) + (-393.51 \text{ kJ}\cdot\text{mol}^{-1})]$$
$$= -16.2 \text{ kJ}\cdot\text{mol}^{-1}$$

$$\Delta S°_r = S°_m(CH_3COOH, l) - [S°_m(CH_4, g) + S°_m(CO_2, g)]$$
$$= 159.8 \text{ J}\cdot\text{K}^{-1}\cdot\text{mol}^{-1} - [186.26 \text{ J}\cdot\text{K}^{-1}\cdot\text{mol}^{-1} + 213.74 \text{ J}\cdot\text{K}^{-1}\cdot\text{mol}^{-1}]$$
$$= -240.2 \text{ J}\cdot\text{K}^{-1}\cdot\text{mol}^{-1}$$

$$\Delta G°_r = \Delta G°_f(CH_3COOH, l) - [\Delta G°_f(CH_4, g) + \Delta G°_f(CO_2, g)]$$
$$= (-389.9 \text{ kJ}\cdot\text{mol}^{-1}) - [(-50.72 \text{ kJ}\cdot\text{mol}^{-1}) + (-394.36 \text{ kJ}\cdot\text{mol}^{-1})]$$
$$= +55.2 \text{ kJ}\cdot\text{mol}^{-1}$$

It is clear from these numbers that the second reaction, the oxidation of ethanol, is by far the most favorable thermodynamically. The addition of CO to methanol is favorable but less so than the oxidation of ethanol. The addition of carbon dioxide to methane is not thermodynamically favored.

7.80 (a) $\frac{1}{2} N_2(g) + \frac{3}{2} H_2(g) \longrightarrow NH_3(g)$

(b) The $\Delta H°_r$ of this reaction is simply the standard heat of formation of ammonia gas, which is $-46.11 \text{ kJ}\cdot\text{mol}^{-1}$.

$$\Delta S°_r = S°(NH_3, g) - [\tfrac{1}{2}S°(N_2, g) + \tfrac{3}{2}S°(H_2, g)]$$
$$= 192.45\ J \cdot K^{-1} \cdot mol^{-1}$$
$$- [\tfrac{1}{2}(191.61\ J \cdot K^{-1} \cdot mol^{-1}) + \tfrac{3}{2}(130.68\ J \cdot K^{-1} \cdot mol^{-1})]$$
$$= -99.38\ J \cdot K^{-1} \cdot mol^{-1}$$

(c) These values are calculated using the relationship $\Delta G°_r = \Delta H°_r - T\Delta S°_r$. The following values of $\Delta G°_r$ are calculated at the different temperatures:

150 K −31.20 kJ

298 K −16.49 kJ

350 K −11.33 kJ

(d) The thermodynamics of this process show that as the temperature is raised, the reaction becomes less favorable. Raising the temperature will increase the rate but will also reduce the amount of products present at equilibrium. The actual reaction conditions employed are a compromise between the rate and thermodynamic considerations.

7.82 For any reaction, $\Delta G°_r = \Delta H°_r - T\Delta S°_r$. If a liquid boils above room temperature, $\Delta G°_r$ for this phase change must be positive at room temperature. $\Delta S°_r$ for a vaporization reaction will generally be positive due to the increase in disorder leading to a negative $-T\Delta S°_r$ term. Therefore, for $\Delta G°_r$ to be positive, $\Delta H°_r$ must be greater than $T\Delta S°_r$ or:

$$\Delta H°_r\left(kJ \cdot mol^{-1}\right) > (298\ K)\left(\Delta S°_r\left(J \cdot mol^{-1} \cdot K^{-1}\right)\right)\left(\tfrac{1\ kJ}{1000\ J}\right)$$
$$\cong 0.3 \cdot \Delta S°_r\left(J \cdot mol^{-1} \cdot K^{-1}\right)$$

Therefore, for the liquid to have a boiling point above room temperature, $\Delta H°_r$ in kJ mol^{-1} must be greater than $\Delta S°_r$ in J mol^{-1} K^{-1}.

7.84 Adenosine triphosphate is a nucleotide that can be viewed as being composed of three subunits: the base adenine, the sugar ribose, and the triphosphate group, as shown below. The conversion of ATP to ADP involves the removal of one of the phosphate groups by hydrolysis. The $\Delta G°$ of this reaction is -30 kJ. In some parts of the cell where energy-producing reactions occur (such as the oxidation of glucose), ADP is converted into ATP (the energy produced is used to drive the hydrolysis reaction in the reverse direction). ATP can then be transferred to other parts of the cell where it can undergo hydrolysis, thereby releasing the energy stored in the bonding of the phosphate to the diphosphate group. These reactions are catalyzed by enzymes that bring the necessary molecules together in order to cause the overall desired reaction.

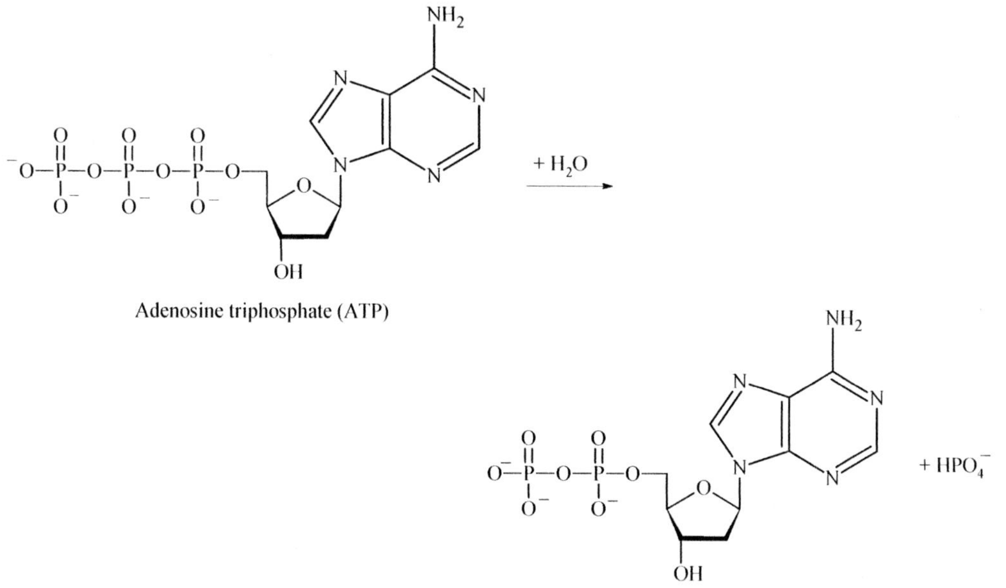

7.86 The hydrogenation reaction corresponds to the general reaction:

$$H_2C{=}CH(R) + H_2(g) \longrightarrow H_3C{-}CH_2(R)$$

where R is H, CH_3, CH_2CH_3, $CH_2CH_2CH_3$, or C_6H_5 for this problem. The calculations of $\Delta G°_r$ and $\Delta H°_r$ are based on the general equations:

$$\Delta G°_r = \Delta G°_f(H_3C{-}CH_2(R)) - \Delta G°_f(H_2C{=}CH(R))$$

and $\Delta H°_r = \Delta H°_f(H_3C{-}CH_2(R)) - \Delta H°_f(H_2C{=}CH(R))$

Using these equations and numbers obtained from the literature, we can calculate the following values:

Compound	$\Delta G^\circ_f\,(\mathrm{kJ \cdot mol^{-1}})$	$\Delta H^\circ_f\,(\mathrm{kJ \cdot mol^{-1}})$
ethene	68.15	52.26
ethane	−32.82	−84.68
propene	62.78	20.42
propane	−23.49	−103.85
1-butene	71.30	−0.13
butane	−17.03	−126.15
1-pentene	79.12	−20.92
pentane	−8.20	−146.44
styrene	213.8	147.8
ethyl benzene	130.6	29.8

ΔS°_r is calculated from $\Delta G^\circ_f = \Delta H^\circ_f(\mathrm{kJ \cdot mol^{-1}}) - T\Delta S^\circ_r$

Hydrogenation reaction	ΔG°_f $(\mathrm{kJ \cdot mol^{-1}})$	ΔH°_f $(\mathrm{kJ \cdot mol^{-1}})$	ΔS°_r $(\mathrm{J \cdot K^{-1} \cdot mol^{-1}})$
ethene	−100.97	−136.94	−121
propene	−86.27	−124.27	−127
1-butene	−88.33	−126.02	−126
1-pentene	−87.32	−125.52	−128
styrene	−83.80	−118.0	−115

The data are reasonably consistent but there are some differences. The molecules that are most similar in structure give the closest agreement (R = H vs. R = alkyl group vs. R = aromatic ring). According to bond enthalpy calculations, we would have expected all of these numbers to be the same, because the same number and type of bonds are broken and formed in each case. It is not unusual to find an error of 10 to 15 percent in bond enthalpy calculations. Note also, that the entropy changes are all negative as expected, based on the loss of one mole of gas as a result of going from reactants to products.

7.88 The reactions (2) and (3) in this problem are used to drive the regeneration of ATP. For every three molecules of NADH that undergo reaction:

$$3\ \mathrm{NADH(aq)} \rightarrow \mathrm{NAD^+(aq)} + \mathrm{H^+(aq)} + 2e^-$$

$$\Delta G^\circ_r = 3(-158.2\ \mathrm{kJ})$$

$$+ \quad \tfrac{1}{2}\mathrm{O_2(g)} + 2\mathrm{H^+(aq)} + 2e^- \rightarrow \mathrm{H_2O(l)}$$

$$\Delta G^\circ_r = 3(-61.9\ \mathrm{kJ})$$

$$3\ \mathrm{NADH(aq)} + \tfrac{1}{2}\mathrm{O_2(g)} + 2\mathrm{H^+(aq)} \rightarrow \mathrm{H_2O(l)} + \mathrm{NAD^+(aq)} + \mathrm{H^+(aq)}$$

$$\Delta G^\circ_r = -660.3\ \mathrm{kJ}$$

Therefore, for every three NADH molecules, 660.3 kJ / 30.5 kJ = 21.6 ATP molecules would be regenerated if all of the free energy released were used to drive reaction (1).

CHAPTER 8
PHYSICAL EQUILIBRIA

8.2 The volume of the supply room (neglecting any contents of the room) is $3.0\ \text{m} \times 2.0\ \text{m} \times 2.0\ \text{m} = 12\ \text{m}^3$ or $12\ \text{m}^3 \times 1000\ \text{L} \cdot \text{m}^{-3} = 12\ 000\ \text{L}$. The ideal gas law can be used to calculate the mass of mercury present:

$PV = nRT$

let m = mass of mercury

$$\left(\frac{0.227\ \text{Pa}}{1.01325 \times 10^5\ \text{Pa} \cdot \text{atm}^{-1}}\right)(12\ 000\ \text{L})$$

$$= \left(\frac{m}{200.59\ \text{g} \cdot \text{mol}^{-1}}\right)(0.082\ 06\ \text{L} \cdot \text{atm} \cdot \text{K}^{-1} \cdot \text{mol}^{-1})(298\ \text{K})$$

$$m = \frac{(0.227\ \text{Pa})(12\ 000\ \text{L})(200.59\ \text{g} \cdot \text{mol}^{-1})}{(1.0135 \times 10^5\ \text{Pa} \cdot \text{atm}^{-1})(0.082\ 06\ \text{L} \cdot \text{atm} \cdot \text{K}^{-1} \cdot \text{mol}^{-1})(298\ \text{K})}$$

$m = 0.22\ \text{g}$

8.4 (a) about 72°C; (b) about 58°C

8.6 (a) The quantities of $\Delta H°_{\text{vap}}$ and $\Delta S°_{\text{vap}}$ can be calculated using the relationship

$$\ln P = -\frac{\Delta H°_{\text{vap}}}{R} \cdot \frac{1}{T} + \frac{\Delta S°_{\text{vap}}}{R}$$

Because we have two temperatures with corresponding vapor pressures, we can set up two equations with two unknowns and solve for $\Delta H°_{\text{vap}}$ and $\Delta S°_{\text{vap}}$. If the equation is used as is, P must be expressed in atm, which is the standard reference state. Remember that the value used for P is really activity that, for pressure, is P divided by the reference state of 1 atm so that the quantity inside the ln term is dimensionless.

$$8.314\ \text{J}\cdot\text{K}^{-1}\cdot\text{mol}^{-1}\times\ln\frac{67\ \text{Torr}}{760\ \text{Torr}}=-\frac{\Delta H^\circ_{\text{vap}}}{273.2\ \text{K}}+\Delta S^\circ_{\text{vap}}$$

$$8.314\ \text{J}\cdot\text{K}^{-1}\cdot\text{mol}^{-1}\times\ln\frac{222\ \text{Torr}}{760\ \text{Torr}}=-\frac{\Delta H^\circ_{\text{vap}}}{298.2\ \text{K}}+\Delta S^\circ_{\text{vap}}$$

which give, upon combining terms,

$$-20.19\ \text{J}\cdot\text{K}^{-1}\cdot\text{mol}^{-1}=-0.003\,660\ \text{K}^{-1}\times\Delta H^\circ_{\text{vap}}+\Delta S^\circ_{\text{vap}}$$

$$-10.23\ \text{J}\cdot\text{K}^{-1}\cdot\text{mol}^{-1}=-0.003\,353\ \text{K}^{-1}\times\Delta H^\circ_{\text{vap}}+\Delta S^\circ_{\text{vap}}$$

Subtracting one equation from the other will eliminate the $\Delta S^\circ_{\text{vap}}$ term and allow us to solve for

$\Delta H^\circ_{\text{vap}}$:

$$-9.96\ \text{J}\cdot\text{K}^{-1}\cdot\text{mol}^{-1}=-0.000\,307\times\Delta H^\circ_{\text{vap}}$$

$$\Delta H^\circ_{\text{vap}}=+32.4\ \text{kJ}\cdot\text{mol}^{-1}$$

(b) We can then use $\Delta H^\circ_{\text{vap}}$ to calculate $\Delta S^\circ_{\text{vap}}$ using either of the two equations:

$$-20.19\ \text{J}\cdot\text{K}^{-1}\cdot\text{mol}^{-1}=-0.003\,660\ \text{K}^{-1}\times(+32\,400\ \text{J}\cdot\text{mol}^{-1})+\Delta S^\circ_{\text{vap}}$$

$$\Delta S^\circ_{\text{vap}}=98.4\ \text{J}\cdot\text{K}^{-1}\cdot\text{mol}^{-1}$$

$$-10.23\ \text{J}\cdot\text{K}^{-1}\cdot\text{mol}^{-1}=-0.003\,353\ \text{K}^{-1}\times(+32\,400\ \text{J}\cdot\text{mol}^{-1})+\Delta S^\circ_{\text{vap}}$$

$$\Delta S^\circ_{\text{vap}}=98.4\ \text{J}\cdot\text{K}^{-1}\cdot\text{mol}^{-1}$$

(c) The $\Delta G^\circ_{\text{vap}}$ is calculated using $\Delta G^\circ_{\text{r}}=\Delta H^\circ_{\text{r}}-T\Delta S^\circ_{\text{r}}$

$$\Delta G^\circ_{\text{r}}=+32.4\ \text{kJ}\cdot\text{mol}^{-1}-(298\ \text{K})(98.4\ \text{J}\cdot\text{K}^{-1}\cdot\text{mol}^{-1})/(1000\ \text{J}\cdot\text{kJ}^{-1})$$

$$\Delta G^\circ_{\text{r}}=+3.08\ \text{kJ}\cdot\text{mol}^{-1}$$

(d) The boiling point can be calculated using one of several methods. The easiest to use is the method developed in the last chapter:

$$\Delta G^\circ_{\text{vap}}=\Delta H^\circ_{\text{vap}}-T_{\text{B}}\Delta S^\circ_{\text{vap}}=0$$

$$\Delta H^\circ_{\text{vap}}=T_{\text{B}}\Delta S^\circ_{\text{vap}}\ \text{or}\ T_{\text{B}}=\frac{\Delta H^\circ_{\text{vap}}}{\Delta S^\circ_{\text{vap}}}$$

$$T_B = \frac{32.4\ \text{kJ}\cdot\text{mol}^{-1} \times 1000\ \text{J}\cdot\text{kJ}^{-1}}{98.4\ \text{J}\cdot\text{K}^{-1}\cdot\text{mol}^{-1}} = 329\ \text{K or } 56°\text{C}$$

Alternatively, we could use the relationship $\ln \frac{P_2}{P_1} = -\frac{\Delta H°_{vap}}{R}\left[\frac{1}{T_2} - \frac{1}{T_1}\right]$.

Here, we would substitute, in one of the known vapor pressure points, the value of the enthalpy of vaporization and the condition that $P = 1$ atm at the normal boiling point.

8.8 (a) The quantities $\Delta H°_{vap}$ and $\Delta S°_{vap}$ can be calculated using the relationship $\ln P = -\frac{\Delta H°_{vap}}{R}\cdot\frac{1}{T} + \frac{\Delta S°_{vap}}{R}$

Because we have two temperatures with corresponding vapor pressures (we know that the vapor pressure = 1 atm at the boiling point), we can set up two equations with two unknowns and solve for $\Delta H°_{vap}$ and $\Delta S°_{vap}$. If the equation is used as is, P must be expressed in atm, which is the standard reference state. Remember that the value used for P is really activity that, for pressure, is P divided by the reference state of 1 atm so that the quantity inside the ln term is dimensionless.

$$8.314\ \text{J}\cdot\text{K}^{-1}\cdot\text{mol}^{-1} \times \ln 1 = -\frac{\Delta H°_{vap}}{326.1\ \text{K}} + \Delta S°_{vap}$$

$$8.314\ \text{J}\cdot\text{K}^{-1}\cdot\text{mol}^{-1} \times \ln \frac{96\ \text{Torr}}{760\ \text{Torr}} = -\frac{\Delta H°_{vap}}{273.2\ \text{K}} + \Delta S°_{vap}$$

which give, upon combining terms,

$$0\ \text{J}\cdot\text{K}^{-1}\cdot\text{mol}^{-1} = -0.003\ 067\ \text{K}^{-1} \times \Delta H°_{vap} + \Delta S°_{vap}$$

$$-17.2\ \text{J}\cdot\text{K}^{-1}\cdot\text{mol}^{-1} = -0.003\ 660\ \text{K}^{-1} \times \Delta H°_{vap} + \Delta S°_{vap}$$

Subtracting one equation from the other will eliminate the $\Delta S°_{vap}$ term and allow us to solve for $\Delta H°_{vap}$:

$$+17.2\ \text{J}\cdot\text{K}^{-1}\cdot\text{mol}^{-1} = +0.000\ 593\ \text{K}^{-1} \times \Delta H°_{vap}$$

$$\Delta H°_{vap} = +29.0\ \text{kJ}\cdot\text{mol}^{-1}$$

(b) We can then use ΔH°_{vap} to calculate ΔS°_{vap} using either of the two equations:

$$0 = -0.003\ 066\ \text{K}^{-1} \times (+29\ 000\ \text{J}\cdot\text{mol}^{-1}) + \Delta S^\circ_{vap}$$

$$\Delta S^\circ_{vap} = 88.9\ \text{J}\cdot\text{K}^{-1}\cdot\text{mol}^{-1}$$

$$-17.2\ \text{J}\cdot\text{K}^{-1}\cdot\text{mol}^{-1} = -0.003\ 660\ \text{K}^{-1} \times (+29\ 000\ \text{J}\cdot\text{mol}^{-1}) + \Delta S^\circ_{vap}$$

$$\Delta S^\circ_{vap} = 88.9\ \text{J}\cdot\text{K}^{-1}\cdot\text{mol}^{-1}$$

(c) The vapor pressure at another temperature is calculated using

$$\ln\frac{P_2}{P_1} = -\frac{\Delta H^\circ_{vap}}{R}\left[\frac{1}{T_2} - \frac{1}{T_1}\right]$$

We need to insert the calculated value of the enthalpy of vaporization and one of the known vapor pressure points:

$$\ln\frac{P_{\text{at }35^\circ\text{C}}}{1\ \text{atm}} = -\frac{29\ 000\ \text{J}\cdot\text{mol}^{-1}}{8.314\ \text{J}\cdot\text{K}^{-1}\cdot\text{mol}^{-1}}\left[\frac{1}{308\ \text{K}} - \frac{1}{326.1\ \text{K}}\right]$$

$$P_{\text{at }35^\circ\text{C}} = 0.63\ \text{atm or } 4.8\times10^2\ \text{Torr}$$

8.10 (a) The quantities ΔH°_{vap} and ΔS°_{vap} can be calculated using the relationship $\ln P = -\dfrac{\Delta H^\circ_{vap}}{R}\cdot\dfrac{1}{T} + \dfrac{\Delta S^\circ_{vap}}{R}$

Because we have two temperatures with corresponding vapor pressures, we can set up two equations with two unknowns and solve for ΔH°_{vap} and ΔS°_{vap}. If the equation is used as is, P must be expressed in atm, which is the standard reference state. Remember that the value used for P is really activity that, for pressure, is P divided by the reference state of 1 atm so that the quantity inside the ln term is dimensionless.

$$8.314\ \text{J}\cdot\text{K}^{-1}\cdot\text{mol}^{-1}\times\ln\frac{155\ \text{Torr}}{760\ \text{Torr}} = -\frac{\Delta H^\circ_{vap}}{250.40\ \text{K}} + \Delta S^\circ_{vap}$$

$$8.314\ \text{J}\cdot\text{K}^{-1}\cdot\text{mol}^{-1}\times\ln\frac{485\ \text{Torr}}{760\ \text{Torr}} = -\frac{\Delta H^\circ_{vap}}{273.2\ \text{K}} + \Delta S^\circ_{vap}$$

which give, upon combining terms,

$$-13.22\ \text{J}\cdot\text{K}^{-1}\cdot\text{mol}^{-1} = -0.003\,994\ \text{K}^{-1} \times \Delta H^\circ_{\text{vap}} + \Delta S^\circ_{\text{vap}}$$

$$-3.734\ \text{J}\cdot\text{K}^{-1}\cdot\text{mol}^{-1} = -0.003\,660\ \text{K}^{-1} \times \Delta H^\circ_{\text{vap}} + \Delta S^\circ_{\text{vap}}$$

Subtracting one equation from the other will eliminate the $\Delta S^\circ_{\text{vap}}$ term and allow us to solve for $\Delta H^\circ_{\text{vap}}$:

$$-9.49\ \text{J}\cdot\text{K}^{-1}\cdot\text{mol}^{-1} = -0.000\,334\ \text{K}^{-1} \times \Delta H^\circ_{\text{vap}}$$

$$\Delta H^\circ_{\text{vap}} = +28.4\ \text{kJ}\cdot\text{mol}^{-1}$$

(b) We can then use $\Delta H^\circ_{\text{vap}}$ to calculate $\Delta S^\circ_{\text{vap}}$ using either of the two equations:

$$-13.22\ \text{J}\cdot\text{K}^{-1}\cdot\text{mol}^{-1} = -0.003\,994\ \text{K}^{-1} \times (+28\,400\ \text{J}\cdot\text{mol}^{-1}) + \Delta S^\circ_{\text{vap}}$$

$$\Delta S^\circ_{\text{vap}} = 1.00 \times 10^2\ \text{J}\cdot\text{K}^{-1}\cdot\text{mol}^{-1}$$

$$-3.734\ \text{J}\cdot\text{K}^{-1}\cdot\text{mol}^{-1} = -0.003\,660\ \text{K}^{-1} \times (+28\,400\ \text{J}\cdot\text{mol}^{-1}) + \Delta S^\circ_{\text{vap}}$$

$$\Delta S^\circ_{\text{vap}} = 1.00 \times 10^2\ \text{J}\cdot\text{K}^{-1}\cdot\text{mol}^{-1}$$

(c) The $\Delta G^\circ_{\text{vap}}$ is calculated using $\Delta G^\circ_{\text{r}} = \Delta H^\circ_{\text{r}} - T\Delta S^\circ_{\text{r}}$

$$\Delta G^\circ_{\text{r}} = +28.4\ \text{kJ}\cdot\text{mol}^{-1} - (298\ \text{K})(100\ \text{J}\cdot\text{K}^{-1}\cdot\text{mol}^{-1})/(1000\ \text{J}\cdot\text{kJ}^{-1})$$

$$\Delta G^\circ_{\text{r}} = -1.4\ \text{kJ}\cdot\text{mol}^{-1}$$

Notice that the standard $\Delta G^\circ_{\text{r}}$ is negative, so the vaporization of ClO_2 is spontaneous as expected; under those conditions it is a gas at room temperature.

(d) The boiling point can be calculated using one of several methods. The easiest to use is the one developed in the last chapter:

$$\Delta G^\circ_{\text{vap}} = \Delta H^\circ_{\text{vap}} - T_{\text{B}}\Delta S^\circ_{\text{vap}} = 0$$

$$\Delta H^\circ_{\text{vap}} = T_{\text{B}}\Delta S^\circ_{\text{vap}} \text{ or } T_{\text{B}} = \frac{\Delta H^\circ_{\text{vap}}}{\Delta S^\circ_{\text{vap}}}$$

$$T_{\text{B}} = \frac{28.4\ \text{kJ}\cdot\text{mol}^{-1} \times 1000\ \text{J}\cdot\text{kJ}^{-1}}{100\ \text{J}\cdot\text{K}^{-1}\cdot\text{mol}^{-1}} = 284\ \text{K or } 11^\circ\text{C}$$

Alternatively, we could use the relationship $\ln \dfrac{P_2}{P_1} = -\dfrac{\Delta H^\circ_{\text{vap}}}{R}\left[\dfrac{1}{T_2} - \dfrac{1}{T_1}\right]$.

Here, we would substitute, in one of the known pressure points, the value of the enthalpy of vaporization and the condition that $P = 1$ atm at the normal boiling point.

8.12 (a) The quantities ΔH°_{vap} and ΔS°_{vap} can be calculated using the relationship

$$\ln P = -\frac{\Delta H^\circ_{vap}}{R} \cdot \frac{1}{T} + \frac{\Delta S^\circ_{vap}}{R}$$

Because we have two temperatures with corresponding vapor pressures (we know that the vapor pressure = 1 atm at the boiling point), we can set up two equations with two unknowns and solve for ΔH°_{vap} and ΔS°_{vap}. If the equation is used as is, P must be expressed in atm, which is the standard reference state. Remember that the value used for P is really activity that, for pressure, is P divided by the reference state of 1 atm so that the quantity inside the ln term is dimensionless.

$$8.314\ \text{J}\cdot\text{K}^{-1}\cdot\text{mol}^{-1} \times \ln 1 = -\frac{\Delta H^\circ_{vap}}{311.6\ \text{K}} + \Delta S^\circ_{vap}$$

$$8.314\ \text{J}\cdot\text{K}^{-1}\cdot\text{mol}^{-1} \times \ln \frac{13\ \text{Torr}}{760\ \text{Torr}} = -\frac{\Delta H^\circ_{vap}}{227.94\ \text{K}} + \Delta S^\circ_{vap}$$

which give, upon combining terms,

$$0\ \text{J}\cdot\text{K}^{-1}\cdot\text{mol}^{-1} = -0.003\,209\ \text{K}^{-1} \times \Delta H^\circ_{vap} + \Delta S^\circ_{vap}$$

$$-33.9\ \text{J}\cdot\text{K}^{-1}\cdot\text{mol}^{-1} = -0.004\,387\,1\ \text{K}^{-1} \times \Delta H^\circ_{vap} + \Delta S^\circ_{vap}$$

Subtracting one equation from the other will eliminate the ΔS°_{vap} term and allow us to solve for ΔH°_{vap}:

$$+33.9\ \text{J}\cdot\text{K}^{-1}\cdot\text{mol}^{-1} = +0.001\,178\ \text{K}^{-1} \times \Delta H^\circ_{vap}$$

$$\Delta H^\circ_{vap} = +28.8\ \text{kJ}\cdot\text{mol}^{-1}$$

(b) We can then use ΔH°_{vap} to calculate ΔS°_{vap} using either of the two equations:

$$0 = -0.003\,209\ \mathrm{K^{-1}} \times (+28\,800\ \mathrm{J \cdot mol^{-1}}) + \Delta S^\circ_{\mathrm{vap}}$$

$$\Delta S^\circ_{\mathrm{vap}} = 92.4\ \mathrm{J \cdot K^{-1} \cdot mol^{-1}}$$

$$-33.9\ \mathrm{J \cdot K^{-1} \cdot mol^{-1}} = -0.004\,387\,1\ \mathrm{K^{-1}} \times (+28\,800\ \mathrm{J \cdot mol^{-1}}) + \Delta S^\circ_{\mathrm{vap}}$$

$$\Delta S^\circ_{\mathrm{vap}} = 92.4\ \mathrm{J \cdot K^{-1} \cdot mol^{-1}}$$

(c) The vapor pressure at another temperature is calculated using

$$\ln \frac{P_2}{P_1} = -\frac{\Delta H^\circ_{\mathrm{vap}}}{R}\left[\frac{1}{T_2} - \frac{1}{T_1}\right]$$

We need to insert the calculated value of the enthalpy of vaporization and one of the known vapor pressure points:

$$\ln \frac{P_{\mathrm{at\ 15.0^\circ C}}}{1\ \mathrm{atm}} = -\frac{28\,800\ \mathrm{J \cdot mol^{-1}}}{8.314\ \mathrm{J \cdot K^{-1} \cdot mol^{-1}}}\left[\frac{1}{288.2\ \mathrm{K}} - \frac{1}{311.6\ \mathrm{K}}\right]$$

$$P_{\mathrm{at\ 25.0^\circ C}} = 0.41\ \mathrm{atm\ or\ } 3.1 \times 10^2\ \mathrm{Torr}$$

8.14 Table 6.2 contains the enthalpy of vaporization and the boiling point of ammonia (at which the vapor pressure = 1 atm). Using this data and the equation

$$\ln \frac{P_2}{P_1} = -\frac{\Delta H^\circ_{\mathrm{vap}}}{R}\left[\frac{1}{T_2} - \frac{1}{T_1}\right]$$

$$\ln \frac{P_{215\ \mathrm{K}}}{1} = -\frac{23\,400\ \mathrm{J \cdot mol^{-1}}}{8.314\ \mathrm{J \cdot K^{-1} \cdot mol^{-1}}}\left[\frac{1}{215\ \mathrm{K}} - \frac{1}{239.7\ \mathrm{K}}\right]$$

$$P_{215\ \mathrm{K}} = 0.26\ \mathrm{atm\ or\ } 2.0 \times 10^2\ \mathrm{Torr}$$

8.16 (a) solid; (b) vapor; (c) liquid; (d) equilibrium between solid, liquid, and vapor (triple point)

8.18 (a) close to 10^5 atm; (b) approximately 3600 K; (c) approximately 10^5 atm; (d) The phase diagram indicates that diamonds are not thermodynamically stable (see Chapter 6) under normal conditions; they appear to be so because the rate of conversion is very slow. We say that diamonds are *kinetically inert.*

8.20 (a) graphite $\longrightarrow$ diamond $\longrightarrow$ liquid; (b) The diamond/graphite interface line has a positive slope; hence diamond is denser than graphite. The diamond/liquid line has a negative slope, hence diamond is less dense than liquid carbon. The order is graphite < diamond < liquid.

8.22 (a)

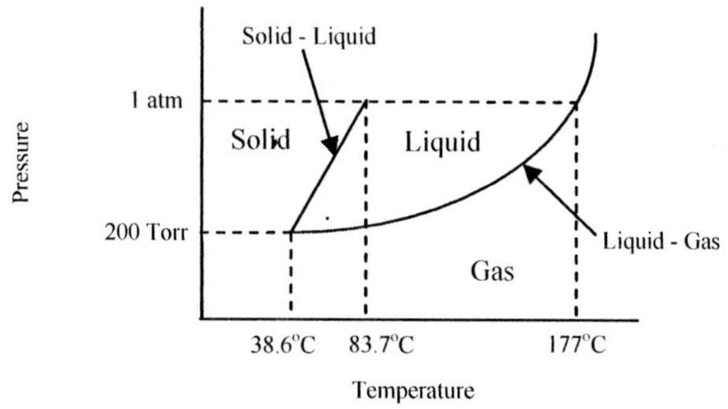

(b) The cooling curve for a sample of this material will resemble this sketch (not to scale):

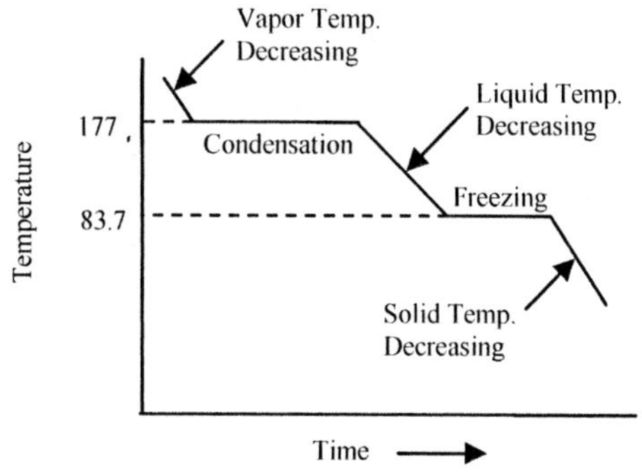

8.24 In the previous problem, we are given the relationship:

$$\frac{dP}{dT} = \frac{\Delta H_{fus}}{T \cdot \Delta V}$$

Therefore, substituting in the values given in the problem:

$$\frac{317 \times 10^5\ \text{Pa}}{1.7\ \text{K}} = \frac{2290\ \text{J} \cdot \text{mol}^{-1}}{235.4\ \text{K} \cdot \Delta V}$$

Solving for ΔV we find $\Delta V = -5.217 \times 10^{-7}\ \text{m}^3\cdot\text{mol}^{-1}$ or $-0.5217\ \text{cm}^3\cdot\text{mol}^{-1}$. Converting to a per gram quantity:

$$\frac{-0.5217\ \text{cm}^3 \cdot \text{mol}^{-1}}{200.6\ \text{g} \cdot \text{mol}^{-1}} = -2.60 \times 10^{-3}\ \text{cm}^3 \cdot \text{g}^{-1}$$

Given the density of liquid mercury, the volume of one gram of liquid mercury is

$$\frac{1\ \text{g}}{13.60\ \text{g} \cdot \text{cm}^{-3}} = 0.07353\ \text{cm}^3$$

The volume of 1 g of solid mercury under the conditions given is then:

$0.07353\ \text{cm}^3 - 2.60 \times 10^{-3}\ \text{cm}^3 = 0.07093\ \text{cm}^3$

giving a density for the solid:

$$\frac{1\ \text{g}}{0.07093\ \text{cm}^3} = 14.1\ \text{g} \cdot \text{cm}^{-3}$$

8.26 (a) water, (b) water, (c) tetrachloromethane

8.28 (a) hydrophilic, hydrogen bonding; (b) hydrophobic, nonpolar; (c) hydrophilic, hydrogen bonding, and dipole-dipole interactions; (d) Cl could be hydrophilic because of possible dipole-dipole interactions (depending on the electronegativity of the atom to which the Cl atom is bonded, as well as the symmetry or asymmetry of the molecule in which it is found). Otherwise, London forces would predominate and Cl would be hydrophobic.

8.30 (a) The solubility of air in water is $7.9 \times 10^{-4}\ \text{mol} \cdot \text{L}^{-1} \cdot \text{atm}^{-1}$ and the molar mass of air (average) = $28.97\ \text{g} \cdot \text{mol}^{-1}$ (see Table 5.1).

solubility = $k_{\text{H}} \times P$

solubility

$$\begin{aligned}(\text{g} \cdot \text{L}^{-1}) &= 1.0\ \text{atm} \times 7.9 \times 10^{-4}\ \text{mol} \cdot \text{L}^{-1} \cdot \text{atm}^{-1} \\ &\quad \times 28.97\ \text{g} \cdot \text{mol}^{-1} \times 10^3\ \text{mg} \cdot \text{g}^{-1} \\ &= 23\ \text{mg} \cdot \text{L}^{-1}\end{aligned}$$

(b) The solubility of He is $3.7 \times 10^{-4}\ \text{mol} \cdot \text{L}^{-1} \cdot \text{atm}^{-1}$.

solubility

$$(\text{g} \cdot \text{L}^{-1}) = 1.0 \text{ atm} \times 3.7 \times 10^{-4} \text{ mol} \cdot \text{L}^{-1} \cdot \text{atm}^{-1}$$
$$\times 4.00 \text{ g} \cdot \text{mol}^{-1} \times 10^{3} \text{ mg} \cdot \text{g}^{-1}$$
$$= 1.48 \text{ mg} \cdot \text{L}^{-1}$$

(c) solubility

$$(\text{g} \cdot \text{L}^{-1}) = \frac{25 \text{ kPa}}{101.325 \text{ kPa} \cdot \text{atm}^{-1}} \times 3.7 \times 10^{-4} \text{ mol} \cdot \text{L}^{-1} \cdot \text{atm}^{-1} \times 4.00 \text{ g} \cdot \text{mol}^{-1}$$

$$\times 10^{3} \text{ mg} \cdot \text{g}^{-1} = 0.37 \text{ mg}$$

8.32 This answer can be calculated from the solubility data, which will give an answer in $\text{mol} \cdot \text{L}^{-1}$. We then use the solution component of the blood volume to determine the total number of moles present, which can be converted to volume using the ideal gas expression:

volume of plasma = 0.45×6.00 L

$$S = k_{\text{H}} \times P = 5.8 \times 10^{-4} \text{ mol} \cdot \text{L}^{-1} \cdot \text{atm}^{-1} \times 10.00 \text{ atm}$$

$$n_{\text{N}_2} = S \times 2.7 \text{ L}$$

$$PV = nRT \text{ or } V = \frac{nRT}{P}$$

$$V_{\text{N}_2} = (5.8 \times 10^{-4} \text{ mol} \cdot \text{L}^{-1} \cdot \text{atm}^{-1})(10.0 \text{ atm})(0.45)(6.00 \text{ L})$$
$$\times \frac{(0.082\,06 \text{ L} \cdot \text{atm} \cdot \text{K}^{-1} \cdot \text{mol}^{-1})(310 \text{K})}{1 \text{ atm}}$$
$$= 0.40 \text{ L}$$

8.34 $CO_2(g) \longrightarrow CO_2(aq)$ + heat (exothermic reaction)

(a) If the pressure of CO_2 is increased, more $CO_2(g)$ will be forced into solution under pressure (Henry's law). The concentration of CO_2 in solution will thus increase. The factor by which the concentration will increase cannot be predicted, however, because the amount of CO_2 present above the solution is unknown, as is the amount of water. (b) If the temperature is raised, the concentration of CO_2 in solution will decrease; the heat added will favor the escape of CO_2 from solution—the reverse process is exothermic. Further, addition of heat increases the speed and energy of the CO_2 molecules, allowing more of them to break out into the gas phase.

8.36 (a) endothermically: a positive enthalpy of solution means enthalpy is absorbed by the system during the dissolution process; (b) $NH_4NO_3(s)$ + heat $\rightleftharpoons NH_4^+(aq) + NO_3^-(aq)$; (c) Given that $\Delta H_L + \Delta H_{hydration} = \Delta H$ of solution, the endothermic values of solution enthalpy result from systems with lattice enthalpies greater than their enthalpies of hydration. Therefore, we expect that for NH_4NO_3 the lattice enthalpy will be greater than the enthalpy of hydration.

8.38 The molar enthalpies of solution are given in Table 8.6. Multiplying these numbers by the number of moles of solid dissolved will give the amount of heat released. The change in temperature will be given by dividing the heat released by the specific heat capacity of the solution.

(a) enthalpy of solution of KCl $= +17.2\ \text{kJ}\cdot\text{mol}^{-1}$

$$\Delta T = -\frac{\dfrac{10.0\ \text{g KCl}}{74.55\ \text{g}\cdot\text{mol}^{-1}\ \text{KCl}} \times (+17\,200\ \text{J}\cdot\text{mol}^{-1})}{(4.18\ \text{J}\cdot\text{K}^{-1}\cdot\text{g}^{-1})(100.0\ \text{g})} = -5.52\ \text{K or } -5.52°\text{C}$$

(b) enthalpy of solution of $MgBr_2 = -185.6\ \text{kJ}\cdot\text{mol}^{-1}$

$$\Delta T = -\frac{\dfrac{10.0\ \text{g MgBr}_2}{184.13\ \text{g}\cdot\text{mol}^{-1}\ \text{MgBr}_2} \times (-185\,600\ \text{J}\cdot\text{mol}^{-1})}{(4.18\ \text{J}\cdot\text{K}^{-1}\cdot\text{g}^{-1})(100.0\ \text{g})}$$

$= +24.1\ \text{K or } +24.1°\text{C}$

(c) enthalpy of solution of $KNO_3 = +34.9\ \text{kJ}\cdot\text{mol}^{-1}$

$$\Delta T = -\frac{\dfrac{10.0\ \text{g KNO}_3}{101.11\ \text{g}\cdot\text{mol}^{-1}\ \text{KNO}_3} \times (+34\,900\ \text{J}\cdot\text{mol}^{-1})}{(4.18\ \text{J}\cdot\text{K}^{-1}\cdot\text{g}^{-1})(100.0\ \text{g})}$$

$= -8.26\ \text{K or } -8.26°\text{C}$

(d) enthalpy of solution of NaOH $= -44.5\ \text{kJ}\cdot\text{mol}^{-1}$

$$\Delta T = -\frac{\dfrac{10.0\text{ g NaOH}}{40.00\text{ g}\cdot\text{mol}^{-1}\text{ NaOH}} \times (-44\,500\text{ J}\cdot\text{mol}^{-1})}{(4.18\text{ J}\cdot\text{K}^{-1}\cdot\text{g}^{-1})(100.0\text{ g})}$$

$$= +26.6\text{ K or } +26.6°\text{C}$$

8.40 The data for the heats of solution are (in $\text{kJ}\cdot\text{mol}^{-1}$) NaF, $1.9\text{ kJ}\cdot\text{mol}^{-1}$; NaCl, $+3.9\text{ kJ}\cdot\text{mol}^{-1}$; NaBr, -0.6; NaI, -7.5. $\Delta H_{\text{soln}} = \Delta H_{\text{L}} + \Delta H_{\text{hyd}}$. As the size of the anion increases, ΔH_{L} decreases (see Table 6.3), but ΔH_{hyd} increases (becomes less negative, Table 8.8), so ΔH_{sol} would be expected to decrease, as the data partly indicate. The decrease in ΔH_{L} outweighs the increase in ΔH_{hyd}, and ΔH_{sol} generally does decrease on proceeding down the group of halides. The exception to the trend, NaF, has an unusually high ΔH_{L}, indicating a reluctance to dissolve. It is difficult to say precisely why ionic size has a slightly greater effect on ΔH_{L} than on ΔH_{hyd}. We are effectively taking the difference of two large quantities (recall ΔH_{hyd} is negative), and that difference cannot be precisely related to ionic size.

8.42 (a) $$m_{\text{KOH}} = \frac{\left(\dfrac{13.72\text{ g KOH}}{56.11\text{ g}\cdot\text{mol}^{-1}\text{ KOH}}\right)}{0.0750\text{ kg}} = 3.26\ m$$

(b) $$\frac{\left(\dfrac{\text{mass}_{\text{ethylene glycol}}}{62.07\text{ g}\cdot\text{mol}^{-1}}\right)}{1.5\text{ kg}} = 0.44\ m$$

$$\text{mass}_{\text{ethylene glycol}} = 41\text{ g}$$

(c) 1.00 kg of solution will contain 38.9 g HCl and 961.1 g H_2O.

$$\frac{\left(\dfrac{38.9\text{ g HCl}}{36.46\text{ g}\cdot\text{mol}^{-1}\text{ HCl}}\right)}{0.9611\text{ kg}} = 1.11\ m$$

8.44 (a) $\dfrac{\left(\dfrac{13.63 \text{ g sucrose}}{342.29 \text{ g}\cdot\text{mol}^{-1} \text{ sucrose}}\right)}{0.612 \text{ kg}} = 0.0650\ m$

(b) 1 kg of 10.00% CsCl will contain 100.0 g CsCl and 900.0 g H_2O.

$$\frac{\left(\dfrac{100.0 \text{ g CsCl}}{168.36 \text{ g}\cdot\text{mol}^{-1} \text{ CsCl}}\right)}{0.9000 \text{ kg}} = 0.6600\ m$$

(c) The solution contains 0.235 mol of acetone for 0.765 mol H_2O.

$$\frac{0.197 \text{ mol acetone}}{(0.765 \text{ mol } H_2O)(18.02 \text{ g}\cdot\text{mol}^{-1})\left(\dfrac{1 \text{ kg}}{1000 \text{ g}}\right)} = 14.3\ m$$

8.46 (a) If x_{FeCl_3} is 0.0205, then there are 0.0205 mol $FeCl_3$ for every 0.9795 mol H_2O. The mass of water will be

$18.02 \text{ g}\cdot\text{mol}^{-1} \times 0.9795 \text{ mol} = 17.65 \text{ g or } 0.01765 \text{ kg}.$

$$m_{OH^-} = \frac{\left(\dfrac{3\ Cl^-}{FeCl_3}\right)(0.0205 \text{ mol } FeCl_3)}{0.01765 \text{ kg solvent}} = 3.48\ m$$

(b) $m_{Cl^-} = \dfrac{\left(\dfrac{2 \text{ mol } OH^-}{1 \text{ mol Ba(OH)}_2}\right)\left(\dfrac{9.25 \text{ g Ba(OH)}_2}{171.36 \text{ g}\cdot\text{mol}^{-1} \text{ Ba(OH)}_2}\right)}{0.183 \text{ kg solvent}} = 0.590\ m$

(c) 1.000 L of 12.00 M NH_3(aq) will contain 12.00 mol with a mass of $12.00 \times 17.03 \text{ g}\cdot\text{mol}^{-1} = 204.4$ g. The density of the 1.000 L of solution is $0.9519 \text{ g}\cdot\text{cm}^{-3}$, so the total mass in the solution is 951.9 g. This leaves $951.9 \text{ g} - 204.4 \text{ g} = 747.5$ g as water.

$$\frac{12.00 \text{ mol } NH_3}{0.7475 \text{ kg solvent}} = 16.05\ m$$

8.48 (a) $X \times 1.07 \text{ g}\cdot\text{cm}^{-3} \times \dfrac{0.100 \text{ g } H_2SO_4}{1 \text{ g}} = 8.37 \text{ g } H_2SO_4$

$$X = \frac{8.37\text{ g H}_2\text{SO}_4}{1.07\text{ g}\cdot\text{cm}^{-3} \times \dfrac{0.100\text{ g H}_2\text{SO}_4}{1\text{ g}}}$$

$$= 78.2\text{ mL}$$

(b) 100 g of solution contains 10.0 g of H_2SO_4 and 90.0 g of water. $10.0\text{ g} \div 98.07\text{ g}\cdot\text{mol}^{-1} = 0.102$ mol H_2SO_4. 90.0 g equals 0.0900 kg of solvent,

$$\text{so molality} = \frac{0.102\text{ mol}}{0.0900\text{ kg}} = 1.13\text{ mol}\cdot\text{kg}^{-1}$$

(c)

$$250\text{ mL} \times 1.07\text{ g}\cdot\text{cm}^{-3} \times 0.100\text{ g H}_2\text{SO}_4\cdot(\text{g solution})^{-1} = 26.8\text{ g H}_2\text{SO}_4$$

8.50 (a) The vapor pressure of water at 80°C is 355.26 Torr. If the mole fraction of glucose is 0.050, then the mole fraction of the solvent water will be $1.000 - 0.050 = 0.950$.

$$P = x_{\text{solvent}} \times P_{\text{pure solvent}}$$
$$P = 0.950 \times 355.26\text{ Torr} = 338\text{ Torr}$$

(b) At 25°C, the vapor pressure of water is 23.76 Torr. The molality of the urea solution must be converted to mole fraction. A 0.10 *m* solution will contain 0.10 mol urea per 1000 g H_2O.

$$x_{\text{H}_2\text{O}} = \frac{n_{\text{H}_2\text{O}}}{n_{\text{H}_2\text{O}} + n_{\text{urea}}} = \frac{\dfrac{1000\text{ g}}{18.02\text{ g}\cdot\text{mol}^{-1}}}{\dfrac{1000\text{ g}}{18.02\text{ g}\cdot\text{mol}^{-1}} + 0.10\text{ mol}} = 0.9982$$

$$P = 0.9982 \times 23.76\text{ Torr} = 23.72\text{ Torr}$$

8.52 (a) The vapor pressure of pure water at 40°C is 55.34 Torr. If the mole fraction of fructose in solutions is 0.11, then the mole fraction of the solvent will be 0.81.

$$P = x_{\text{solvent}} \times P_{\text{pure solvent}}$$
$$P = 0.81 \times 55.34\text{ Torr} = 45\text{ Torr}$$

$\Delta P = 55.34 \text{ Torr} - 45 \text{ Torr} = 10 \text{ Torr}.$

(b) The vapor pressure of pure water at 20°C is 17.54 Torr. When MgF_2 dissolves, we will assume it dissociates completely into Mg^{+2} and F^- ions:

$MgF_2(s) \longrightarrow Mg^{2+}(aq) + 2\ F^-(aq)$

Note that the F^- ion concentration will be $2 \times$ that of the Mg^{2+} concentration:

$$x_{H_2O} = \frac{n_{H_2O}}{n_{H_2O} + n_{Mg^{2+}} + n_{F^-}}$$

$$= \frac{\dfrac{100\text{ g}}{18.02\text{ g}\cdot\text{mol}^{-1}}}{\dfrac{100\text{ g}}{18.02\text{ g}\cdot\text{mol}^{-1}} + \dfrac{0.008\text{ g}}{62.31\text{ g}\cdot\text{mol}^{-1}} + \left(2\times\dfrac{0.008\text{ g}}{62.31\text{ g}\cdot\text{mol}^{-1}}\right)}$$

$$\approx 1.00$$

$P = 1.00 \times 17.54 \text{ Torr} = 17.5 \text{ Torr}$

(c) The vapor pressure of pure water is 4.58 Torr at 0°C.

Assume that the $Fe(NO_3)_3$ undergoes complete dissociation is solution:

$Fe(NO_3)_3(s) \longrightarrow Fe^{3+}(aq) + 3\ NO_3^-(aq)$

The concentration must be converted from molality to mole fraction:

$$x_{H_2O} = \frac{n_{H_2O}}{n_{H_2O} + n_{Fe^{3+}} + n_{NO_3^-}}$$

$$= \frac{\dfrac{1000\text{ g}}{18.02\text{ g}\cdot\text{mol}^{-1}}}{\dfrac{1000\text{ g}}{18.02\text{ g}\cdot\text{mol}^{-1}} + 0.025\text{ mol} + (3\times 0.025\text{ mol})}$$

$$\approx 1.00$$

$P = 1.00 \times 4.58 \text{ Torr} = 4.58 \text{ Torr}$

8.54 (a) From the relationship $P = x_{\text{solvent}} \times P_{\text{pure solvent}}$ we can calculate the mole fraction of the solvent, making use of the fact that at its normal boiling point, the vapor pressure of any liquid will be 760.00 Torr:

$740 \text{ Torr} = x_{\text{solvent}} \times 760.00 \text{ Torr}$

$x_{\text{solvent}} = 0.974$

The mole fraction of the unknown compound will be

$1.000 - 0.974 = 0.026$.

(b) The molar mass can be calculated using the definition of mole fraction for either the solvent or the solute. In this case, the math is slightly easier if the definition of mole fraction of the solvent is used:

$$x_{\text{solvent}} = \frac{n_{\text{solvent}}}{n_{\text{unknown}} + n_{\text{solvent}}}$$

$$0.974 = \frac{\dfrac{100\text{ g}}{46.07\text{ g}\cdot\text{mol}^{-1}}}{\dfrac{100\text{ g}}{46.07\text{ g}\cdot\text{mol}^{-1}} + \dfrac{9.15\text{ g}}{M_{\text{unknown}}}}$$

$$M_{\text{unknown}} = \frac{9.15\text{ g}}{\left[\left(\dfrac{100\text{ g}}{46.07\text{ g}\cdot\text{mol}^{-1}}\right)\Big/ 0.974\right] - \left(\dfrac{100\text{ g}}{46.07\text{ g}\cdot\text{mol}^{-1}}\right)} = 158\text{ g}\cdot\text{mol}^{-1}$$

8.56 (a) $\Delta T_{\text{b}} = ik_{\text{b}}m$

For $CaCl_2$, $i = 3$

$$\Delta T_{\text{b}} = 3 \times 0.51\text{ K}\cdot\text{kg}\cdot\text{mol}^{-1} \times 0.22\text{ mol}\cdot\text{kg}^{-1} = 0.34\text{ K or }0.34°\text{C}$$

The boiling point will be 100.00°C + 0.34°C = 100.34°C.

(b) $\Delta T_{\text{b}} = ik_{\text{b}}m$

For Li_2CO_3, $i = 3$

$$\Delta T_{\text{b}} = 3 \times 0.51\text{ K}\cdot\text{kg}\cdot\text{mol}^{-1} \times \frac{\left(\dfrac{0.72\text{ g}}{73.89\text{ g}\cdot\text{mol}^{-1}}\right)}{0.100\text{ kg}} = 0.15\text{ K or }0.15°\text{C}$$

The boiling point will be 100.00°C + 0.15°C = 100.15°C.

(c) $\Delta T_{\text{b}} = ik_{\text{b}}m$

Because urea is a nonelectrolyte, $i = 1$.

A 1.7% solution of urea will contain 1.7 g of urea per 98.3 g of water.

$$\Delta T_b = 0.51\ \text{K}\cdot\text{kg}\cdot\text{mol}^{-1} \times \frac{\left(\dfrac{1.7\ \text{g}}{60.06\ \text{g}\cdot\text{mol}^{-1}}\right)}{0.0983\ \text{kg}} = 0.15\ \text{K or } 0.15°\text{C}$$

The boiling point will be 100.00°C + 0.15°C or 100.15°C.

8.58 (a) The molality of the solution can be calculated, knowing the freezing point; this value can, in turn, be used to calculate the boiling point.

$\Delta T_f = k_f m$

Because the solvent is water with a normal freezing point of 0°C, the freezing point of the solution is also the ΔT_f.

$1.04\ \text{K} = 1.86\ \text{K}\cdot\text{kg}\cdot\text{mol}^{-1} \times \text{molality}$

$\text{molality} = 0.559\ \text{mol}\cdot\text{kg}^{-1}$

$$\begin{aligned}\Delta T_b &= k_b m\\ &= 0.51\ \text{K}\cdot\text{kg}\cdot\text{mol}^{-1} \times 0.559\ \text{mol}\cdot\text{kg}^{-1}\\ &= 0.29\ \text{K or } 0.29°\text{C}\end{aligned}$$

boiling point = 100.00°C + 0.29°C = 100.29°C

(b) $\Delta T_f = k_f m$

$\Delta T_f = 5.5°\text{C} - 2.0°\text{C} = 3.5°\text{C or } 3.5\ \text{K}$

$3.5\ \text{K} = 5.12\ \text{K}\cdot\text{kg}\cdot\text{mol}^{-1} \times \text{molality}$

$\text{molality} = 0.68\ \text{mol}\cdot\text{kg}^{-1}$

$$\begin{aligned}\Delta T_b &= k_b m\\ &= 2.53\ \text{K}\cdot\text{kg}\cdot\text{mol}^{-1} \times 0.68\ \text{mol}\cdot\text{kg}^{-1}\\ &= 1.7\ \text{K or } 1.7°\text{C}\end{aligned}$$

boiling point = 80.1°C + 1.7°C = 81.8°C

8.60 $\Delta T_b = 0.481°\text{C or } 0.481\ \text{K}$

$\Delta T_b = k_b \times \text{molality}$

$$0.481\ \text{K} = 2.79\ \text{K} \cdot \text{kg} \cdot \text{mol}^{-1} \times \text{molality}$$

$$0.481\ \text{K} = 2.79\ \text{K} \cdot \text{kg} \cdot \text{mol}^{-1} \times \frac{\left(\dfrac{2.25\ \text{g}}{M_{\text{unknown}}}\right)}{0.150\ \text{kg}}$$

$$\frac{0.150\ \text{kg} \times 0.481\ \text{K}}{2.79\ \text{K} \cdot \text{kg} \cdot \text{mol}^{-1}} = \frac{2.25\ \text{g}}{M_{\text{unknown}}}$$

$$M_{\text{unknown}} = \frac{2.25\ \text{g} \times 2.79\ \text{K} \cdot \text{kg} \cdot \text{mol}^{-1}}{0.150\ \text{kg} \times 0.481\ \text{K}} = 87.0\ \text{g} \cdot \text{mol}^{-1}$$

8.62 (a) $\Delta T_{\text{f}} = ik_{\text{f}}m$

For $CaCl_2$, $i = 3$

$$\Delta T_{\text{f}} = 3 \times 1.86\ \text{K} \cdot \text{kg} \cdot \text{mol}^{-1} \times 0.22\ \text{mol} \cdot \text{kg}^{-1} = 1.2\ \text{K or } 1.2°\text{C}$$

The freezing point will be $0.00°\text{C} - 1.2°\text{C} = -1.2°\text{C}$.

(b) $\Delta T_{\text{f}} = ik_{\text{f}}m$

For Li_2CO_3, $i = 3$

$$\Delta T_{\text{f}} = 3 \times 1.86\ \text{K} \cdot \text{kg} \cdot \text{mol}^{-1} \times \frac{\left(\dfrac{0.001\ 54\ \text{g}}{73.89\ \text{g} \times \text{mol}^{-1}}\right)}{0.100\ \text{kg}}$$

$$= 1.16 \times 10^{-3}\ \text{K or } 1.16 \times 10^{-3}°\text{C}$$

The boiling point will be $0.00°\text{C} - 1.16 \times 10^{-3}°\text{C} = -1.16 \times 10^{-3}°\text{C}$.

(c) $\Delta T_{\text{f}} = ik_{\text{f}}m$

Because urea is a nonelectrolyte, $i = 1$

A 1.7% solution of urea will contain 1.7 g of urea per 98.3 g of water.

$$\Delta T_{\text{f}} = 1.86\ \text{K} \cdot \text{kg} \cdot \text{mol}^{-1} \times \frac{\left(\dfrac{1.7\ \text{g}}{60.06\ \text{g} \cdot \text{mol}^{-1}}\right)}{0.0983\ \text{kg}} = 0.54\ \text{K or } 0.54°\text{C}$$

The freezing point will be $0.00°\text{C} - 0.54°\text{C}$ or $-0.54°\text{C}$.

8.64 $\Delta T_{\text{f}} = k_{\text{f}}m$

$$1.454 \text{ K} = (7.27 \text{ K}\cdot\text{kg}\cdot\text{mol}^{-1})\, m$$

$$1.454 \text{ K} = (7.27 \text{ K}\cdot\text{kg}\cdot\text{mol}^{-1})\frac{\left(\dfrac{1.32 \text{ g}}{M_{\text{unknown}}}\right)}{0.0500 \text{ kg}}$$

$$\frac{0.0500 \text{ kg} \times 1.454 \text{ K}}{7.27 \text{ K}\cdot\text{kg}\cdot\text{mol}^{-1}} = \frac{1.32 \text{ g}}{M_{\text{unknown}}}$$

$$M_{\text{unknown}} = \frac{1.32 \text{ g} \times 7.27 \text{ K}\cdot\text{kg}\cdot\text{mol}^{-1}}{0.0500 \text{ kg} \times 1.454 \text{ K}} = 132 \text{ g}\cdot\text{mol}^{-1}$$

8.66 (a) We use the vapor pressure to calculate the mole fraction of benzene and then convert this quantity to molality, which in turn is used to calculate the freezing point depression.

$$P_{\text{solvent}} = x_{\text{solvent}} \times P_{\text{pure solvent}}$$

$$740 \text{ Torr} = x_{\text{solvent}} \times 760 \text{ Torr}$$

(Remember that the vapor pressure of a liquid at its boiling point is 760 Torr by definition.)

$x_{\text{solvent}} = 0.974$

Because the absolute amount of solvent is not important here, we can assume that the total number of moles = 1.

$$x_{\text{solvent}} = 0.974 = \frac{n_{\text{solvent}}}{n_{\text{solvent}} + n_{\text{solute}}} = \frac{n_{\text{solvent}}}{1}$$

$$n_{\text{solvent}} = 0.974,\ n_{\text{solute}} = 0.026$$

$$\text{molality of solute} = \frac{0.026 \text{ mol}}{\left(\dfrac{0.974 \text{ mol} \times 78.11 \text{ g}\cdot\text{mol}^{-1}}{1000 \text{ g}\cdot\text{kg}^{-1}}\right)} = 0.34 \text{ mol}\cdot\text{kg}^{-1}$$

$$\Delta T_{\text{f}} = k_{\text{f}} m$$

$$= 5.12 \text{ K}\cdot\text{kg}\cdot\text{mol}^{-1} \times 0.34 \text{ mol}\cdot\text{kg}^{-1} = 1.7°\text{C}$$

The freezing point will be $5.12°\text{C} - 1.7°\text{C} = 3.4°\text{C}$.

(b) $\Delta T_f = k_f m$

$$4.02\ \text{K} = 1.86\ \text{K}\cdot\text{kg}\cdot\text{mol}^{-1} \times \text{molality}$$

$$4.02\ \text{K} = 1.86\ \text{K}\cdot\text{kg}\cdot\text{mol}^{-1} \times \frac{n_{CO(NH_2)_2}}{\text{kg solvent}} = 1.86\ \text{K}\cdot\text{kg}\cdot\text{mol}^{-1} \times \frac{n_{CO(NH_2)_2}}{1.200\ \text{kg}}$$

$$n_{CO(NH_2)_2} = 2.59\ \text{mol}$$

$$\Delta T_f = k_f m$$

$$\Delta T_f = 1.86\ \text{K}\cdot\text{kg}\cdot\text{mol}^{-1} \times \frac{n_{\text{protein}}}{\text{kg solvent}}$$

$$= 1.86\ \text{K}\cdot\text{kg}\cdot\text{mol}^{-1} \times \frac{\left(\dfrac{1.0\ \text{g}}{1.0\times10^5\ \text{g}\cdot\text{mol}^{-1}}\right)}{1.0\ \text{kg}}$$

$$= 1.9\times10^{-5}\ \text{K or } 1.9\times10^{-5}\,^{\circ}\text{C}$$

(c) freezing point $\approx$ 0.0°C

8.68 (a) A 1.00% aqueous solution of $MgSO_4$ will contain 1.00 g of $MgSO_4$ for 99.0 g of water. To use the freezing point depression equation, we need the molality of the solution.

$$\text{molality} = \frac{\left(\dfrac{1.00\ \text{g}}{120.37\ \text{g}\cdot\text{mol}^{-1}}\right)}{0.0990\ \text{kg}} = 0.0839\ \text{mol}\cdot\text{kg}^{-1}$$

$$\Delta T_f = i k_f m$$

$$\Delta T_f = i\,(1.86\ \text{K}\cdot\text{kg}\cdot\text{mol}^{-1})(0.0839\ \text{mol}\cdot\text{kg}^{-1}) = 0.192\ \text{K}$$

$$i = 1.23$$

(b) molality of all solute species (undissociated $MgSO_4$(aq) plus Mg^{2+}(aq) + SO_4^{2-}(aq)) $= 1.23 \times 0.0839\ \text{mol}\cdot\text{kg}^{-1} = 0.103\ \text{mol}\cdot\text{kg}^{-1}$

(c) If all the $MgSO_4$ had dissociated, the total molality in solution would have been $0.168\ \text{mol}\cdot\text{kg}^{-1}$, giving an i value equal to 2. If no dissociation had taken place, the molality in solution would have equaled $0.0839\ \text{mol}\cdot\text{kg}^{-1}$.

$$MgSO_4(aq) \rightleftharpoons Mg^{2+}(aq) + SO_4^{2-}(aq)$$

$$0.0839\ \text{mol}\cdot\text{kg}^{-1} - x \qquad x \qquad x$$

$0.0839 \text{ mol} \cdot \text{kg}^{-1} - x + x + x = 0.103 \text{ mol} \cdot \text{kg}^{-1}$

$0.0839 \text{ mol} \cdot \text{kg}^{-1} + x = 0.103 \text{ mol} \cdot \text{kg}^{-1}$

$x = 0.019 \text{ mol} \cdot \text{kg}^{-1}$

$$\% \text{ dissociation} = \frac{0.019 \text{ mol} \cdot \text{kg}^{-1}}{0.0839 \text{ mol} \cdot \text{kg}^{-1}} \times 100 = 23\%$$

8.70 First calculate the van't Hoff i factor:

$\Delta T_f = ik_f m$

$0.423 \text{ K} = i \times 1.86 \text{ K} \cdot \text{kg} \cdot \text{mol}^{-1} \times 0.124 \text{ mol} \cdot \text{kg}^{-1}$

$i = 1.83$

The molality of all solute species (undissociated $CCl_3COOH(aq)$ plus $CCl_3COO^-(aq) + H^+(aq)) = 1.83 \times 0.124 \text{ mol} \cdot \text{kg}^{-1} = 0.227 \text{ mol} \cdot \text{kg}^{-1}$

If all the $CCl_3COOH(aq)$ had dissociated, the total molality in solution would have been $0.248 \text{ mol} \cdot \text{kg}^{-1}$, giving an i value equal to 2. If no dissociation had taken place, the molality in solution would have equaled $0.124 \text{ mol} \cdot \text{kg}^{-1}$.

$CCl_3COOH(aq)$	$\rightleftharpoons$	$H^+(aq)$	+	$CCl_3COO^-(aq)$
$0.124 \text{ mol} \cdot \text{kg}^{-1} - x$		x		x

$0.124 \text{ mol} \cdot \text{kg}^{-1} - x + x + x = 0.227 \text{ mol} \cdot \text{kg}^{-1}$

$0.124 \text{ mol} \cdot \text{kg}^{-1} + x = 0.227 \text{ mol} \cdot \text{kg}^{-1}$

$x = 0.103 \text{ mol} \cdot \text{kg}^{-1}$

$$\% \text{ ionization} = \frac{0.103 \text{ mol} \cdot \text{kg}^{-1}}{0.124 \text{ mol} \cdot \text{kg}^{-1}} \times 100 = 83.1\%$$

8.72 First, calculate the osmotic pressure of each solution from $\Pi = iRT \times \text{molarity}$.

(a) KCl is an ionic compound that dissociates into 2 ions, so $i = 2$.

$\Pi = 2 \times 0.082\,06 \text{ L} \cdot \text{atm} \cdot \text{K}^{-1} \cdot \text{mol}^{-1} \times 323 \text{ K} \times 0.10 \text{ mol} \cdot \text{L}^{-1}$

$= 5.3 \text{ atm}$

(b) urea, $CO(NH_2)_2$ is a nonelectrolyte, so $i = 1$.

$$\Pi = 1 \times 0.082\ 06\ \text{L}\cdot\text{atm}\cdot\text{K}^{-1}\cdot\text{mol}^{-1} \times 323\ \text{K} \times 0.60\ \text{mol}\cdot\text{L}^{-1}$$
$$= 15.9\ \text{atm}$$

(c) K_2SO_4 is an ionic solid that dissolves in solution to produce 3 ions, so $i = 3$.

$$\Pi = 3 \times 0.082\ 06\ \text{L}\cdot\text{atm}\cdot\text{K}^{-1}\cdot\text{mol}^{-1} \times 323\ \text{K} \times 0.30\ \text{mol}\cdot\text{L}^{-1}$$
$$= 24\ \text{atm}$$

Solution (c) has the highest osmotic pressure.

8.74 Insulin is a nonelectrolyte, so $i = 1$.

$$\Pi = iRT \times \text{molarity}$$

$$\Pi = \frac{2.30\ \text{Torr}}{760\ \text{Torr}\cdot\text{atm}^{-1}} = 1 \times 0.082\ 06\ \text{L}\cdot\text{atm}\cdot\text{K}^{-1}\cdot\text{mol}^{-1} \times 298\ \text{K} \times \frac{\left(\dfrac{0.10\ \text{g}}{M_{\text{unknown}}}\right)}{0.200\ \text{L}}$$

$$M_{\text{unknown}} = \frac{0.082\ 06\ \text{L}\cdot\text{atm}\cdot\text{K}^{-1}\cdot\text{mol}^{-1}}{2.30\ \text{Torr} \times 0.200\ \text{L}} \times \frac{293\ \text{K} \times 0.10\ \text{g} \times 760\ \text{Torr}\cdot\text{atm}^{-1}}{2.30\ \text{Torr} \times 0.200\ \text{L}}$$
$$= 4.0 \times 10^3\ \text{g}\cdot\text{mol}^{-1}$$

8.76 We assume the polymer to be a nonelectrolyte, so $i = 1$.

$$\Pi = iRT \times \text{molarity}$$

$$\Pi = \frac{0.582\ \text{Torr}}{760\ \text{Torr}\cdot\text{atm}^{-1}} = 1 \times 0.082\ 06\ \text{L}\cdot\text{atm}\cdot\text{K}^{-1}\cdot\text{mol}^{-1} \times 293\ \text{K} \times \frac{\left(\dfrac{0.50\ \text{g}}{M_{\text{unknown}}}\right)}{0.200\ \text{L}}$$

$$M_{\text{unknown}} = \frac{0.082\ 06\ \text{L}\cdot\text{atm}\cdot\text{K}^{-1}\cdot\text{mol}^{-1}\times 293\ \text{K}\times}{0.582\ \text{Torr}\times 0.200\ \text{L}}$$

$$\times\frac{0.50\ \text{g}\times 760\ \text{Torr}\cdot\text{atm}^{-1}}{0.582\ \text{Torr}\times 0.200\ \text{L}}$$

$$= 7.8\times 10^{4}\ \text{g}\cdot\text{mol}^{-1}$$

8.78 (a) Assume that $C_6H_{12}O_6$ is a nonelectrolyte, so $i = 1$.

$$\Pi = iRT \times \text{molarity}$$
$$= 1\times 0.082\ 06\ \text{L}\cdot\text{atm}\cdot\text{K}^{-1}\cdot\text{mol}^{-1}\times 293\ \text{K}\times 3.0\times 10^{-3}\ \text{mol}\cdot\text{L}^{-1}$$
$$= 0.072\ \text{atm}$$

(b) $CaCl_2$ is an ionic compound that will dissolve in solution to give 3 ions, so $i = 3$.

$$\Pi = iRT \times \text{molarity}$$
$$= 3\times 0.082\ 06\ \text{L}\cdot\text{atm}\cdot\text{K}^{-1}\cdot\text{mol}^{-1}\times 293\ \text{K}\times 2.0\times 10^{-3}\ \text{mol}\cdot\text{L}^{-1}$$
$$= 0.14\ \text{atm}$$

(c) K_2SO_4 is an ionic compound that will dissolve into 3 ions in solution, $i = 3$.

$$\Pi = iRT \times \text{molarity}$$
$$= 3\times 0.082\ 06\ \text{L}\cdot\text{atm}\cdot\text{K}^{-1}\cdot\text{mol}^{-1}\times 293\ \text{K}\times 0.010\ \text{mol}\cdot\text{L}^{-1}$$
$$= 0.72\ \text{atm}$$

8.80 $5\%\ \text{glucose} = \dfrac{5.0\ \text{g glucose}}{1.0\times 10^{2}\ \text{g solution}}$

assume the solution density $\approx 1\ \text{g}\cdot\text{mL}^{-1}$

Then $\left(\dfrac{5.0\ \text{g glucose}}{1.0\times 10^{2}\ \text{mL}}\right)\left(\dfrac{1\ \text{mL}}{10^{-3}\ \text{L}}\right)\left(\dfrac{1\ \text{mol glucose}}{180.16\ \text{g}\cdot\text{mol}^{-1}}\right) = 0.28\ \text{mol}\cdot\text{L}^{-1}$

and

$$\Pi = iRTM = (1)(0.082\ 06\ \text{L}\cdot\text{atm}\cdot\text{K}^{-1}\cdot\text{mol}^{-1})(310\ \text{K})(0.28\ \text{mol}\cdot\text{L}^{-1})$$
$$= 7.1\ \text{atm}$$

8.82 (a) To determine the vapor pressure of the solution, we need to know the mole fraction of each component.

$$x_{\text{hexane}} = \frac{0.25\ \text{mol}}{0.25\ \text{mol} + 0.65\ \text{mol}} = 0.28$$

$$x_{\text{cyclohexane}} = 1 - x_{\text{hexane}} = 0.72$$

$$P_{\text{total}} = (0.28 \times 151\ \text{Torr}) + (0.72 \times 98\ \text{Torr}) = 113\ \text{Torr}$$

The vapor phase composition will be given by

$$x_{\text{hexane in vapor phase}} = \frac{P_{\text{hexane}}}{P_{\text{total}}} = \frac{0.28 \times 151\ \text{Torr}}{113\ \text{Torr}} = 0.37$$

$$x_{\text{cyclohexane in vapor phrase}} = 1 - 0.37 = 0.63$$

The vapor is richer in the more volatile cyclohexane, as expected.

(b) The procedure is the same as in (a) but the number of moles of each component must be calculated first:

$$n_{\text{hexane}} = \frac{10.0\ \text{g}}{86.18\ \text{g} \cdot \text{mol}^{-1}} = 0.116$$

$$n_{\text{cyclohexane}} = \frac{10.0\ \text{g}}{84.16\ \text{g} \cdot \text{mol}^{-1}} = 0.119$$

$$x_{\text{hexane}} = \frac{0.116\ \text{mol}}{0.116\ \text{mol} + 0.119\ \text{mol}} = 0.494$$

$$x_{\text{cyclohexane}} = 1 - x_{\text{hexane}} = 0.506$$

$$P_{\text{total}} = (0.494 \times 151\ \text{Torr}) + (0.506 \times 98\ \text{Torr}) = 124\ \text{Torr}$$

The vapor phase composition will be given by

$$x_{\text{hexane in vapor phase}} = \frac{P_{\text{hexane}}}{P_{\text{total}}} = \frac{0.494 \times 151\ \text{Torr}}{124\ \text{Torr}} = 0.602$$

$$x_{\text{cyclohexane in vapor phase}} = 1 - 0.602 = 0.398$$

8.84 We are given the following information:

$$P_{\text{butanone, pure}} = 100.\ \text{Torr at } 25^{\circ}\text{C},\ 350.0\text{g} \quad (72.107\ \text{g} \cdot \text{mol}^{-1} \quad \text{M.W.})$$

$$P_{\text{propanone, pure}} = 222\ \text{Torr at } 25^{\circ}\text{C} \quad (58.080\ \text{g} \cdot \text{mol}^{-1} \quad \text{M.W.})$$

$$P_{\text{total}} = 135\ \text{Torr}$$

and Raoult's Law states:

$$P_{\text{total}} = X_{\text{butanone}} P_{\text{butanone, pure}} + X_{\text{propanone}} P_{\text{propanone, pure}}$$

$$= \frac{n_{\text{butanone}}}{n_{\text{total}}} P_{\text{butanone, pure}} + \frac{n_{\text{propanone}}}{n_{\text{total}}} P_{\text{propanone, pure}}$$

given $n_{\text{tot}}al = n_{\text{butanone}} + n_{\text{propanone}}$ we can write:

$$P_{\text{tot}} = \frac{n_{\text{butanone}}}{n_{\text{total}}} P_{\text{butanone, pure}} + \frac{n_{\text{total}} - n_{\text{butanone}}}{n_{\text{total}}} P_{\text{propanone, pure}}$$

Rearranging to solve for n_{total}:'

$$n_{\text{total}} = \frac{\left(n_{\text{butanone}} \cdot P_{\text{butanone, pure}}\right) - \left(n_{\text{butanone}} \cdot P_{\text{propanone, pure}}\right)}{P_{\text{total}} - P_{\text{propanone, pure}}}$$

$$= \frac{(4.8539 \text{ mol} \cdot 100. \text{ Torr}) - (4.8539 \text{ mol} \cdot 222 \text{ Torr})}{135 \text{ Torr} - 222 \text{ Torr}}$$

$$= 6.81 \text{ mol}$$

Therefore,

$$n_{\text{propanone}} = n_{\text{total}} - n_{\text{butanone}} = 6.81 \text{ mol} - 4.85 \text{ mol} = 1.95 \text{ mol}$$

and

$$mass_{\text{propanone}} = 1.95 \text{ mol} \cdot 58.080 \text{ g} \cdot \text{mol}^{-1} = 113 \text{ g}$$

8.86 Raoult's Law applies to the vapor pressure of the mixture, so a positive deviation means that the vapor pressure is higher than expected for an ideal solution. Negative deviation means that the vapor pressure is lower than expected for an ideal solution. Negative deviation will occur when the interactions between the different molecules are somewhat stronger than the interactions between molecules of the same kind.

(a) For HBr and H_2O, the possibility of intermolecular hydrogen bonding between water and HBr would suggest that negative deviation would be observed, which is the case. HBr and H_2O form an azeotrope that boils at $126\,°\text{C}$, which is higher than the boiling point of either HBr (−67°C) or water.

(b) Because formic acid is a very polar molecule with hydrogen bonding and benzene is nonpolar, we would expect a positive deviation, which is observed. Benzene and formic acid form an azeotrope that boils at $71°C$, which is well below the boiling point of either benzene ($80.1°C$) or formic acid ($101°C$).

(c) Because cyclohexane and cyclopentane are both nonpolar hydrocarbons of similar size and with similar intermolecular forces, we would expect them to form an ideal solution.

8.88 (a) Given:

$P_{A,\text{ pure}} = 1.55 \text{ atm}, \quad P_{B,\text{ pure}} = 0.650 \text{ atm}, \quad \text{and } P_{\text{total}} = 1.00 \text{ atm}$

Raoult's Law states:

$P_{\text{tot}} = X_A P_{A,\text{ pure}} + X_B P_{B,\text{ pure}}$

To solve for the mole fractions of A and B we use:

$X_A + X_B = 1$, and therefore: $X_B = 1 - X_A$.

Substituting into the equation above we obtain:

$P_{\text{tot}} = X_A P_{A,\text{ pure}} + (1 - X_A) P_{B,\text{ pure}}$.

Solving for X_A :

$$X_A = \frac{P_{\text{tot}} - P_{B,\text{ pure}}}{P_{A,\text{ pure}} - P_{B,\text{ pure}}} = \frac{1.00 \text{ atm} - 0.650 \text{ atm}}{1.55 \text{ atm} - 0.650 \text{ atm}} = 0.389$$

and $X_B = 1 - X_A = 1 - 0.389 = 0.611$

(b) The gas-phase mole fraction is given by:

$P_A = X_{A,\text{ liquid}} P_{A,\text{ pure}} = 0.389 \cdot 1.55 \text{ atm} = 0.603 \text{ atm}$

$P_B = X_{B,\text{ liquid}} P_{B,\text{ pure}} = 0.611 \cdot 0.650 \text{ atm} = 0.397 \text{ atm}$

$P_{\text{tot}} = P_A + P_B = 0.603 \text{ atm} + 0.397 \text{ atm} = 1.00 \text{ atm}$

and

$$X_{A,\text{ gas}} = \frac{0.603 \text{ atm}}{1.00 \text{ atm}} = 0.603, \quad X_{B,\text{ gas}} = \frac{0.397 \text{ atm}}{1.00 \text{ atm}} = 0.397$$

(c) Qualitatively, as the boiling proceeds, the liquid becomes enriched in component B, the less volatile component. The gas phase, on the other hand, starts out highly enriched in the more volatile component, component A, but will slowly gain more of component B as the boiling process proceeds.

8.90 H_2O_2 has a greater molar mass than H_2O, which allows for greater London forces. Hydrogen bonding should occur for both molecules.

8.92 (a) Vapor pressure increases due to the increased kinetic energy of the molecules at higher temperatures. (b) No effect on the vapor pressure as such, which is determined only by the temperature, but the rate of evaporation increases. (c) No effect on the vapor pressure, which is determined only by the temperature, but additional liquid evaporates. (d) Very little effect. Adding air above the liquid could increase the external pressure on the liquid, but pressure changes have only a small and usually negligible effect on vapor pressure.

8.94 (a) At 0°C and 2 atm, the system exists at the ice/liquid boundary. Decreased pressure brings it to the ice/vapor boundary, when the solid sublimes. Sublimation is complete upon further pressure decrease: the system now contains the vapor alone. (b) At 50°C, water begins in the liquid phase. The vapor pressure of water at 50°C is greater than 5 Torr (between 55 and 150 Torr). As the pressure is lowered to 5 Torr, the water will begin to boil.

8.96 (a) In a sense, nothing (the air mass simply warms up and the percent humidity is reduced). (b) It condenses to fog or freezes to frost.

8.98 Partial pressure of $H_2O = \dfrac{64\% \times 39.90 \text{ Torr}}{100\%} = 26 \text{ Torr}$

The vapor pressure of H_2O at 25°C is 23.76 Torr; therefore, fog or dew will form.

8.100 Boiling point elevation and freezing point depression both arise because dissolving a solute in a solvent increases the entropy of the solvent, thereby decreasing its free energy. For the vapor pressure curve, the lines representing the free energies of the liquid solution and the vapor intersect at a higher temperature than for the pure solvent, so the boiling point is higher in the presence of a solute (see Fig. 8.33b). Similarly for freezing point depression, the lines representing the free energies of the liquid and solid phases of the solvent intersect at a lower temperature than for the pure solvent, so the freezing point is lower in the presence of the solute (see Fig. 8.34). This is explained in detail in section 8.17.

8.102 (a) $\Delta T_{\mathrm{f}} = 1.20°\mathrm{C} - 5.5°\mathrm{C} = -4.3°\mathrm{C}$

$$\Delta T_{\mathrm{f}} = ik_f m = ik_f \frac{\left(\frac{m_{\mathrm{solute}}}{M_{\mathrm{solute}}}\right)}{\mathrm{kg\ solvent}}$$

$$M_{\mathrm{solute}} = \frac{\mathrm{ik}_f \mathrm{m}_{\mathrm{solute}}}{(\mathrm{kg\ solvent})(\Delta T_{\mathrm{f}})} = \frac{(1)(5.12\ \mathrm{K \cdot kg \cdot mol^{-1}})(10.0\ \mathrm{g})}{(0.0800\ \mathrm{kg})(4.3°\mathrm{C})}$$
$$= 1.5 \times 10^2\ \mathrm{g \cdot mol^{-1}}$$

(b) The empirical formula mass $= 73.4\ \mathrm{g \cdot mol^{-1}}$; the experimental formula mass is roughly double this value, so the molecular formula is $C_6H_4Cl_2$.

(c) molar mass $= 146.99\ \mathrm{g \cdot mol^{-1}}$

8.104 (a) The elemental analysis yields the following:

element	% by mass	mol in 100 g	ratio to smallest moles (0.542 for N)
C	59.0	4.91	9.04
O	26.2	1.64	3.02
H	7.10	7.04	13.0
N	7.60	0.542	1

The empirical formula is $C_9H_{13}O_3N$ (formula mass $\sim 183\ \mathrm{g \cdot mol^{-1}}$).

(b) and (c)

$$\Delta T_{\text{f}} = ik_f m = ik_f \frac{\left(\dfrac{\text{mass}_{\text{solute}}}{M_{\text{solute}}}\right)}{\text{kg solvent}}$$

$$M_{\text{solute}} = \frac{ik_f \text{mass}_{\text{solute}}}{(\text{kg solvent})(\Delta T_{\text{f}})} = \frac{(1)(5.12\ \text{K}\cdot\text{kg}\cdot\text{mol}^{-1})(0.64\ \text{g})}{(0.036\ \text{kg})(0.50°\text{C})}$$
$$= 1.8\times 10^{2}\ \text{g}\cdot\text{mol}^{-1}$$

The molecular formula is the same as the empirical formula, $C_9H_{13}O_3N$.

Molar mass of epinephrine $=183.20\ \text{g}\cdot\text{mol}^{-1}$.

8.106 (a) The partial pressure remains the same. Some of the ethanol will condense to return the ethanol(l) $\rightleftharpoons$ ethanol(g) reaction to the equilibrium point. (b) The total pressure will be the sum of the pressures due to the ethanol vapor and to the air. The air will not condense, so its pressure should follow the ideal gas law. If the total pressure initially is 750 Torr and 58.9 Torr is due to ethanol vapor, the 750 Torr – 58.9 Torr = 691 Torr will be due to the air. If the volume is halved at constant temperature, then the pressure due to the air will be doubled, or 2×691 Torr $=1.38\times 10^{3}$ Torr. The total pressure will then be $1.38\times 10^{3}\ \text{Torr} + 58.9\ \text{Torr} = 1.44\times 10^{3}\ \text{Torr}$.

8.108 (a) Given $d = 1.00\ \text{g}\cdot\text{cm}^{-3} = 1.00\ \text{g}\cdot\text{mL}^{-1}$,

$$\therefore \frac{1.54\ \text{g}}{100\ \text{mL solution}} = 15.4\ \text{g}\cdot\text{L}^{-1}$$

$$15.4\ \text{g}\cdot\text{L}^{-1}\times\frac{1\ \text{mol Li}_2\text{CO}_3}{73.89\ \text{g Li}_2\text{CO}_3} = 0.208\ \text{mol}\cdot\text{L}^{-1}$$

$$\Pi = iRT\times\text{molarity}$$
$$= 3\times 0.082\,06\ \text{L}\cdot\text{atm}\cdot K^{-1}\cdot\text{mol}^{-1}\times 273\ K\times 0.208\ \text{mol}\cdot\text{L}^{-1} = 14.0\ \text{atm}$$

(b) $P_{\text{solution}} = x_{\text{solvent}}\times P_{\text{pure solvent}}$

$751\ \text{Torr} = x_{\text{solvent}}\times 760\ \text{Torr}$

$x_{\text{solvent}} = 0.988$

$x_{\text{solute}} = 1 - 0.988 = 0.012$

Consider a solution of 0.012 mol solute in 0.988 mol H_2O. Assume that the volume of solution equals the volume of H_2O. Then

$$V = 0.988 \text{ mol } H_2O \times \frac{18.02 \text{ g } H_2O}{1.00 \text{ mol } H_2O} \times \frac{1.00 \text{ mL } H_2O}{1.00 \text{ g } H_2O}$$

$$= 17.8 \text{ mL} = 0.0178 \text{ L}$$

$$\text{molarity of solution} = \frac{0.012 \text{ mol}}{0.0178 \text{ L}} = 0.67 \text{ mol} \cdot \text{L}^{-1}$$

Assuming that $i = 1$,

$$\Pi = 1 \times 0.082\,06 \text{ L} \cdot \text{atm} \cdot \text{K}^{-1} \cdot \text{mol}^{-1} \times 373 \text{ K} \times 0.67 \text{ mol} \cdot \text{L}^{-1} = 21 \text{ atm}$$

(c) $\Delta T = ik_b \times \text{molality}$, assume $i = 1$, then

$$\text{molality} = \frac{\Delta T_b}{k_b} = \frac{1 \text{ K}}{0.51 \text{ K} \cdot \text{kg} \cdot \text{mol}^{-1}} = 2.0 \text{ mol} \cdot \text{kg}^{-1}$$

Given that $d_{\text{solution}} = d_{\text{water}} = 1.00 \text{ g} \cdot \text{cm}^{-3} = 1.00 \text{ g} \cdot \text{mL}^{-1}$, we have

$$2.0 \text{ mol} \cdot \text{kg}^{-1} \times \frac{1 \text{ kg}}{1000 \text{ g}} \times \frac{1.00 \text{ g}}{1.00 \text{ mL}} \times \frac{1000 \text{ mL}}{1.00 \text{ L}} = 2.0 \text{ mol} \cdot \text{L}^{-1}$$

Then $\Pi = iRT \times \text{molarity}$

$$= 1 \times 0.082\,06 \text{ L} \cdot \text{atm} \cdot \text{K}^{-1} \cdot \text{mol}^{-1} \times 374 \text{ K} \times 2.0 \text{ mol} \cdot \text{L}^{-1}$$

$$= 61 \text{ atm}$$

8.110 (a) Because the osmotic pressure is calculated from $\Pi = iRT \times \text{molarity}$ (both compounds are nonelectrolytes, so $i = 1$), the solution with the higher concentration will have the higher osmotic pressure, which in this case is the solution of urea, $CO(NH_2)_2$.

(b) The more concentrated solution, $CO(NH_2)_2$, will become more dilute with the passage of H_2O molecules through the membrane.

(c) Pressure must be applied to the more concentrated solution, $CO(NH_2)_2$, to equilibrate the water flow.

(d) $\Pi = iRT \times \Delta M$ (where ΔM is the difference in molar concentration between the two solutions):

$\Pi = 1 \times 0.082\,06\ \text{L} \cdot \text{atm} \cdot \text{K}^{-1} \cdot \text{mol}^{-1} \times 298\ \text{K} \times 0.038\ \text{mol} \cdot \text{L}^{-1} = 0.73\ \text{atm}$

8.112 As illustrated in example 8.8, the molar mass is determined by first finding the concentration of the prepared solution:

$RTc = gdh$, and, therefore, $c = \dfrac{gdh}{RT}$

$$= \frac{\left(9.80665\,\frac{\text{m}}{\text{s}^2}\right)\left(1000\,\frac{\text{cm}}{\text{m}}\right)\left(1000\,\frac{\text{cm}^3}{\text{L}}\right)\left(0.79\,\frac{\text{g}}{\text{cm}^3}\right)\left(\frac{1\ \text{kg}}{1000\ \text{g}}\right)(32.5\ \text{cm})}{\left(8.31447\,\frac{\text{kg m}^2}{\text{s}^2\ \text{K mol}}\right)\left(\frac{100\ \text{cm}}{\text{m}}\right)^2(298\ \text{K})}$$

$= 1.02 \times 10^{-3}\ \text{M}$

The moles of protein in the solution is then:

$n_{\text{protein}} = \left(1.02 \times 10^{-3}\ \text{M}\right)\left(0.010\ \text{L}\right) = 1.02 \times 10^{-5}\ \text{mol}$

and the molar mass is:

$$\text{M.M.} = \frac{0.155\ \text{g}}{1.02 \times 10^{-5}\ \text{mol}} = 15300\,\frac{\text{g}}{\text{mol}}$$

8.114 (a) The data in Table 6.2 can be used to obtain the heat of vaporization and boiling point of benzene.

$C_6H_6(l) \rightleftharpoons C_6H_6(g)$

$\Delta H^\circ_{\text{vap}} = 30.8\ \text{KJ} \cdot \text{mol}^{-1}$

The boiling point is $353.2\ \text{K}$ or $80.0°\text{C}$, at which the vapor pressure = 1 atm.

To derive the general equation, we start with the expression that $\Delta G^\circ_{\text{vap}} = -RT \ln P$, where P is the vapor pressure of the solvent. Because $\Delta G^\circ_{\text{vap}} = \Delta H^\circ_{\text{vap}} - T\Delta S^\circ_{\text{vap}}$, this is the relationship to use to determine the temperature dependence of ln P:

$\Delta H^\circ_{\text{vap}} - T\Delta S^\circ_{\text{vap}} = -RT \ln P$

This equation can be rearranged to give:

$$\ln P = -\frac{\Delta H^\circ_{vap}}{R} \times \frac{1}{T} + \frac{\Delta S^\circ_{vap}}{R}$$

Because we do not have enough data to calculate ΔS°_{vap} from the table or appendix, we can use the alternate form of the equation, which relates the vapor pressure at two points to the corresponding temperatures. This equation is obtained by subtracting one specific point from another as shown:

$$\ln P_2 = -\frac{\Delta H^\circ_{vap}}{R} \times \frac{1}{T_2} + \frac{\Delta S^\circ_{vap}}{R}$$

$$-\left[\ln P_1 = -\frac{\Delta H^\circ_{vap}}{R} \times \frac{1}{T_1} + \frac{\Delta S^\circ_{vap}}{R}\right]$$

$$\ln \frac{P_2}{P_1} = -\frac{\Delta H^\circ_{vap}}{R}\left[\frac{1}{T_2} - \frac{1}{T_1}\right]$$

Because we know ΔH°_{vap} and one point, we can introduce those values:

$$\ln \frac{P}{1} = -\frac{30\,800 \text{ J}\cdot\text{mol}^{-1}}{8.314 \text{ J}\cdot\text{K}^{-1}\cdot\text{mol}^{-1}}\left[\frac{1}{T} - \frac{1}{353.2 \text{ K}}\right]$$

$$\ln P = -\frac{3.70 \times 10^3 \text{ K}}{T} + 10.5$$

(b) The appropriate quantities to plot to get a straight line are ln P versus $\frac{1}{T}$.

(c) The solution will boil when the vapor pressure equals the external pressure.

Simply substitute the numbers once the equation has been derived:

$$\ln P = -\frac{3.70 \times 10^3 \text{ K}}{T} + 10.5$$

$$\ln 0.655 = -\frac{3.70 \times 10^3 \text{ K}}{T} + 10.5$$

$T = 339$ K or 66°C

(d) The equation of the line is written in such a form that the constant 10.5 is

equal to $\frac{\Delta S^\circ_{vap}}{R}$.

$$\frac{\Delta S^\circ_{vap}}{R} = 10.5$$

$$\Delta S^\circ_{vap} = 10.5 \times 8.314\ \text{J}\cdot\text{K}^{-1}\cdot\text{mol}^{-1} = 87.3\ \text{J}\cdot\text{K}^{-1}\cdot\text{mol}^{-1}$$

Because $\Delta S^\circ_{vap} = S^\circ_m(\text{benzene, g}) - S^\circ_m(\text{benzene, l})$

we can now calculate the standard molar entropy of benzene(g)

$$87.3\ \text{J}\cdot\text{K}^{-1}\cdot\text{mol}^{-1} = S^\circ_m(\text{benzene, g}) - 173.3\ \text{J}\cdot\text{K}^{-1}\cdot\text{mol}^{-1}$$

$$S^\circ_m(\text{benzene, g}) = 260.6\ \text{J}\cdot\text{K}^{-1}\ \text{mol}^{-1}$$

8.116 The equation to give vapor pressure as a function of temperature is

$$\ln P = -\frac{\Delta H^\circ_{vap}}{R} \times \frac{1}{T} + \frac{\Delta S^\circ_{vap}}{R}$$

$$\ln P = \frac{57\,814\ \text{J}\cdot\text{mol}^{-1}}{8.314\ \text{J}\cdot\text{K}^{-1}\cdot\text{mol}^{-1}} \times \frac{1}{T} + \frac{124\ \text{J}\cdot\text{mol}^{-1}}{8.314\ \text{J}\cdot\text{K}^{-1}\cdot\text{mol}^{-1}}$$

$$\ln P = -\frac{6954\ \text{K}}{T} + 14.9$$

In order for the substance to boil, we need to reduce the external pressure on the sample to its vapor pressure at that temperature. Thus, the problem becomes simply a matter of calculating the vapor pressure at 80°C:

$$\ln P = -\frac{6954\ \text{K}}{353\ \text{K}} + 14.9 = -4.80$$

$P = 0.008\ 23$ atm or 6.26 Torr

The pressure needs to be reduced to 6.26 Torr.

8.118 The critical pressures and temperatures of the compounds are as follows:

Compound	T_C (°C)	P_C (atm)
Methane, CH_4	−82.1	45.8
Methyl amine, CH_3NH_2	156.9	40.2
Ammonia, NH_3	132.5	112.5
Tetrafluoromethane, CF_4	−45.7	41.4

In order to access the supercritical fluid state, we must have conditions in excess of the critical temperature and pressure. Given the rating of the autoclave, ammonia would not be suitable because one could not access the supercritical state due to the pressure limitation. Methylamine would not be suitable for a room temperature extraction because its T_C is too high. Either methane or tetrafluoromethane would be suitable for this application.

8.120 (a) If there is enough diethyl ether present, then the pressure will be due to the vapor pressure of diethyl ether at that temperature. If there is insufficient diethyl ether present, then it will all convert to gas and the pressure will be determined from the ideal gas law. We can calculate the amount of diethyl ether necessary to achieve the vapor pressure from the ideal gas equation, using the vapor pressure as the pressure of the gas.

$$\left(\frac{57\ \text{Torr}}{760\ \text{Torr}\cdot\text{atm}^{-1}}\right)(1.00\ \text{L}) = \left(\frac{m_{\text{diethyl ether}}}{74.12\ \text{g}\cdot\text{mol}^{-1}\ \text{diethyl ether}}\right)$$
$$(0.082\,06\ \text{L}\cdot\text{atm}\cdot\text{K}^{-1}\cdot\text{mol}^{-1})\,(228\ \text{K})$$
$$m_{\text{diethyl ether}} = 0.297\ \text{g}$$

Because there are 1.50 g of diethyl ether, the vapor pressure will be achieved, so the answer to (a) is 57 Torr.

(b)

$$\left(\frac{535\ \text{Torr}}{760\ \text{Torr}\cdot\text{atm}^{-1}}\right)(1.00\ \text{L}) = \left(\frac{m_{\text{diethyl ether}}}{74.12\ \text{g}\cdot\text{mol}^{-1}\ \text{diethyl ether}}\right)$$
$$(0.082\,06\ \text{L}\cdot\text{atm}\cdot\text{K}^{-1}\cdot\text{mol}^{-1})(298\ \text{K})$$
$$m_{\text{diethyl ether}} = 2.13\ \text{g}$$

In this case, there is not enough diethyl ether to achieve equilibrium so all of the ether will vaporize. The pressure will be

$$(P)(1.00\text{ L}) = \left(\frac{1.50\text{ g diethyl ether}}{74.12\text{ g}\cdot\text{mol}^{-1}\text{ diethyl ether}}\right)(0.082\,06\text{ L}\cdot\text{atm}\cdot\text{K}^{-1}\cdot\text{mol}^{-1})(298\text{ K})$$

$P = 0.495$ atm or 376 Torr

(c) The system is at the same temperature as in (a) but the volume is now doubled. This volume will accommodate twice as much diethyl ether or $2 \times 0.297\text{ g} = 0.594\text{ g}$. There is still sufficient diethyl ether for this to occur so that the pressure will again be equal to the vapor pressure of diethyl ether, or 57 Torr. (d) If flask B is cooled by liquid nitrogen, its temperature will approach −196°C. The vapor pressure of ether at that temperature is negligible and the ether will solidify in flask B. As the ether condenses in flask B, the liquid ether in flask A will continue to vaporize and will do so until all the ether has condensed in flask B. At the end of this, there will be no ether left in flask A and the pressure in the apparatus will be 0.

8.122

$$0.50 = \frac{\chi_{\text{pentane, liquid}} P^{\circ}_{\text{pentane}}}{\chi_{\text{pentane, liquid}} P^{\circ}_{\text{pentane}} + (1 - \chi_{\text{pentane, liquid}}) P^{\circ}_{\text{hexane}}}$$

$$\chi_{\text{pentane, gas}} = 0.50 = \frac{\chi_{\text{pentane, liquid}} P^{\circ}_{\text{pentane}}}{\chi_{\text{pentane, liquid}} P^{\circ}_{\text{pentane}} + \chi_{\text{hexane, liquid}} P^{\circ}_{\text{hexane}}}$$

$$0.50 = \frac{(\chi_{\text{pentane, liquid}})(512\text{ Torr})}{(\chi_{\text{pentane, liquid}})(512\text{ Torr}) + (1 - \chi_{\text{pentane, liquid}})(151\text{ Torr})}$$

$$\chi_{\text{pentane, liquid}} = 0.228;\ \chi_{\text{hexane, liquid}} = 0.772$$

8.124 (a) The vapor pressures over the two solutions are different. The volatile component, ethanol, will transfer from one solution to the other until the vapor pressures (and therefore concentrations) are equal. The vapor pressure is lower over the more concentrated solution so ethanol will condense on that side of the apparatus. As this takes place, there will be a net transfer of ethanol from the less concentrated side to the more concentrated side of the apparatus.

(b) The vapor pressure will be determined by the concentration of the solution (the solutions in both sides of the apparatus will have the same concentration) once the system reaches equilibrium.

For the 0.15 *m* solution, there will be 0.15 mol sucrose $(342.29\ \text{g}\cdot\text{mol}^{-1})$ for 1000 g ethanol $(46.07\ \text{g}\cdot\text{mol}^{-1})$. This corresponds to 51 g of sucrose for a total solution mass of 1051 g.

In 15.0 g of solution there will be 0.73 g sucrose and 14.27 g of ethanol. Similarly, for the 0.050 *m* solution, there will be 0.050 mol sucrose for 1000 g ethanol, corresponding to 17 g sucrose and a total solution mass of 1017 g. In 15.0 g of solution there will be 0.25 g sucrose and 14.75 g ethanol.

At equilibrium, the concentrations must be equal. In order to calculate the vapor pressure, we need to know the concentration in terms of mole fractions. The concentrations will be made equal by transferring some ethanol from one solution to the other. This can be expressed mathematically by the following relationship:

$$\frac{\left(\dfrac{0.73\ \text{g}}{342.29\ \text{g}\cdot\text{mol}^{-1}}\right)}{\left(\dfrac{14.27+x\ \text{g}}{46.07\ \text{g}\cdot\text{mol}^{-1}}\right)+\left(\dfrac{0.73\ \text{g}}{342.29\ \text{g}\cdot\text{mol}^{-1}}\right)} = \frac{\left(\dfrac{0.25\ \text{g}}{342.29\ \text{g}\cdot\text{mol}^{-1}}\right)}{\left(\dfrac{14.75-x\ \text{g}}{46.07\ \text{g}\cdot\text{mol}^{-1}}\right)+\left(\dfrac{0.25\ \text{g}}{342.29\ \text{g}\cdot\text{mol}^{-1}}\right)}$$

Solving this equation gives $x = 7.3$ g. The mole fraction of sucrose for the solution will be 0.0045, and the mole fraction of ethanol will be 1 – 0.0045 = 0.9955. The vapor pressure will be

$P = \chi_{\text{ethanol}} P^{\circ}_{\text{ethanol}} = (0.9955)(60\ \text{Torr}) \cong 60\ \text{Torr}$. A 0.0045 mole fraction corresponds to a 0.098 *m* solution of sucrose.

8.126 (a) The polymer concentration is given by:

$$c = \frac{\Pi}{iRT} = \frac{0.325 \text{ atm}}{(1)(0.08206 \frac{\text{L atm}}{\text{K mol}})(298 \text{ K})} = 0.0133 \text{ M}$$

The moles of polymer in solution are:

$$n_{\text{polymer}} = (0.0133 \text{ M})(0.500 \text{ L}) = 0.00664 \text{ mol}$$

and the molar mass of the polymer is:

$$\text{M.M.} = \frac{47.7 \text{ g}}{0.00664 \text{ mol}} = 7180 \frac{\text{g}}{\text{mol}}$$

(b) If a monomer in the polymer is $-CH_2CH(CN)-$, then the molar mass of a monomer is 53.06 g/mol and the average number of monomers in a polymer is: $(7180\text{g/mol})/(53.06 \text{ g/mol}) = 135$ monomers.

(c) The pressure of $H_2O(g)$ above the mixture is given by:

$$P_{H_2O} = X_{H_2O} \cdot P_{H_2O,\text{ pure}}$$

$$\text{in 100 mL, } n_{H_2O} = \frac{(100 \text{ mL})(1.00 \frac{\text{g}}{\text{mL}})}{18.02 \frac{\text{g}}{\text{mol}}} = 5.55 \text{ mol}$$

and, $n_{\text{polymer}} = (0.100 \text{ L})(0.01328 \frac{\text{mol}}{\text{L}}) = 0.00133$ mol.

Therefore,

$$X_{H_2O} = \frac{5.55 \text{ mol}}{5.55 \text{ mol} + 0.00133 \text{ mol}} = 0.9998 \quad \text{and}$$

$$P_{H_2O} = (0.9998)(23.76 \text{ Torr}) = 23.75 \text{ Torr}$$

(d) Measuring the change in osmotic pressure proves to be a better method in this case. The osmotic pressure developed by the resulting polymer solution is readily measured while the change in partial pressure of $H_2O(g)$ changes by less than 0.1 % upon addition of the polymer.

8.128 Compound (a). The stationary phase is more polar than the liquid phase, so the more polar compound of (a) and (b) should be attracted more strongly to the stationary phase and should remain on the column longer. Because compound (a) has two carboxylic acid units (COOH), it will be more polar and will have the greater value of *k*.

8.130 We can use the initial data to calculate an experimental freezing-point depression constant for naphthalene:

Freezing-point depression = $k_f \times$ molality

$$k_f = \frac{\text{Freezing-point depression}}{\text{molality}} = \frac{7.0\ \text{K}}{1\ \text{mol} \cdot \text{kg}^{-1}} = 7.0\ \text{K} \cdot \text{mol}^{-1} \cdot \text{kg}$$

The molecular weight of the S_x is found using:

Freezing-point depression = $k_f \times$ molality

$$0.7 = \left(7.0\ \text{K} \cdot \text{mol}^{-1} \cdot \text{kg}\right)\left(\frac{\frac{14.8\ \text{g S}}{M_{S_x}}}{0.575\ \text{kg}}\right)$$

$$M_{S_x} = 257\ \text{g} \cdot \text{mol}^{-1}$$

The molecular formula for the sulfur compound is:

Freezing-point depression = $k_f \times$ molality

$$\frac{257\ \text{g S}_x \cdot \text{mol}^{-1}}{32.06\,\text{g S} \cdot \text{mol}^{-1}} = 8$$

Therefore, the molecular formula is S_8.

8.132 All three molecules possess the same weak London forces holding them in the condensed phase. One would expect that neopentane would have the highest vapor pressure as the molecules are rigid and unable to "flex" to increase favorable intermolecular interactions. Octane would have the lowest vapor pressure as it is the most massive of the three options.

8.134 (a) The molar mass of L-carnitine is found as in problem 8.130 above:

$$\Pi = iRT \times \text{molarity}$$

$$0.501\ \text{atm} = 1 \times 0.082\,06\ \text{L} \cdot \text{atm} \cdot \text{K}^{-1} \cdot \text{mol}^{-1} \times 305.2\ \text{K} \times \left(\frac{\frac{0.322\ \text{g}}{M}}{0.100\ \text{kg}}\right)$$

$$M = 161\,\text{g} \cdot \text{mol}^{-1}$$

(b) The empirical formula is found first using the observed composition of the compound:

Assuming 100 g of compound:

$$\text{C: } \frac{52.16 \text{ g}}{12.011 \text{ g} \cdot \text{mol}^{-1}} = 4.343 \text{ mol} \qquad (7)$$

$$\text{H: } \frac{9.38 \text{ g}}{1.00794 \text{ g} \cdot \text{mol}^{-1}} = 9.31 \text{ mol} \qquad (15)$$

$$\text{N: } \frac{8.69 \text{ g}}{14.01 \text{ g} \cdot \text{mol}^{-1}} = 0.620 \text{ mol} \qquad (1)$$

$$\text{O: } \frac{29.78 \text{ g}}{15.9994 \text{ g} \cdot \text{mol}^{-1}} = 1.861 \text{ mol} \qquad (3)$$

(the whole numbers on the right were obtained by dividing the number of moles of each element by the moles of N present.)

The empirical formula is, therefore, $C_7H_{15}O_3N$.

The molecular mass of this formula is 161.2 g mol^{-1}, which matches the molecular mass determined using the osmotic pressure above indicating that $C_7H_{15}O_3N$ is the molecular formula for L-carnitine.

8.136 (a) CH_3OH will be more soluble as it has a polar –OH group, which gives this molecule some polar character; CH_4 is completely nonpolar. (b) KI would be more soluble as it is less polar than KCl, a highly polar molecule. (c) CsBr would be more soluble, again because CsBr is less polar than highly polar LiBr.

CHAPTER 9
CHEMICAL EQUILIBRIA

9.2 (a) True

(b) False. Changing the rate of a reaction will not affect the value of the equilibrium constant; it merely changes how fast one gets to equilibrium.

(c) True

(d) False. The *standard* reaction free energy ΔG_r° is not 0 at equilibrium. The reaction free energy ΔG_r, which is dependent upon the concentrations of the products and reactants, is 0 at equilibrium.

9.4 Cl_2 decreases from 2.15 to 2.13 bar, PCl_3 decreases from 1.28 to 1.26 bar, and PCl_5 increases from 0.02 to 0.04 bar. The shapes for the curves can only be determined accurately if the rate law for the reaction is known (see Chapter 13).

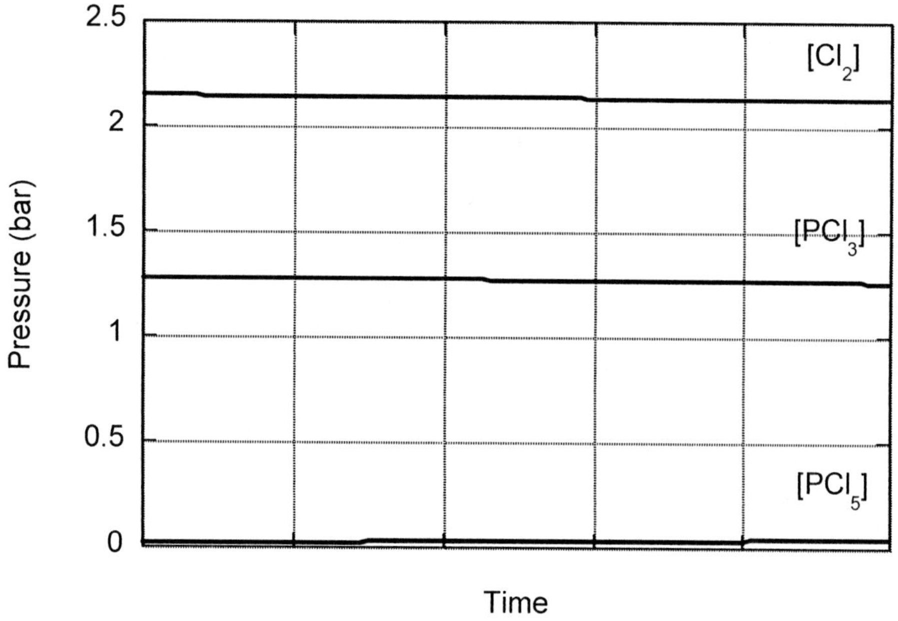

9.6 (a) $K = \frac{P_{NO_2}^{\ 2}}{P_{NO}^{\ 2}\ P_{O_2}}$; (b) $K = \frac{P_{SbCl_3}\ P_{Cl_2}}{P_{SbCl_5}}$; (c) $K = \frac{P_{N_2H_4}}{P_{N_2}\ P_{H_2}^{\ 2}}$

9.8 All values should be the same because the same amounts of the substances are present at equilibrium. It doesn't matter whether we begin with reactants or with products; the equilibrium composition will be the same if the same amounts of materials are used. If different amounts had been used, only (e) would be the same in the two containers. A more detailed analysis follows:

$H_2(g) + Br_2(g) \rightarrow 2\ HBr(g)$

The equilibrium constant expression for this system is

$$K_C = \frac{[HBr]^2}{[H_2][Br_2]}$$

In terms of the change in concentration, x, of H_2 and Br_2 that has come about at equilibrium, we may write

$$K_C = \frac{[2\,x]^2}{[0.05 - x][0.05 - x]} = \frac{[2\,x]^2}{[0.05 - x]^2} \qquad (1)$$

in the first container, and in terms of the change in concentration, y, of HBr that has come about at equilibrium in the second container, we may write

$$K_C = \frac{[0.10 - y]^2}{\left[\frac{y}{2}\right]\left[\frac{y}{2}\right]} = \frac{[0.10 - y]^2}{\left(\frac{y}{2}\right)^2} \qquad (2)$$

Because K_C is a constant, and because the relative amounts of starting materials are in the ratio of their stoichiometric factors, we must have

$$\frac{[2\,x]^2}{[0.05 - x]^2} = \frac{[0.10 - y]^2}{\left(\frac{y}{2}\right)^2}$$

$$\frac{[2\,x]}{[0.05 - x]} = \frac{[0.10 - y]}{\left(\frac{y}{2}\right)}$$

Cross multiplying:

$$xy = (0.05 - x)(0.10 - y)$$
$$xy = 0.005 - 0.05\,y - 0.10\,x + xy$$
$$y = 0.10 - 2\,x$$

This may also be seen by solving quadratic equations for Eqs. 1 and 2 for x and y to obtain any value of K_C.

(a) $[Br_2] = 0.05 - x$ in the first container, $[Br_2] = y/2$ in the second container. Because $y/2 = 0.05 - x$ satisfies the conditions of Eq. 3, the concentrations and hence the amounts of Br_2 are the same in the two cases.

(b) $[H_2] = 0.05 - x = y/2$, as above; hence the concentrations of H_2 are the same in both systems.

(c) Because all concentrations are the same in both cases, this ratio will also be the same.

(d) For the same reason as in part (c), this ratio is the same in both cases.

(e) This ratio is the equilibrium constant, so it must be the same for both systems.

(f) Because all the concentrations and amounts are the same in both cases, and because the volumes and temperatures are the same, the total pressure must be the same:

$$P = \frac{(n_{H_2} + n_{Br_2} + n_{HBr})\,RT}{V}$$

9.10 $K = P_{NH_3}\,P_{H_2S}$

For condition 1, $K = 0.307 \times 0.307 = 0.0942$
For condition 2, $K = 0.364 \times 0.258 = 0.0939$
For condition 3, $K = 0.539 \times 0.174 = 0.0938$

9.12 (a) $\dfrac{[Cl^-]^2[ClO_3^-]}{[ClO^-]^3}$

(b) $[CO_2]$

(c) $\dfrac{[H^+][CH_3COO^-]}{[CH_3COOH]}$

9.14 (a) $2\ CH_4(g) + S_8(s) \rightarrow 2\ CS_2(l) + 4\ H_2S(g)$

$$\begin{aligned}\Delta G^\circ_r &= 2 \times \Delta G^\circ_f(CS_2, l) + 4 \times \Delta G^\circ_f(H_2S, g) \\ &\quad - [2 \times \Delta G^\circ_f(CH_4, g)] \\ &= 2(65.27\ kJ \cdot mol^{-1}) + 4\,(-33.56\ kJ \cdot mol^{-1}) \\ &\quad - [2\,(-50.72\ kJ \cdot mol^{-1})] \\ &= +97.74\ kJ \cdot mol^{-1}\end{aligned}$$

$$\Delta G^\circ = -RT \ln K$$

or

$$\ln K = -\frac{\Delta G^\circ}{RT}$$

$$\ln K = -\frac{+97\,740\ J \cdot mol^{-1}}{(8.314\ J \cdot K^{-1} \cdot mol^{-1})(298\ K)} = -39.4$$

$$K = 8 \times 10^{-18}$$

(b) $CaC_2(s) + 2\ H_2O(l) \rightarrow Ca(OH)_2(s) + C_2H_2(g)$

$$\begin{aligned}\Delta G^\circ_r &= \Delta G^\circ_f(Ca(OH)_2, s) + \Delta G^\circ_f(C_2H_2, g) \\ &\quad - [\Delta G^\circ_f(CaC_2, s) + 2 \times \Delta G^\circ_f(H_2O, l)] \\ &= (-898.49\ kJ \cdot mol^{-1}) + (+209.20\ kJ \cdot mol^{-1}) \\ &\quad - [(-64.9\ kJ \cdot mol^{-1}) + 2(-237.13\ kJ \cdot mol^{-1})] \\ &= -150.1\ kJ \cdot mol^{-1}\end{aligned}$$

$$\ln K = -\frac{-150\,100\ J \cdot mol^{-1}}{(8.314\ J \cdot K^{-1} \cdot mol^{-1})(298\ K)} = +60.6$$

$$K = 2 \times 10^{26}$$

(c) $4\ NH_3(g) + 5\ O_2(g) \rightarrow 4\ NO(g) + 6\ H_2O(l)$

$$\begin{aligned}\Delta G^\circ_r &= 4\Delta G^\circ_f(NO, g) + 6\Delta G^\circ_f(H_2O, l) - [4\Delta G^\circ_f(NH_3, g)] \\ &= 4(86.55\ kJ \cdot mol^{-1}) + 6(-237.13\ kJ \cdot mol^{-1}) - [4(-16.45\ kJ \cdot mol^{-1})] \\ &= -1010.8\ kJ \cdot mol^{-1}\end{aligned}$$

$$\ln K = -\frac{-1010.8 \times 10^3\ J \cdot mol^{-1}}{(8.314\ J \cdot K^{-1} \cdot mol^{-1})(298\ K)} = +408$$

$$K \cong 10^{177}$$

(d) $CO_2(g) + 2\ NH_3(g) \rightarrow CO(NH_2)_2(s) + H_2O(l)$

$$\begin{aligned}\Delta G^\circ_r &= \Delta G^\circ_f(CO(NH_2)_2, s) + \Delta G^\circ_f(H_2O, l) \\ &\quad - [\Delta G^\circ_f(CO_2, g) + 2 \times \Delta G^\circ_f(NH_3, g)] \\ &= (-197.33\ kJ \cdot mol^{-1}) + (-237.13\ kJ \cdot mol^{-1}) \\ &\quad - [(-394.36\ kJ \cdot mol^{-1}) + 2(-16.45\ kJ \cdot mol^{-1})] \\ &= -7.20\ kJ \cdot mol^{-1}\end{aligned}$$

$$\ln K = -\frac{-7.20 \times 10^3\ J}{(8.314\ J \cdot K^{-1} \cdot mol^{-1})(298\ K)} = +2.91$$

$$K = 18$$

9.16 (a) $\Delta G^\circ_r = -RT \ln K$

$$= -(8.314\ J \cdot K^{-1} \cdot mol^{-1})(700\ K) \ln 54$$
$$= -23.2\ kJ \cdot mol^{-1}$$

(b) $\Delta G^\circ_r = -RT \ln K$

$$= -(8.314\ J \cdot K^{-1} \cdot mol^{-1})(298\ K) \ln 0.30$$
$$= +2.98\ kJ \cdot mol^{-1}$$

9.18 $\Delta G^\circ_r = \Delta G^\circ_f(CO_2, g)$

$$= -394.36\ kJ \cdot mol^{-1}$$

$$\Delta G^\circ = -RT \ln K$$

$$\ln K = -\frac{\Delta G^\circ}{RT}$$

$$\ln K = -\frac{-394.36 \times 10^3\ J \cdot mol^{-1}}{(8.314\ J \cdot K^{-1} \cdot mol^{-1})(298\ K)} = +159.17$$

$$K = 1.3 \times 10^{69}$$

In practice, no K will be so precise. A better estimate would be 1×10^{69}.

Because $Q < K$, the reaction will tend to proceed to produce products.

9.20 The free energy at a specific set of conditions is given by

$$\Delta G_r = \Delta G°_r + RT \ln Q$$
$$\Delta G_r = -RT \ln K + RT \ln Q$$
$$\Delta G_r = -RT \ln K + RT \ln \frac{p_{Cl_5}}{p_{Cl_3} \cdot p_{Cl_2}}$$
$$= -(8.314\ J \cdot K^{-1} \cdot mol^{-1})(503\ K) \ln 49$$
$$+ (8.314\ J \cdot K^{-1} \cdot mol^{-1})(503\ K) \ln \frac{(1.33)}{(0.22)(0.41)}$$
$$= -5.0\ kJ \cdot mol^{-1}$$

Because ΔG_r is negative, the reaction will be spontaneous to form products.

9.22 The free energy at a specific set of conditions is given by

$$\Delta G_r = \Delta G°_r + RT \ln Q$$
$$\Delta G_r = -RT \ln K + RT \ln Q$$
$$\Delta G_r = -RT \ln K + RT \ln \frac{[HI]^2}{[H_2][I_2]}$$
$$= -(8.314\ J \cdot K^{-1} \cdot mol^{-1})(700\ K) \ln 54$$
$$+ (8.314\ J \cdot K^{-1} \cdot mol^{-1})(700\ K) \ln \frac{(2.17)^2}{(0.16)(0.25)}$$
$$= 4.5\ kJ \cdot mol^{-1}$$

Because ΔG_r is positive, the reaction will proceed to form reactants.

9.24 (a) $K = \dfrac{P_{SO_3}{}^2}{P_{SO_2}{}^2\ P_{O_2}} = 3.4$

$$K = \left(\frac{T}{12.027\ K}\right)^{\Delta n} K_C$$

$$K_C = \left(\frac{12.027\ K}{T}\right)^{\Delta n} K = \left(\frac{12.027\ K}{1000\ K}\right)^{(2-3)} 3.4 = 2.8 \times 10^2$$

(b) $K = P_{NH_3} P_{H_2S} = 9.4 \times 10^{-2}$

$$K_C = \left(\frac{12.02\text{ K}}{T}\right)^{\Delta n} K = \left(\frac{12.027\text{ K}}{297\text{ K}}\right)^{(2-0)} 9.4\times10^{-2} = 1.5\times10^{-4}$$

9.26 For the equation written $2\,SO_2(g) + O_2(g) \rightarrow 2\,SO_3(g)$ Eq. 1

$$K = \frac{P_{SO_3}{}^2}{P_{SO_2}{}^2\,P_{O_2}} = 2.5\times10^{10}$$

(a) For the equation written $SO_2(g) + \frac{1}{2}O_2(g) \rightarrow SO_3(g)$ Eq. 2

$$K_{\text{Eq. 2}} = \frac{P_{SO_3}}{P_{SO_2}P_{O_2}{}^{1/2}} = \sqrt{K_{\text{Eq. 1}}} = \sqrt{2.5\times10^{10}} = 1.6\times10^5$$

(b) For the equation written $SO_3(g) \rightarrow SO_2(g) + \frac{1}{2}O_2(g)$ Eq. 3

This equation is the reverse of Eq. 2, so $K_{\text{Eq. 3}} = \dfrac{1}{K_{\text{Eq. 2}}} = \dfrac{1}{\sqrt{K_{\text{Eq. 1}}}}$

$$K_{\text{Eq. 2}} = \frac{P_{SO_2}P_{O_2}{}^{1/2}}{P_{SO_3}} = \frac{1}{\sqrt{K_{\text{Eq. 1}}}} = \frac{1}{\sqrt{2.5\times10^{10}}} = 6.3\times10^{-6}$$

(c) For the equation written $3\,SO_2(g) + \frac{3}{2}O_2(g) \rightarrow 3\,SO_3(g)$ Eq. 4

This equation is $\frac{3}{2}\times$ Eq. 1, so $K_{\text{Eq. 4}} = K_{\text{Eq. 1}}{}^{3/2}$

$$K_{\text{Eq. 2}} = \frac{P_{SO_3}{}^3}{P_{SO_2}{}^3P_{O_2}{}^{3/2}} = K_{\text{Eq. 1}}{}^{3/2} = (2.5\times10^{10})^{3/2} = 4.0\times10^{15}$$

9.28 $H_2(g) + Cl_2(g) \rightarrow 2\,HCl(g)$ $K_C = 5.1\times10^8$

$$K_C = \frac{[HCl]^2}{[H_2][Cl_2]}$$

$$5.1\times10^8 = \frac{(1.45\times10^{-3})^2}{[H_2](2.45\times10^{-3})}$$

$$[H_2] = \frac{(1.45\times10^{-3})^2}{(5.1\times10^8)(2.45\times10^{-3})}$$

$$[H_2] = 1.7\times10^{-12}\text{ mol}\cdot\text{L}^{-1}$$

9.30 $K = \frac{P_{SbCl_3} P_{Cl_2}}{P_{SbCl_5}}$

$$3.5 \times 10^{-4} = \frac{(5.02 \times 10^{-3}) P_{Cl_2}}{0.072}$$

$$P_{Cl_2} = \frac{(3.5 \times 10^{-4})(0.072)}{5.02 \times 10^{-3}} = 5.0 \times 10^{-3} \text{ bar}$$

9.32 (a) $K_C = \frac{[NH_3]^2}{[N_2][H_2]^3} = 0.278$

$$Q_C = \frac{[0.122]^2}{[0.417][0.524]^3} = 0.248$$

(b) $Q_C \neq K_C$; therefore the system is not at equilibrium.

(c) Because $Q_C < K_C$, more products will be formed.

9.34 (a) $K_C = \frac{[NH_3]^2}{[N_2][H_2]^3} = 62$

$$Q_C = \frac{[1.12 \times 10^{-4}]^2}{[2.23 \times 10^{-3}][1.24 \times 10^{-3}]^3} = 2.95 \times 10^3$$

(b) Because $Q_C > K_C$, ammonia will decompose to form reactants.

9.36 $\frac{1.00 \text{ g } I_2}{253.8 \text{ g} \cdot \text{mol}^{-1}} = 0.003\,94 \text{ mol } I_2$; $\frac{0.830 \text{ g } I_2}{253.8 \text{ g} \cdot \text{mol}^{-1}} = 0.003\,27 \text{ mol } I_2$

$I_2(g) \rightleftharpoons 2\,I(g)$

$0.003\,94 \text{ mol} - x \qquad 2x$

$0.003\,94 \text{ mol} - x = 0.003\,27 \text{ mol}$

$x = 0.000\,67 \text{ mol}$; $2x = 0.0013 \text{ mol}$

$$K_C = \frac{\left[\frac{0.0013}{1.00}\right]^2}{\left[\frac{0.003\,27}{1.00}\right]} = 5.2 \times 10^{-4}$$

9.38

Pressure (Torr)	$CO(g)$ +	$H_2O(g)$ $\rightleftharpoons$	$CO_2(g)$ +	$H_2(g)$
initial	200	200	0	0
final	200 – 88	200 – 88	88	88

Note: Because pressure is directly proportional to the number of moles of a substance, the pressure changes can be used directly in calculating the reaction stoichiometry. Technically, to achieve the correct standard state condition, the Torr must be converted to bar $(750.1 \text{ Torr} \cdot \text{bar}^{-1})$; however, in this case those conversion factors will cancel because there are equal numbers of moles of gas on both sides of the equation.

$$K = \frac{P_{CO_2} P_{H_2}}{P_{CO} P_{H_2O}} = \frac{\left(\frac{88}{750.1}\right)\left(\frac{88}{750.1}\right)}{\left(\frac{112}{750.1}\right)\left(\frac{112}{750.1}\right)} = 0.62$$

9.40 (a) The balanced equation is $Cl_2(g) \rightleftharpoons 2\,Cl(g)$

The initial concentration of $Cl_2(g)$ is $\dfrac{0.0050 \text{ mol } Cl_2}{2.0 \text{ L}} = 0.0025 \text{ mol} \cdot \text{L}^{-1}$

Concentration $(\text{mol} \cdot \text{L}^{-1})$	$Cl_2(g)$ $\rightleftharpoons$	$2\,Cl(g)$
initial	0.0025	0
change	$-x$	$+2\,x$
equilibrium	$0.0025 - x$	$+2\,x$

$$K_C = \frac{[Cl]^2}{[Cl_2]} = \frac{(2\,x)^2}{(0.0025 - x)} = 1.7 \times 10^{-3}$$

$$4\,x^2 = (1.7 \times 10^{-3})(0.0025 - x)$$

$$4\,x^2 + (1.7 \times 10^{-3})x - (4.25 \times 10^{-6}) = 0$$

$$x = \frac{-(1.7 \times 10^{-3}) \pm \sqrt{(1.7 \times 10^{-3})^2 - 4(4)(-4.3 \times 10^{-6})}}{2 \times 4}$$

$$x = \frac{-(1.7 \times 10^{-3}) \pm 8.47 \times 10^{-3}}{8}$$

$$x = -1.3 \times 10^{-3} \text{ or } +8.5 \times 10^{-4}$$

The negative answer is not meaningful, so we choose $x =$ 8.5×10^{-4} mol·L^{-1}. The concentration of Cl_2 is $0.0025 - 8.5 \times 10^{-4}$ $= 0.0017$. The concentration of Cl atoms is $2 \times (8.5 \times 10^{-4})$ $= 1.7 \times 10^{-4}$ mol·L^{-1}. The percentage decomposition of Cl_2 is given by

$$\frac{8.5 \times 10^{-4}}{0.0025} \times 100 = 34\%$$

(b) The balanced equation is $Br_2(g) \rightleftharpoons 2\ Br(g)$

The initial concentration of $Br_2(g)$ is $\dfrac{5.0 \text{ mol } Br_2}{2.0 \text{ L}} = 2.5 \text{ mol} \cdot L^{-1}$

Concentration (mol·L^{-1})	$Br_2(g)$	$\rightleftharpoons$	2 Br(g)
initial	2.5		0
change	$-x$		$+2x$
equilibrium	$2.5 - x$		$+2x$

$$K_C = \frac{[Br]^2}{[Br_2]} = \frac{(2x)^2}{(2.5 - x)} = 1.7 \times 10^{-3}$$

$$4x^2 = (1.7 \times 10^{-3})(2.5 - x)$$

$$4x^2 + (1.7 \times 10^{-3})x - (4.25 \times 10^{-3}) = 0$$

$$x = \frac{-(1.7 \times 10^{-3}) \pm \sqrt{(1.7 \times 10^{-3})^2 - 4(4)(-4.25 \times 10^{-3})}}{2 \times 4}$$

$$x = \frac{-(1.7 \times 10^{-3}) \pm 2.608 \times 10^{-1}}{8}$$

$$x = -3.3 \times 10^{-2} \text{ or } +3.2 \times 10^{-2}$$

The negative answer is not meaningful, so we choose $x =$ 3.2×10^{-2} mol·L^{-1}. The concentration of Br_2 is $2.5 - 3.2 \times 10^{-2} \approx 2.5$. The concentration of Br atoms is $2 \times (3.2 \times 10^{-2}) =$ 6.4×10^{-2} mol·L^{-1}. The percentage decomposition of Br_2 is given by

$$\frac{3.2 \times 10^{-2}}{2.5} \times 100 = 1.3\%$$

(c) At this temperature, Cl_2 and Br_2 are equally stable since their equilibrium constants are the same. If we had used the same initial amounts of Cl_2 and Br_2 in parts (a) and (b), then the percent decomposition would have been the same as well.

9.42

Pressure (bar)	$2\ BrCl(g)$	$\rightleftharpoons$	$Br_2(g)$ +	$Cl_2(g)$
initial	1.4×10^{-3}		0	0
change	$-2\ x$		$+x$	$+x$
final	$1.4 \times 10^{-3} - 2x$		$+x$	$+x$

$$K = \frac{p_{Br_2} \cdot p_{Cl_2}}{p_{BrCl}^{\ 2}}$$

$$32 = \frac{(x)(x)}{(1.4 \times 10^{-3} - 2\ x)^2} = \frac{x^2}{(1.4 \times 10^{-3} - 2\ x)^2}$$

$$\sqrt{32} = \sqrt{\frac{x^2}{(1.4 \times 10^{-3} - 2\ x)^2}}$$

$$\frac{x}{(1.4 \times 10^{-3} - 2\ x)} = \sqrt{32}$$

$$x = (\sqrt{32})(1.4 \times 10^{-3} - 2\ x)$$

$$x + 2\sqrt{32}\ x = (\sqrt{32})(1.4 \times 10^{-3})$$

$$(1 + 2\sqrt{32})\ x = (\sqrt{32})(1.4 \times 10^{-3})$$

$$x = \frac{(\sqrt{32})(1.4 \times 10^{-3})}{(1 + 2\sqrt{32})}$$

$$x = 6.4 \times 10^{-4}$$

$$p_{Br_2} = p_{Cl_2} = 6.4 \times 10^{-4}\ \text{bar} = 0.64\ \text{mbar}$$

$$p_{BrCl} = 1.4 \times 10^{-3}\ \text{bar} - 2\ (6.4 \times 10^{-4}\ \text{bar}) = 1.2 \times 10^{-4}\ \text{bar} = 0.12\ \text{mbar}$$

The percentage decomposition is given by

$$\frac{2\ (6.4 \times 10^{-4}\ \text{bar})}{1.4 \times 10^{-3}\ \text{bar}} \times 100 = 91\%$$

9.44 (a) concentration of PCl_5 initially =

$$\frac{\left(\frac{2.0\text{ g PCl}_5}{208.22\text{ g}\cdot\text{mol}^{-1}\text{ PCl}_5}\right)}{0.300\text{ L}} = 0.032\text{ mol}\cdot\text{L}^{-1}$$

Concentration ($mol \cdot L^{-1}$)	$PCl_5(g)$	$\rightleftharpoons$	$PCl_3(g)$	+	$Cl_2(g)$
initial	0.032		0		0
change	$-x$		$+x$		$+x$
final	$0.032 - x$		$+x$		$+x$

$$K_C = \frac{[PCl_3][Cl_2]}{[PCl_5]} = \frac{(x)(x)}{(0.032 - x)} = \frac{x^2}{(0.032 - x)}$$

$$\frac{x^2}{(0.032 - x)} = 0.61$$

$$x^2 = (0.61)(0.032 - x)$$

$$x^2 + (0.61)x - 0.020 = 0$$

$$x = \frac{-(0.61) \pm \sqrt{(0.61)^2 - (4)(1)(-0.020)}}{2 \cdot 1}$$

$$x = \frac{-(0.61) \pm 0.67}{2 \cdot 1} = +0.03 \text{ or } -0.64$$

The negative root is not meaningful, so we choose $x = 0.03\text{ mol}\cdot\text{L}^{-1}$.

$[PCl_3] = [Cl_2] = 0.03\text{ mol}\cdot\text{L}^{-1}$;

$[PCl_5] = 0.032 - 0.03\text{ mol}\cdot\text{L}^{-1} = 0.002\text{ mol}\cdot\text{L}^{-1}$. The solution of this problem again points up the problems with following the significant figure conventions for calculations of this type, because we would use only one significant figure as imposed by the subtraction in the quadratic equation solution. The fact that the equilibrium constant is given to two significant figures, however, suggests that the concentrations could be determined to that level of accuracy. Plugging the answers back into the equilibrium expression gives a value of 0.45, which seems somewhat off from the starting value of 0.61. The closest agreement to the equilibrium expression comes from using $[PCl_3] = [Cl_2] = 0.0305\text{ mol}\cdot\text{L}^{-1}$ and

$[PCl_5] = 0.0015 \text{ mol} \cdot L^{-1}$; this gives an equilibrium constant of 0.62. Rounding these off to two significant figures gives $[PCl_3] = [Cl_2] = 0.030 \text{ mol} \cdot L^{-1}$ and $[PCl_5] = 0.0015$, which in the equilibrium expression produces a value of 0.60. The percentage decomposition is given by

$$\frac{0.030}{0.032} \times 100\% = 94\%$$

9.46 Starting concentration of $NH_3 = \dfrac{0.200 \text{ mol}}{2.00 \text{ L}} = 0.100 \text{ mol} \cdot L^{-1}$

Concentration $(\text{mol} \cdot L^{-1})$	$NH_4HS(s)$	$\rightleftharpoons$	$NH_3(g)$	+	$H_2S(g)$
initial	—		0.100		0
change	—		$+x$		$+x$
final	—		$0.100 + x$		$+x$

$$K_C = [NH_3][H_2S] = (0.100 + x)(x)$$

$$1.6 \times 10^{-4} = (0.100 + x)(x)$$

$$x^2 + 0.100\,x - 1.6 \times 10^{-4} = 0$$

$$x = \frac{-(+0.100) \pm \sqrt{(+0.100)^2 - (4)(1)(-1.6 \times 10^{-4})}}{2 \cdot 1}$$

$$x = \frac{-0.100 \pm 0.1031}{2 \cdot 1} = +0.002 \text{ or } -0.102$$

The negative root is not meaningful, so we choose $x = 2 \times 10^{-3} \text{ mol} \cdot L^{-1}$.

$$[NH_3] = +0.100 \text{ mol} \cdot L^{-1} + 2 \times 10^{-3} \text{ mol} \cdot L^{-1}$$
$$= 0.102 \text{ mol} \cdot L^{-1}$$

$$[H_2S] = 2 \times 10^{-3} \text{ mol} \cdot L^{-1}$$

Alternatively, we could have assumed that $x << 0.100$, in which case $0.100\,x = 1.6 \times 10^{-4}$ or $x = 1.6 \times 10^{-3}$.

9.48 The initial concentrations of PCl_3 and Cl_2 are calculated as follows:

$$[PCl_3] = \frac{0.200 \text{ mol}}{8.00 \text{ L}} = 0.0250 \text{ mol} \cdot L^{-1}; [Cl_2] = \frac{0.600 \text{ mol}}{8.00 \text{ L}}$$

$$= 0.0750 \text{ mol} \cdot L^{-1}$$

Concentrations $(\text{mol} \cdot L^{-1})$	PCl_5	$\rightleftharpoons$ $PCl_3(g)$	+ $Cl_2(g)$
initial	0	0.0250	0.0750
change	$+x$	$-x$	$-x$
final	$+x$	$0.0250 - x$	$0.0750 - x$

$$K_C = \frac{[PCl_3][Cl_2]}{[PCl_5]} = \frac{(0.0250 - x)(0.0750 - x)}{(+x)} = 33.3$$

$$x^2 - 0.100\,x + 0.001\,875 = 33.3\,x$$

$$x^2 - 33.40\,x + 0.001\,875 = 0$$

$$x = \frac{+33.40 \pm \sqrt{(-33.40)^2 - (4)(1)(0.001\,875)}}{(2)(1)} = \frac{+33.40 \pm 33.399\,89}{2}$$

$$= +5.5 \times 10^{-5} \text{ or } +33.4$$

The root +33.4 has no physical meaning because it is greater than the starting concentrations of PCl_3 and Cl_2, so it can be discarded.

$[PCl_5] = 5.5 \times 10^{-5} \text{ mol} \cdot L^{-1}; [PCl_3] = 0.0250 \text{ mol} \cdot L^{-1}$

$- 5.5 \times 10^{-5} \text{ mol} \cdot L^{-1} = 0.0249 \text{ mol} \cdot L^{-1};$

$[Cl_2] = 0.0750 \text{ mol} \cdot L^{-1} - 5.5 \times 10^{-5} \text{ mol} \cdot L^{-1} = 0.0749 \text{ mol} \cdot L^{-1}$. Note that the normal conventions concerning significant figures were ignored in order to obtain a meaningful answer.

9.50 The initial concentrations of N_2 and O_2 are

$$[N_2] = \frac{0.0140 \text{ mol}}{10.0 \text{ L}} = 0.001\,40 \text{ mol} \cdot L^{-1}; [O_2] = \frac{0.240 \text{ mol}}{10.0 \text{ L}}$$

$$= 0.0214 \text{ mol} \cdot L^{-1}$$

Concentrations $(\text{mol} \cdot L^{-1})$	$N_2(g)$ +	$O_2(g)$ $\rightleftharpoons$	$2\,NO(g)$
initial	0.001 40	0.0214	0
change	$-x$	$-x$	$+2x$
final	$0.001\,40 - x$	$0.0214 - x$	$+2x$

$$K_C = \frac{[NO]^2}{[N_2][O_2]} = \frac{(2\,x)^2}{(0.001\,40 - x)(0.0214 - x)}$$

$$1.00 \times 10^{-5} = \frac{(2\,x)^2}{(0.001\,40 - x)(0.0214 - x)}$$

$$1.00 \times 10^{-5} = \frac{4\,x^2}{x^2 - 0.0228\,x + 3.0 \times 10^{-5}}$$

$$4\,x^2 = (1.00 \times 10^{-5})(x^2 - 0.0228x + 3.0 \times 10^{-5})$$

$$4\,x^2 = 1.00 \times 10^{-5}\,x^2 - 2.28 \times 10^{-7}\,x + 3.0 \times 10^{-10}$$

$$4\,x^2 + 2.28 \times 10^{-7}\,x - 3.0 \times 10^{-10} = 0$$

$$x = \frac{-2.28 \times 10^{-7} \pm \sqrt{(2.28 \times 10^{-7})^2 - (4)(4)(-3.0 \times 10^{-10})}}{(2)(4)}$$

$$x = \frac{-2.28 \times 10^{-7} \pm 6.93 \times 10^{-5}}{8} = 8.6 \times 10^{-6} \text{ or } -8.7 \times 10^{-6}$$

The negative root can be discarded because it has no physical meaning.

$[NO] = 2\,x = 2(8.6 \times 10^{-6}) = 1.7 \times 10^{-5}\,\text{mol} \cdot \text{L}^{-1}$; the concentrations of N_2 and O_2 remain essentially unchanged at $0.001\,39\ \text{mol} \cdot \text{L}^{-1}$ and $0.0214\ \text{mol} \cdot \text{L}^{-1}$, respectively.

9.52 The initial concentrations of N_2 and H_2 are

$$[N_2] = [H_2] = \frac{0.20\ \text{mol}}{25.0\ \text{L}} = 0.0080\ \text{mol} \cdot \text{L}^{-1}.$$

At equilibrium, 5.0 % of the N_2 had reacted, so 95.0 % of the N_2 remains:

$$[N_2] = (0.950)(0.0080\ \text{mol} \cdot \text{L}^{-1}) = 0.0076\ \text{mol} \cdot \text{L}^{-1}$$

If 5.0 % reacted, then

$$0.050 \times 0.200\ \text{mol } N_2 \times \frac{2\ \text{mol } NH_3}{\text{mol } N_2} = 0.020\ \text{mol } NH_3 \text{ formed.}$$

$$\text{The concentration of } NH_3 \text{ formed} = \frac{0.020\ \text{mol}}{25.0\ \text{L}} = 8.0 \times 10^{-4}\ \text{mol} \cdot \text{L}^{-1}.$$

The amount of

$$H_2 \text{ reacted} = 0.050 \times 0.200\ \text{mol } N_2 \times \frac{3\ \text{mol } H_2}{\text{mol } N_2} = 0.030\ \text{mol } H_2 \text{ used.}$$

Concentration of H_2 present at equilibrium

$$= \frac{0.200\text{ mol} - 0.030\text{ mol}}{25.0\text{ L}} = 0.0068\text{ mol}\cdot\text{L}^{-1}.$$

$$K_C = \frac{[\text{NH}_3]^2}{[\text{N}_2][\text{H}_2]^3} = \frac{(8.0\times10^{-4})^2}{(0.0076)(0.0068)^3} = 2.7\times10^2$$

9.54 Note: The volume of the system is not used because we are given pressures and K.

Pressures (bar)	$N_2(g)$ +	$3\ H_2(g)$ ⇌	$2\ NH_3(g)$
initial	0.025	0.015	0
change	$-x$	$-3x$	$+2x$
final	$0.025 - x$	$0.015 - 3x$	$+2x$

$$K = \frac{P_{\text{NH}_3}{}^2}{P_{\text{N}_2}\,P_{\text{H}_2}{}^3} = \frac{(2x)^2}{(0.025 - x)(0.015 - 3x)^3} = 0.036$$

Solving this explicitly will lead to a high-order equation, so first check to see if the assumption that $3x \ll 0.015$ can be used to simplify the math:

$$\frac{(2x)^2}{(0.025)(0.015)^3} = 0.036$$

$$4x^2 = 3.04\times10^{-9}$$

$$x = 2.8\times10^{-5}$$

Comparing x to 0.015, we see that the approximation was justified.

At equilibrium, $P_{\text{NH}_3} = 2\times2.8\times10^{-5}\text{ bar} = 5.6\times10^{-5}\text{ bar}$; the pressures of N_2 and H_2 remain essentially unchanged.

9.56 Concentrations

$(\text{mol}\cdot\text{L}^{-1})$	CH_3COOH +	C_2H_5OH $\rightleftharpoons$	$CH_3COOC_2H_5$ +	H_2O
initial	0.024	0.059	0	0.015
change	$-x$	$-x$	$+x$	$+x$
final	$0.024 - x$	$0.059 - x$	$+x$	$0.015 + x$

$$K_C = \frac{[CH_3COOC_2H_5][H_2O]}{[CH_3COOH][C_2H_5OH]} = \frac{(x)(0.015 + x)}{(0.024 - x)(0.059 - x)}$$

$$= \frac{x^2 + 0.015\,x}{x^2 - 0.083\,x + 0.0014}$$

$$4.0 = \frac{x^2 + 0.015\,x}{x^2 - 0.083\,x + 0.001\,42}$$

$$4.0\,x^2 - 0.332\,x + 0.005\,68 = x^2 + 0.015\,x$$

$$3.0\,x^2 - 0.347\,x + 0.005\,68 = 0$$

$$x = \frac{-(-0.347) \pm \sqrt{(-0.347)^2 - (4)(3.0)(0.005\,68)}}{(2)(3.0)} = \frac{+0.347 \pm 0.228}{6.0}$$

$x = 0.0958$ or 0.0198

The root 0.0958 is meaningless because it is larger than the initial concentration of acetic acid and ethanol, so the value 0.0198 is chosen. The equilibrium concentration of the product ester is, therefore, $0.0198\ \text{mol}\cdot\text{L}^{-1}$. The numbers can be confirmed by placing them into the equilibrium expression:

$$K_C = \frac{[CH_3COOC_2H_5][H_2O]}{[CH_3COOH][C_2H_5OH]} = \frac{(0.0198)(0.015 + 0.0198)}{(0.024 - 0.0198)(0.059 - 0.0198)} = 4.1$$

This is reasonably good agreement, given the nature of the calculation. Given that the K_C value is reported to only two significant figures, the best report of the concentration of ester will be $0.020\ \text{mol}\cdot\text{L}^{-1}$.

9.58 $$K = \frac{P_{PCl_5}}{P_{PCl_3}P_{Cl_2}}$$

$$3.5 \times 10^4 = \frac{1.3 \times 10^2}{(9.56)P_{Cl_2}}$$

$$P_{Cl_2} = \frac{1.3 \times 10^2}{(9.56)(3.5 \times 10^4)} = 3.9 \times 10^{-4} \text{ bar}$$

9.60 We use the reaction stoichiometry to calculate the amounts of substances present at equilibrium:

Amounts (mol)	$CO(g)$ +	$H_2O(g)$ $\rightleftharpoons$	$CO_2(g)$ +	$H_2(g)$
initial	1.000	1.000	0	0
change	$-x$	$-x$	$+x$	$+x$
final	$1.000 - x$	$1.000 - x$	0.665	$+x$

(a) Because $x = 0.665$ mol, there will be $(1.000 - 0.665)$ mol = 0.335 mol CO; 0.335 mol H_2O; 0.665 mol H_2. The concentrations are easy to calculate because $V = 10.00$ L:

$$[CO] = [H_2O] = 0.0335 \text{ mol} \cdot L^{-1}; [CO_2] = [H_2] = 0.0665 \text{ mol} \cdot L^{-1}$$

(b) $$K_C = \frac{[CO_2][H_2]}{[CO][H_2O]} = \frac{(0.0665)^2}{(0.0335)^2} = 3.94$$

9.62 The initial concentration of $H_2S = \frac{0.100 \text{ mol}}{10.0 \text{ L}} = 0.0100 \text{ mol} \cdot L^{-1}$

The final concentration of $H_2 = \frac{0.0285 \text{ mol}}{10.0 \text{ L}} = 0.002\,85 \text{ mol} \cdot L^{-1}$

Concentrations ($mol \cdot L^{-1}$)	$2\,H_2S(g)$ $\rightleftharpoons$	$2\,H_2(g)$ +	$S_2(g)$
initial	0.0100	0	0
change	$-2\,x$	$+2\,x$	$+x$
final	$0.0100 - 2\,x$	0.002 85	$+x$

Thus, $2\,x = 0.002\,85 \text{ mol} \cdot L^{-1}$ or $x = 0.001\,42 \text{ mol} \cdot L^{-1}$

At equilibrium:

$[H_2S] = 0.0100\ \text{mol} \cdot L^{-1} - 0.002\,85\ \text{mol} \cdot L^{-1} = 0.0072\ \text{mol} \cdot L^{-1}$

$[H_2] = 0.002\,85\ \text{mol} \cdot L^{-1}$

$[S_2] = 0.001\,42\ \text{mol} \cdot L^{-1}$

$$K_C = \frac{(0.002\,85)^2(0.001\,42)}{(0.0072)^2} = 2.22 \times 10^{-4}$$

9.64 The initial concentration of $PCl_5 = \dfrac{0.865\ \text{mol}}{0.500\ \text{L}} = 1.73\ \text{mol} \cdot L^{-1}$

Concentrations ($\text{mol} \cdot L^{-1}$)	$PCl_5(g)$ $\rightleftharpoons$	$PCl_3(g)$ +	$Cl_2(g)$
initial	1.73	0	0
change	$-x$	$+x$	$+x$
final	$1.73 - x$	$+x$	$+x$

$$K_C = \frac{[PCl_3][Cl_2]}{[PCl_5]}$$

$$1.80 = \frac{(x)(x)}{(1.73 - x)} = \frac{x^2}{(1.73 - x)}$$

$$(1.80)(1.73 - x) = x^2$$

$$x^2 + 1.80\,x - 3.114 = 0$$

$$x = \frac{-1.80 \pm \sqrt{(1.80)^2 - (4)(1 - 3.114)}}{(2)(1)} = \frac{-1.80 \pm 3.96}{2}$$

$$x = +1.08 \text{ or } -2.88$$

The negative root is not physically meaningful and can be discarded. The concentrations of PCl_3 and Cl_2 are, therefore, 1.08 $\text{mol} \cdot L^{-1}$ at equilibrium, and the concentration of PCl_5 is 1.73 $\text{mol} \cdot L^{-1}$ $- 1.08\ \text{mol} \cdot L^{-1}$ $= 0.65\ \text{mol} \cdot L^{-1}$. These numbers can be checked by substituting back into the equilibrium expression:

$$K_C = \frac{[PCl_3][Cl_2]}{[PCl_5]}$$

$$\frac{(1.08)^2}{(0.65)} = 1.79$$

which compares well to K_C (1.80).

9.66 We use the ideal gas relationship to find the initial concentration of HCl(g) at 25°C:

$$n = \frac{PV}{RT}$$

$$n = \frac{\left(1.00 \text{ bar} \times \dfrac{1 \text{ atm}}{1.013\ 25 \text{ bar}}\right)(4.00 \text{ L})}{(0.082\ 06 \text{ L} \cdot \text{atm} \cdot \text{K}^{-1} \cdot \text{mol}^{-1})(273 \text{ K})} = 0.176 \text{ mol}$$

$$[\text{HCl}] = \frac{0.176 \text{ mol}}{12.00 \text{ L}} = 0.0147 \text{ mol} \cdot \text{L}^{-1}$$

Concentrations ($\text{mol} \cdot \text{L}^{-1}$)	2 HCl(g) +	I_2(s) $\rightleftharpoons$	2 HI(g) +	Cl_2(g)
initial	0.0147	—	0	0
change	$-2\,x$	—	$+2\,x$	$+x$
final	$0.147 - 2\,x$	—	$+2\,x$	$+x$

$$K_C = \frac{[\text{HI}]^2[\text{Cl}_2]}{[\text{HCl}]^2}$$

$$1.6 \times 10^{-34} = \frac{(2\,x)^2(x)}{(0.0147 - 2\,x)^2}$$

Because the equilibrium constant is very small, we will assume that $x <<$ 0.0147:

$$1.6 \times 10^{-34} = \frac{4\,x^3}{(0.0147)^2}$$

$$x^3 = \frac{(1.6 \times 10^{-34})(0.0147)^2}{4}$$

$$x = \sqrt[3]{\frac{(1.6 \times 10^{-34})(0.0147)^2}{4}} = 2.1 \times 10^{-13}$$

At equilibrium:

$$[\text{HI}] = 2 \times 2.1 \times 10^{-13} = 4.2 \times 10^{-13}$$

$$[\text{Cl}_2] = 2.1 \times 10^{-13}$$

$$[\text{HCl}] = 0.0147 \text{ mol} \cdot \text{L}^{-1}$$

9.68 initial $[NH_3] = \dfrac{\left(\dfrac{25.6\text{ g}}{17.03\text{ g}\cdot\text{mol}^{-1}}\right)}{5.00\text{ L}} = 0.301\text{ mol}\cdot\text{L}^{-1}$

Concentrations ($\text{mol}\cdot\text{L}^{-1}$)	$2\,NH_3(g) \rightleftharpoons$	$N_2(g)$ +	$3\,H_2(g)$
initial	0.301	0	0
change	$-2x$	$+x$	$+3x$
final	$0.301 - 2\,x$	$+x$	$+3x$

$$K_C = \frac{[N_2][H_2]^3}{[NH_3]^2}$$

$$\frac{(x)(3\,x)^3}{(0.301 - 2\,x)^2} = 0.395$$

$$\frac{27\,x^4}{(0.301 - 2\,x)^2} = 0.395$$

$$\sqrt{\frac{27\,x^4}{(0.301 - 2\,x)^2}} = \sqrt{0.395}$$

$$\frac{3\sqrt{3}\,x^2}{(0.301 - 2\,x)} = 0.628$$

$$3\sqrt{3}\,x^2 = (0.628)(0.301 - 2\,x) = 0.189 - 1.26\,x$$

$$5.20\,x^2 + 1.26\,x - 0.189 = 0$$

$$x = \frac{-1.26 \pm \sqrt{(1.26)^2 - (4)(5.20)(-0.189)}}{(2)(5.20)} = \frac{-1.26 \pm 2.34}{10.4}$$

$$x = +0.105 \text{ or } -0.396$$

The negative root is discarded because it is not physically meaningful.

Thus, at equilibrium we should have

$[N_2] = 0.105\text{ mol}\cdot\text{L}^{-1}$; $[H_2] = 0.315\text{ mol}\cdot\text{L}^{-1}$; $[NH_3] = 0.091\text{ mol}\cdot\text{L}^{-1}$.

9.70 $2\,HCl(g) \rightarrow H_2(g) + Cl_2(g)$

$$K = \frac{P_{H_2}P_{Cl_2}}{{P_{HCl}}^2} = 3.2 \times 10^{-34} \qquad \text{(Eq. 1)}$$

At equilibrium, $P_{Total} = P_{H_2} + P_{Cl_2} + P_{HCl} = 3.0\text{ bar}$. (Eq. 2)

And, since the atomic ratio Cl:H is 1:3,

$$\frac{\text{moles Cl}}{\text{moles H}} = \frac{1}{3} = \frac{n_{\text{HCl}} + 2n_{\text{Cl}_2}}{n_{\text{HCl}} + 2n_{\text{H}_2}}$$

$$n_{\text{HCl}} + 2n_{\text{H}_2} = 3(n_{\text{HCl}} + 2n_{\text{Cl}_2})$$

$$n_{\text{H}_2} = n_{\text{HCl}} + 3n_{\text{Cl}_2}$$

$P_{gas} \propto n_{gas}$ at T,V constant so

$$P_{\text{H}_2} = P_{\text{HCl}} + 3P_{\text{Cl}_2} \qquad \text{(Eq. 3)}$$

We now have three equations in three unknowns, so we can rearrange and substitute to find each partial pressure.

Substituting equation 3 into equation 2 gives

$$P_{Total} = 4P_{\text{Cl}_2} + 2P_{\text{HCl}} = 3.0 \text{ bar}$$

$$P_{\text{HCl}} = \frac{3.0 \text{ bar}}{2} - 2P_{\text{Cl}_2}$$

Substitution back into equation 3 gives

$$P_{\text{H}_2} = \frac{3.0 \text{ bar}}{2} + P_{\text{Cl}_2}$$

Using these two expressions for partial pressures in equation 1 gives

$$K = \frac{\left(\frac{3.0}{2} + P_{\text{Cl}_2}\right) \cdot P_{\text{Cl}_2}}{\left(\frac{3.0}{2} - 2P_{\text{Cl}_2}\right)^2} = 3.2 \times 10^{-34}$$

Since the equilibrium constant is so small, we would expect the partial pressure of chlorine to be very low at equilibrium, so we can neglect it to simplify this expression.

$$K \approx \frac{\left(\frac{3.0}{2}\right) \cdot P_{\text{Cl}_2}}{\left(\frac{3.0}{2}\right)^2} = 3.2 \times 10^{-34}$$

$$P_{\text{Cl}_2} = (1.5)(3.2 \times 10^{-34}) = 4.8 \times 10^{-34} \text{ bar}$$

$$P_{\text{HCl}} = P_{\text{H}_2} = 1.5 \text{ bar}$$

9.72 (a) According to Le Chatelier's principle, an increase in the partial pressure CO_2 will shift the equilibrium to the left, increasing the partial pressure of CH_4.

(b) According to Le Chatelier's principle, a decrease in the partial pressure of CH_4 will shift the equilibrium to the left, decreasing the partial pressure of CO_2.

(c) The equilibrium constant for the reaction is unchanged, because it is unaffected by any change in concentration.

(d) According to Le Chatelier's principle, a decrease in the concentration of H_2O will shift the equilibrium to the right, increasing the concentration of CO_2.

9.74 The questions can all be answered qualitatively using Le Chatelier's principle:

(a) Adding a reactant will promote the formation of products; the amount of HI should increase.

(b) I_2 is solid and already present in excess in the original equilibrium. Adding more solid will not affect the equilibrium so the amount of Cl_2 will not change.

(c) Removing a product will shift the reaction toward the formation of more products; the amount of Cl_2 should increase.

(d) Removing a product will shift the reaction toward the formation of more products; the amount of HCl should decrease.

(e) The equilibrium constant will be unaffected by changes in the concentrations of any of the species present.

(f) Removing the reactant HCl will cause the reaction to shift toward the production of more reactants; the amount of I_2 should increase.

(g) As in (e), the equilibrium constant will be unaffected by the changes to the system.

9.76 Increasing the total pressure on the system by decreasing its volume will shift the equilibrium toward the side of the reaction with fewer numbers of moles of gaseous components. If the total number of moles of gas is the same on the product and reactant sides of the balanced chemical equation, then changing the pressure will have little or no effect on the equilibrium distribution of species present. (a) decrease in the amount of NO_2; (b) increase in the amount and concentration of NO; (c) decrease in the amount of HI; (d) decrease in the amount of SO_2; (e) increase in the amount and concentration of NO_2. (Note that it is not possible to predict whether or not the concentration will increase, decrease, or remain the same in cases where the amount of the substance is decreasing. More specific information about the extent of the changes in the amount and the volume is needed in those cases.)

9.78 (a) The partial pressure of SO_3 will decrease when the partial pressure of SO_2 is decreased. According to Le Chatelier's principle, a decrease in the partial pressure of a reactant will shift the equilibrium to the left, decreasing the partial pressure of the products, in this case SO_3.
(b) When the partial pressure of SO_2 increases, the partial pressure of O_2 will decrease. According to Le Chaltelier's principle, an increase in the partial pressure of a reactant shifts the equilibrium toward products, decreasing the partial pressure of the other reactant, O_2.

9.80 If a reaction is exothermic, raising the temperature will tend to shift the reaction toward reactants, whereas if the reaction is endothermic, a shift toward products will be observed. For the specific reactions given, raising the temperature should favor products in (a) and reactants in (b) and (c).

9.82 Even though numbers are given, we do not need to do a calculation to answer this qualitative question. Because the equilibrium constant is larger at lower temperatures (4.0×10^{24} at 298 K vs. 2.5×10^{10} at 500 K, see

Table 9.2, page 334), more products will be present at the lower temperature. Thus we expect more SO_3 to be present at 25°C than at 500 K, assuming that no other changes occur to the system (the volume is fixed and no reactants or products are added or removed from the vessel).

9.84 To answer this question, we must calculate *Q:*

$$Q = \frac{[\text{ClF}]^2}{[\text{Cl}_2][\text{F}_2]} = \frac{(0.92)^2}{(0.18)(0.31)} = 15$$

Because $Q \neq K$, the system is not at equilibrium, and because $Q < K$, the reaction will proceed to produce more products so ClF will tend to form.

9.86 (a) $\text{HgO(s)} \rightleftharpoons \text{Hg(l)} + \frac{1}{2}\,\text{O}_2\text{(g)}$

$$\Delta H°_r = -[\Delta H°_f(\text{HgO, s})]$$
$$\Delta H°_r = -[-90.83\ \text{kJ}\cdot\text{mol}^{-1}]$$
$$\Delta H°_r = +90.83\ \text{kJ}\cdot\text{mol}^{-1}$$
$$\Delta S°_r = S°\,(\text{Hg, l}) + \tfrac{1}{2}\times S°\,(\text{O}_2, \text{g}) - [S°\,(\text{HgO, s})]$$
$$\Delta S°_r = 76.02\ \text{J}\cdot\text{K}^{-1}\cdot\text{mol}^{-1} + \tfrac{1}{2}\times 205.14\ \text{J}\cdot\text{K}^{-1}\cdot\text{mol}^{-1}$$
$$-[70.29\ \text{J}\cdot\text{K}^{-1}\cdot\text{mol}^{-1}]$$
$$\Delta S°_r = 108.30\ \text{J}\cdot\text{K}^{-1}\cdot\text{mol}^{-1}$$

At 298 K:

$$\Delta G°_{r(298\ \text{K})} = 90.83\ \text{kJ} - (298\ \text{K})(108.30\ \text{J}\cdot\text{K}^{-1})/(1000\ \text{J}\cdot\text{kJ}^{-1})$$
$$= 58.56\ \text{kJ}\cdot\text{mol}^{-1}$$
$$\Delta G°_{r(298\ \text{K})} = -RT\ \ln K$$

$$\ln K = -\frac{\Delta G°_{r(298\ \text{K})}}{RT}$$
$$= -\frac{58\,560\ \text{J}}{(8.314\ \text{J}\cdot\text{K}^{-1})(298\ \text{K})} = -23.6$$

$$K = 6\times 10^{-11}$$

At 373 K:

$$\Delta G°_{r(373\ \text{K})} = 90.83\ \text{kJ} - (373\ \text{K})(108.3\ \text{J}\cdot\text{K}^{-1})/(1000\ \text{J}\cdot\text{kJ}^{-1})$$
$$= 50.4\ \text{kJ}\cdot\text{mol}^{-1}$$

$$\ln K = -\frac{50\,400\text{ J}}{(8.314\text{ J}\cdot\text{K}^{-1})(373\text{ K})} = -16.3$$

$$K = 8 \times 10^{-8}$$

(b) propene (C_3H_6, g) $\rightarrow$ cyclopropane (C_3H_6, g)

$$\Delta H^\circ_r = \Delta H^\circ_f(\text{cyclopropane, g}) - [\Delta H^\circ_f(\text{propene, g})]$$

$$\Delta H^\circ_r = 53.30\text{ kJ}\cdot\text{mol}^{-1} - [(20.42\text{ kJ}\cdot\text{mol}^{-1})]$$

$$\Delta H^\circ_r = +32.88\text{ kJ}\cdot\text{mol}^{-1}$$

$$\Delta S^\circ_r = S^\circ(\text{cyclopropane, g}) - [S^\circ(\text{propene, g})]$$

$$\Delta S^\circ_r = 237.4\text{ J}\cdot\text{K}^{-1}\cdot\text{mol}^{-1} - [266.6\text{ J}\cdot\text{K}^{-1}\cdot\text{mol}^{-1}]$$

$$\Delta S^\circ_r = -29.2\text{ J}\cdot\text{K}^{-1}\cdot\text{mol}^{-1}$$

At 298 K:

$$\begin{aligned}\Delta G^\circ_{r(298\text{ K})} &= 32.88\text{ kJ}\cdot\text{mol}^{-1} \\ &\quad - (298\text{ K})(-29.2\text{ J}\cdot\text{K}^{-1}\cdot\text{mol}^{-1})/(1000\text{ J}\cdot\text{kJ}^{-1}) \\ &= 41.58\text{ kJ}\cdot\text{mol}^{-1}\end{aligned}$$

$$\Delta G^\circ_{r(298\text{ K})} = -RT\ln K$$

$$\begin{aligned}\ln K &= -\frac{\Delta G^\circ_{r(298\text{ K})}}{RT} \\ &= -\frac{41\,580\text{ J}}{(8.314\text{ J}\cdot\text{K}^{-1})(298\text{ K})} = -16.8\end{aligned}$$

$$K = 5 \times 10^{-8}$$

At 373 K:

$$\begin{aligned}\Delta G^\circ_{r(373\text{ K})} &= 32.88\text{ kJ}\cdot\text{mol}^{-1} \\ &\quad - (373\text{ K})(-29.2\text{ J}\cdot\text{K}^{-1}\cdot\text{mol}^{-1})/(1000\text{ J}\cdot\text{kJ}^{-1}) \\ &= 43.8\text{ kJ}\cdot\text{mol}^{-1}\end{aligned}$$

$$\ln K = -\frac{43\,800\text{ J}}{(8.314\text{ J}\cdot\text{K}^{-1})(373\text{ K})} = -14.1$$

$$K = 7 \times 10^{-7}$$

9.88 We can think of the vaporization of a liquid as an equilibrium reaction, e.g., for substance A, $A(l) \rightleftharpoons A(g)$. This reaction has an equilibrium constant expression $K = P_A$. Since P_A is the pressure of the gas above a liquid, it is the equilibrium vapor pressure of that liquid. Therefore, the vapor pressure varies with temperature in the same way that this equilibrium constant varies with temperature.

The van't Hoff equation, $\ln\dfrac{K_2}{K_1}=\dfrac{\Delta H_r^\circ}{R}\left(\dfrac{1}{T_1}-\dfrac{1}{T_2}\right)$, (see equation 16, page 354), gives the temperature dependence of K. To make this equation fit the special case of vapor pressure all we need to do is substitute P for K and ΔH_{vap}° for ΔH_r°.

$$\ln\frac{P_2}{P_1}=\frac{\Delta H_{vap}^\circ}{R}\left(\frac{1}{T_1}-\frac{1}{T_2}\right)$$

This relationship is known as the Clausius-Clapeyron equation (see equation 4, page 285). Since $\Delta H_{vap}^\circ > 0$, vapor pressure increases with temperature.

9.90 Recall $\Delta G_r^\circ = -RT\ln K = \Delta H_r^\circ - T\Delta S_r^\circ$ and $K = 10^{-pK}$

and $\ln\dfrac{K_2}{K_1}=\dfrac{\Delta H_r^\circ}{R}\left(\dfrac{1}{T_1}-\dfrac{1}{T_2}\right)$, the van't Hoff equation (see equation 16, page 354).

Rearranging and plugging in values,

$$\Delta H_r^\circ = \left(\ln\frac{K_2}{K_1}\right)\frac{R}{\left(\dfrac{1}{T_1}-\dfrac{1}{T_2}\right)} = \left(\ln\frac{10^{-13.8330}}{10^{-15.136}}\right)\left(\frac{8.314\ \text{J}\cdot\text{K}^{-1}\cdot\text{mol}^{-1}}{\dfrac{1}{293\ \text{K}}-\dfrac{1}{303\ \text{K}}}\right)$$

$$= (3.000)(7.381\times10^4\ \text{J}\cdot\text{mol}^{-1})$$

$$= 221.4\ \text{kJ}\cdot\text{mol}^{-1}$$

$$\Delta S_r^\circ = R\ln K + \frac{\Delta H_r^\circ}{T}$$

$$= (8.314\ \text{J}\cdot\text{K}^{-1}\cdot\text{mol}^{-1})\ln(10^{-15.136}) + \frac{221.4\times10^3\ \text{J}\cdot\text{mol}^{-1}}{293\ \text{K}}$$

$$= 465.9\ \text{J}\cdot\text{K}^{-1}\cdot\text{mol}^{-1}$$

The sign on ΔS_r° is positive because of the randomization that occurs as deuterons dissociate and redistribute throughout the liquid.

From the results above, the "autoprotolysis" constant of heavy water at 25°C is

$$\ln\frac{K_2}{K_1}=\frac{\Delta H_r^\circ}{R}\left(\frac{1}{T_1}-\frac{1}{T_2}\right)$$

$$\begin{aligned}\ln K_2 &= \left(\frac{221.4\times10^3\ \text{J}\cdot\text{mol}^{-1}}{8.314\ \text{J}\cdot\text{K}^{-1}\cdot\text{mol}^{-1}}\right)\left(\frac{1}{293\ \text{K}}-\frac{1}{298\ \text{K}}\right)+\ln 10^{-15.136}\\ &= -33.327\end{aligned}$$

$K_{\text{D, 298 K}} = 3.359\times10^{-15}$ or $pK_{\text{D, 298 K}} = 14.474$

Since $pK_{\text{H, 298 K}} = 14.00$ for regular water $< pK_{\text{D, 298 K}} = 14.474$, substitution of D for H causes reactants, i.e., the associated molecules, to be favored more in heavy water than in regular water even though we expect D and H to be chemically equivalent. We can suggest that this observation is another example of the kinetic isotope effect where the zero-point energy of the O-D bond is lower than that of the O-H bond (see the solutions to Exercises 17.55 and 17.85 for further discussion). If bond dissociation is the rate-limiting step in the mechanism, and if only the vibrational energy of the ground state of the water molecules is changed appreciably, then the activation barrier for dissociation is bigger in heavy water since the zero-point energy is lower due to the heavier mass of D. This increase in the activation energy for bond dissociation causes the rate of dissociation to decrease while the rate of association is not changed appreciably. The position of equilibrium then favors the associated D_2O molecules more than it does in the H_2O molecules.

9.92 (a) $$\begin{aligned}\Delta H^\circ_{\text{r}} &= -[2\Delta H^\circ_{\text{f}}(\text{O}_3,\text{g})]\\ &= -[2(142.7\ \text{kJ}\cdot\text{mol}^{-1})]\\ &= -285.4\ \text{kJ}\cdot\text{mol}^{-1}\end{aligned}$$

$$\Delta S^\circ_r = 3S^\circ(O_2, g) - [2S^\circ(O_3, g)]$$
$$= 3(205.14\ J \cdot K^{-1} \cdot mol^{-1}) - [2(238.93\ J \cdot K^{-1} \cdot mol^{-1})]$$
$$= 137.56\ J \cdot K^{-1} \cdot mol^{-1}$$

At 298 K:

$$\Delta G^\circ_{r(298\ K)} = -285.4\ kJ \cdot mol^{-1}$$
$$- (298\ K)(137.56\ J \cdot K^{-1} \cdot mol^{-1})/(1000\ J \cdot kJ^{-1})$$
$$= -326.4\ kJ \cdot mol^{-1}$$

or directly from ΔG°_f values

$$\Delta G^\circ_r = -2\Delta G^\circ_f(O_3, g)$$
$$= -2(163.2\ kJ \cdot mol^{-1})$$
$$= -326.4\ kJ \cdot mol^{-1}$$

(b) $\Delta G^\circ_{r(298\ K)} = -RT \ln K$

$$\ln K = -\frac{\Delta G^\circ_{r(298\ K)}}{RT}$$
$$= -\frac{-326\ 390\ J}{(8.314\ J \cdot K^{-1})(298\ K)} = +132$$
$$K = 10^{57}$$

The equilibrium constant for the decomposition of ozone to oxygen is extremely large, making this process extremely favorable. Note: The presence of ozone in the upper atmosphere is largely due to kinetic factors that inhibit the decomposition; however, in the presence of a suitable catalyst, this process should occur extremely readily.

9.94 $[N_2O_4] = \dfrac{\left(\dfrac{2.50\ g}{92.02\ g \cdot mol^{-1}}\right)}{2.00\ L} = 0.0136\ mol \cdot L^{-1}; [NO_2] = \dfrac{\left(\dfrac{0.330\ g}{46.01\ g \cdot mol^{-1}}\right)}{2.00\ L}$

$$= 0.003\ 59\ mol \cdot L^{-1}$$

Concentration ($mol \cdot L^{-1}$)	$N_2O_4(g)$	$\rightleftharpoons$	$2\ NO_2(g)$
initial	0.0136		0.003 59
change	$-x$		$+2x$
final	$0.0136 - x$		$0.003\ 59 + 2x$

$$K_{\Psi} = \frac{[NO_2]^2}{[N_2O_4]}$$

$$4.66 \times 10^{-3} = \frac{(0.003\,59 + 2\,x)^2}{0.0136 - x}$$

$$(4.66 \times 10^{-3})(0.0136 - x) = (0.003\,59 + 2\,x)^2$$

$$6.34 \times 10^{-5} - 4.66 \times 10^{-3}\,x = 4\,x^2 + 1.44 \times 10^{-2}\,x + 1.29 \times 10^{-5}$$

$$4\,x^2 + 1.91 \times 10^{-2}\,x - 5.05 \times 10^{-5} = 0$$

Solving by using the quadratic equation gives $x = 1.89 \times 10^{-3}$.

$$[N_2O_4] = 0.0136\ \text{mol} \cdot \text{L}^{-1} - 1.89 \times 10^{-3}\ \text{mol} \cdot \text{L}^{-1} = 0.012\ \text{mol} \cdot \text{L}^{-1}$$

$$[NO_2] = 0.003\,59\ \text{mol} \cdot \text{L}^{-1} + 2\,(1.89 \times 10^{-3}\ \text{mol} \cdot \text{L}^{-1}) = 0.0074\ \text{mol} \cdot \text{L}^{-1}$$

These numbers can be checked by substituting them into the equilibrium expression:

$$\frac{(0.0074)^2}{0.012} \stackrel{?}{=} 4.66 \times 10^{-3}$$

$$4.6 \times 10^{-3} \stackrel{\checkmark}{=} 4.66 \times 10^{-3}$$

9.96 We can write an equilibrium expression for the interconversion of each pair of isomers:

2-methyl propene $\rightleftharpoons$ *cis*-2-butene K_1

2-methyl propene $\rightleftharpoons$ *trans*-2-buteneK_2

cis-2-butene $\rightleftharpoons$ *trans*-2-butene K_3

We need to calculate only two of these values in order to determine the ratios.

The K's can be obtained from the ΔG°_r's for each interconversion:

$$\Delta G^\circ_1 = 65.86\ \text{kJ} \cdot \text{mol}^{-1} - 58.07\ \text{kJ} \cdot \text{mol}^{-1} = 7.79\ \text{kJ} \cdot \text{mol}^{-1}$$

$$\Delta G^\circ_2 = 62.97\ \text{kJ} \cdot \text{mol}^{-1} - 58.07\ \text{kJ} \cdot \text{mol}^{-1} = 4.90\ \text{kJ} \cdot \text{mol}^{-1}$$

$$\Delta G^\circ_3 = 62.97\ \text{kJ} \cdot \text{mol}^{-1} - 65.86\ \text{kJ} \cdot \text{mol}^{-1} = -2.89\ \text{kJ} \cdot \text{mol}^{-1}$$

$$\Delta G^\circ_r = -RT \ln K; \ln K = \frac{\Delta G^\circ_r}{-RT}$$

$$\ln K_1 = \frac{\Delta G^\circ_1}{-RT} = -\frac{7790 \text{ J} \cdot \text{mol}^{-1}}{(8.314 \text{ J} \cdot \text{K}^{-1} \cdot \text{mol}^{-1})(298 \text{ K})} = -3.14; K_1 = 0.043$$

$$\ln K_2 = \frac{\Delta G^\circ_2}{-RT} = -\frac{4900 \text{ J} \cdot \text{mol}^{-1}}{(8.314 \text{ J} \cdot \text{K}^{-1} \cdot \text{mol}^{-1})(298 \text{ K})} = -1.98; K_2 = 0.14$$

$$\ln K_3 = \frac{\Delta G^\circ_3}{-RT} = -\frac{-2890 \text{ J} \cdot \text{mol}^{-1}}{(8.314 \text{ J} \cdot \text{K}^{-1} \cdot \text{mol}^{-1})(298 \text{ K})} = +1.17; K_3 = 3.2$$

Each K gives a ratio of two of the isomers:

$$K_1 = \frac{[\textit{cis}\text{-2-butene}]}{[\text{2-methylpropene}]}; K_2 = \frac{[\textit{trans}\text{-2-butene}]}{[\text{2-methylpropene}]}; K_3 = \frac{[\textit{trans}\text{-2-butene}]}{[\textit{cis}\text{-2-butene}]}$$

We will choose to use the first two K's to calculate the final answer:

$$\frac{[\textit{cis}\text{-2-butene}]}{[\text{2-methylpropene}]} = 0.043 \qquad \frac{[\textit{trans}\text{-2-butene}]}{[\text{2-methylpropene}]} = 0.14$$

If we let the number of moles of 2-methylpropene = 1, then there will be 0.14 mol *trans*-2-butene and 0.043 mol *cis*-2-butene. The total number of moles will be 1 mol + 0.14 mol + 0.043 mol = 1.18 mol. The percentage of each will be

$$\%\text{ 2-methylpropene} = \frac{1 \text{ mol}}{1.18 \text{ mol}} \times 100\% = 85\%$$

$$\%\ \textit{cis}\text{-2-butene} = \frac{0.043 \text{ mol}}{1.18 \text{ mol}} \times 100\% = 4\%$$

$$\%\ \textit{trans}\text{-2-butene} = \frac{0.14 \text{ mol}}{1.18 \text{ mol}} \times 100\% = 12\%$$

(The sum of these should be 100% but varies by 1%, due to limitations in the use of significant figures/rounding conventions.)

The relative amounts are what we would expect, based upon the free energies of formation (the more negative or less positive value corresponding to the thermodynamically more stable compound), which indicate that the most stable compound is the 2-methylpropene followed by *trans*-2-butene with *cis*-2-butene being the least stable isomer. We can use K_3 as a check of these numbers:

$$K_3 = 3.2 \stackrel{?}{=} \frac{[trans\text{-2-butene}]}{[cis\text{-2-butene}]} = \frac{\frac{0.14}{V}}{\frac{0.043}{V}}$$

$3.2 \cong 3.3$

9.98

	A	$\rightleftharpoons$ 2 B	+ 3 C
initial	10.00 atm	0 atm	0 atm
change	$-x$	$+2x$	$+3x$
final	$10.00 - x$	$+2x$	$+3x$

The equilibrium expression is $K = \dfrac{P_B^{\ 2} \times P_C^{\ 3}}{P_A}$

We can also write $P_{\text{total}} = 10.00 - x + 2x + 3x = 10.00 + 4x = 15.76$

$x = 1.44$ atm

$P_A = 8.56$ atm; $P_B = 2.88$ atm; $P_C = 4.32$ atm

$$K = \frac{(2.88)^2(4.32)^3}{8.56} = 78.1$$

$$\Delta G° = -RT \ln K = -(8.314\ \text{J}\cdot\text{K}^{-1}\cdot\text{mol}^{-1})(298\ \text{K})\ln(78.1)$$
$$= -10.8\ \text{kJ}\cdot\text{mol}^{-1}$$

9.100 (a) First we calculate the initial pressure of the gas at 127°C using the ideal gas relationship:

$$PV = nRT$$

$$PV = \frac{m}{M}RT$$

$$P = \frac{mRT}{MV}$$
$$= \frac{(10.00\ \text{g})(0.082\ 06\ \text{L}\cdot\text{atm}\cdot\text{K}^{-1}\cdot\text{mol}^{-1})(298\ \text{K})}{(17.03\ \text{g}\cdot\text{mol}^{-1})(4.00\ \text{L})}$$
$$= 3.59\ \text{atm}$$

$$\frac{P_1}{T_1} = \frac{P_2}{T_2}$$

$$\frac{3.59 \text{ atm}}{298 \text{ K}} = \frac{P_2}{400 \text{ K}}$$

P_2 = 4.82 atm or 4.88 bar (1 atm = 1.01325 bar)

From Table 9.1 we find that the equilibrium constant is 41 for the equation

$N_2(g) + 3\,H_2(g) \rightleftharpoons 2\,NH_3(g)$

Although this exercise technically begins with pure ammonia and allows it to dissociate into $N_2(g)$ and $H_2(g)$, we can use the same expression.

pressure (bar)	$N_2(g)$ +	$3\,H_2(g) \rightleftharpoons$	$2\,NH_3(g)$
initial	0	0	4.88
change	$+x$	$+3x$	$-2x$
final	$+x$	$+3x$	$4.88 - 2x$

$$\frac{(4.88 - 2\,x)^2}{(x)(3\,x)^3} = 41$$

$$\frac{(4.88 - 2\,x)^2}{27\,x^4} = 41$$

$$\frac{4.88 - 2\,x}{3\sqrt{3}\,x^2} = \sqrt{41}$$

$$4.88 - 2\,x = 3\sqrt{123}\,x^2$$

$$33.3\,x^2 + 2\,x - 4.88 = 0$$

$$x = 0.354$$

P_{NH_3} = 4.17 bar or 4.11 atm; P_{N_2} = 0.354 bar or 0.349 atm; P_{H_2} = 1.06 bar or 1.05 atm

P_{total} = 4.11 atm + 0.349 atm + 1.05 atm = 5.51 atm

(b) The equilibrium constant can be calculated from $\Delta G°$ at 400 K, which can be obtained from $\Delta H°$ and $\Delta S°$. The values are

$$\Delta H° = 2(-46.11\ \text{kJ}\cdot\text{mol}^{-1}) = -92.22\ \text{kJ}\cdot\text{mol}^{-1}$$

$$\begin{aligned}\Delta S° &= 2(192.45\ \text{J}\cdot\text{K}^{-1}\cdot\text{mol}^{-1}) - [(191.61\ \text{J}\cdot\text{K}^{-1}\cdot\text{mol}^{-1}) \\ &\quad + 3(130.68\ \text{J}\cdot\text{K}^{-1}\cdot\text{mol}^{-1})] \\ &= -198.75\ \text{J}\cdot\text{K}^{-1}\cdot\text{mol}^{-1}\end{aligned}$$

$$\begin{aligned}\Delta G° &= (-92.22\ \text{kJ}\cdot\text{mol}^{-1})(1000\ \text{J}\cdot\text{K}^{-1}) \\ &\quad - (400\ \text{K})(-198.75\ \text{J}\cdot\text{K}^{-1}\cdot\text{mol}^{-1}) \\ &= -12.72\ \text{kJ}\cdot\text{mol}^{-1} = -12720\ \text{J}\cdot\text{mol}^{-1}\end{aligned}$$

$$K = e^{-\Delta G°/RT} = e^{+12720\ \text{J}\cdot\text{mol}^{-1}/(8.314\ \text{J}\cdot\text{K}^{-1}\cdot\text{mol}^{-1})(400\text{K})} = 46$$

This is reasonably good agreement considering the logarithmic nature of these calculations. For comparison, the $\Delta G°$ value calculated from $K = 41$ at 400 K is 12.34 $\text{kJ}\cdot\text{mol}^{-1}$.

9.102 (a)

pressure (bar)	$2\ AsH_3(g)$	$\rightleftharpoons$	$2\ As(s)$	+	$3\ H_2(g)$
initial	0.52				
change	$-2x$				$+3x$
final	$0.52 - 2x$				$+3x$

At equilibrium, total pressure = 0.64 bar.

$$0.64 = 0.52 - 2x + 3x = 0.52 + x$$

$$x = 0.12\ \text{bar}$$

$$P_{AsH_3} = 0.52\ \text{bar} - 2(0.12\ \text{bar}) = 0.28\ \text{bar}$$

$$P_{H_2} = 3(0.12\ \text{bar}) = 0.36\ \text{bar}$$

(b) Find the number of moles of AsH_3 that decomposed and relate it to the mass of As.

$$n = \frac{PV}{RT} = \frac{(0.24\ \text{bar})(1.00\ \text{L})}{(0.0831\ \text{L}\cdot\text{bar}\cdot\text{K}^{-1}\cdot\text{mol}^{-1})(298\ \text{K})}$$

$$= 9.69\times10^{-3}\ \text{mol AsH}_3\ \text{decomposed}$$

$$\text{mass As} = (9.69\times10^{-3}\ \text{mol AsH}_3)\left(\frac{2\ \text{mol As}}{2\ \text{mol AsH}_3}\right)\left(\frac{74.9\ \text{g As}}{1\ \text{mol As}}\right)$$

$$= 0.73\ \text{g As}$$

(c) $K = \frac{P_{H_2}^3}{P_{AsH_3}^2} = \frac{(0.36)^3}{(0.28)^2} = 0.60$

9.104 The free energy change for the reaction is calculated from the equation

$\Delta G = \Delta G° + RT \ln Q = -RT \ln K + RT \ln Q$

In order to determine the range of Cl_2 pressures that we need to examine, we will first calculate the equilibrium pressures of the gases present.

	$Br_2(g)$ +	$Cl_2(g)$ ⇌	2 $BrCl(g)$
initial	1.00 bar	1.00 bar	0
change	$-x$	$-x$	$2x$
final	$1.00 - x$	$1.00 - x$	$2x$

$$0.2 = \frac{P_{BrCl}^{\ 2}}{P_{Br_2}\,P_{Cl_2}} = \frac{(2x)^2}{(1.00 - x)^2}$$

$$\sqrt{0.2} = \frac{2x}{1.00 - x}$$

$$x = 0.183 \approx 0.2$$

In other words, equilibrium will be reached when $P_{BrCl} = 2x = 0.36$ bar, at the point on the graph where $\Delta G = 0$.

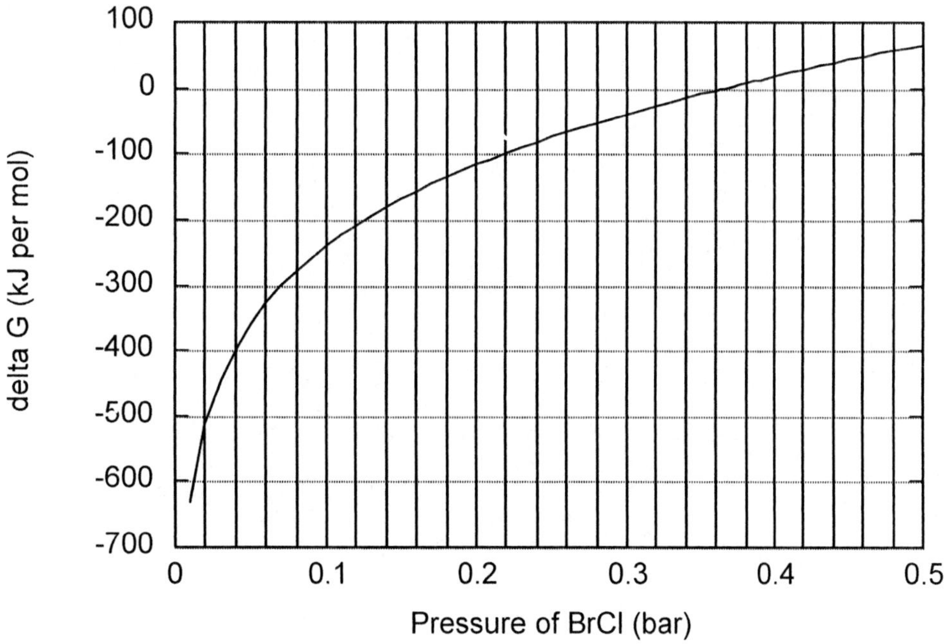

Notice that the free energy change for the reaction is most negative the farther the system is from equilibrium and that the value approaches 0 as equilibrium is attained.

9.106 The graph is generated for ln K vs. $1/T$ according to the relationship

$$\ln K = -\frac{\Delta H°}{R} \cdot \frac{1}{T} + \frac{\Delta S°}{R}$$

where $-\frac{\Delta H°}{R}$ is the slope and $\frac{\Delta S°}{R}$ is the intercept.

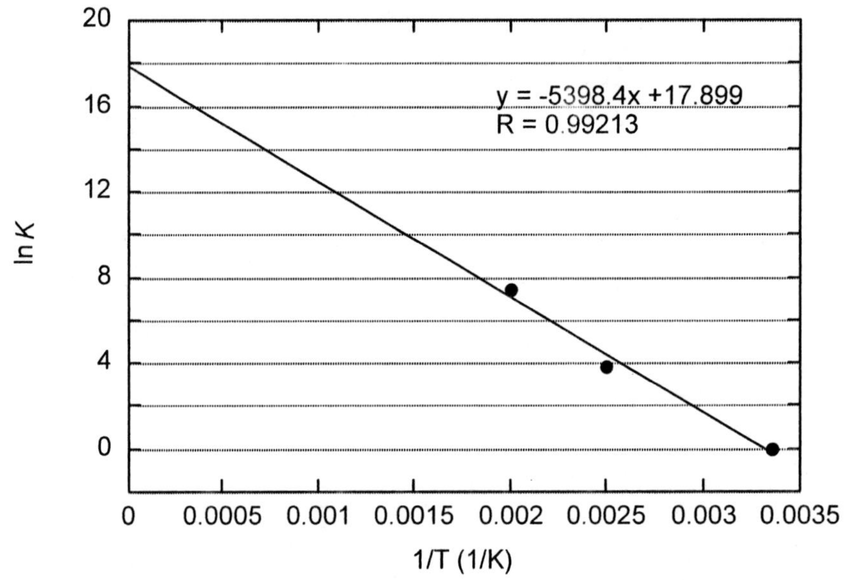

$$-\frac{\Delta H°}{R} = -5398 \text{ K}$$

$$\Delta H° = 44.9 \text{ kJ} \cdot \text{mol}^{-1}$$

$$\frac{\Delta S°}{R} = 17.9$$

$$\Delta S° = 149 \text{ J} \cdot \text{K}^{-1} \cdot \text{mol}^{-1}$$

Because the reaction involves breaking only the N—N bond, the enthalpy of reaction should be an approximation of the N—N bond strength. Notice that this value is considerably less than the average 163 $\text{kJ} \cdot \text{mol}^{-1}$ value for N—N bonds given in Table 2.2. Therefore it is a very weak N—N bond.

9.108 (a) Since we are only using two temperatures, it really is not necessary to make graphs in order to determine $\Delta H°$ and $\Delta S°$. We can easily substitute into the van't Hoff equation, $\ln\dfrac{K_2}{K_1}=\dfrac{\Delta H_r^\circ}{R}\left(\dfrac{1}{T_1}-\dfrac{1}{T_2}\right)$, for each halogen to get $\Delta H°$, then use it to find $\Delta S°$ from $\Delta G° = \Delta H° - T\Delta S = -RT\ln K$. A spread sheet will work well.

(b) & (c) Results from our spread sheet based on the relationships given above and data from Table 9.2 are given below. The $\Delta S°$ values are for the reactions $X_2(g) \rightleftharpoons 2\ X(g)$, so we can calculate the standard molar entropies for the atomic species using data from Appendix 2A and the relationship $\Delta S° = 2S°_m\ (X, g) - S°_m\ (X_2, g)$.

Halogen	$\Delta H°$ $kJ\cdot mol^{-1}$	$\Delta S°$ $J\cdot K^{-1}\cdot mol^{-1}$	$S°(X_2, g)$ $J\cdot K^{-1}\cdot mol^{-1}$	$S°(X, g)$ $J\cdot K^{-1}\cdot mol^{-1}$
Fluorine	164	126	202.78	164
Chlorine	256	160	223.07	192
Bromine	195	110	245.46	178
Iodine	163	152	260.69	206

9.110 The general form of this type of equation is

$$\ln K = -\frac{\Delta H°_r}{R}\left(\frac{1}{T}\right)+\frac{\Delta S°_r}{R}$$

which is easily derived from the relationships $\Delta G°_r = -RT\ \ln K$ and $\Delta G°_r = \Delta H°_r - T\Delta S°_r$. By inspection, we can see that for the original expression to be valid,

$$-\frac{\Delta H°_r}{R} = -21\ 700\text{, which gives } \Delta H°_r$$

$$= +(21\ 700\ K)(8.314\ J\cdot K^{-1}\cdot mol^{-1})/(1000\ J\cdot kJ^{-1}) = 180\ kJ\cdot mol^{-1}$$

9.112 (a) $K = \dfrac{1}{P_{PH_3} \cdot P_{BCl_3}}$

(b) $K = K_c(RT)^{\Delta n}$

$$K_c = \frac{K}{(RT)^{\Delta n}} = \frac{19.2}{[(0.08314\ L \cdot bar \cdot K^{-1} \cdot mol^{-1})(333\ K)]^{-2}}$$

$$= 1.48 \times 10^4$$

(c) The initial pressure of PH_3 is:

$$P_{PH_3} = \frac{nRT}{V} = \frac{(0.0128\ mol)(0.08314\ L \cdot bar \cdot K^{-1}mol^{-1})(333\ K)}{0.500\ L}$$

$= 0.7088$ bar

pressure (bar)	$PH_3(g)$	+	$BCl_3(g)$	$\rightleftharpoons$	$PH_3BCl_3(s)$
initial	0.7088		0		
change	$+x$		$+x$		
equilibrium	$0.7088 + x$		x		

$$K = 19.2 = \frac{1}{P_{PH_3} \cdot P_{BCl_3}} = \frac{1}{(0.7088 + x)x}$$

$$x^2 + 0.7088x - 0.05208 = 0$$

$$x = \frac{-0.7088 \pm \sqrt{(0.7088)^2 + 4(0.05208)}}{2}$$

$= 0.06712$ (negative root isn't physically possible)

So $P_{PH_3} = 0.7088 + 0.06712 = 0.7759$ bar

and $c_{PH_3} = \dfrac{P_{PH_3}}{RT} = \dfrac{0.7759\ bar}{(0.08314\ L \cdot bar \cdot K^{-1}mol^{-1})(333\ K)} = 0.0280\ mol \cdot L^{-1}$

(d) *K* increases as T increases, so the equilibrium position is shifting toward products with the addition of heat. The reaction is endothermic.

(e) $K = K_c(RT)^{\Delta n}$

$$K_c = \frac{K}{(RT)^{\Delta n}} = \frac{26.2}{[(0.08314\ L \cdot bar \cdot K^{-1} \cdot mol^{-1})(343\ K)]^{-2}}$$

$$= 2.13 \times 10^4$$

(f) PH_3 has a lone pair of electrons to donate in order to form an adduct with BCl_3, which can be an electron acceptor. Therefore PH_3 is a Lewis base while BCl_3 is a Lewis acid.

9.114 (a) The values of the standard free energies of formation are

Compound	$\Delta G°_f\ (\text{kJ}\cdot\text{mol}^{-1})$
$BCl_3(g)$	−387.969
$BBr_3(g)$	−236.914
$BCl_2Br(g)$	−338.417
$BClBr_2(g)$	−287.556

(b) A number of equilibrium reactions are possible including

$2\ BCl_2Br(g) \rightleftharpoons BCl_3(g) + BClBr_2(g)$

$2\ BClBr_2(g) \rightleftharpoons BBr_3(g) + BCl_2Br(g)$

$2\ BCl_3(g) + BBr_3(g) \rightleftharpoons 3\ BCl_2Br(g)$

$BCl_3(g) + 2\ BBr_3(g) \rightleftharpoons 3\ BClBr_2(g)$

(c) BCl_3 because it is the most stable as determined by its $\Delta G°_f$ value.

(d) No. Although the equilibrium reaction given in the exercise would require that these two partial pressures be equal, there are other equilibria occurring as in (b) that will cause the values to be unequal.

(e) We will solve this problem graphically rather than by writing a computer program to solve a system of simultaneous equations. Standard free energies of formation are only used to get equilibrium constants for two essential reactions from $K = e^{\frac{-\Delta G_r^\circ}{RT}}$ and

$\Delta G_r^\circ = \sum n_{prods} \Delta G_f^\circ - \sum m_{reacts} \Delta G_f^\circ$.

In order to solve this part of the problem, first we need to figure out the independent reactions between the four species. If we consider BCl_3 and BBr_3 as pure species in terms of the halogens atoms, then the mixed species $BClBr_2$ and BCl_2Br can be made from the pure species by these two reactions:

(1) $BCl_3 + 2\ BBr_3 \rightleftharpoons 3\ BClBr_2$ $\quad K_1 = \exp\left(\dfrac{871}{8.314 \times 298.15}\right) = 1.421$

(2) $2\ BCl_3 + BBr_3 \rightleftharpoons 3\ BCl_2Br$ $\quad K_2 = \exp\left(\dfrac{2399}{8.314 \times 298.15}\right) = 2.632$

All other reactions between the species can be derived from these two essential reactions, so there are only two independent reactions for the system.

Let $\quad P_{BCl_3} = x$, and $P_{BBr_3} = y$ then $K_1 = \dfrac{P_{BClBr_2}{}^3}{xy^2}$ and $K_2 = \dfrac{P_{BCl_2Br}{}^3}{x^2 y}$

so $\quad P_{BClBr_2} = K_1{}^{1/3} x^{1/3} y^{2/3}$, and $P_{BCl_2Br} = K_2{}^{1/3} x^{2/3} y^{1/3}$

Atom balances are determined by the initial composition of the system. Balancing B atoms we get

$$P^o_{BCl_3} + P^o_{BBr_3} = P_{BCl_3} + P_{BCl_2Br} + P_{BClBr_2} + P_{BBr_3}$$

where $P^o_{BCl_3} = P^o_{BBr_3} = 1$ bar are the initial pressures. Balancing Cl atoms and Br atoms gives us

$$3P^o_{BCl_3} = 3P_{BCl_3} + 2P_{BCl_2Br} + P_{BClBr_2}$$

$$3P^o_{BBr_3} = P_{BCl_2Br} + 2P_{BClBr_2} + 3P_{BBr_3}$$

(Notice that the B balance is not an independent condition because it is implied by the Cl and Br balances.) In terms of *x* and *y*, the Cl balance and Br balance can be written as

Cl balance: $\quad 3x + 2K_2{}^{1/3} x^{2/3} y^{1/3} + K_1{}^{1/3} x^{1/3} y^{2/3} = 3$

Br balance: $\quad K_2{}^{1/3} x^{2/3} y^{1/3} + 2K_1{}^{1/3} x^{1/3} y^{2/3} + 3y = 3$

Subtracting the second equation from the first equation, dividing the resulting equation by *y*, and also using this new definition

$$\left(\frac{x}{y}\right)^{1/3} \equiv z$$

we arrive at following equation in terms of *z*:

$$3z^3 + K_2{}^{1/3} z^2 - K_1{}^{1/3} z - 3 = 0$$

$3z^3 + 1.381z^2 - 1.124z - 3 = 0$

This cubic equation can be solved graphically by noting where $f(z) = 3z^3 + 1.381z^2 - 1.124z - 3$ crosses 0. The only real root is $z = 0.975$.

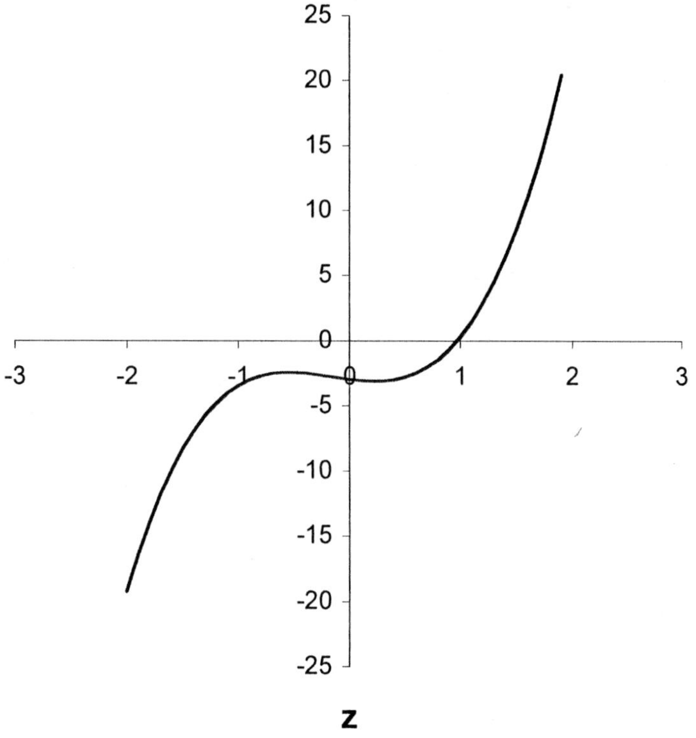

Substituting $x = z^3 y$ back into the Cl balance equation gives

$$y = \frac{3}{3z^3 + 2K_2^{1/3}z^2 + K_1^{1/3}z}$$

$$= \frac{3}{3(0.975)^3 + 2(1.381)(0.975)^2 + (1.124)(0.975)}$$

$$= 0.461$$

Then $x = z^3 y = (0.975)^3(0.461) = 0.428$

And finally:

$$P_{BClBr_2} = \left((1.421)(0.428)(0.461)^2\right)^{1/3} = 0.506$$

$$P_{BCl_2Br} = \left((2.632)(0.428)^2(0.461)\right)^{1/3} = 0.606$$

The results can be summarized as:

$P_{BCl_3} = 0.428$ bar

$P_{BBr_3} = 0.461$ bar

$P_{BCl_2Br} = 0.606$ bar

$P_{BClBr_2} = 0.506$ bar

Note that in part (c) we predicted that BCl_3 is the most abundant species in the equilibrated mixture. However, this calculation reveals that it is the least abundant species at equilibrium given the stated initial conditions. The result is simply non-intuitive. What we need to recognize is that the formation reactions and their corresponding values of free energy of reaction describe equilibria between the elements B(s), Cl_2(g), and Br_2(l) and the boron trihalides under standard conditions. In our mixture the compounds are certainly not in equilibrium with their elements in their standard states, so the prediction in part (c) fails.

Also note that in this system, B(s) and Br_2(l) cannot exist at same time, because the equilibrium constant of B(s) + 1.5 Br_2(l) $\rightleftharpoons$ BBr_3(g) is very large, while the actual pressure of BBr_3(g) is way below this equilibrium constant. B(s) may exist from a microscopic amount of decomposition of BCl_3(g), so Br_2(l) can not exist. This condition further prevents BCl_2Br and $BClBr_2$ from decomposing into their elements, so only BCl_3(g) may decompose to a small extent. Clearly, this small amount of Cl_2(g) is far below standard conditions.

9.116 (a) Find Q to see if the reaction is at equilibrium.

cyclohexane (C) $\rightleftharpoons$ methylcyclohexane (M)

$$Q = \frac{[M]}{[C]} = \frac{0.100}{0.0200} = 5.00 > K = 0.140$$

The reaction will proceed to the left to form more reactant.

(b)

	[C]	[M]
initial	0.0200	0.100
change	$+x$	$-x$
equil	$0.0200+x$	$0.100-x$

$$K=\frac{[M]}{[C]}=\frac{0.100-x}{0.0200+x}=0.140$$

$$0.100-x=(0.140)(0.0200+x)$$

$$1.140x=9.72\times10^{-2}$$

$$x=8.53\times10^{-2}$$

[M] = 0.100 – 0.0853 = 0.0147 = 0.015 M

[C] = 0.0200 + 0.0853 = 0.1053 M

(c) At equilibrium, [C] = 0.100 M. The total concentration of all species is 0.120 M. Therefore, [M] = 0.020 at equilibrium.

$$K_{50^{\circ}C}=\frac{[M]}{[C]}=\frac{0.100}{0.020}=5.0$$

(d) As the temperature increased, K also increased resulting in the reaction forming more products: C + heat $\rightleftharpoons$ M. Therefore, the reaction is endothermic.

9.118 Refer to Graham's Law of Effusion, equation 19, page 144, which states that the ratio of the rates of effusion of two gases, both at the same temperature, is equal to the square-root of the inverse ratio of their molar masses.

$$\frac{\text{Rate of effusion of } H_2}{\text{Rate of effusion of HI}}=\sqrt{\frac{M_{HI}}{M_{H_2}}}$$

$$=\sqrt{\frac{127.9\ g\cdot mol^{-1}}{2.016\ g\cdot mol^{-1}}}$$

$$=7.97$$

This treatment suggests that H_2 effuses ~8 times faster than HI. However, it is also important to note that the partial pressure of each gas affects the collision frequency of the molecules with the pinhole. Therefore, an expression that takes this dependence into account is

$$\frac{\text{Rate of effusion of H}_2}{\text{Rate of effusion of HI}} = \frac{P_{\mathrm{H_2}}}{P_{\mathrm{HI}}}\sqrt{\frac{M_{\mathrm{HI}}}{M_{\mathrm{H_2}}}}$$

The partial pressure of hydrogen is 0.7 atm, or 0.69 bar. The partial pressure of HI can be found from

$$K = \frac{P_{\mathrm{HI}}^2}{P_{\mathrm{H_2}}}$$

$$P_{\mathrm{HI}} = \sqrt{K \cdot P_{\mathrm{H_2}}} = (0.345 \times 0.69)^{1/2} = 0.49 \text{ bar}$$

$$\begin{aligned}\frac{\text{Rate of effusion of H}_2}{\text{Rate of effusion of HI}} = \frac{P_{\mathrm{H_2}}}{P_{\mathrm{HI}}}\sqrt{\frac{M_{\mathrm{HI}}}{M_{\mathrm{H_2}}}} &= \left(\frac{0.69 \text{ bar}}{0.49 \text{ bar}}\right)\sqrt{\frac{127.9 \text{ g} \cdot \text{mol}^{-1}}{2.016 \text{ g} \cdot \text{mol}^{-1}}} \\ &= 11 \\ &= 10 \text{ (to 1 sig fig)}\end{aligned}$$

So hydrogen effuses ten times as fast as hydrogen iodide under these conditions.

CHAPTER 10

ACIDS AND BASES

10.2 (a) H_3O^+ (b) H_2O (c) $C_6H_5NH_3^+$ (d) HS^- (e) PO_4^{3-}

(f) ClO_4^-

10.4 (a)

conjugate

$H_2O(l) + CN^-(aq) \rightleftharpoons HCN(l) + OH^-(aq)$

acid$_1$ base$_2$ acid$_2$ base$_1$

conjugate

(b)

conjugate

$H_2O(l) + NH_2NH_2(aq) \rightleftharpoons NH_2NH_3^+(aq) + OH^-(aq)$

acid$_1$ base$_2$ acid$_2$ base$_1$

conjugate

(c)

conjugate

$H_2O(l) + CO_3^{2-}(aq) \rightleftharpoons HCO_3^-(aq) + OH^-(aq)$

acid$_1$ base$_2$ acid$_2$ base$_1$

conjugate

(d)

conjugate

$H_2O(l) + HPO_4^{2-}(aq) \rightleftharpoons H_2PO_4^-(aq) + OH^-(aq)$

acid$_1$ base$_2$ acid$_2$ base$_1$

conjugate

(e)

conjugate

$H_2O(l) + NH_2CONH_2(aq) \rightleftharpoons NH_2CONH_3^+(aq) + OH^-(aq)$

acid$_1$ base$_2$ acid$_2$ base$_1$

conjugate

10.6 (a) Brønsted acid: NH_4^+

Brønsted base: HSO_3^-

(b) Conjugate base to NH_4^+ : NH_3

Conjugate acid to HSO_3^- : H_2SO_3

10.8 (a) as an acid: $H_2PO_4^-(aq) + H_2O(l) \rightleftharpoons H_3O^+(aq) + HPO_4^{2-}(aq)$

Conjugate acid/base pairs: $H_2PO_4^-(aq)$ / $HPO_4^{2-}(aq)$

$H_3O^+(aq)$ / $H_2O(l)$

as a base: $H_2PO_4^-(aq) + H_2O(l) \rightleftharpoons OH^-(aq) + H_3PO_4(aq)$

Conjugate acid/base pairs: $H_3PO_4(aq)$ / $H_2PO_4^-(aq)$

$H_2O(l)$ / $OH^-(aq)$

(c) as an acid: $HC_2O_4^-(aq) + H_2O(l) \rightleftharpoons H_3O^+(aq) + C_2O_4^{2-}(aq)$

Conjugate acid/base pairs: $HC_2O_4^-(aq)$ / $C_2O_4^{2-}(aq)$

$H_3O^+(aq)$ / $H_2O(l)$

as a base: $HC_2O_4^-(aq) + H_2O(l) \rightleftharpoons OH^-(aq) + H_2C_2O_4(aq)$

Conjugate acid/base pairs: $H_2C_2O_4(aq)$ / $HC_2O_4^-(aq)$

$H_2O(l)$ / $OH^-(aq)$

10.10 (a) acidic; (b) basic; (c) acidic; (d) amphoteric

10.12 (a) $2\ OH^-(aq) + SiO_2(s) \rightarrow SiO_3^{2-}(aq) + H_2O(l)$

(b)

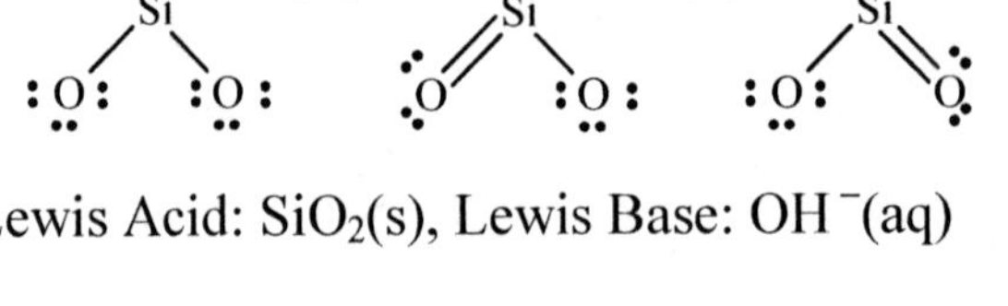

(c) Lewis Acid: $SiO_2(s)$, Lewis Base: $OH^-(aq)$

10.14 In each case use $K_w = [H_3O^+][OH^-] = 1.0 \times 10^{-14}$, then

$$[H_3O^+] = \frac{K_w}{[OH^-]} = \frac{1.0 \times 10^{-14}}{[OH^-]}$$

(a) $$[H_3O^+] = \frac{1.0 \times 10^{-14}}{0.012} = 8.3 \times 10^{-13}$$

(b) $$[H_3O^+] = \frac{1.0 \times 10^{-14}}{6.2 \times 10^{-5}} = 1.6 \times 10^{-10}$$

(c) $$[H_3O^+] = \frac{1.0 \times 10^{-14}}{2.3 \times 10^{-3}} = 4.3 \times 10^{-12}$$

10.16 (a) $[H_3O^+] = [OH^-] = 3.9 \times 10^{-8}\ \text{mol} \cdot \text{L}^{-1}$

$$K_w = [H_3O^+][OH^-] = (3.9 \times 10^{-8})^2 = 1.5 \times 10^{-15}$$
$$pK_w = -\log K_w = -\log(1.5 \times 10^{-15}) = 14.82$$

(b) $\text{pH} = -\log(3.9 \times 10^{-8}) = 7.41 = \text{pOH}$

10.18 $[KNH_2]_0$ = nominal concentration of KNH_2

$$[KNH_2]_0 = \left(\frac{0.5\ \text{g}\ KNH_2}{0.250\ \text{L}}\right)\left(\frac{1\ \text{mol}\ KNH_2}{55.13\ \text{g}\ KNH_2}\right) = 0.036\ \text{mol} \cdot \text{L}^{-1}$$

Because KNH_2 is a soluble salt, $[KNH_2]_0 = [K^+] = [NH_2^-]_0$ (where $[NH_2^-]_0$ is the nominal concentration of NH_2^-); thus $[K^+]$ and $[NH_2^-]$ = $0.036\ \text{mol} \cdot \text{L}^{-1}$.

NH_2^- reacts with water:

$$NH_2^-(aq) + H_2O(l) \longrightarrow NH_3(aq) + OH^-(aq)$$

Because NH_2^- is a strong base, this reaction goes essentially to completion; therefore

$$[NH_2^-]_0 = [OH^-] = 0.036\ \text{mol} \cdot \text{L}^{-1}$$

$$[H_3O^+] = \frac{K_w}{[OH^-]} = \frac{1.0 \times 10^{-14}}{0.036} = 2.8 \times 10^{-13}\ \text{mol} \cdot \text{L}^{-1}$$

10.20 $[H_3O^+] = 10^{-pH}\ mol \cdot L^{-1}$ (antilog pH, $mol \cdot L^{-1}$)

(a) $[H_3O^+] = \text{antilog}(-5) = 1 \times 10^{-5}\ mol \cdot L^{-1}$

(b) $[H_3O^+] = 10^{-2.3}\ mol \cdot L^{-1}$; antilog $(-2.3) = 5 \times 10^{-3}\ mol \cdot L^{-1}$

(c) $[H_3O^+] = 10^{-7.4}\ mol \cdot L^{-1}$; antilog $(-7.4) = 4 \times 10^{-8}\ mol \cdot L^{-1}$

(d) $[H_3O^+] = 10^{-10.5}\ mol \cdot L^{-1}$; antilog $(-10.5) = 3 \times 10^{-11}\ mol \cdot L^{-1}$

Acidity increases as pH decreases. The order is thus:
milk of magnesia < blood < urine < lemon juice

10.22 (a) $pH = -\log(0.0356) = 1.448$

$pOH = 14.00 - 1.448 = 12.55$

(b) $pH = -\log(0.0725) = 1.12$

$pOH = 14.00 - 1.12 = 12.86$

(c) $Ba(OH)_2 \longrightarrow Ba^{2+} + 2\ OH^-$

$[OH^-] = 2 \times 3.46 \times 10^{-3}\ mol \cdot L^{-1} = 6.92 \times 10^{-3}\ mol \cdot L^{-1}$

$pOH = -\log(6.92 \times 10^{-3}) = 2.160$

$pH = 14.00 - 2.160 = 11.84$

(d) $$\frac{10.9 \times 10^{-3}\ g}{56.11\ g \cdot mol^{-1}} = 1.94 \times 10^{-4}\ mol\ KOH$$

$$[OH^-] = \frac{1.94 \times 10^{-4}\ mol}{0.0100\ L} = 1.94 \times 10^{-2}\ mol \cdot L^{-1}$$

$pOH = -\log(1.94 \times 10^{-2}) = 1.712$

$pH = 14.00 - 1.712 = 12.29$

(e) $$[OH^-] = \left(\frac{10.0\ mL}{2500\ mL}\right) \times (5.00\ mol \cdot L^{-1}) = 0.0200\ mol \cdot L^{-1}$$

$pOH = -\log(0.200) = 1.70$

$pH = 14.00 - 0.699 = 12.30$

(f) $$[H_3O^+] = \left(\frac{5.0\ mL}{25.0\ mL}\right) \times (3.5 \times 10^{-4}\ mol \cdot L^{-1}) = 7.0 \times 10^{-5}\ mol \cdot L^{-1}$$

$pH = -\log(7.0 \times 10^{-5}) = 4.15$

$pOH = 14.00 - 4.15 = 9.85$

10.24

Base	K_b	pK_b
(a) NH_3	1.8×10^{-5}	4.74
(b) ND_3	1.1×10^{-5}	4.96
(c) NH_2NH_2	1.7×10^{-6}	5.77
(d) NH_2OH	1.1×10^{-8}	7.96

(e) $NH_2OH < NH_2NH_2 < ND_3 < NH_3$

10.26 (a) $(CH_3)_2NH(aq) + H_2O(l) \rightleftharpoons (CH_3)_2NH_2^+(aq) + OH^-(aq)$

$$K_b = \frac{[(CH_3)_2NH_2^+][OH^-]}{[(CH_3)_2NH]}$$

$(CH_3)_2NH_2^+(aq) + H_2O(l) \rightleftharpoons H_3O^+(aq) + (CH_3)_2NH(aq)$

$$K_a = \frac{[H_3O^+][(CH_3)_2NH]}{[(CH_3)_2NH_2^+]}$$

(b) $C_{14}H_{10}N_2(aq) + H_2O(l) \rightleftharpoons C_{14}H_{10}N_2H^+(aq) + OH^-(aq)$

$$K_b = \frac{[C_{14}H_{10}N_2H^+][OH^-]}{[C_{14}H_{10}N_2]}$$

$C_{14}H_{10}N_2H^+(aq) + H_2O(l) \rightleftharpoons H_3O^+(aq) + C_{14}H_{10}N_2(aq)$

$$K_a = \frac{[H_3O^+][C_{14}H_{10}N_2]}{[C_{14}H_{10}N_2H^+]}$$

(c) $C_6H_5NH_2(aq) + H_2O(l) \rightleftharpoons C_6H_5NH_3^+(aq) + OH^-(aq)$

$$K_b = \frac{[C_6H_5NH_3^+][OH^-]}{[C_6H_5NH_2]}$$

$C_6H_5NH_3^+(aq) + H_2O(l) \rightleftharpoons H_3O^+(aq) + C_6H_5NH_2(aq)$

$$K_a = \frac{[H_3O^+][C_6H_5NH_2]}{[C_6H_5NH_3^+]}$$

10.28 Decreasing pK_a will correspond to increasing acid strength because $pK_a = -\log K_a$. The pK_a values (given in parentheses) determine the following ordering:

$(CH_3)_3NH^+ (14.00 - 4.19 = 9.71) < N_2H_5^+ (14.00 - 5.77 = 8.23)$

$< HCOOH\ (3.75) < HF\ (3.45)$

Remember that the pK_a for the conjugate acid of a weak base will be given by

$pK_a + pK_b = 14$.

10.30 Decreasing pK_b will correspond to increasing base strength because $pK_b = -\log K_b$. The pK_b values (given in parentheses) determine the following ordering:

$$N_2H_4 (5.77) < BrO^- (14.00 - 8.69 = 5.31) < CN^- (14.00 - 9.31 = 4.69)$$
$$< (C_2H_5)_3N (2.99)$$

Remember that the pK_b for the conjugate base of a weak acid will be given by

$pK_a + pK_b = 14$.

10.32 Any base whose conjugate acid lies below water in Table 10.3 will be a strong base, that is, the conjugate acid of the base will be a weaker acid than water, and so water will preferentially protonate the base. Based upon this information, we obtain the following analysis: (a) O^{2-}, strong; (b) Br^-, weak; (c) HSO_4^-, weak; (d) HCO_3^-, weak; (e) CH_3NH_2, weak; (f) H^-, strong; (g) CH_3^-, strong.

10.34 In oxoacids with the same number of oxygen atoms attached to the central atom, the greater the electronegativity of the central atom, the more the electrons of the O—H bond are withdrawn, making the bond more polar. This allows the hydrogen of the OH group to be more readily donated as a proton to H_2O, due to the stronger hydrogen bonds that it forms with the oxygen of water. Therefore, HClO is the stronger acid, with the lower pK_a.

10.36 (a) H_3PO_4 is stronger; it has the more electronegative central atom.

(b) $HBrO_3$ is stronger; there are more O atoms attached to the central atom in $HBrO_3$, making the H—O bond in $HBrO_3$ more polar than in HBrO.

(c) We would predict H_3PO_4 to be a stronger acid due to more oxygens on the central atom.

(d) H_2Te is the stronger acid, because the H—Te bond is weaker than the H—Se bond.

(e) HCl is the stronger acid. Within a period, the acidities of the binary acids are controlled by the bond polarity rather than the bond strength, and HCl has the greater bond polarity, due to the greater electronegativity of Cl relative to S.

(f) HClO is stronger because Cl has a greater electronegativity than I.

10.38 (a) Methylamine is CH_3NH_2, ammonia is NH_3. Methylamine can be thought of as being formed from NH_3 by replacing one H atom with CH_3. Because CH_3 is less electron withdrawing than H, CH_3NH_2 is a weaker acid and therefore a stronger base.

(b) Hydroxylamine is $HONH_2$, hydrazine is $H_2N—NH_2$. The former can be thought of as being derived from NH_3 by replacement of one H atom with OH; the latter by replacement of one H atom with NH_2. Because the hydroxyl group is more electron withdrawing than the amino group, NH_2, hydroxylamine is the stronger acid and therefore a weaker base.

10.40 The solution of 0.10 M H_2SO_4 would have the higher pH (would be the weaker acid) because the conjugate base, HSO_4^-, is less electronegative than the conjugate base of hydrobromic acid, namely Br^-.

10.42 The smaller the value of pK_b, the stronger the base; hence, aniline is the stronger base. 4-Chloroaniline is the stronger acid due to the presence of the electron-withdrawing Cl atom, making it the weaker base; and again we see that aniline is the stronger base.

10.44 The higher the pK_a of an acid, the stronger the corresponding conjugate base; therefore, the order is

3-hydroxyaniline < aniline < 2-hydroxyaniline < 4-hydroxyaniline

No simple pattern exists, but the position of the —OH group does affect the basicity.

10.46 (a) $K_a = 8.4 \times 10^{-4} = \dfrac{[H_3O^+][CH_3CH(OH)CO_2^-]}{[CH_3CH(OH)COOH]} = \dfrac{x^2}{0.12 - x} \approx \dfrac{x^2}{0.12}$

Here we have assumed that x is small enough to neglect it relative to 0.12 $mol \cdot L^{-1}$. This is a borderline case. We will also solve this exercise without making this approximation and compare the results below.

$x = [H_3O^+] = 0.010\ mol \cdot L^{-1}$

$pH = -\log(0.010) = 2.00$

$pOH = 14.00 - 2.00 = 12.00$

Without the approximation, the quadratic equation that must be solved is

$$8.4 \times 10^{-4} = \frac{x^2}{0.12 - x}$$

or $x^2 + 8.4 \times 10^{-4} x - 1.01 \times 10^{-4} = 0$

$$x = \frac{-8.4 \times 10^{-4} \pm \sqrt{(8.4 \times 10^{-4})^2 - 4(-1.01 \times 10^{-4})}}{2}$$

$x = 0.0096, -0.018.$

$x = [H_3O^+] = 0.0096$

$pH = -\log(0.0096) = 2.02$

(b) $K_a = 8.4 \times 10^{-4} = \dfrac{x^2}{1.2 \times 10^{-3} - x}$

or $x^2 + 8.4 \times 10^{-4} x - 1.0 \times 10^{-6} = 0$

$x = -1.5 \times 10^{-3}, 1.1 \times 10^{-3}$

The negative root can be eliminated:

$x = [H_3O^+] = 1.1 \times 10^{-3}$

$pH = -\log(1.1 \times 10^{-3}) = 2.96$

$pOH = 14.00 - 2.96 = 11.04$

(c) $8.4\times10^{-4} = \dfrac{x^2}{1.2\times10^{-5} - x}$

or $x^2 + 8.4\times10^{-4}x - 1.0\times10^{-8} = 0$

$x = 1.2\times10^{-5}, -8.5\times10^{-4}$

The negative root can be eliminated.

$[H_3O^+] = 1.2\times10^{-5}$

$pH = -\log(1.2\times10^{-5})$

$pH = 4.92$

$pOH = 14.00 - 4.92 = 9.08$

10.48 (a) $K_b(\text{pyridine}) = 1.8\times10^{-9}$

$C_6H_5N + H_2O \rightleftharpoons C_6H_5NH^+ + OH^-$

Concentration ($mol\cdot L^{-1}$)	C_6H_5N +	H_2O ⇌	$C_6H_5NH^+$ +	OH^-
initial	0.075	—	0	0
change	$-x$	—	$+x$	$+x$
final	$0.075 - x$	—	$+x$	$+x$

$K_b = 1.8\times10^{-9} = \dfrac{x^2}{0.075 - x} \approx \dfrac{x^2}{0.075}$

$x = [OH^-] = 1.2\times10^{-5}\ mol\cdot L^{-1}$

$pOH = -\log(1.2\times10^{-5}) = 4.92$

$pH = 14.00 - 4.92 = 9.00$

Percentage ionized $= \dfrac{1.2\times10^{-5}}{0.075}\times100\% = 1.6\times10^{-2}\%$

(b) The setup is similar to that in (a).

$K_b = 1.0\times10^{-6} = \dfrac{x^2}{0.0112 - x} \approx \dfrac{x^2}{0.0112}\quad x = [OH^-]$

$x = 1.1 \times 10^{-4}\ \text{mol} \cdot \text{L}^{-1}$

$\text{pOH} = -\log(1.1 \times 10^{-4}) = 3.97$

$\text{pH} = 14.00 - 3.97 = 10.03$

$$\text{Percentage protonation} = \frac{3.2 \times 10^{-4}}{0.0122} \times 100\% = 2.6\%$$

(c) $\text{p}K_\text{b} = 14.00 - 8.52 = 5.48$

$K_\text{b} = 3.3 \times 10^{-6}$

$\text{quinine} + H_2O \rightleftharpoons \text{quinineH}^+ + OH^-$

$$K_\text{b} = 3.3 \times 10^{-6} = \frac{x^2}{0.021 - x} \approx \frac{x^2}{0.021}$$

$x = [OH^-] = 2.6 \times 10^{-4}\ \text{mol} \cdot \text{L}^{-1}$

$\text{pOH} = -\log(2.6 \times 10^{-4}) = 3.58$

$\text{pH} = 14.00 - 3.58 = 10.42$

$$\text{Percentage protonation} = \frac{2.6 \times 10^{-4}}{0.021} \times 100\% = 1.2\%$$

(d) $\text{strychnine} + H_2O \rightleftharpoons \text{strychnineH}^+ + OH^-$

$x = [OH^-]$

$$K_\text{b} = \frac{K_\text{w}}{K_\text{a}} = \frac{1.00 \times 10^{-14}}{5.49 \times 10^{-9}} = 1.82 \times 10^{-6}$$

$$K_\text{b} = 1.82 \times 10^{-6} = \frac{x^2}{0.059 - x} \approx \frac{x^2}{0.059}$$

$x = [OH^-] = 3.3 \times 10^{-4}\ \text{mol} \cdot \text{L}^{-1}$

$\text{pOH} = -\log(3.3 \times 10^{-4}) = 3.48$

$\text{pH} = 14.00 - 3.48 = 10.52$

$$\text{Percentage protonation} = \frac{3.3 \times 10^{-4}}{0.059} \times 100\% = 0.56\%$$

10.50 (a) $HNO_2 + H_2O \rightleftharpoons H_3O^+ + NO_2^-$

$[H_3O^+] = [NO_2^-] = 10^{-2.63} = \text{antilog}(-2.63) = 2.3 \times 10^{-3}\ \text{mol} \cdot \text{L}^{-1}$

$$K_a = \frac{(2.3 \times 10^{-3})^2}{0.015 - 2.3 \times 10^{-3}} = 4.2 \times 10^{-4}$$

$pK_a = -\log(4.3 \times 10^{-4}) = 3.38$

(b) $C_4H_9NH_2 + H_2O \rightleftharpoons C_4H_9NH_3^+ + OH^-$

$pOH = 14.00 - 12.04 = 1.96$

$[C_4H_9NH_3^+] = [OH^-] = 10^{-1.96} = \text{antilog}(-1.96) = 0.011\ \text{mol} \cdot \text{L}^{-1}$

$$K_b = \frac{[C_4H_9NH_3^+][OH^-]}{[C_4H_9NH_2]} = \frac{(0.011)^2}{0.10 - 0.011} = 1.4 \times 10^{-3}$$

$pK_b = -\log(1.4 \times 10^{-3}) = 2.85$

10.52 (a) $K_a(\text{HCN}) = 4.9 \times 10^{-10}$

$[H_3O^+] = 10^{-pH} = 10^{-5.3} = \text{antilog}(-5.3) = 5 \times 10^{-6}\ \text{mol} \cdot \text{L}^{-1}$

Let x = nominal concentration of HCN, then

Concentration ($\text{mol} \cdot \text{L}^{-1}$)	HCN	+ H_2O	$\rightleftharpoons$ H_3O^+	+ CN^-
nominal	x	—	0	0
equilibrium	$x - 5 \times 10^{-6}$	—	5×10^{-6}	5×10^{-6}

$$4.9 \times 10^{-10} = \frac{(5 \times 10^{-6})^2}{x - 5 \times 10^{-6}}$$

Solve for x: $x = 5 \times 10^{-2}\ \text{mol} \cdot \text{L}^{-1} = 0.05\ \text{mol} \cdot \text{L}^{-1}$

(b) $K_b(\text{pyridine}) = 1.8 \times 10^{-9}$

$pOH = 14.00 - 8.8 = 5.2$

$[OH^-] = 10^{-pOH} = 10^{-5.2} = \text{antilog}(-5.2) = 6 \times 10^{-6}\ \text{mol} \cdot \text{L}^{-1}$

Let x = nominal concentration of C_5H_5N, then

Concentration ($\text{mol} \cdot \text{L}^{-1}$)	C_5H_5N	+ H_2O	$\rightleftharpoons$ $C_5H_6N^+$	+ OH^-
nominal	x			
equilibrium	$x - 6 \times 10^{-6}$		6×10^{-6}	6×10^{-6}

$$1.8\times10^{-9} = \frac{(6\times10^{-6})^2}{x - 6\times10^{-6}}$$

Solve for x: $x = 2\times10^{-2}\ \text{mol}\cdot\text{L}^{-1} = 0.02\ \text{mol}\cdot\text{L}^{-1}$

10.54 veronal + $H_2O \rightleftharpoons H_3O^+$ + veronalate ion

The equilibrium concentrations are

$$[\text{veronal}] = 0.020 - 0.0014\times0.020 = 0.020\ \text{mol}\cdot\text{L}^{-1}$$

$$[H_3O^+] = [\text{veronalate ion}] = 0.0014\times0.020 = 2.8\times10^{-5}\ \text{mol}\cdot\text{L}^{-1}$$

$$\text{pH} = -\log(2.8\times10^{-5}) = 4.55$$

$$K_a = \frac{[H_3O^+][\text{veronalate ion}]}{[\text{veronal}]} = \frac{(2.8\times10^{-5})^2}{0.020} = 3.9\times10^{-8}$$

10.56 cacodylic acid + $H_2O \rightleftharpoons H_3O^+$ + cacodylate ion

The equilibrium concentrations are

$$[\text{cacodylic acid}] = 0.0110 - (0.0077\times0.0110) = 0.0110 - 0.000\,08 = 0.109\ \text{mol}\cdot\text{L}^{-1}$$

$$[H_3O^+] = [\text{cacodylate ion}] = 0.0077\times0.0110 = 8.5\times10^{-5}\ \text{mol}\cdot\text{L}^{-1}$$

$$\text{pH} = -\log(8.5\times10^{-5}) = 4.07$$

$$K_a = \frac{[H_3O^+][\text{cacodylate ion}]}{[\text{cacodylic acid}]} = \frac{(8.5\times10^{-5})^2}{0.0109} = 6.6\times10^{-7}$$

10.58 (a)

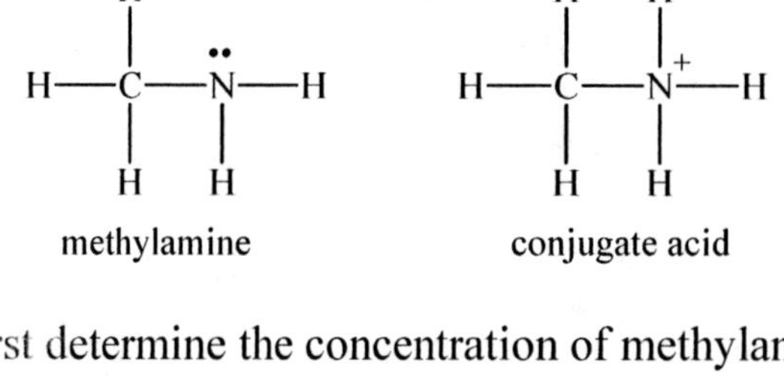

methylamine conjugate acid

(b) First determine the concentration of methylamine following dilution.

Moles of methylamine are found by:

$(50 \text{ mL})\cdot(0.85 \text{ g mL}^{-1}) = 42.5 \text{ g of solution}$

$(42.5 \text{ g})\cdot(0.35) = 14.9 \text{ g of methylamine}$

$$\frac{(14.9 \text{ g})}{(31.06 \text{ g}\cdot\text{mol}^{-1})} = 0.479 \text{ mol of methylamine}$$

diluted to 1.000 L, the resulting solution is 0.479 M in methylamine

Concentration ($\text{mol}\cdot\text{L}^{-1}$)	CH_3NH_2 +	H_2O ⇌	$CH_3NH_3^+$ +	OH^-
initial	0.479	—	0	0
equilibrium	$0.479 - x$	—	x	x

$$K_b = 3.6\times10^{-4} = \frac{[CH_3NH_3^+][OH^-]}{[CH_3NH_2]} = \frac{x^2}{0.479 - x}$$

solving for x we find:

$x = 1.3\times10^{-2} \text{ M} = [OH^-]$

$\text{pOH} = -\log(1.3\times10^{-2}) = 1.9$

$\text{pH} = 14.0 - 1.9 = 12.1$

10.60 (a) pH > 7, basic: $H_2O(l) + C_2O_4^{2-}(aq) \rightleftharpoons HC_2O_4^-(aq) + OH^-(aq)$

(b) pH = 7, neutral: Ca^{2+} is not an acid and NO_3^- is not a base

(c) pH < 7, acidic: $CH_3NH_3^+(aq) + H_2O(l) \rightleftharpoons H_3O^+(aq) + CH_3NH_2(aq)$

(d) pH > 7, basic: $H_2O(l) + PO_4^{3-}(aq) \rightleftharpoons HPO_4^{2-}(aq) + OH^-(aq)$

(e) pH < 7, acidic: $Fe(H_2O)_6^{3+}(aq) + H_2O(l) \rightleftharpoons$

$H_3O^+(aq) + Fe(H_2O)_5OH^{2+}(aq)$

(f) pH < 7, acidic: $C_5H_5NH^+(aq) + H_2O(l) \rightleftharpoons H_3O^+(aq) + C_5H_5N(aq)$

10.62 (a)

Concentration ($\text{mol}\cdot\text{L}^{-1}$)	$CH_3NH_3^+$	+ $H_2O(l)$ ⇌	CH_3NH_2 +	H_3O^+
initial	0.25	—	0	0
change	$-x$	—	$+x$	$+x$
equilibrium	$0.25 - x$	—	x	x

(See Table 10.2 for K_b of CH_3NH_2, the conjugate base of $CH_3NH_3^+$.)

$$K_a = \frac{K_w}{K_b} = \frac{1.00 \times 10^{-14}}{3.6 \times 10^{-4}} = 2.8 \times 10^{-11} = \frac{[CH_3NH_2][H_3O^+]}{[CH_3NH_3^+]}$$

$$2.8 \times 10^{-11} = \frac{x^2}{0.25 - x} \approx \frac{x^2}{0.25}$$

$x = 2.6 \times 10^{-6}\ \text{mol} \cdot \text{L}^{-1} = [H_3O^+]$ and $\text{pH} = -\log(2.6 \times 10^{-6}) = 5.58$

(b) Concentration

$(\text{mol} \cdot \text{L}^{-1})$	SO_3^{2-}	+ $H_2O(l)$ $\rightleftharpoons$	HSO_3^-	+ OH^-
initial	0.13	—	0	0
change	$-x$	—	$+x$	$+x$
equilibrium	$0.13 - x$	—	x	x

[See Table 10.9 for $K_a(HSO_3^-) = K_{a2}(H_2SO_3)$.]

$$K_b = \frac{K_w}{K_{a2}} = \frac{1.00 \times 10^{-14}}{1.2 \times 10^{-7}} = 8.3 \times 10^{-8} = \frac{[HSO_3^-][OH^-]}{[SO_3^{2-}]} = \frac{x^2}{0.13 - x} \approx \frac{x^2}{0.13}$$

$$8.3 \times 10^{-8} = \frac{x^2}{0.13}$$

$x = 1.0 \times 10^{-4}\ \text{mol} \cdot \text{L}^{-1} = [OH^-]$

$\text{pOH} = -\log(1.0 \times 10^{-4}) = 4.00$

$\text{pH} = 14.00 - \text{pOH} = 10.00$

(c) Concentration

$(\text{mol} \cdot \text{L}^{-1})$	$Fe(H_2O)_6^{3+}(aq)$	+ $H_2O(l)$ $\rightleftharpoons$	$H_3O^+(aq)$	+ $Fe(H_2O)_5OH^+(aq)$
initial	0.071	—	0	0
change	$-x$	—	$+x$	$+x$
equilibrium	$0.071 - x$	—	x	x

$$K_a = \frac{[H_3O^+][Fe(H_2O)_5OH^+(aq)]}{[Fe(H_2O)_6^{3+}]} = 3.5 \times 10^{-3} = \frac{x^2}{0.071 - x}$$

$x^2 + 3.5 \times 10^{-3} x \qquad 2.5 \times 10^{-4} = 0$

$x = 0.014, \qquad -0.018$

The negative root can be discarded

$[H_3O^+] = 0.014$, $\text{pH} = -\log 0.014 = .85$

10.64 (a) Initial concentration of $C_6H_5NH_3^+$ is:

$$\frac{(7.8\text{ g})\left(\dfrac{1}{94.133\text{ g mol}^{-1}}\right)}{0.350\text{ L}} = 0.24\text{ M}$$

Given this initial concentration and the Ka for this acid found in Table 10.7, the percent deprotonation is found as follows:

Concentration $(\text{mol}\cdot\text{L}^{-1})$	$C_6H_5NH_3^+$ +	$H_2O(l)$	$\rightleftharpoons$ $C_6H_5NH_2^{2+}$ +	H_3O^+
initial	0.24	—	0	0
change	$-x$	—	$+x$	$+x$
equilibrium	$0.24-x$	—	x	x

$$K_a = 2.3\times10^{-5} = \frac{[C_6H_5NH_2^{2+}][H_3O^+]}{[C_6H_5NH_3^+]}$$

$$2.3\times10^{-5} = \frac{x^2}{0.24-x} \approx \frac{x^2}{0.24}$$

$$x = 2.3\times10^{-3}\text{ mol}\cdot\text{L}^{-1} = [H_3O^+]$$

$$\text{percent dissociation} = \frac{2.3\times10^{-3}\text{ mol}\cdot\text{L}^{-1}}{0.23\text{ mol}\cdot\text{L}^{-1}}\cdot100\% = 1.0\%$$

10.66 (a) HSO_3^- can act as both an acid and a base. Both actions need to be considered simultaneously:

$$HSO_3^- + H_2O \rightleftharpoons H_3O^+ + SO_3^{2-}$$

$$HSO_3^- + H_2O \rightleftharpoons H_2SO_3 + OH^-$$

Summing,

$$HSO_3^- + HSO_3^- \rightleftharpoons H_2SO_3 + SO_3^{2-}$$

The simplest approach is to recognize that there are two conjugate acid-base pairs and to use the Henderson-Hasselbalch equation twice, once for each pair.

$$pH = pK_{a1} + \log\left(\frac{[HSO_3^-]}{[H_2SO_3]}\right)$$

$$pH = pK_{a2} + \log\left(\frac{[SO_3^{2-}]}{[HSO_3^-]}\right)$$

Summing,

$$2pH = pK_{a1} + pK_{a2} + \log\left(\frac{[HSO_3^-]}{[H_2SO_3]} \times \frac{[SO_3^{2-}]}{[HSO_3^-]}\right)$$

Because $[H_2SO_3] = [SO_3^{2-}]$, $2pH = pK_{a1} + pK_{a2}$

Therefore, $pH = \frac{1}{2}(pK_{a1} + pK_{a2}) = \frac{1}{2}(1.81 + 6.91) = 4.36$

Note that it is not necessary to know the concentration of HSO_3^-, because it cancels out of the equation. At extremely low concentrations of HSO_3^-, however, the approximations upon which the use of this equation are based are no longer valid.

(b) Neither silver ion nor nitrate ion is acidic or basic in aqueous solution. Therefore, the pH is that of neutral water, 7.00.

10.68 (a) $\frac{0.250\ \text{mol}\cdot\text{L}^{-1}\ \text{KCN} \times 0.0350\ \text{L}}{0.1000\ \text{L}} = 0.0875\ \text{mol}\cdot\text{L}^{-1}\ \text{KCN} = [CN^-]_0$

Concentration $(\text{mol}\cdot\text{L}^{-1})$	$H_2O(l)$	$+ CN^-(aq)$	$\rightleftharpoons$	$HCN(aq) +$	$OH^-(aq)$
initial	—	0.0875		0	0
change	—	$-x$		$+x$	$+x$
equilibrium	—	$0.0875 - x$		x	x

$$K_b = \frac{K_w}{K_a} = \frac{1.00 \times 10^{-14}}{4.9 \times 10^{-10}} = 2.0 \times 10^{-5} = \frac{[HCN][OH^-]}{[CN^-]}$$

$$2.0 \times 10^{-5} = \frac{x^2}{0.0875 - x} \approx \frac{x^2}{0.0875}$$

$x = 1.3 \times 10^{-3}\ \text{mol}\cdot\text{L}^{-1} = [HCN]$

(b) $\dfrac{1.59\text{ g NaHCO}_3}{84.01\text{ g NaHCO}_3/\text{mol NaHCO}_3} \times \dfrac{1}{0.200\text{ L}}$

$= 9.46 \times 10^{-2}\ \text{mol}\cdot\text{L}^{-1}\ \text{NaHCO}_3$

Concentration $(\text{mol}\cdot\text{L}^{-1})$	$H_2O(l)$ +	$HCO_3^-(aq)$ $\rightleftharpoons$	$H_2CO_3(aq)$ +	$OH^-(aq)$
initial	—	9.46×10^{-2}	0	0
change	—	$-x$	$+x$	$+x$
equilibrium	—	$9.46 \times 10^{-2} - x$	x	x

$$K_{b1} = \frac{[H_2CO_3][OH^-]}{[HCO_3^-]} = \frac{K_w}{K_{a1}} = \frac{1.00 \times 10^{-14}}{4.3 \times 10^{-7}} = 2.3 \times 10^{-8}$$

$$2.3 \times 10^{-8} = \frac{x^2}{9.46 \times 10^{-2} - x} \approx \frac{x^2}{9.46 \times 10^{-2}},\ x = 4.7 \times 10^{-5}\ \text{mol}\cdot\text{L}^{-1}$$

$$= [OH^-]$$

$\text{pOH} = -\log(4.7 \times 10^{-5}) = 4.33,\ \text{pH} = 14.00 - 4.33 = 9.67$

10.70 The reaction of interest is:

$C_6H_5CH_2(CH_3)NH_3^+(aq) + H_2O(l) \rightleftharpoons C_6H_5CH_2(CH_3)NH_2(aq) + H_3O^+(aq)$

K_a for this reaction is: $K_a = \dfrac{1\times10^{-14}}{7.8\times10^{-4}} = 1.3\times10^{-11}$

The initial concentration of $C_6H_5CH_2(CH_3)NH_3^+(aq)$ is:

$$\frac{(6.48\text{ g})\cdot(216.12\text{ g mol}^{-1})}{0.200\text{ L}} = 0.150\text{ mol L}^{-1}$$

Concentration $(\text{mol}\cdot\text{L}^{-1})$	$H_2O(l)$ +	$C_6H_5CH_2(CH_3)NH_3^+(aq)$ $\rightleftharpoons$	$C_6H_5CH_2(CH_3)NH_2(aq)$ +	$H_3O^+(aq)$
initial	—	9.46×10^{-2}	0	0
change	—	$-x$	$+x$	$+x$
equilibrium	—	$9.46 \times 10^{-2} - x$	x	x

$$1.3\times10^{-11}=\frac{x^2}{0.145-x}\approx\frac{x^2}{0.145},\ x=1.4\times10^{-6}\ \text{mol}\cdot\text{L}^{-1}$$

$$=[H_3O^+]$$

$$\text{pH}=-\log(1.4\times10^{-6})=5.9$$

10.72 (a) $H_3PO_4(aq)+H_2O(l)\rightleftharpoons H_3O^+(aq)+H_2PO_4^-(aq)$

$H_2PO_4^-(aq)+H_2O(l)\rightleftharpoons H_3O^+(aq)+HPO_4^{2-}(aq)$

$HPO_4^{2-}(aq)+H_2O(l)\rightleftharpoons H_3O^+(aq)+PO_4^{3-}(aq)$

(b)

$(CH_2)_4(COOH)_2(aq)+H_2O(l)\rightleftharpoons H_3O^+(aq)+(CH_2)_4(COOH)CO_2^-(aq)$

$(CH_2)_4(COOH)CO_2^-(aq)+H_2O(l)\rightleftharpoons H_3O^+(aq)+(CH_2)_4(CO_2)_2^{2-}(aq)$

(c)

$(CH_2)_2(COOH)_2(aq)+H_2O(l)\rightleftharpoons H_3O^+(aq)+(CH_2)_2(COOH)CO_2^-(aq)$

$(CH_2)_2(COOH)CO_2^-(aq)+H_2O(l)\rightleftharpoons H_3O^+(aq)+(CH_2)_2(CO_2)_2^{2-}(aq)$

10.74 The reaction is (after the first, essentially complete ionization)

$HSeO_4^- + H_2O \rightleftharpoons H_3O^+ + SeO_4^{2-}$

The initial concentrations of $HSeO_4^-$ and H_3O^+ are both 0.010 $\text{mol}\cdot\text{L}^{-1}$ due to the complete ionization of H_2SeO_4 in the first step. The second ionization is incomplete.

Concentration ($\text{mol}\cdot\text{L}^{-1}$)	$HSeO_4^-$	+	H_2O	$\rightleftharpoons$	H_3O^+	+	SeO_4^{2-}
initial	0.010		—		0.010		0
change	$-x$		—		$+x$		$+x$
equilibrium	$0.010-x$		—		$0.010+x$		x

$$K_{a2} = 1.2 \times 10^{-2} = \frac{[H_3O^+][SeO_4^{2-}]}{[HSeO_4^-]} = \frac{(0.010 + x)(x)}{0.010 - x}$$

$$x^2 + 0.022x - 1.2 \times 10^{-4} = 0$$

$$x = \frac{-0.022 + \sqrt{(0.022)^2 - (4)(-1.2 \times 10^{-4})}}{2} = 4.5 \times 10^{-3}$$

$$[H_3O^+] = 0.010 + x = (0.010 + 4.5 \times 10^{-3})\ mol \cdot L^{-1} = 1.5 \times 10^{-2}\ mol \cdot L^{-1}$$

$$pH = -\log(1.5 \times 10^{-2}) = 1.82$$

10.76 (a) Second ionization is ignored because $K_{a2} << K_{a1}$

$$H_2S + H_2O \rightleftharpoons H_3O^+ + HS^-$$

$$K_{a1} = 1.3 \times 10^{-7} = \frac{[H_3O^+][HS^-]}{[H_2S]} = \frac{x^2}{0.10 - x} \approx \frac{x^2}{0.10}$$

$$x = [H_3O^+] = 1.1 \times 10^{-4}\ mol \cdot L^{-1}$$

$$pH = -\log(1.1 \times 10^{-4}) = 3.96$$

(b) This is a situation where it may not be justified to ignore the second ionization. $K_{a1} = 6.0 \times 10^{-4}$, $K_{a2} = 1.5 \times 10^{-5}$, and $K_{a2}/K_{a1} = 40$. This is a marginal case; we can work it both ways, first without ignoring the second ionization. Adopt the following notation:

$H^+ = H_3O^+$

$[H^+] = [H_3O^+]$ = equilibrium concentration of H_3O^+

H_2A = tartaric acid

c_0 = solute molarity = $0.15\ mol \cdot L^{-1}$

$$c_0 = [H_2A] + [HA^-] + [A^{2-}] \qquad (1)$$

$$H = \text{total H present} = [H^+] + 2[H_2A] + [HA^-] = 2c_0 \qquad (2)$$

The following equilibria occur:

$$H_2A \rightleftharpoons H^+ + HA^- \qquad K_{a1} = \frac{[H^+][HA^-]}{[H_2A]} \qquad (3)$$

$$HA^- \rightleftharpoons H^+ + A^{2-} \qquad K_{a2} = \frac{[H^+][A^{2-}]}{[HA^-]} \qquad (4)$$

Examination of Eqs. 1 to 4 shows that there are four unknowns,

$[H^+]$, $[H_2A]$, $[HA^-]$, and $[A^{2-}]$, to be determined. However, these four simultaneous equations allow for their determination.

The unknowns $[H_2A]$, $[HA^-]$, and $[A^{2-}]$ may all be expressed in terms of one unknown, $[H^+]$.

According to Eq. 1, $c_0 = [H_2A] + [HA^-] + [A^{2-}]$

From Eq. 3, $[H_2A] = \dfrac{[H^+][HA^-]}{K_{a1}}$

From Eq. 4, $[HA^-] = \dfrac{[H^+][A^{2-}]}{K_{a2}}$

Substituting, $[H_2A] = \dfrac{[H^+]^2[A^{2-}]}{K_{a1}\,K_{a2}}$

$$\text{Then Eq. 1 becomes } c_0 = [A^{2-}]\left\{1 + \frac{[H^+]}{K_{a2}} + \frac{[H^+]^2}{K_{a1}K_{a2}}\right\} \qquad (5)$$

Now, subtract 2 × Eq. 1 from Eq. 2:

$$2c_0 = [H^+] + 2[H_2A] + [HA^-] \qquad \text{Eq. 2}$$

$$\frac{2c_0 = 2[H_2A] + 2[HA^-] + 2[A^{2-}]}{[H^+] - [HA^-] - 2[A^{2-}] = 0} \qquad 2 \times \text{Eq. 1}$$

$$[A^{2-}],\ [A^{2-}] = \frac{[H^+]}{\dfrac{[H^+]}{K_{a2}} + 2}$$

Substitute this into Eq. 5 to give

$$c_0\left(2 + \frac{[H^+]}{K_{a2}}\right) = [H^+] + \frac{[H^+]^2}{K_{a2}} + \frac{[H^+]^3}{K_{a1}\,K_{a2}}$$

Rearranging into standard cubic form gives

$$\frac{[H^+]^3}{K_{a1}\,K_{a2}} + \frac{[H^+]^2}{K_{a2}} + [H^+]\left(1 - \frac{c_0}{K_{a2}}\right) - 2c_0 = 0$$

Now, put in the numerical values for c_0, K_{a1}, and K_{a2}:

$$1.1 \times 10^8[H^+]^3 + 6.7 \times 10^4[H^+]^2 - 1.0 \times 10^4[H^+] - 0.30 = 0$$

Solution of this cubic by standard methods gives

$$[H^+] = [H_3O^+] = 9.2 \times 10^{-3} \text{ mol} \cdot \text{L}^{-1}$$

$$\text{pH} = -\log(9.2 \times 10^{-3}) = 2.04$$

It is left as an exercise for the reader to show that, if the second ionization is ignored,

$$[H^+] = 9.5 \times 10^{-3} \text{ mol} \cdot \text{L}^{-1} \text{ and pH} = 2.02$$

The difference is small but, perhaps, not within experimental error.

(c) The second ionization can be ignored because $K_{a2} << K_{a1}$.

$$H_2TeO_4 + H_2O \rightleftharpoons H_3O^+ + HTeO_4^- \subsetneq$$

$$K_{a1} = 2.1 \times 10^{-8} = \frac{[H_3O^+][HTeO_4^-]}{[H_2TeO_4]} = \frac{x^2}{1.1 \times 10^{-3} - x} \approx \frac{x^2}{1.1 \times 10^{-3}}$$

$$x = [H_3O^+] = 4.8 \times 10^{-6} \text{ mol} \cdot \text{L}^{-1}$$

$$\text{pH} = -\log(4.8 \times 10^{-6}) = 5.32$$

10.78 (a) The pH is given by $\text{pH} = \frac{1}{2}(\text{p}K_{a1} + \text{p}K_{a2})$. From Table 10.9, we find

$K_{a1} = 4.3 \times 10^{-7}$ $\qquad$ $\text{p}K_{a1} = 6.37$

$K_{a2} = 5.6 \times 10^{-11}$ $\qquad$ $\text{p}K_{a2} = 10.25$

$$\text{pH} = \tfrac{1}{2}(6.37 + 10.25) = 8.31$$

(b) The nature of the spectator counter ion does not affect the equilibrium and the pH of a salt solution of a polyprotic acid is independent of the concentration of the salt, therefore pH = 8.31.

10.80 The pH is given by $\text{pH} = \frac{1}{2}(\text{p}K_{a1} + \text{p}K_{a2})$.

$$\text{pH} = \tfrac{1}{2}(2.46 + 7.31) = 4.89$$

10.82 The equilibrium reactions of interest are

$$H_2SO_3(aq) + H_2O(l) \rightleftharpoons H_3O^+(aq) + HSO_3^-(aq) \qquad K_{a1} = 1.5 \times 10^{-2}$$

$$HSO_3^-(aq) + H_2O(l) \rightleftharpoons H_3O^+(aq) + SO_3^{2-}(aq) \qquad K_{a2} = 1.2 \times 10^{-7}$$

Because the second ionization constant is much smaller than the first, we can assume that the first step dominates:

Concentration (mol · L^{-1})	H_2SO_3(aq) +	H_2O(l) $\rightleftharpoons$	H_3O^+(aq)	+ HSO_3^-(aq)
initial	0.125	—	0	0
change	$-x$	—	$+x$	$+x$
final	$0.125 - x$	—	$+x$	$+x$

$$K_{a1} = \frac{[H_3O^+][HSO_3^-]}{[H_2SO_3]}$$

$$1.5 \times 10^{-2} = \frac{(x)(x)}{0.125 - x} = \frac{x^2}{0.125 - x}$$

Assume that $x << 0.125$, then

$$x^2 = (1.5 \times 10^{-2})(0.125)$$
$$x = 0.043$$

Because $x > 5\%$ of 0.0456, the assumption was not valid, and the full expression must be evaluated using the quadratic:

$$x^2 + 1.5 \times 10^{-2}\ x - (1.5 \times 10^{-2})(0.125) = 0$$

Solving with the quadratic equation gives $x = 0.036\ \text{mol} \cdot \text{L}^{-1}$.

$$x = [H_3O^+] = [HSO_3^-] = 0.036\ \text{mol} \cdot \text{L}^{-1}$$
$$[H_2SO_3] = 0.125\ \text{mol} \cdot \text{L}^{-1} - 0.036\ \text{mol} \cdot \text{L}^{-1} = 0.089\ \text{mol} \cdot \text{L}^{-1}$$

We can then use the other equilibria to determine the remaining concentrations:

$$K_{a2} = \frac{[H_3O^+][SO_3^{2-}]}{[HSO_3^-]}$$

$$1.2 \times 10^{-7} = \frac{(0.036)[SO_3^{2-}]}{(0.036)}$$

$$[SO_3^{2-}] = 1.2 \times 10^{-7}\ \text{mol} \cdot \text{L}^{-1}$$

Because $1.2 \times 10^{-7} << 0.036$, the initial assumption that the first dissociation would dominate is valid. To calculate $[OH^-]$, we use the K_w relationship:

$K_w = [H_3O^+][OH^-]$

$$[OH^-] = \frac{K_w}{[H_3O^+]} = \frac{1.00 \times 10^{-14}}{0.036} = 2.8 \times 10^{-13}\ \text{mol} \cdot \text{L}^{-1}$$

In summary,

$[H_2SO_3] = 0.089\ \text{mol} \cdot \text{L}^{-1}, [H_3O^+] = [HSO_3^-] = 0.036\ \text{mol} \cdot \text{L}^{-1}, [SO_3^{2-}]$

$= 1.2 \times 10^{-7}\ \text{mol} \cdot \text{L}^{-1}, [OH^-] = 2.8 \times 10^{-13}\ \text{mol} \cdot \text{L}^{-1}$

10.84 The equilibrium reactions of interest are now the base forms of the carbonic acid equilibria, so K_b values should be calculated for the following changes:

$SO_3^{2-}(aq) + H_2O(l) \rightleftharpoons HSO_3^-(aq) + OH^-(aq)$

$$K_{b1} = \frac{K_w}{K_{a2}} = \frac{1.00 \times 10^{-14}}{1.2 \times 10^{-7}} = 8.3 \times 10^{-8}$$

$HSO_3^-(aq) + H_2O(l) \rightleftharpoons H_2SO_3(aq) + OH^-(aq)$

$$K_{b2} = \frac{K_w}{K_{a2}} = \frac{1.00 \times 10^{-14}}{1.5 \times 10^{-2}} = 6.7 \times 10^{-13}$$

Because the second hydrolysis constant is much smaller than the first, we can assume that the first step dominates:

Concentration ($\text{mol} \cdot \text{L}^{-1}$)	$SO_3^{2-}(aq)+$	$H_2O(l)$ $\rightleftharpoons$	$HSO_3^-(aq)$ +	$OH^-(aq)$
initial	0.125	—	0	0
change	$-x$	—	$+x$	$+x$
final	$0.125 - x$	—	$+x$	$+x$

$$K_{b1} = \frac{[HSO_3^-][OH^-]}{[SO_3^{2-}]}$$

$$8.3 \times 10^{-8} = \frac{(x)(x)}{0.125 - x} = \frac{x^2}{0.125 - x}$$

Assume that $x << 0.125$, then

$x^2 = (8.3 \times 10^{-8})(0.125)$

$x = 1.0 \times 10^{-4}$

Because x < 1% of 0.125, the assumption was valid.

$x = [HSO_3^-] = [OH^-] = 1.0 \times 10^{-4}\ mol \cdot L^{-1}$

Therefore, $[SO_3^{2-}] = 0.125\ mol \cdot L^{-1} - 1.0 \times 10^{-4}\ mol \cdot L^{-1} \cong 0.125\ mol \cdot L^{-1}$

We can then use the other equilibria to determine the remaining concentrations:

$$K_{b2} = \frac{[H_2SO_3][OH^-]}{[HCO_3^-]}$$

$$6.7 \times 10^{-13} = \frac{[H_2SO_3](1.0 \times 10^{-4})}{(1.0 \times 10^{-4})}$$

$$[H_2SO_3] = 6.7 \times 10^{-13}\ mol \cdot L^{-1}$$

Because $6.7 \times 10^{-13} << 1.0 \times 10^{-4}$, the initial assumption that the first hydrolysis would dominate is valid.

To calculate $[H_3O^+]$, we use the K_w relationship:

$$K_w = [H_3O^+][OH^-]$$

$$[H_3O^+] = \frac{K_w}{[OH^-]} = \frac{1.00 \times 10^{-14}}{1.0 \times 10^{-14}} = 1.0 \times 10^{-10}\ mol \cdot L^{-1}$$

In summary, $[H_2SO_3] = 6.7 \times 10^{-13}\ mol \cdot L^{-1}, [OH^-]$
$= [HSO_3^-] = 1.0 \times 10^{-4}\ mol \cdot L^{-1}, [SO_3^{2-}]$
$= 0.125\ mol \cdot L^{-1}, [H_3O^+] = 1.0 \times 10^{-10}\ mol \cdot L^{-1}$

10.86 (a) tartaric acid: The two pK_a values are 3.22 and 4.82. Because pH = 5.0 lies above both of these values, the major form present will be the doubly deprotonated ion A^{2-}. (b) hydrosulfuric acid: The two pK_a values are 6.89 and 14.15. Because the pH of the solution lies below both of these values, the dominant form will be the doubly protonated H_2A. (c) phosphoric acid: The three pK_a values are 2.12, 7.21, and 12.68. The pH of the solution lies between the first and second ionization, so the predominant species should be the singly deprotonated ion $H_2PO_4^-$.

10.88 The equilibria present in the solution are:

$$H_2(aq) + H_2O(l) \rightleftharpoons H_3O^+(aq) + HS^-(aq) \quad K_{a1} = 1.3 \times 10^{-7}$$
$$HS^-(aq) + H_2O(l) \rightleftharpoons H_3O^+(aq) + S^{2-}(aq) \quad K_{a2} = 7.1 \times 10^{-15}$$

The calculation of the desired concentrations follows exactly after the method derived in Eq. 25, substituting H_2S for H_2CO_3, HS^- for HCO_3^-, and S^{2-} for CO_3^{2-}. First, calculate the quantity

f (at pH = 9.35 $[H_3O^+] = 10^{-9.35} = 4.5 \times 10^{-10}\ mol \cdot L^{-1}$):

$$f = [H_3O^+]^2 + [H_3O^+]K_{a1} + K_{a1}\ K_{a2}$$
$$= (4.5 \times 10^{-10})^2 + (4.5 \times 10^{-10})(1.3 \times 10^{-7}) + (1.3 \times 10^{-7})(7.1 \times 10^{-15})$$
$$= 5.9 \times 10^{-17}$$

The fractions of the species present are then given by

$$\alpha(H_2S) = \frac{[H_3O^+]}{f} = \frac{(4.5 \times 10^{-10})^2}{5.9 \times 10^{-17}} = 3.4 \times 10^{-3}$$

$$\alpha(HS^-) = \frac{[H_3O^+]K_{a1}}{f} = \frac{(4.5 \times 10^{-10})(1.3 \times 10^{-7})}{5.9 \times 10^{-17}} = 0.99$$

$$\alpha(S^{2-}) = \frac{K_{a1}K_{a2}}{f} = \frac{(1.3 \times 10^{-17})(7.1 \times 10^{-15})}{5.9 \times 10^{-17}} = 1.6 \times 10^{-15}$$

Thus, in a solution at pH 9.35, the dominant species will be HS^- with a concentration of $(0.250\ mol \cdot L^{-1})(0.99) \cong 0.25\ mol \cdot L^{-1}$. The concentration of H_2S will be

$(3.4 \times 10^{-3})(0.250\ mol \cdot L^{-1}) = 8.5 \times 10^{-4}\ mol \cdot L^{-1}$, and the concentration of S^{2-} will be $(1.6 \times 10^{-5})(0.250\ mol \cdot L^{-1}) = 4.0 \times 10^{-6}\ mol \cdot L^{-1}$.

10.90 For the first ionization of $(COOH)_2$—or $H_2C_2O_4$—we write

$$H_2C_2O_4 + H_2O \rightleftharpoons H_3O^+ + HC_2O_4^-,\ K_{a1} = 5.9 \times 10^{-2}$$

Concentration ($mol \cdot L^{-1}$)	$H_2C_2O_4$	+ H_2O	$\rightleftharpoons$	H_3O^+ +	$HC_2O_4^-$
initial	0.10	—		0	0
change	$-x$	—		$+x$	$+x$
equilibrium	$0.10 - x$	—		x	x

$K_{a1} = 5.9 \times 10^{-2}$, $K_{a2} = 6.5 \times 10^{-5}$

Because $K_{a2} << K_{a1}$, the second ionization can safely be ignored in the calculation of $[H_3O^+]$.

$$K_{a1} = 5.9 \times 10^{-2} = \frac{x^2}{0.10 - x}$$

$$x^2 + 0.059x - 0.0059 = 0$$

$$x = \frac{-0.059 + \sqrt{(0.059)^2 + (4)(1)(0.0059)}}{2} = 0.053 \text{ mol} \cdot L^{-1} = [H_3O^+]$$

$$[OH^-] = \frac{1.0 \times 10^{-14}}{0.053} = 1.9 \times 10^{-13} \text{ mol} \cdot L^{-1}$$

$$[H_2C_2O_4] = 0.10 - 0.053 = 0.05 \text{ mol} \cdot L^{-1}$$

Concentration ($mol \cdot L^{-1}$)	$HC_2O_4^-$	+ H_2O	$\rightleftharpoons$	H_3O^+ +	$C_2O_4^{2-}$
initial	0.053	—		0.053	0
change	$-x$	—		$+x$	$+x$
equilibrium	$0.053 - x$	—		$0.053 + x$	x

$$K_{a2} = 6.5 \times 10^{-5} = \frac{(0.053 + x)(x)}{0.053 - x} \approx x \text{ (because } x \text{ is small)}$$

or $x = [C_2O_4^{2-}] = 6.5 \times 10^{-5}$ mol $\cdot$ L^{-1}, and

$$[HC_2O_4^-] = 0.053 - x = 0.053 - 0.000\,065 = 0.053 \text{ mol} \cdot L^{-1}$$

10.92 The equilibria involved are:

$H_2CO_3(aq) + H_2O(l) \rightleftharpoons HCO_3^-(aq) + H_3O^+$ $\quad K_{a1} = 4.3 \times 10^{-7}$

$HCO_3^-(aq) + H_2O(l) \rightleftharpoons CO_3^{2-}(aq) + H_3O^+$ $\quad K_{a2} = 5.6 \times 10^{-11}$

Given these two equilibria, two equations can be written:

$$K_{a1} = \frac{[HCO_3^-][H_3O^+]}{[H_2CO_3]}$$

$$K_{a2} = \frac{[CO_3^{2-}][H_3O^+]}{[HCO_3^-]}$$

also, enough information is provided to construct two other equations:

$$[H_2CO_3]+[HCO_3^-]+[CO_3^{2-}]=4.5\times10^{-3}$$

$$[H_3O^+]=10^{-4.8}=1.6\times10^{-5}$$

Solving this set of simultaneous equations for the three unknown concentrations, one finds:

$$[H_2CO_3]=4.38\times10^{-3}$$

$$[HCO_3^-]=1.19\times10^{-4}$$

$$[CO_3^{2-}]=4.23\times10^{-10}$$

10.94 We can use the relationship derived in the text:

$$[H_3O^+]^2-[HA]_{initial}[H_3O^+]-K_w=0$$

$$[H_3O^+]^2-(7.49\times10^{-8})[H_3O^+]-(1.00\times10^{-14})=0$$

Solving using the quadratic equation gives

$[H_3O^+]=1.44\times10^{-7}$; pH = 6.842

This value is lower than the value calculated, based on the acid concentration alone $(pH=-\log(7.49\times10^{-8})=7.126)$.

10.96 We can use the relationship derived in the text:

$[H_3O^+]^2+[B]_{initial}[H_3O^+]-K_w=0$, in which B is any strong base.

$$[H_3O^+]^2+(8.23\times10^{-7})[H_3O^+]-(1.00\times10^{-14})=0$$

Solving using the quadratic equation gives

$[H_3O^+]=1.20\times10^{-8}$, pH = 7.922

This value is slightly higher than the value calculated, based on the base concentration alone

$(pOH=-\log(8.23\times10^{-7})=6.084$, $pH=14.00-6.084=7.916)$.

10.98 (a) In the absence of a significant effect due to the autoprotolysis of water, the pH values of the 2.50×10^{-4} M and 2.50×10^{-6} M C_6H_5OH solutions can be calculated as described earlier.

For 2.50×10^{-4} $mol \cdot L^{-1}$:

Concentration $(mol \cdot L^{-1})$	$C_6H_5OH(aq)$	$+ H_2O(l) \rightleftharpoons$	$H_3O^+(aq) +$	$C_6H_5O^-(aq)$
initial	2.50×10^{-4}	—	0	0
change	$-x$	—	$+x$	$+x$
final	$2.50 \times 10^{-4} - x$	—	$+x$	$+x$

$$K_a = \frac{[H_3O^+][C_6H_5O^-]}{[C_6H_5OH]}$$

$$1.3 \times 10^{-10} = \frac{(x)(x)}{2.50 \times 10^{-4} - x} = \frac{x^2}{2.50 \times 10^{-4} - x}$$

Assume $x << 2.50 \times 10^{-4}$

$$x^2 = (1.3 \times 10^{-10})(2.50 \times 10^{-4})$$
$$x = 1.8 \times 10^{-7}$$

Because $x < 1\%$ of 2.50×10^{-4}, the assumption was valid. Given this value, the pH is then calculated to be $-\log(1.8 \times 10^{-8}) = 6.74$.

For 2.50×10^{-6} $mol \cdot L^{-1}$:

Concentration $(mol \cdot L^{-1})$	$C_6H_5OH(aq)$	$+ H_2O(l) \rightleftharpoons$	$H_3O^+(aq)$	$+ C_6H_5O^-(aq)$
initial	2.50×10^{-6}	—	0	0
change	$-x$	—	$+x$	$+x$
final	$2.50 \times 10^{-6} - x$	—	$+x$	$+x$

$$K_a = \frac{[H_3O^+][C_6H_5O^-]}{[C_6H_5OH]}$$

$$1.3 \times 10^{-10} = \frac{(x)(x)}{2.50 \times 10^{-6} - x} = \frac{x^2}{2.50 \times 10^{-6} - x}$$

Assume $x << 2.50 \times 10^{-6}$

$x^2 = (1.3 \times 10^{-10})(2.50 \times 10^{-6})$
$x = 1.8 \times 10^{-8}$

Because $x < 1\%$ of 2.50×10^{-6}, the assumption is valid and the pH should be 7.74. However, this number does not make sense because an acid was added to the water.

(b) To calculate the value, taking into account the autoprotolysis of water, we can use Eq. 22:

$x^3 + K_a x^2 - (K_w + K_a \cdot [HA]_{initial})x - K_w \cdot K_a = 0$, where $x = [H_3O^+]$.

To solve the expression, you substitute the values of $K_w = 1.00 \times 10^{-14}$, the initial concentration of acid, and $K_a = 1.3 \times 10^{-10}$ into this equation and then solve the expression either by trial and error or, preferably, using a graphing calculator. Alternatively, you can use a computer program designed to solve simultaneous equations. Because the unknowns include $[H_3O^+]$, $[OH^-]$, [HClO], and $[ClO^-]$, you will need four equation. As seen in the text, the pertinent equations are

$$K_a = \frac{[H_3O^+][C_6H_5O^-]}{[C_6H_5OH]}$$
$$K_w = [H_3O^+][OH^-]$$
$$[H_3O^+] = [OH^-] + [C_6H_5O^-]$$
$$[C_6H_5OH]_{initial} = [C_6H_5OH] + [C_6H_5O^-]$$

Using either method should produce the same result.

The values obtained for $2.50 \times 10^{-4}\ mol \cdot L^{-1}$ are

$[H_3O^+] = 2.1 \times 10^{-7}\ mol \cdot L^{-1}, pH = 6.68$ (compare to 6.74 obtained in (a))

$[C_6H_5O^-] = 1.6 \times 10^{-7}\ mol \cdot L^{-1}$
$[C_6H_5OH] \cong 2.5 \times 10^{-4}\ mol \cdot L^{-1}$
$[OH^-] = 4.8 \times 10^{-8}\ mol \cdot L^{-1}$

Similarly, for $[C_6H_5OH]_{initial} = 2.50 \times 10^{-6}$:

$[H_3O^+] = 1.0 \times 10^{-7}\ mol \cdot L^{-1}, pH = 7.00$ (compare to 7.74 obtained in (a))

$[C_6H_5O^-] = 3.2 \times 10^{-9}\ mol \cdot L^{-1}$

$[C_6H_5OH] \cong 2.5 \times 10^{-6}\ mol \cdot L^{-1}$

$[OH^-] = 9.8 \times 10^{-8}\ mol \cdot L^{-1}$

Note that for the more concentrated solution, the effect of the autoprotolysis of water is very small. Notice also that the less concentrated solution is more acidic, due to the autoprotolysis of water, than would be predicted if this effect were not operating.

10.100 (a) In the absence of a significant effect due to the autoprotolysis of water, the pH values of the 1.89×10^{-5} M and 9.64×10^{-7} M HClO solutions can be calculated as described earlier.

For $1.89 \times 10^{-5}\ mol \cdot L^{-1}$:

Concentration $(mol \cdot L^{-1})$	$HClO(aq)$	+ $H_2O(l)$	$\rightleftharpoons H_3O^+(aq)$	+ $ClO^-(aq)$
initial	1.89×10^{-5}	—	0	0
change	$-x$	—	$+x$	$+x$
final	$1.89 \times 10^{-5} - x$	—	$+x$	$+x$

$$K_a = \frac{[H_3O^+][ClO^-]}{[HClO]}$$

$$3.0 \times 10^{-8} = \frac{(x)(x)}{1.89 \times 10^{-5} - x} = \frac{x^2}{1.89 \times 10^{-5} - x}$$

Assume $x << 1.89 \times 10^{-5}$

$x^2 = (3.0 \times 10^{-8})(1.89 \times 10^{-5})$

$x = 7.5 \times 10^{-7}$

Because $x < 5\%$ of 1.89×10^{-5}, the assumption was valid. Given this value, the pH is then calculated to be $-\log(7.5 \times 10^{-7}) = 6.12$.

For $9.64 \times 10^{-7}\ mol \cdot L^{-1}$:

Concentration (mol · L^{-1})	HClO(aq) +	H_2O(l)	$\rightleftharpoons H_3O^+$(aq) +	ClO^-(aq)
initial	9.64×10^{-7}	—	0	0
change	$-x$	—	$+x$	$+x$
final	$9.64 \times 10^{-7} - x$	—	$+x$	$+x$

$$K_a = \frac{[H_3O^+][ClO^-]}{[HClO]}$$

$$3.0 \times 10^{-8} = \frac{(x)(x)}{9.64 \times 10^{-7} - x} = \frac{x^2}{9.64 \times 10^{-7} - x}$$

Assume $x << 9.64 \times 10^{-7} - x$

$$x^2 = (3.0 \times 10^{-8})(9.64 \times 10^{-7} - x)$$
$$x = 1.7 \times 10^{-7}$$

x is approximately 10% of 9.64×10^{-7}; the assumption is not reasonable, and so you must calculate the value explicitly for the following expression, using the quadratic equation.

$$x^2 + 3.0 \times 10^{-8} x - (3.0 \times 10^{-8})(9.64 \times 10^{-7}) = 0$$

Upon solving the quadratic equation, a value of $x = 1.6 \times 10^{-7}$ is obtained, yielding pH = 6.80.

(b) To calculate the value, taking into account the autoprotolysis of water, we can use Eq. 21:

$x^3 + K_a x^2 - (K_w + K_a \cdot [HA]_{initial})x - K_w \cdot K_a = 0$, where $x = [H_3O^+]$.

To solve the expression, you substitute the values of $K_w = 1.00 \times 10^{-14}$, the initial concentration of acid, and $K_a = 3.0 \times 10^{-8}$ into this equation and then solve the expression either by trial and error or, preferably, using a graphing calculator. Alternatively, you can use a computer program designed to solve simultaneous equations. Because the unknowns include $[H_3O^+]$, $[OH^-]$, [HClO], and $[ClO^-]$, you will need four equations. As seen in the text, the pertinent equations are

$$K_a = \frac{[H_3O^+][ClO^-]}{[HClO]}$$

$$K_w = [H_3O^+][OH^-]$$

$$[H_3O^+] = [OH^-] + [ClO^-]$$

$$[HClO]_{initial} = [HClO] + [ClO^-]$$

Using either method should produce the same result.

The values obtained for 1.89×10^{-5} mol·L^{-1} are

$[H_3O^+] = 7.4 \times 10^{-7}$ mol·L^{-1}, pH = 6.13 (compare to 6.12 obtained in (a))

$[ClO^-] = 7.3 \times 10^{-7}$ mol·L^{-1}

$[HClO] \cong 1.8 \times 10^{-5}$ mol·L^{-1}

$[OH^-] = 1.3 \times 10^{-8}$ mol·L^{-1}

Similarly, for $[HClO]_{initial} = 9.64 \times 10^{-7}$:

$[H_3O^+] = 1.9 \times 10^{-7}$ mol·L^{-1}, pH = 6.72 (compare to 6.80 obtained in (a))

$[ClO^-] = 1.3 \times 10^{-7}$ mol·L^{-1}

$[HClO] \cong 8.3 \times 10^{-7}$ mol·L^{-1}

$[OH^-] = 5.3 \times 10^{-8}$ mol·L^{-1}

Note than for the more concentrated solution, the effect of the autoprotolysis of water is very small. Notice also that the less concentrated solution is more acidic, due to the autoprotolysis of water, than would be predicted if this effect were not operating.

10.102 Because the process is only 90% efficient, to remove 50.0 kg of SO_2 one must supply enough $CaCO_3$ to remove 55.6 kg of SO_2 (50.0 kg is 90% of 55.6 kg). The moles of SO_2 to be removed is given by:

$$\frac{55600 \text{ g}}{65.06 \text{ g mol}^{-1}} = 867 \text{ mol } SO_2$$

1 mole of SO_2 is consummed by 1 mole of $CaCO_3$. Therefore,

$$(867 \text{ mol } CaCO_3)(100.09 \text{ g mol}^{-1}) = 86.8 \text{ kg}$$

10.104 (1) $2\,H_2O(l) \rightleftharpoons H_3O^+(aq) + OH^-(aq)$ $\quad K_w = 1.00 \times 10^{-14}$

(2) $NH_3(aq) + H_2O(l) \rightleftharpoons NH_4^+(aq) + OH^-(aq)$ $\quad K_b(NH_3) = 1.8 \times 10^{-5}$

or

(3)

$NH_4^+(aq) + H_2O(l) \rightleftharpoons NH_3(aq) + H_3O^+(l)$

$$K_a(NH_4^+) = K_w / K_b(NH_3) = 5.56 \times 10^{-10}$$

(4) $CH_3COOH(aq) + H_2O(l) \rightleftharpoons H_3O^+(l) + CH_3COO^-(aq)$

$$K_a(CH_3COOH) = 1.8 \times 10^{-5}$$

or

(5)

$CH_3COO^-(aq) + H_2O(l) \rightleftharpoons CH_3COOH(aq) + OH^-(aq)$

$$K_b(CH_3COO)^- = K_w / K_a(CH_3COOH)$$
$$= 5.56 \times 10^{-10}$$

(6)

$NH_4^+(aq) + CH_3COO^-(aq) \rightleftharpoons NH_3(aq) + CH_3COOH(aq)$

$$K(NH_4CH_3COO) = \frac{[NH_3][CH_3COOH]}{[NH_4^+][CH_3COO^-]}$$

This equation is obtained by adding Equations (3) and (5) and subtracting Equation (1):

$+[NH_4^+(aq) + H_2O(l) \rightleftharpoons NH_3(aq) + H_3O^+(l)]$

$K_w / K_b(NH_3)$

$+[CH_3COO^-(aq) + H_2O(l) \rightleftharpoons CH_3COOH(aq) + OH(aq)]$

$K_w / K_a(CH_3COOH)$

$-[2\,H_2O(l) \rightleftharpoons H_3O^+(aq) + OH^-(aq)]\ K_w$

Because we are starting with pure ammonium acetate, we will use the last relationship to calculate the concentrations of

NH_4^+, CH_3COO^-, NH_3, and CH_3COOH in solution.

Concentration

$(mol \cdot L^{-1})$	$NH_4^+(aq)$ +	$CH_3COO^-(aq) \rightleftharpoons$	$NH_3(aq)$ +	$CH_3COOH(aq)$
initial	0.100	0.100	0	0
change	$-x$	$-x$	$+x$	$+x$
equilibrium	$0.100 - x$	$0.100 - x$	$+x$	$+x$

For the initial calculation, we will assume that the subsequent deprotonation of acetic acid by water or the protonation of ammonia by water will be small compared to the reaction of the ammonium ion with the acetate ion.

$$\frac{[NH_3][CH_3COOH]}{[NH_4^+][CH_3COO^-]} = 3.09 \times 10^{-5}$$

$$\frac{x^2}{(0.100 - x)^2} = 3.09 \times 10^{-5}$$

$$\frac{x}{0.100 - x} = \sqrt{3.09 \times 10^{-5}}$$

$$x = (0.100 - x)\sqrt{3.09 \times 10^{-5}}$$

$$(1 + \sqrt{3.09 \times 10^{-5}})x = (0.100)\sqrt{3.09 \times 10^{-5}}$$

$$x = 5.52 \times 10^{-4}$$

By using this as the value of the concentrations of NH_3 and CH_3COOH, we can calculate the concentrations of NH_4^+, CH_3COO^-, H_3O^+, and OH^- in solution using the above equilibrium relationship. The values calculated are

$[NH_4^+] = [CH_3COOH] = 0.099$

$[H_3O^+] = [OH^-] = 1.00 \times 10^{-7}$

If we substitute these numbers back into each of the equilibrium constants expressions, we get very good agreement. This justifies the assumption that the subsequent hydrolysis reactions of the NH_3 and CH_3COOH were small compared to the main reaction.

Alternatively (had this not been the case), we could have solved the system of equations using a graphing calculator or suitable computer software package by setting up a system of simultaneous equations. The answer is the same in any case. This problem is simplified by the fact that the tendency for subsequent hydrolysis of NH_3 and CH_3COOH have the same magnitude.

10.106 According to Table 10.3, the amide ion that is formed from the autoionization of ammonia, $2\,NH_3 \rightleftharpoons NH_4^+ + NH_2^-$, is a stronger base than OH^-. Therefore, carbonic acid (and other weak acids) is expected to be stronger in liquid ammonia than in water. Furthermore, carbonic acid is a stronger acid than NH_4^+; therefore, it will donate a proton to NH_3 to form NH_4^+. It can then be concluded that carbonic acid will be leveled in liquid NH_3 and will behave as a strong acid.

10.108 (a) The pH is given by $\text{pH} = \frac{1}{2}(\text{p}K_{a1} + \text{p}K_{a2})$. The first and second dissociations of H_3AsO_4 are the pertinent values for NaH_2AsO_4:

$$\text{pH} = \tfrac{1}{2}(2.25 + 6.77) = 4.51$$

(b) For Na_2HAsO_4 the second and third acid dissociation constants must be used:

$$\text{pH} = \tfrac{1}{2}(6.77 + 11.60) = 9.19$$

10.110 (a) $NH_3 + NH_3 \rightleftharpoons NH_4^+ + NH_2^-$

(b) acid $= NH_4^+$, base $= NH_2^-$

(c) $\text{p}K_{am} = -\log K_{am} = -\log(1 \times 10^{-33}) = 33.0$

(d) $\text{p}K_{am} = [NH_4^+][NH_2^-] = x^2 \; (x = [NH_4^+] = [NH_2^-])$

$[NH_4^+] = \sqrt{1 \times 10^{-33}} = 3 \times 10^{-17}\ \text{mol} \cdot \text{L}^{-1}$

(e) $\text{pNH}_4^+ = \text{pNH}_2^- = -\log(3.2 \times 10^{-17}) = 17$

(f) $\text{pNH}_4^+ = \text{pNH}_2^- = \text{p}K_{am} \approx 33.0$

10.112 The osmotic pressure of 0.10 H_2SO_4 will be slightly higher than that of 0.10 M HCl. When H_2SO_4 is added to solution it completely deprotonates into HSO_4^-, a small fraction of which deprotonates further to SO_4^{2-}. Therefore, each mole of H_2SO_4 gives rise to slightly more than 2 moles of dissolved ions. HCl, on the other hand, is a strong monoprotic acid giving exactly two moles of dissolved ions per mole of HCl molecules.

To calculate the moles of dissolved ions in a 0.10 M solution of H_2SO_4 one assumes that H_2SO_4 is completely deprotonated, and solves the equilibrium problem:

Concentration ($mol \cdot L^{-1}$)	$HSO_4^-(aq)$ +	$H_2O(l)$ ⇌	$H_3O^+(aq)$ +	$SO_4^{2-}(aq)$
initial	0.10	—	0.10	0
change	$-x$	—	$+x$	$+x$
final	$0.10 - x$	—	$0.10 + x$	$+x$

$$K_a = \frac{[H_3O^+][SO_4^{2-}]}{[HSO_4^-]}$$

$$1.2 \times 10^{-2} = \frac{(0.10 + x)(x)}{0.10 - x}$$

Using the quadratic equation one finds that $x = 9.8 \times 10^{-3}$ and, therefore: $[HSO_4^-] = 0.0902$ M, $[H_3O^+] = 0.1098$ M, and $[SO_4^{2-}] = 0.0098$ M giving a total dissolved ion concentration of 0.210 M (compared to a total dissolved ion concentration of 0.20 for the 0.10 M HCl solution).

Assuming a temperature of 298 K, the osmotic pressure of the 0.10 M H_2SO_4 solution is:

$$\Pi = RTc = \left(0.082057\ L \cdot atm \cdot K^{-1}\ mol^{-1}\right)\left(298\ K\right)\left(0.210\ mol \cdot L^{-1}\right)$$
$$= 5.13\ atm$$

While that of the HCl solution is:

$$\Pi = RTc = \left(0.082057\ L \cdot atm \cdot K^{-1} \cdot mol^{-1}\right)\left(298\ K\right)\left(0.200\ mol \cdot L^{-1}\right)$$
$$= 4.89\ atm$$

10.114 $\Delta H° = +57 \text{ kJ} \cdot \text{mol}^{-1}$ (where 1 mol refers to the reaction as written)

$$\ln \frac{K_1}{K_2} = -\frac{\Delta H°}{R}\left(\frac{1}{T_1} - \frac{1}{T_2}\right)$$

$$2.303 \log \frac{K_1}{K_2} = -\frac{\Delta H°}{R}\left(\frac{1}{T_1} - \frac{1}{T_2}\right)$$

$$-2.303 \log \frac{K_1}{K_2} = \frac{\Delta H°}{R}\left(\frac{1}{T_1} - \frac{1}{T_2}\right)$$

$$\text{p}K_1 - \text{p}K_2 = \frac{\Delta H°}{2.303R}\left(\frac{1}{T_1} - \frac{1}{T_2}\right)$$

Let condition 2 be 25°C, where $\text{p}K_\text{w} = 14$:

$$\text{p}K_1 - 14 = \frac{57\,000 \text{ J}}{2.303 \cdot 8.314 \text{ J} \cdot \text{K}^{-1} \cdot \text{mol}^{-1}}\left(\frac{1}{T_1} - \frac{1}{298 \text{ K}}\right)$$

$$\text{p}K_1 = 14 + \frac{57\,000 \text{ J}}{2.303 \cdot 8.314 \text{ J} \cdot \text{K}^{-1} \cdot \text{mol}^{-1}}\left(\frac{1}{T_1} - \frac{1}{298 \text{ K}}\right)$$

We will not define K_1 as the K_w value at some unknown temperature T:

$$\text{p}K_\text{w} = 14 + \frac{57\,000 \text{ J} \cdot \text{mol}^{-1}}{2.303 \cdot 8.314 \text{ J} \cdot \text{K}^{-1} \cdot \text{mol}^{-1}}\left(\frac{1}{T} - \frac{1}{298 \text{ K}}\right)$$

$$\text{p}K_\text{w} = \frac{3.0 \times 10^3 \text{ K}}{\text{T}} + 4.00$$

Substituting the value of T = 373 K gives $\text{p}K_\text{w}$ = 12.04. If the solution is neutral, the pH = pOH = 6.02.

10.116 (a) The structures of alanine, glycine, phenylalanine, and cysteine are

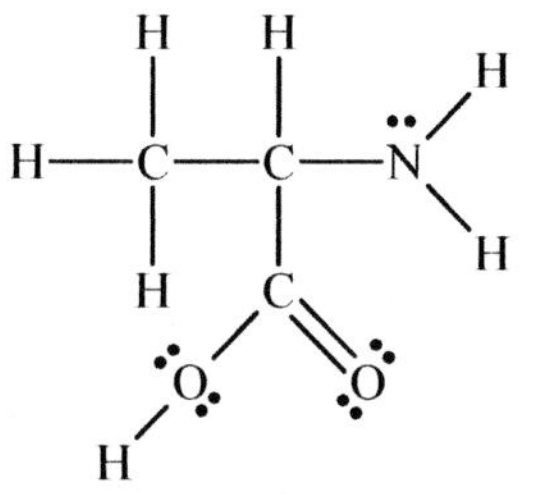

alanine

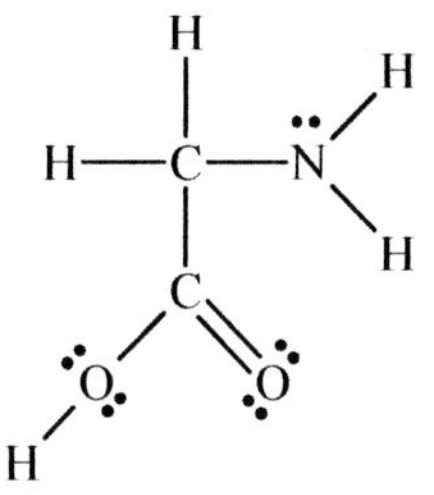

glycine

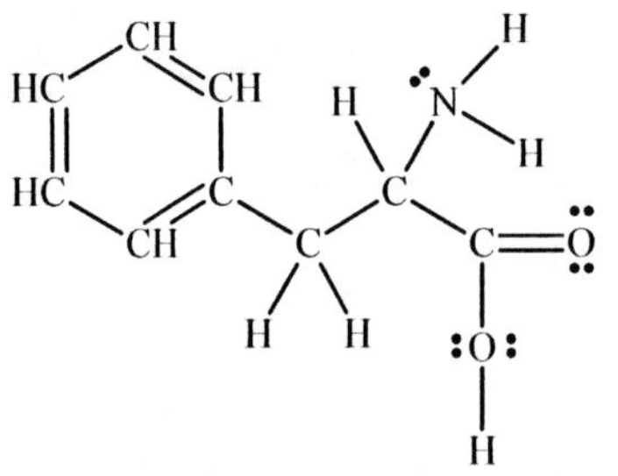

phenylalanine

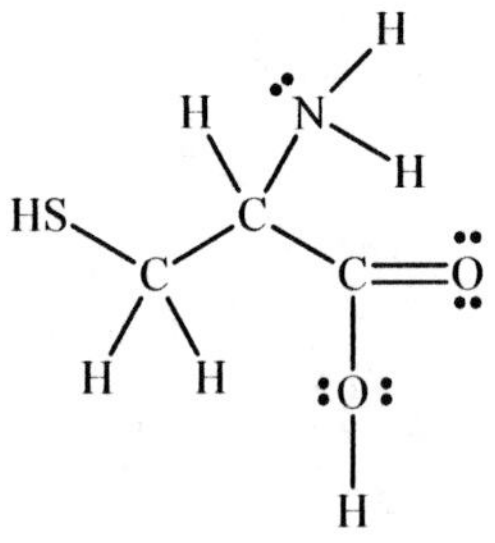

cysteine

All the amino acids have an amine —NH_2 function as well as a carboxylic acid —COOH group.

(b) The pK_a value of the —COOH group of alanine is 2.34 and the pK_b of the —NH_2 group is 4.31. Often, we find instead of the pK_b value, the pK_a value of the conjugate acid of the —NH_2 group. The relationship $pK_a \cdot pK_b = pK_w$ is used to convert to the pK_b value.

(c) To find the equilibrium constant of the reaction of the acid function, with the base function in the amino acid, we first write the known equilibrium reactions and corresponding K values:

$$R-COOH(aq) + H_2O(l) \rightleftharpoons H_3O^+(aq) + R-COO^-(aq) \quad pK_a = 2.34$$

$$K_a = 4.6 \times 10^{-3}$$

$$R-NH_2(aq) + H_2O(l) \rightleftharpoons RNH_3^+(aq) + OH^-(aq) \quad pK_b = 4.31$$

$$K_b = 4.9 \times 10^{-5}$$

If we sum these reactions, we obtain

$$R-COOH(aq) + R-NH_2(aq) + 2\,H_2O(l) \rightleftharpoons H_3O^+(aq) + OH^-(aq)$$

$$+ R-COO^-(aq) + RNH_3^+(aq) \quad K_a \cdot K_b = 2.3 \times 10^{-7}$$

The presence of the H_2O, H_3O^+, and OH^- can be eliminated by subtracting the autoprotolysis reaction of water to give the desired final equation:

$$2\,H_2O(l) \rightleftharpoons H_3O^+(aq) + OH^-(aq) \quad K_w = 1.00 \times 10^{-14}$$

$$R-COOH(aq) + R-NH_2(aq) = R-COO^-(aq) + RNH_3^+(aq)$$

$$K = K_a \cdot K_b \cdot K_w^{-1}$$
$$K = 2.3 \times 10^7$$

(d) A zwitterion is a compound that contains both a positive ion and a negative ion in the same molecule. Overall, the molecule is neutral, but it is ionic in that it possesses both positive and negative ions within itself. This is a common occurrence for the amino acids in which the acid function reacts to protonate the base site. Given the large value of the equilibrium constant found in (c), it is more appropriate to write an amino acid as $R(NH_3^+)(COO^-)$ then as $R(NH_2)(COOH)$.

10.118 (a)

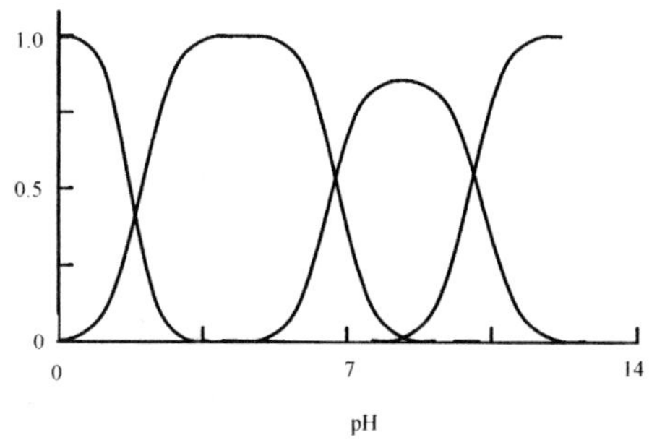

(b) The major species at pH 7.5 is the doubly deprotonated form.

(c) pH = 5.97

(d) at pH values greater than about 11.4

(e) from approximately pH = 7.0 to 8.4

CHAPTER 11

AQUEOUS EQUILIBRIA

11.2 (a) When solid sodium dihydrogen phosphate is added to a solution of phosphoric acid, the following equilibrium

$H_3PO_4(aq) + H_2O(l) \rightleftharpoons H_3O^+(aq) + H_2PO_4^-(aq)$

shifts to the left to relieve the stress resulting from the increased $H_2PO_4^-$. Consequently, $[H_3O^+]$ decreases and the pH increases.

(b) When HBr, a strong acid, is added to a solution of HCN, the percentage of HCN that is deprotonated decreases because the following equilibrium

$HCN(aq) + H_2O(l) \rightleftharpoons H_3O^+(aq) + CN^-(aq)$

shifts to the left to relieve the stress imposed by the increased H_3O^+ supplied by the HBr.

(c) When pyridinium chloride (C_5H_5NHCl) is added to a solution of pyridine, the following equilibrium

$C_5H_5N(aq) + H_2O(l) \rightleftharpoons C_5H_5NH^+(aq) + OH^-(aq)$

shifts to the left to relieve the stress resulting from the increased $C_5H_5NH^+$. Consequently, the $[OH^-]$ decreases. Because $[H_3O^+][OH^-]$ is constant, $[H_3O^+]$ must increase and the pH consequently decreases.

11.4 (a) The reaction is $HCN(aq) + H_2O(l) \rightleftharpoons H_3O^+(aq) + CN^-(aq)$.

Concentration

$(\text{mol}\cdot\text{L}^{-1})$	$HCN(aq)$ +	$H_2O(l)$	$\rightleftharpoons$ $H_3O^+(aq)$ +	$CN^-(aq)$
initial	0.075	—	0	0.045
change	$-x$	—	$+x$	$+x$
equilibrium	$0.075-x$	—	x	$0.045+x$

$$K_a = \frac{[H_3O^+][CN^-]}{[HCN]} = \frac{(x)(0.045+x)}{(0.075-x)}$$

$$4.9\times10^{-10} \approx \frac{[H_3O^+](0.045)}{(0.075)}$$

$$[H_3O^+] \approx 8.2\times10^{-10}\ \text{mol}\cdot\text{L}^{-1}$$

(b) 0.50 M NaCl will have no effect:

Concentration

$(\text{mol}\cdot\text{L}^{-1})$	$NH_2NH_2(aq)$ +	$H_2O(l)$	$\rightleftharpoons$ $NH_2NH_3^+(aq)$ +	$OH^-(aq)$
initial	0.20	0	0	
change	$-x$	—	$+x$	$+x$
equilibrium	$0.20-x$	—	x	x

$$K_b = \frac{[NH_2NH_3^+][OH^-]}{[NH_2NH_2]}$$

$$1.7\times10^{-6} = \frac{[OH^-]^2}{0.20}$$

$$[OH^-] = \sqrt{3.4\times10^{-7}} = 5.8\times10^{-4}\ \text{mol}\cdot\text{L}^{-1}$$

$$[H_3O^+] = \frac{K_w}{[OH^-]} = \frac{1.00\times10^{-14}}{5.8\times10^{-4}} = 1.7\times10^{-11}\ \text{mol}\cdot\text{L}^{-1}$$

(c) Setup is similar to part (a).

$$K_a = 4.9\times10^{-10} = \frac{[H_3O^+][CN^-]}{[HCN]} = \frac{(x)(0.030+x)}{(0.015-x)}$$

$$4.9\times10^{-10} \approx \frac{[H_3O^+][0.030]}{(0.015)}$$

$$[H_3O^+] = 2.5\times10^{-10}\ \text{mol}\cdot\text{L}^{-1}$$

(d) When the concentrations of a weak base and its conjugate acid are equal, the pOH equals the pK_b. Therefore, the pOH of hydrazine = pK_b = 5.77, and

$$\text{pH} = 14.00 - \text{pOH} = 14.00 - 5.77 = 8.23. [\text{H}_3\text{O}^+] = 10^{-8.23}$$
$$= 5.9 \times 10^{-9}\ \text{mol} \cdot \text{L}^{-1}$$

11.6 In each case, the equilibrium involved is $HPO_4^{2-}(aq) + H_2O(l) \rightleftharpoons H_3O^+(aq) + PO_4^{3-}(aq)$. $HPO_4^{2-}(aq)$ and $PO_4^{3-}(aq)$ are conjugate acid and base; therefore, the pH calculation is most easily performed with the Henderson-Hasselbalch equation:

$$\text{pH} = \text{p}K_a + \log\left(\frac{[\text{PO}_4^{3-}]}{[\text{HPO}_4^{2-}]}\right)$$

For H_3PO_4, $\text{p}K_{a3} = 12.68$

(a)

$$\text{pH} = 12.68 + \log\left(\frac{0.25\ \text{mol} \cdot \text{L}^{-1}}{0.17\ \text{mol} \cdot \text{L}^{-1}}\right) = 12.85;\ \text{pOH} = 14.00 - 12.85 = 1.15$$

(b)

$$\text{pH} = 12.68 + \log\left(\frac{0.42\ \text{mol} \cdot \text{L}^{-1}}{0.66\ \text{mol} \cdot \text{L}^{-1}}\right) = 12.48;\ \text{pOH} = 14.00 - 12.48 = 1.52$$

(c)

$$\text{pH} = 12.68 + \log\left(\frac{0.12\ \text{mol} \cdot \text{L}^{-1}}{0.12\ \text{mol} \cdot \text{L}^{-1}}\right) = 12.68;\ \text{pOH} = 14.00 - 12.68 = 1.32$$

11.8 molarity of NaBrO $= \left(\frac{7.50\ \text{g NaBrO}}{118.90\ \text{g/mol}}\right)\left(\frac{1}{0.100\ \text{L}}\right) = 0.631\ \text{mol} \cdot \text{L}^{-1}$

Concentration $(mol \cdot L^{-1})$	$H_2O(l)$ +	$HBrO(aq)$ $\rightleftharpoons$	$H_3O^+(aq)$ +	$BrO^-(aq)$
initial	—	0.50	0	0.631
change	—		$+x$	$+x$
equilibrium	—	$0.50 - x$	x	$0.631 + x$

$$K_a = 2.0 \times 10^{-9} = \frac{[H_3O^+][BrO^-]}{[HBrO]} = \frac{x(0.631 + x)}{0.50 - x} \approx \frac{0.631x}{0.50}$$

$$x = [H_3O^+] = 1.6 \times 10^{-9}\ mol \cdot L^{-1}$$

$$pH = -\log[H_3O^+] = -\log(1.6 \times 10^{-9}) = 8.80$$

Change in pH is $8.80 - 4.50 = 4.30$

11.10 (a) $(CH_3)_2NH(aq) + H_2O(l) \rightleftharpoons (CH_3)_2NH_2^+(aq) + OH^-(aq)$

total volume 400 mL = 0.400 L

moles of $(CH_3)_2NH = (0.100\ L)(0.020\ mol \cdot L^{-1})$

$= 2.0 \times 10^{-3}\ mol\ (CH_3)_2NH$

moles of $(CH_3)_2NH_2Cl = (0.300\ L)(0.030\ mol \cdot L^{-1})$

$= 9.0 \times 10^{-3}\ mol\ (CH_3)_2NH_2Cl$

$$\text{initial } [(CH_3)_2NH] = \frac{2.0 \times 10^{-3}\ mol}{0.400\ L} = 5.0 \times 10^{-3}\ mol \cdot L^{-1}$$

$$\text{initial } [(CH_3)_2NH_2Cl] = \frac{9.0 \times 10^{-3}\ mol}{0.400\ L} = 2.3 \times 10^{-2}\ mol \cdot L^{-1}$$

Concentration $(mol \cdot L^{-1})$	$H_2O(l)$ +	$(CH_3)_2NH(aq)$ $\rightleftharpoons$	$(CH_3)_2NH_2^+(aq)$ +	$OH^-(aq)$
initial	—	5.0×10^{-3}	2.3×10^{-2}	0
change	—	$-x$	$+x$	$+x$
equilibrium	—	$5.0 \times 10^{-3} - x$	$2.3 \times 10^{-2} + x$	x

↑ HH eqn   need pKb = 3.27

pKa = 10.73 for $(CH_3)_2NH_2^+$

HHeqn

pH = pKa + log $\frac{[3.5\times10^{-3}]}{[2.3\times10^{-2}]}$

10.07 = 10.73 − 0.66

$$K_b = \frac{[(CH_3)_2NH_2^+][OH^-]}{[(CH_3)_2NH]} = \frac{(2.3\times10^{-2}+x)(x)}{(5.0\times10^{-3}-x)} \cong \frac{(2.3\times10^{-2})(x)}{(5.0\times10^{-3})}$$

$$= 5.4\times10^{-4}$$

$x = 1.2\times10^{-4}\ \text{mol}\cdot\text{L}^{-1} \cong [OH^-]$ 1.17×10^{-4}

$\text{pOH} = -\log[OH^-] = -\log(1.2\times10^{-4}) = 3.92$

$\text{pH} = 14.00 - 3.92 = 10.08$

(b) $(CH_3)_2NH(aq) + H_2O(l) \rightleftharpoons (CH_3)_2NH_2^+(aq) + OH^-(aq)$

Total volume 75 mL = 0.0750 L

moles of $(CH_3)_2NH$ = $(0.065\ \text{L})(0.010\ \text{mol}\cdot\text{L}^{-1})$

$= 6.5\times10^{-4}$ mol $(CH_3)_2NH$

moles of $(CH_3)_2NH_2Cl$ = $(0.010\ \text{L})(0.150\ \text{mol}\cdot\text{L}^{-1})$

$= 1.5\times10^{-3}$ mol $(CH_3)_2NH_2Cl$

$$\text{initial } [(CH_3)_2NH_2Cl] = \frac{1.5\times10^{-3}\ \text{mol}}{0.0750\ \text{L}} = 2.0\times10^{-2}\ \text{mol}\cdot\text{L}^{-1}$$

$$\text{and } [(CH_3)_2NH] = \frac{6.5\times10^{-4}\ \text{mol}}{0.0750\ \text{L}} = 8.7\times10^{-3}\ \text{mol}\cdot\text{L}^{-1}$$

As in part (a),

$$K_b = \frac{[(CH_3)_2NH_2^+][OH^-]}{[(CH_3)_2NH]} = \frac{(2.0\times10^{-2}+x)(x)}{(8.7\times10^{-3}-x)} \approx \frac{(2.0\times10^{-2})(x)}{(8.7\times10^{-3})}$$

$$= 5.4\times10^{-4}$$

$x = 2.3\times10^{-4}\ \text{mol}\cdot\text{L}^{-1} \cong [OH^-]$

$\text{pOH} = -\log[OH^-] = -\log(2.3\times10^{-4}) = 3.64$

$\text{pH} = 14.00 - 3.64 = 10.36$

(c) $(CH_3)_2NH(aq) + H_2O(l) \rightleftharpoons (CH_3)_2NH_2^+(aq) + OH^-(aq)$

Total volume = 175 mL = 0.175 L

moles of $(CH_3)_2NH = (0.050\ \text{L})(0.015\ \text{mol}\cdot\text{L}^{-1})$

$= 7.5\times10^{-4}$ mol $(CH_3)_2NH$

moles of $(CH_3)_2NH_2Cl = (0.125\ \text{L})(0.015\ \text{mol}\cdot\text{L}^{-1})$

$$= 1.9 \times 10^{-3} \text{ mol } (CH_3)_2NH_2Cl$$

$$\text{initial } [(CH_3)_2NH_2Cl] = \frac{1.9 \times 10^{-3} \text{ mol}}{0.175 \text{ L}} = 1.1 \times 10^{-2} \text{ mol} \cdot L^{-1}$$

$$\text{and } [(CH_3)_2NH] = \frac{7.5 \times 10^{-4} \text{ mol}}{0.175 \text{ L}} = 4.3 \times 10^{-3} \text{ mol} \cdot L^{-1}$$

$$K_b = \frac{[(CH_3)_2NH_2^+][OH^-]}{[(CH_3)_2NH]} = \frac{(1.1 \times 10^{-2} + x)(x)}{(4.3 \times 10^{-3} - x)} = \frac{(1.1 \times 10^{-2})(x)}{(4.3 \times 10^{-3})}$$

$$= 5.4 \times 10^{-4}$$

$$x = 2.1 \times 10^{-4} \text{ mol} \cdot L^{-1} \cong [OH^-]$$

$$pOH = -\log[OH^-] = -\log(2.1 \times 10^{-4}) = 3.68$$

$$pH = 14.00 - 3.68 = 10.32$$

11.12 HA = acetylsalicylic acid

A^- = conjugate base of acetylsalicylic acid

$$HA(aq) + H_2O(l) \rightleftharpoons H_3O^+(aq) + A^-(aq) \qquad K_a = 3.2 \times 10^{-4}$$

We will use the following formula to evaluate the ratio of $[A^-]$ to [HA] at equilibrium:

$$pH = pK_a + \log\frac{[A^-]}{[HA]}$$

The formula is derived in this way:

$$K_a = \frac{[H_3O^+][A^-]}{[HA]}$$

$$[H_3O^+] = \frac{K_a[HA]}{[A^-]}$$

$$pH = p\left(\frac{K_a[HA]}{[A^-]}\right)$$

$$= pK_a - \log\frac{[HA]}{[A^-]}$$

$$= pK_a + \log\frac{[A^-]}{[HA]}$$

Given that pH = 4.67 and $K_a = 3.2 \times 10^{-4}$ ($pK_a = 3.49$),

$$4.67 = 3.49 + \log\frac{[A^-]}{[HA]}$$

$$1.18 = \log\frac{[A^-]}{[HA]}$$

Taking antilogs of both sides of this equation gives

$$15 = \frac{[A^-]}{[HA]}$$

11.14 (a) $pK_a = 3.37$, pH range $\approx$ 2 to 4

(b) $pK_a = 3.75$, pH range $\approx$ 3 to 5

(c) $pK_{a2} = 10.25$, pH range $\approx$ 9 to 11

(d) $pK_b = 4.75$, $pK_a = 9.25$, pH range $\approx$ 8 to 10

(e) $pK_b = 8.75$, $pK_a = 5.25$, pH range $\approx$ 4 to 6

11.16 (a) C_6H_5COOH and $NaC_6H_5CO_2$, $pK_a = 4.19$

(b) NH_4Cl and NH_3, $pK_b = 4.75$, $pK_a = 9.25$

(c) CH_3COOH and $NaCH_3CO_2$, $pK_a = 4.75$

(d) $(C_2H_5)_3$ NHCl and $(C_2H_5)_3N$, $pK_b = 2.99$, $pK_a = 11.01$

11.18 $HPO_4^{2-}(aq) + H_2O(l) \rightleftharpoons H_3O^+(aq) + PO_4^{3-}(aq)$

$$K_{a3} = \frac{[H_3O^+][PO_4^{3-}]}{[HPO_4^{2-}]}$$

$$pH = pK_{a3} + \log\frac{[PO_4^{3-}]}{[HPO_4^{2-}]} \text{(see Exercise 11.12)}$$

$$\log\frac{[PO_4^{3-}]}{[HPO_4^{2-}]} = pH - pK_{a3};\ pK_{a3} = 12.68$$

(a) $\log\frac{[PO_4^{3-}]}{[HPO_4^{2-}]} = 12.00 - 12.68 = -0.68$

$$\frac{[PO_4^{3-}]}{[HPO_4^{2-}]} = 0.21$$

(b)

$$\text{molarity of } PO_4^{3-} = 0.21 \times \text{molarity of } HPO_4^{2-} = 0.21 \times 0.100 \text{ mol} \cdot L^{-1}$$
$$= 2.1 \times 10^{-2} \text{ mol} \cdot L^{-1}$$

$$\text{moles of } PO_4^{3-} = \text{moles of } K_3PO_4 = 2.1 \times 10^{-2} \text{ mol} \cdot L^{-1} \times 1 \text{ L}$$
$$= 2.1 \times 10^{-2} \text{ mol}$$

$$\text{mass of } K_3PO_4 = 2.1 \times 10^{-2} \text{ mol} \times \frac{212.27 \text{ g } K_3PO_4}{1 \text{ mol } K_3PO_4} = 4.46 \text{ g } K_3PO_4$$

(c) $$\frac{[PO_4^{3-}]}{[HPO_4^{2-}]} = 0.21 = \frac{0.100 \text{ mol} \cdot L^{-1}}{x \text{ mol} \cdot L^{-1}} = \frac{0.100 \text{ mol}}{x \text{ mol}} = \frac{0.100 \text{ mol } K_3PO_4}{x \text{ mol } K_2HPO_4}$$

$$x \text{ mol } K_2HPO_4 = \frac{0.100 \text{ mol}}{0.21} = 0.48 \text{ mol } K_2HPO_4$$

$$\text{mass of } K_2HPO_4 = 0.48 \text{ mol } K_2HPO_4 \times \frac{174.18 \text{ g } K_2HPO_4}{1 \text{ mol } K_2HPO_4}$$

$$= 84 \text{ g } K_2HPO_4$$

(d) moles of K_3PO_4 required $= 0.21 \times$ moles of K_2HPO_4

$= 0.21 \times 0.0500 \text{ L}$

$\times 0.100 \text{ mol} \cdot L^{-1} = 1.1 \times 10^{-3} \text{ mol}$

$$\text{volume of } K_3PO_4 \text{ solution} = \frac{1.1 \times 10^{-3} \text{ mol}}{0.150 \text{ mol} \cdot L^{-1}} = 7.3 \times 10^{-3} \text{ L} = 7.3 \text{ mL}$$

11.20 (a) The initial pH is found by:

$$pH = pK_{a1} + \log \frac{[HPO_4^{2-}]}{[H_2PO_4^-]}$$
$$= 7.21 + \log \frac{0.150 \text{ M}}{0.100 \text{ M}}$$
$$= 7.39$$

$(0.0800 \text{ L})(0.0100 \text{ mol} \cdot L^{-1}) = 8.00 \times 10^{-4}$ mol NaOH (strong base),

producing 8.0×10^{-4} mol HPO_4^{2-} from $H_2PO_4^-$

$(0.150\ \text{mol}\cdot\text{L}^{-1})(0.100\ \text{L}) = 1.50\times10^{-2}\ \text{mol HPO}_4{}^{2-}$ (initially)

$(0.100\ \text{mol}\cdot\text{L}^{-1})(0.100\ \text{L}) = 1.00\times10^{-2}\ \text{mol H}_2\text{PO}_4{}^{-}$ (initially)

After adding NaOH:

$$[\text{HPO}_4{}^{2-}] = \frac{(1.50\times10^{-2} + 8.00\times10^{-4})\ \text{mol}}{0.180\ \text{L}} = 8.78\times10^{-2}\ \text{mol}\cdot\text{L}^{-1}$$

$$[\text{H}_2\text{PO}_4{}^{-}] = \frac{(1.00\times10^{-2} - 8.00\times10^{-4})\ \text{mol}}{0.180\ \text{L}} = 5.1\times10^{-2}\ \text{mol}\cdot\text{L}^{-1}$$

$$\text{pH} = \text{p}K_\text{a} + \log\frac{[\text{HPO}_4{}^{-}]}{[\text{H}_2\text{PO}_4{}^{2-}]}$$

$$\text{pH} = 7.21 + \log\frac{(8.78\times10^{-2}\ \text{mol}\cdot\text{L}^{-1})}{(5.1\times10^{-2}\ \text{mol}\cdot\text{L}^{-1})} = 7.21 + 0.24 = 7.45$$

$\Delta\text{pH} = +0.06$

(b) $(0.0100\ \text{L})(1.0\ \text{mol}\cdot\text{L}^{-1}) = 1.0\times10^{-2}\ \text{mol HNO}_3$ (strong acid),

produing $1.0\times10^{-2}\ \text{mol H}_2\text{PO}_4{}^{-}$ from $\text{HPO}_4{}^{2-}$

$$\text{molarity of HPO}_4{}^{2-} = \frac{(1.5\times10^{-2} - 1.0\times10^{-2})\ \text{mol}}{0.110\ \text{L}} = 5\times10^{-2}\ \text{mol}\cdot\text{L}^{-1}$$

$$\text{molarity of H}_2\text{PO}_4{}^{-} = \frac{(1.0\times10^{-2} + 1.0\times10^{-2})\ \text{mol}}{0.110\ \text{L}} = 1.8\times10^{-1}\ \text{mol}\cdot\text{L}^{-1}$$

$$\text{pH} = \text{p}K_\text{a} + \log\frac{[\text{HPO}_4{}^{-}]}{[\text{H}_2\text{PO}_4{}^{2-}]}$$

$$\text{pH} = 7.21 + \log\frac{(5\times10^{-2}\ \text{mol}\cdot\text{L})}{(1.8\times10^{-1}\ \text{mol}\cdot\text{L})} = 7.21 - 0.6 = 6.6$$

$\Delta\text{pH} = 6.6 - 7.39 = -0.8$

11.22

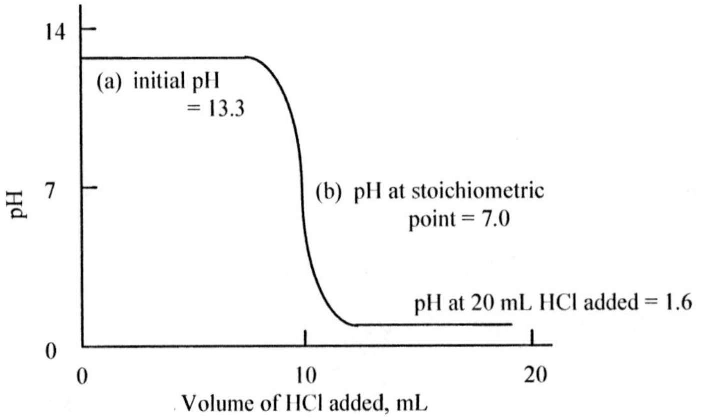

Initial molarity of $OH^- = 2 \times 0.10\ \text{mol} \cdot \text{L}^{-1} = 0.20\ \text{mol} \cdot \text{L}^{-1}$

$\text{pOH} = -\log(0.20) = 0.70,\ \text{pH} = 14.00 - 0.70 = 13.30$

11.24 (a)

$$V_{HCl} = \left(\frac{25.0\ \text{mL}}{2}\right)\left(\frac{10^{-3}\ \text{L}}{1\ \text{mL}}\right)\left(\frac{0.215\ \text{mol KOH}}{1\ \text{L KOH}}\right) \times \left(\frac{1\ \text{mol HCl}}{1\ \text{mol KOH}}\right)\left(\frac{1\ \text{L HCl}}{0.116\ \text{mol HCl}}\right)$$

$$= 0.0232\ \text{L HCl} = 23.2\ \text{mL HCl}$$

(b) $2 \times 0.0232\ \text{L} = 0.0464\ \text{L} = 46.4\ \text{mL HCl}$

(c) In HCl, $[\text{HCl}]_0 = [\text{Cl}^-]$

volume $= (0.0250 + 0.0464)\ \text{L} = 0.0714\ \text{L}$

$$\text{molarity of Cl}^- = (0.0464\ \text{L})\left(\frac{0.116\ \text{mol Cl}^-}{1\ \text{L}}\right)\left(\frac{1}{0.0714\ \text{L}}\right)$$

$$= 0.0754\ \text{mol} \cdot \text{L}^{-1}$$

(d) number of moles of H_3O^+ (from acid) $= (0.0400\ \text{L})(0.116\ \text{mol} \cdot \text{L}^{-1})$

$= 4.64 \times 10^{-3}\ \text{mol}$

number of moles of OH^- (from base) $= (0.0250\ \text{L})(0.215\ \text{mol} \cdot \text{L}^{-1})$

$= 5.38 \times 10^{-3}\ \text{mol}$

excess $OH^- = (5.38 - 4.64) \times 10^{-3}\ \text{mol} = 7.4 \times 10^{-4}\ \text{mol}$

volume solution $= (0.0400 + 0.0250)\ \text{L} = 0.0650\ \text{L}$

$$[\text{OH}^-] = \frac{7.4 \times 10^{-4}\ \text{mol}}{0.0650\ \text{L}} = 0.011\ \text{mol} \cdot \text{L}^{-1}$$

$\text{pOH} = -\log(0.011) = 1.96$

$\text{pH} = 14.00 - 1.96 = 12.04$

11.26 (a)

$$M = \frac{m}{n} = \frac{0.483g}{0.250\text{M} \times 42.0\text{mL}} = \frac{0.483g}{0.250\text{mol}{\times}\text{L}^{-1} \times 0.0420\text{L}} = 46.0g \cdot \text{mol}^{-1}$$

(b) $\text{p}K_a = pH - \log\frac{[A^-]}{[HA]}$ = pH at the midpoint of the titration = 3.75

Comparing to values listed in Table 10.1, the acid is most likely formic acid (HCOOH).

11.28 $Ba(OH)_2 + 2\,HCl \longrightarrow BaCl_2 + 2\,H_2O$

mass of pure $Ba(OH)_2$

$$= (0.0176\text{ L HCl})\left(\frac{0.0935\text{ mol HCl}}{1\text{ L HCl}}\right)\left(\frac{1\text{ mol Ba(OH)}_2}{2\text{ mol HCl}}\right)$$

$$\left(\frac{171.36\text{ g Ba(OH)}_2}{1\text{ mol Ba(OH)}_2}\right)\left(\frac{250\text{ mL}}{35.0\text{ mL}}\right)$$

$$= 1.01\text{ g}$$

$$\text{percentage purity} = \frac{1.01\text{ g}}{1.331\text{ g}} \times 100\% = 75.9\%$$

11.30 The reaction is $HCl + KOH \longrightarrow H_2O + KCl$; thus $HCl \propto KOH$

(a) Initial : $[OH^-] = 0.215\text{ mol}\cdot L^{-1}$

$$pOH = -\log(0.215) = 0.668$$
$$pH = 14.00 - pOH = 13.33$$

(b) After addition of 5.0 mL of $0.116\text{ mol}\cdot L^{-1}$ HCl:

$$(0.0250\text{ L base})\left(\frac{0.215\text{ mol KOH}}{1\text{ L base}}\right) - (0.0050\text{ L})\left(\frac{0.116\text{ mol HCl}}{1\text{ L}}\right)$$

$$= 4.80 \times 10^{-3}\text{ mol KOH unreacted}$$

$$[OH^-] = \frac{4.80 \times 10^{-3}\text{ mol OH}^-}{0.0300\text{ L}} = 0.16\text{ mol}\cdot L^{-1}\text{ OH}^-$$

$$pOH = -\log(0.16) = 0.80$$
$$pH = 14.00 - 0.80 = 13.20$$

(c) After addition of 10.0 mL of $0.116\text{ mol}\cdot L^{-1}$ HCl:

$$(0.0250\text{ L base})\left(\frac{0.215\text{ mol KOH}}{1\text{ L base}}\right) - (0.0100\text{ L})\left(\frac{0.116\text{ mol HCl}}{1\text{ L}}\right)$$

$$= 4.22 \times 10^{-3}\text{ mol KOH unreacted}$$

$$[OH^-] = \frac{4.22 \times 10^{-3} \text{ mol OH}^-}{0.0350 \text{ L}} = 0.121 \text{ mol} \cdot \text{L}^{-1} \text{ OH}^-$$

$$\text{pOH} = -\log(0.121) = 0.917$$

$$\text{pH} = 14.00 - 0.917 = 13.08$$

(d) At the stoichiometric point:

For the reaction of a strong acid with a strong base, the pH at the stoichiometric point is the same as the pH for pure water. Therefore, the pH = 7.0.

Volume HCl at the stoichiometric point:

$$(0.0250 \text{ L KOH})\left(\frac{0.215 \text{ mol KOH}}{1 \text{ L KOH}}\right)\left(\frac{1 \text{ mol HCl}}{1 \text{ mol KOH}}\right)\left(\frac{1 \text{ L HCl}}{0.116 \text{ mol HCl}}\right)$$

$$= 0.0463 \text{ L HCl}$$

(e) After addition of 5.0 mL of acid beyond the stoichiometric point,

$$(0.0050 \text{ L})\left(\frac{0.116 \text{ mol HCl}}{1 \text{ L}}\right)\left(\frac{1}{(0.0463 + 0.0050 + 0.0250) \text{ L}}\right)$$

$$= 0.0076 \text{ mol} \cdot \text{L}^{-1} \text{ H}_3\text{O}^+$$

$$\text{pH} = -\log(0.0076) = 2.12$$

(f) After addition of 10 mL of acid beyond the stoichiometric point,

$$(0.010 \text{ L})\left(\frac{0.116 \text{ mol HCl}}{1 \text{ L}}\right)\left(\frac{1}{(0.0463 + 0.010 + 0.0250) \text{ L}}\right)$$

$$= 0.014 \text{ mol} \cdot \text{L}^{-1} \text{ H}_3\text{O}^+$$

$$\text{pH} = -\log(0.014) = 1.85$$

11.32 (a) Initial pH of 0.20 M C_6H_5COOH ($K_a = 6.5 \times 10^{-5}$)

$$C_6H_5COOH + H_2O(l) \rightleftharpoons H_3O^+ + C_6H_5CO_2^-$$

$$K_a = \frac{[H_3O^+][C_6H_5CO_2^-]}{[C_6H_5COOH]}$$

Concentration

$(mol \cdot L^{-1})$	$C_6H_5COOH(aq)$ +	$H_2O(l) \rightleftharpoons$	$H_3O^+(aq)$ +	$C_6H_5CO_2^-(aq)$
initial	0.20	—	0	0
change	$-x$	—	$+x$	$+x$
equilibrium	$0.20 - x$	—	x	x

$$6.5 \times 10^{-5} = \frac{(x)(x)}{0.20 - x} \cong \frac{x^2}{0.20}$$

$$x = [H_3O^+] = 3.6 \times 10^{-3}\ mol \cdot L^{-1}$$

$$pH = -\log(3.6 \times 10^{-3}) = 2.44$$

(b) initial moles of $C_6H_5COOH = (0.0300\ L)(0.20\ mol \cdot L^{-1})$
$= 6.0 \times 10^{-3}\ mol\ C_6H_5COOH$

moles of KOH $= (0.0150\ L)(0.30\ mol \cdot L^{-1}) = 4.5 \times 10^{-3}\ mol\ KOH$
$= mol\ C_6H_5CO_2^-$

moles C_6H_5OH remaining $= 6.0 \times 10^{-3}\ mol - 4.5 \times 10^{-3}\ mol$
$= 1.5 \times 10^{-3}\ mol$

$$\frac{1.5 \times 10^{-3}\ mol\ C_6H_5COOH}{0.0450\ L} = 3.3 \times 10^{-2}\ mol \cdot L^{-1}\ C_6H_5COOH$$

and $$\frac{4.5 \times 10^{-3}\ mol\ C_6H_5CO_2^-}{0.0450\ L} = 0.10\ mol \cdot L^{-1}\ C_6H_5O_2^-$$

Then, consider the equilibrium

$$C_6H_5COOH(aq) + H_2O(l) \rightleftharpoons H_3O^+(aq) + C_6H_5CO_2^-(aq)$$

Concentration

$(mol \cdot L^{-1})$	$C_6H_5COOH(aq)$ +	$H_2O(l) \rightleftharpoons$	$H_3O^+(aq)$ +	$C_6H_5CO_2^-(aq)$
initial	3.3×10^{-2}	—	0	0.10
change	$-x$	—	$+x$	$+x$
equilibrium	$3.3 \times 10^{-2} - x$	—	x	$0.10 + x$

$$K_a = \frac{[H_3O^+][C_6H_5CO_2^-]}{[C_6H_5COOH]}$$

$$6.5 \times 10^{-5} = \frac{(x)(x + 0.10)}{(3.3 \times 10^{-2} - x)}; + x \text{ and } - x \text{ are negligible}$$

$$[H_3O^+] = x = 2.1 \times 10^{-5}\ \text{mol} \cdot \text{L}^{-1}$$

and $\text{pH} = -\log(2.1 \times 10^{-5}) = 4.68$

(c)

$$0.0300\ \text{L}\left(\frac{0.20\ \text{mol}\ C_6H_5COOH}{\text{L}}\right)\left(\frac{1\ \text{mol KOH}}{1\ \text{mol}\ C_6H_5COOH}\right)$$

$$\times\left(\frac{\text{L}}{0.30\ \text{mol KOH}}\right)$$

$= 0.020$ L or 20 mL

20 mL to the stoichiometric point, so halfway to the stoichiometric point uses 10 mL.

(d) At half-stoichiometric point, $\text{pH} = \text{p}K_a$

and $\text{pH} = -\log(6.5 \times 10^{-5}) = 4.19$

(e) 20 mL, as calculated in part (c).

(f) Concentration of $KC_6H_5O_2$ at the stoichiometric point is

$$\frac{6.0 \times 10^{-3}\ \text{mol}}{0.050\ \text{L}} = 0.12\ \text{mol} \cdot \text{L}^{-1}$$

$$H_2O + C_6H_5CO_2^- \rightleftharpoons C_6H_5COOH + OH^-$$

$$K_b = \frac{K_w}{K_a} = \frac{1.00 \times 10^{-14}}{6.5 \times 10^{-5}} = 1.5 \times 10^{-10}$$

Concentration $(\text{mol} \cdot \text{L}^{-1})$	H_2O +	$C_6H_5CO_2^-$ ⇌	C_6H_5COOH +	OH^-
initial	—	0.10	0	0
change	—	$-x$	$+x$	$+x$
equilibrium	—	$0.10 - x$	x	x

$$K_b = \frac{[C_6H_5COOH][OH^-]}{[C_6H_5CO_2^-]} = 1.5 \times 10^{-10} = \frac{(x)(x)}{0.12 - x} \approx \frac{x^2}{0.12}$$

$$[OH^-] = x = 4.2 \times 10^{-6}\ \text{mol} \cdot \text{L}^{-1}$$

$\text{pOH} = 5.38$; $\text{pH} = 14.00 - 5.38 = 8.62$

11.34 (a) Initial pH of 0.25 M CH_3NH_2

$$CH_3NH_2(aq) + H_2O(l) \rightleftharpoons CH_3NH_3^+(aq) + OH^-(aq),\ K_b = 3.6\times10^{-4}$$

$$K_b = \frac{[CH_3NH_3^+][OH^-]}{[CH_3NH_2]}$$

Concentration

$(mol\cdot L^{-1})$	$CH_3NH_2(aq)$	+ $H_2O(l)$	$\rightleftharpoons$ $CH_3NH_3^+(aq)$	+ $OH^-(aq)$
initial	0.25	—	0	0
change	$-x$	—	$+x$	$+x$
equilibrium	$0.25 - x$	—	x	x

$$3.6\times10^{-4} = \frac{x^2}{0.25 - x} \approx \frac{x^2}{0.25}$$

$$[OH^-] = x = 9.5\times10^{-3}\ mol\cdot L^{-1}$$

$$pOH = -\log(9.5\times10^{-3}) = 2.02$$

$$\therefore pH = 14.00 - 2.02 = 11.98$$

(b) initial moles of $CH_3NH_2 = (0.0500\ L)(0.25\ mol\cdot L^{-1})$
$= 1.3\times10^{-2}\ mol\ CH_3NH_2$

moles of $HCl = (0.0150\ L)(0.35\ mol\cdot L^{-1}) = 5.3\times10^{-3}\ mol\ HCl$

After neutralization:

$$\text{molarity of } CH_3NH_2 = \frac{1.3\times10^{-2}\ mol - 5.3\times10^{-3}\ mol}{0.0650\ L}$$
$$= 0.1\ mol\cdot L^{-1}\ CH_3NH_2$$

$$\text{molarity of } CH_3NH_3^+ = \frac{5.3\times10^{-3}\ mol\ CH_3NH_3^+}{0.0650\ L}$$
$$= 8.2\times10^{-2}\ mol\cdot L^{-1}\ CH_3NH_3^+$$

Then, consider the equilibrium

$$CH_3NH_2(aq) + H_2O(l) \rightleftharpoons CH_3NH_3^+(aq) + OH^-(aq):$$

Concentration ($mol \cdot L^{-1}$)	$CH_3NH_2(aq)$ +	$H_2O(l)$ ⇌	$CH_3NH_3^+(aq)$ +	$OH^-(aq)$
initial	0.1	—	8.2×10^{-2}	0
change	$-x$	—	$+x$	$+x$
equilibrium	$0.1 - x$	—	$8.2 \times 10^{-2} + x$	x

$$K_b = \frac{[CH_3NH_3^+][OH^-]}{[CH_3NH_2]} = 3.6 \times 10^{-4} = \frac{(x)(x + 8.2 \times 10^{-2})}{(0.1 - x)};$$

$+x$ and $-x$ are negligible

$[OH^-] = x = 4 \times 10^{-4}$ $mol \cdot L^{-1}$ and pOH = 3.4

$\therefore$ pH = 14.00 − 3.4 = 10.6

(c) At the stoichiometric point, the volume of acid needed, V_A, is

$$V_A = (0.050 \text{ L base})\left(\frac{0.25 \text{ mol base}}{1 \text{ L base}}\right)\left(\frac{1 \text{ mol acid}}{1 \text{ mol base}}\right)\left(\frac{1 \text{ L acid}}{0.35 \text{ mol acid}}\right)$$

$$= 36 \times 10^{-3} \text{ L HCl} = 36 \text{ mL}$$

Therefore, the half-stoichiometric point $= \frac{1}{2}(36 \text{ mL}) = 18 \text{ mL}$

(d) At the half-stoichiometric point, pOH = pK_b

and $pOH = -\log(3.6 \times 10^{-4}) = 3.44$

pH = 14.00 − 3.44 = 10.56

(e) 36 mL HCl; see part (c)

(f) $V_{total} = V_A + V_B = 36 + 50.0 = 86$ mL

The solution contains

$$(0.050 \text{ L})\left(\frac{0.25 \text{ mol base}}{1 \text{ L base}}\right)\left(\frac{1}{86 \times 10^{-3} \text{ L}}\right)$$

$= 0.15$ $mol \cdot L^{-1}$ $CH_3NH_3^+$ as CH_3NH_3Cl

$CH_3NH_3^+(aq) + H_2O(l) \Lambda CH_3NH_2(aq) + H_3O^+(aq)$

$$K_a = \frac{K_w}{K_b} = \frac{1.00 \times 10^{-14}}{3.6 \times 10^{-4}} = 2.8 \times 10^{-11}$$

Concentration

$(\text{mol}\cdot\text{L}^{-1})$	$CH_3NH_3^+(aq)$	+ $H_2O(l)$	$\rightleftharpoons$ $CH_3NH_2(aq)$	+ $H_3O^+(aq)$
initial	0.15	—	0	0
change	$-x$	—	$+x$	$+x$
equilibrium	$0.15 - x$	—	x	x

$$2.8 \times 10^{-11} = \frac{[CH_3NH_2][H_3O^+]}{[CH_3NH_3^+]} = \frac{x^2}{0.15 - x} \approx \frac{x^2}{0.15}$$

$[H_3O^+] = x = 2.0 \times 10^{-6}\ \text{mol}\cdot\text{L}^{-1}$

$\text{pH} = -\log(2.0 \times 10^{-6}) = 5.69$

11.36 At the stoichiometric point, the volume of solution will have doubled; therefore, the concentration of NH_4^+ will be 0.10 M. The equilibrium is

Concentration

$(\text{mol}\cdot\text{L}^{-1})$	$NH_4^+(aq)$	+ $H_2O(l)$	$\rightleftharpoons$ $NH_3(aq)$	+ $H_3O^+(aq)$
initial	0.10	—	0	0
change	$-x$	—	$+x$	$+x$
equilibrium	$0.10 - x$	—	x	x

$$K_a = \frac{K_w}{K_b} = \frac{1.00 \times 10^{-14}}{1.8 \times 10^{-5}} = 5.6 \times 10^{-10}$$

$$K_a = \frac{[NH_3][H_3O^+]}{[NH_4^+]} = \frac{x^2}{0.10 - x} \approx \frac{x^2}{0.10} = 5.6 \times 10^{-10}$$

$x = 7.5 \times 10^{-6}\ \text{mol}\cdot\text{L}^{-1} = [H_3O^+]$

$\text{pH} = -\log(7.5 \times 10^{-6}) = 5.12$

From Table 11.2, we see that this pH value lies within the range for (a) bromocresol green (3.8 to 5.4), and (b) methyl red (4.8 to 6.0). Phenol red (pH range: 6.6 to 8.0) and thymol blue (ranges of 1.2 to 2.8 for red-yellow change and 8.0 to 9.6 for yellow-blue change) are unsuitable.

11.38 Exercise 11.32; thymol blue; Exercise 11.34: methyl red.

11.40 (a) To reach the first stoichiometric point, we must add enough solution to neutralize one H^+ on the H_2SO_3. To do this, we will require $0.125\ \text{L} \times 0.197\ \text{mol} \cdot \text{L}^{-1} = 0.0246$ mol of OH^-. The volume of base required will be given by the number of moles of base required, divided by the concentration of base solution:

$$\frac{0.125\ \text{L} \times 0.197\ \text{mol} \cdot \text{L}^{-1}}{0.123\ \text{mol} \cdot \text{L}^{-1}} = 0.200\ \text{L or } 2.00 \times 10^2\ \text{mL}.$$

(b) To reach the second stoichiometric point will require double the amount calculated in (a), or 4.00×10^2 mL.

11.42 (a) The exercise begins with the fully deprotonated base PO_4^{3-} and is essentially the opposite situation from beginning with the polyprotic acid H_3PO_4. It will require an equal number of moles of HCl to react with PO_4^{3-} in order to reach the first equivalence point. The value will be given by $\dfrac{0.0888\ \text{L} \times 0.233\ \text{mol} \cdot \text{L}^{-1}}{0.0848\ \text{mol} \cdot \text{L}^{-1}} = 0.244$ L or 244 mL.

(b) To reach the second equivalence point would require double the amount of solution calculated in (a), or 488 mL.

(c) To reach, the third equivalence point would require triple the amount of solution calculated in (a), or 732 mL.

11.44 (a) This value is calculated as described in section 10.12. First we calculate the molarity of the starting oxalic acid solution:

$\dfrac{0.242\ \text{g}}{90.03\ \text{g} \cdot \text{mol}^{-1}} \Big/ 0.0500\ \text{L} = 0.0538\ \text{mol} \cdot \text{L}^{-1}$. We then use the first acid dissociation of oxalic acid as the dominant equilibrium. The K_{a1} is 5.9×10^{-2}. Let H_2Ox represent the fully protonated oxalic acid:

Concentration

$(\text{mol}\cdot\text{L}^{-1})$	$H_2Ox\,(aq)$ + $H_2O(l)$		$\rightleftharpoons$	$HOx^-(aq)$ +	$H_3O^+(aq)$
initial	0.0538	—		0	0
change	$-x$	—		$+x$	$+x$
final	$0.0538 - x$	—		$+x$	$+x$

$$K_{a1} = \frac{[H_3O^+][HOx^-]}{[H_2Ox]} = 5.9 \times 10^{-2}$$

$$5.9 \times 10^{-2} = \frac{x \cdot x}{0.0538 - x} = \frac{x^2}{0.0538 - x}$$

Because the equilibrium constant is reasonably large, the full quadratic solution should be undertaken. The equation is

$x^2 = (5.9 \times 10^{-2})(0.0538 - x)$ or

$x^2 + 5.9 \times 10^{-2} x - 3.17 \times 10^{-3} = 0$

Using the quadratic formula, we obtain $x = 0.034$.

pH = 1.47

(b) First, carry out the reaction between oxalic acid and the strong base to completion:

$H_2Ox(aq) + OH^-(aq) \longrightarrow HOx^-(aq) + H_2O(l)$

moles of $H_2Ox = (0.0538\ \text{mol}\cdot\text{L}^{-1})(0.0500\ \text{L}) = 2.69 \times 10^{-3}$ mol

moles of $OH^- = (0.0150\ \text{L})(0.150\ \text{mol}\cdot\text{L}^{-1}) = 2.25 \times 10^{-3}$ mol

This amount of base will partially neutralize one H^+ present in H_2Ox. At this point, 0.440×10^{-3} mol of H_2O_x and 2.25×10^{-3} mol of HO_x^- are present in solution , so

$$\text{pH} = \text{p}K_{a1} + \log\frac{\left[HO_x^-\right]}{\left[H_2O_x\right]} = 1.23 + \log\frac{2.25}{0.44}$$

$$= 1.94$$

(c) The total amount of OH^- added will be 15.0 mL + 5.00 mL = 20.00 mL. The moles of $OH^- = (0.0200\ \text{L})(0.150\ \text{mol}\cdot\text{L}^{-1}) = 3.00 \times 10^{-3}$ mol.

Because there are 2.69×10^{-3} moles of H_2Ox, the first deprotonation will be complete:

$H_2Ox(aq) + OH^-(aq) \longrightarrow HOx^-(aq) + H_2O(l)$

The excess base will be $(3.00 \times 10^{-3}\ \text{mol}) - (2.69 \times 10^{-3}\ \text{mol}) = 0.31 \times 10^{-3}$ mol and the amount of HOx^- will be 2.69×10^{-3} mol. This base can then react with HOx^- to form Ox^{2-}. Because the stoichiometry is 1 : 1, the amount of HOx^- that reacts is 7.5×10^{-4} mol, giving that number of moles of Ox^{2-} and leaving 3.1×10^{-4} mol of HOx^-.

The pH is then:

$$\text{pH} = \text{p}K_{a2} + \log\frac{[\text{O}_x^{2-}]}{[\text{HO}_x^-]} = 4.19 + \log\frac{0.31}{2.38}$$

$$= 3.30$$

11.46 (a) The reaction of the base Na_2CO_3 with the strong acid will be taken to completion first.

$CO_3^{2-}(aq) + H_3O^+(aq) \longrightarrow HCO_3^- + H_2O(l)$

moles of $CO_3^{2-} = 0.0750\ \text{L} \times 0.0995\ \text{mol}\cdot\text{L}^{-1} = 7.46 \times 10^{-3}$ mol

moles of $H_3O^+ = 0.0250\ \text{L} \times 0.130\ \text{mol}\cdot\text{L}^{-1} = 3.25 \times 10^{-3}$ mol

There is excess carbonate ion, so all of the strong acid will be consumed, generating a starting solution with residual CO_3^{2-} and product HCO_3^- (bicarbonate ion). Because the reaction stoichiometry is 1 : 1, 3.25×10^{-3} mol of HCO_3^- will be formed; this will leave $(7.46 \times 10^{-3}\ \text{mol}) - (3.25 \times 10^{-3}\ \text{mol}) = 4.21 \times 10^{-3}\ \text{mol}\ CO_3^{2-}$. This places the concentrations of the species in the second buffer region of carbonic acid. The concentrations will be

$$[HCO_3^-] = \frac{3.25 \times 10^{-3}\ \text{mol}}{0.100\ \text{L}} = 0.0325\ \text{mol}\cdot\text{L}^{-1};$$

$$[CO_3^{2-}] = \frac{4.21 \times 10^{-3}\ \text{mol}}{0.100\ \text{L}} = 0.0421\ \text{mol}\cdot\text{L}^{-1}.$$

The equilibrium of interest is

Concentration

$(\text{mol}\cdot\text{L}^{-1})$	$CO_3^{2-}(aq)$	$+\ H_2O(l)$	$\rightleftharpoons$ $HCO_3^-(aq)$	$+\ OH^-(aq)$
initial	0.0421	—	0.0325	0
change	$-x$	—	$+x$	$+x$
final	$0.0421 - x$		$0.0325 + x$	$+x$

The K_{b1} for CO_3^{2-} is given by $\dfrac{K_w}{K_{a2}} = \dfrac{1.00\times10^{-14}}{5.6\times10^{-11}} = 1.8\times10^{-4}$, where K_{a2} is the second acid dissociation constant of H_2CO_3.

The solution is set up as follows.

$$K_{b1} = \frac{[HCO_3^-][OH^-]}{[CO_3^{2-}]}$$

$$1.8\times10^{-4} = \frac{(0.0325 + x)x}{0.0421 - x}$$

If $x \ll 0.0325$, then the problem simplifies to

$$1.8\times10^{-4} = \frac{0.0325x}{0.0421}$$

$x = 2.3\times10^{-4}$

Because x is less than 5% of 0.0325, the assumption was valid.

$x = [OH^-]$, so $-\log x = \text{pOH} = 3.64$

$\text{pH} = 14.00 - 3.64 = 10.36$

(b) The reaction proceeds as in (a) but now there is more strong acid:

moles of $H_3O^+ = 0.0550\ \text{L}\times0.130\ \text{mol}\cdot\text{L}^{-1} = 7.15\times10^{-3}$ mol

The moles of CO_3^{2-} initially is the same as in (a) at 7.46×10^{-3} mol.

There is not enough strong acid present to convert all of the CO_3^{2-} to HCO_3^- so some CO_3^{2-} will remain. The setup is the same as in (a):

$$CO_3^{2-}(aq) + H_3O^+(aq) \longrightarrow HCO_3^- + H_2O(l)$$

moles of $CO_3^{2-} = 0.0300\ \text{L}\times0.175\ \text{mol}\cdot\text{L}^{-1} = 7.46\times10^{-3}$ mol

moles of $H_3O^+ = 0.0550\ \text{L}\times0.130\ \text{mol}\cdot\text{L}^{-1} = 7.15\times10^{-3}$ mol

There is excess carbonate ion, so all of the strong acid will be consumed, generating a starting solution with residual CO_3^{2-} and product HCO_3^- (bicarbonate ion). Because the reaction stoichiometry is 1 : 1, 7.15×10^{-3} mol of HCO_3^- will be formed; this will leave $(7.46 \times 10^{-3}\ \text{mol}) - (7.15 \times 10^{-3}\ \text{mol}) = 3.1 \times 10^{-4}\ \text{mol}\ CO_3^{2-}$. This places the concentrations of the species in the second buffer region of carbonic acid. The concentrations will be

$$[CO_3^{2-}] = \frac{3.1 \times 10^{-4}\ \text{mol}}{0.140\ \text{L}} = 2.2 \times 10^{-3}\ \text{mol} \cdot \text{L}^{-1}; [HCO_3^-]$$

$$= \frac{7.15 \times 10^{-3}\ \text{mol}}{0.140\ \text{L}} = 0.0510\ \text{mol} \cdot \text{L}^{-1}$$

The equilibrium of interest is

Concentration ($\text{mol} \cdot \text{L}^{-1}$)	$CO_3^{2-}(aq)$	+ $H_2O(l) \rightleftharpoons$	$HCO_3^-(aq)$	+ $OH^-(aq)$
initial	2.2×10^{-3}	—	0.0510	0
change	$-x$	—	$+x$	$+x$
final	$2.2 \times 10^{-3} - x$		$0.0510 + x$	$+x$

The K_{b1} for CO_3^{2-} is given by $\frac{K_w}{K_{a2}} = \frac{1.00 \times 10^{-14}}{5.6 \times 10^{-11}} = 1.8 \times 10^{-4}$, where K_{a2} is the second acid dissociation constant of H_2CO_3.

The solution is set up as follows:

$$K_{b1} = \frac{[HCO_3^-][OH^-]}{[CO_3^{2-}]}$$

$$1.8 \times 10^{-4} = \frac{(0.0510 + x)(x)}{(2.2 \times 10^{-3} - x)}$$

If $x << 2.2 \times 10^{-3}$, then the problem simplifies to

$$1.8 \times 10^{-4} = \frac{0.0510x}{2.2 \times 10^{-3}}$$

$$x = 7.8 \times 10^{-6}$$

Because x is less than 5% of 2.2×10^{-3}, the assumption was valid.

$x = [\mathrm{OH}^-]$, so $-\log x = \mathrm{pOH} = 5.10$

$\mathrm{pH} = 14.00 - 5.10 = 8.90$

11.48 (a) The solubility equilibrium is

$\mathrm{AgI(s)} \rightleftharpoons \mathrm{Ag^+(aq)} + \mathrm{I^-(aq)}$

$[\mathrm{Ag^+}] = [\mathrm{I^-}] = 9.1 \times 10^{-9}\ \mathrm{mol \cdot L^{-1}} = S$ (molar solubility)

$K_{sp} = [\mathrm{Ag^+}][\mathrm{I^-}] = (9.1 \times 10^{-9})^2 = 8.3 \times 10^{-17}$

(b) The solubility equilibrium is

$\mathrm{Ca(OH)_2(s)} \rightleftharpoons \mathrm{Ca^{2+}(aq)} + 2\ \mathrm{OH^-(aq)}$

$[\mathrm{Ca^{2+}}] = 0.011\ \mathrm{mol \cdot L^{-1}} = S$

$[\mathrm{OH^-}] = 0.022\ \mathrm{mol \cdot L^{-1}} = 2S$

$K_{sp} = [\mathrm{Ca^{2+}}][\mathrm{OH^-}]^2 = (0.011)(0.022)^2 = 5.3 \times 10^{-6}$

(c) The solubility equilibrium is

$\mathrm{Ag_3PO_4(s)} \rightleftharpoons 3\ \mathrm{Ag^+(aq)} + \mathrm{PO_4^{3-}(aq)}$

$[\mathrm{Ag^+}] = 8.1 \times 10^{-6}\ \mathrm{mol \cdot L^{-1}} = 3S$

$[\mathrm{PO_4^{3-}}] = 2.7 \times 10^{-6}\ \mathrm{mol \cdot L^{-1}} = S$

$K_{sp} = [\mathrm{Ag^+}]^3[\mathrm{PO_4^{3-}}] = (8.1 \times 10^{-6})^3(2.7 \times 10^{-6}) = 1.4 \times 10^{-21}$

(d) The solubility equilibrium is

$\mathrm{Hg_2Cl_2(s)} \rightleftharpoons \mathrm{Hg_2^{2+}(aq)} + 2\ \mathrm{Cl^-(aq)}$

$[\mathrm{Hg_2^{2+}}] = 5.2 \times 10^{-7}\ \mathrm{mol \cdot L^{-1}} = S$

$[\mathrm{Cl^-}] = 1.0 \times 10^{-6}\ \mathrm{mol \cdot L^{-1}} = 2S$

$K_{sp} = [\mathrm{Hg_2^{2+}}][\mathrm{Cl^-}]^2 = (5.2 \times 10^{-7})(1.0 \times 10^{-6})^2 = 5.2 \times 10^{-19}$

11.50 (a) S = molar solubility; $\mathrm{PbSO_4(s)} \rightleftharpoons \mathrm{Pb^{2+}(aq)} + \mathrm{SO_4^{2-}(aq)}$

$K_{sp} = [\mathrm{Pb^{2+}}][\mathrm{SO_4^{2-}}] = S \times S = S^2 = 1.6 \times 10^{-8}$

$S = 1.3 \times 10^{-4}\ \mathrm{mol \cdot L^{-1}}$

(b) $\mathrm{Ag_2CO_3(s)} \rightleftharpoons 2\ \mathrm{Ag^+(aq)} + \mathrm{CO_3^{2-}(aq)}$

$K_{sp} = [Ag^+]^2[CO_3^{2-}] = (2S)^2 \times S = 4S^3 = 6.2 \times 10^{-12}$

$S = 1.2 \times 10^{-4}\ mol \cdot L^{-1}$

(c) $Fe(OH)_2(s) \rightleftharpoons Fe^{2+}(aq) + 2\ OH^-(aq)$

$K_{sp} = [Fe^{2+}][OH^-]^2 = S \times (2S)^2 = 4S^3 = 1.6 \times 10^{-14}$

$S = 1.6 \times 10^{-5}\ mol \cdot L^{-1}$

11.52 S = molar solubility The reaction is $Ce(OH)_3(s) \rightleftharpoons Ce^{3+}(aq) + 3\ OH^-(aq)$

$K_{sp} = [Ce^{3+}][OH^-]^3 = S \times (3S)^3 = 27S^4 = 27(5.2 \times 10^{-6})^4 = 2.0 \times 10^{-20}$

11.54 (a) $AgBr(s) \rightleftharpoons Ag^+(aq) + Br^-(aq)$

$S = mol \cdot L^{-1}$ of AgBr that dissolve = $[Ag^+]$

Concentration ($mol \cdot L^{-1}$)	$AgBr(s)$	$\rightleftharpoons Ag^+(aq) +$	$Br^-(aq)$
initial	—	0	0.050
change	—	$+S$	$+S$
equilibrium	—	S	$0.050 + S$

$0.050 + S = [Br^-]$

$K_{sp} = [Ag^+][Br^-]$

$7.7 \times 10^{-13} = (S) \times (0.050 + S) = (S) \times (0.050)$

$S = 1.5 \times 10^{-11}\ mol \cdot L^{-1} = [Ag^+]$ = molar solubility of AgBr in 0.050 M NaBr

(b) $MgCO_3(s) \rightleftharpoons Mg^{2+}(aq) + CO_3^{2-}(aq)$

Concentration ($mol \cdot L^{-1}$)	$MgCO_3(s)$	$\rightleftharpoons Mg^{2+}(aq) +$	$CO_3^{2-}(aq)$
initial	—	0	1.0×10^{-3}
change	—	$+S$	$+S$
equilibrium	—	S	$1.0 \times 10^{-3} + S$

$K_{sp} = [Mg^{2+}][CO_3^{2-}] = (S) \times (1.0 \times 10^{-3} + S) = 1.0 \times 10^{-5}$

$S^2 + 1.0 \times 10^{-3} S - 1.0 \times 10^{-5} = 0$

$S = 2.7 \times 10^{-3}\ mol \cdot L^{-1} = [Mg^{2+}]$

= molar solubility of $MgCO_3$ in 1.0×10^{-3} M Na_2CO_3

(c) $PbSO_4(s) \rightleftharpoons Pb^{2+}(aq) + SO_4^{2-}$

Concentration $(mol \cdot L^{-1})$	$PbSO_4(s)$ $\rightleftharpoons$	$Pb^{2+}(aq)$ +	$SO_4^{2-}(aq)$
initial	—	0	0.25
change	—	$+S$	$+S$
equilibrium	—	S	$0.25 + S$

$K_{sp} = [Pb^{2+}][SO_4^{2-}] = S \times (0.25 + S) = 1.6 \times 10^{-8}$

(assume S in $(0.25 + S)$ is negligible)

$0.25S = 1.6 \times 10^{-8}$

$S = 6.4 \times 10^{-8}\ mol \cdot L^{-1}$ = molar solubility of $PbSO_4$ in 0.25 M Na_2SO_4

(d) $Ni(OH)_2(s) \rightleftharpoons Ni^{2+}(aq) + 2\ OH^-(aq)$

Concentration $(mol \cdot L^{-1})$	$Ni(OH)_2(s)$	$\rightleftharpoons$ $Ni^{2+}(aq)$ +	$OH^-(aq)$
initial	—	0.125	0
change	—	$+S$	$+2S$
equilibrium	—	$0.125 + S$	$2S$

$K_{sp} = [Ni^{2+}][OH^-]^2 = 6.5 \times 10^{-18} = (S + 0.125) \times (2S)^2$

(assume S in $(S + 0.125)$ is negligible)

$S = 3.6 \times 10^{-9}\ mol \cdot L^{-1}$ = molar solubility of $Ni(OH)_2$ in 0.125 M $NiSO_4$

11.56 (a) $Pb^{2+}(aq) + 2\ I^-(aq) \rightleftharpoons PbI_2(s)$

Concentration ($mol \cdot L^{-1}$)	Pb^{2+}	I^-
initial	0.0020	0
change	0	$+x$
equilibrium	0.0020	x [$Pb(NO_3)_2$ is soluble]

$$K_{sp} = [Pb^{2+}][I^-]^2 = 1.4 \times 10^{-8} = (0.0020)(x)^2$$

$$\text{molarity of } I^- = x = \sqrt{\frac{1.4 \times 10^{-8}}{0.002}} = 2.6 \times 10^{-3}\ mol \cdot L^{-1}$$

(b) mass of KI =

$$\left(\frac{2.6 \times 10^{-3}\ \text{mol KI}}{1\ \text{L}}\right)(0.0250\ \text{L})\left(\frac{166.00\ \text{g KI}}{1\ \text{mol KI}}\right) = 1.1 \times 10^{-2}\ \text{g KI}$$

11.58 (a) $2\ Ag^+(aq) + CO_3^{2-}(aq) \rightleftharpoons Ag_2CO_3(s)$

$$K_{sp} = [Ag^+]^2[CO_3^{2-}] = 6.2 \times 10^{-12}$$

$$Q_{sp} = \left[\frac{(1.00)(0.010)}{1.01}\right]^2\left[\frac{(0.0050)(0.10)}{1.01}\right]$$
$$= 4.9 \times 10^{-8}$$

A precipitate will form because $Q_{sp} > K_{sp}$.

(b) $Ag^+(aq) + Cl^-(aq) \rightleftharpoons AgCl(s)$

$$K_{sp} = [Ag^+][Cl^-] = 1.6 \times 10^{-10}$$

$$Q_{sp} = \left[\frac{(0.0033)(1.0)}{0.050}\right]\left[\frac{(0.0049)(0.0030)}{0.050}\right]$$
$$= 1.9 \times 10^{-5}$$

A precipitate will form because $Q_{sp} > K_{sp}$.

11.60 (a) $\frac{1\ \text{mL}}{20\ \text{drops}} \times 7\ \text{drops} = 0.35\ \text{mL} = 3.5 \times 10^{-4}\ \text{L}$

$$(3.5 \times 10^{-4}\ L)(0.0029\ mol \cdot L^{-1}) = 1.0 \times 10^{-6}\ mol\ K_2CO_3$$
$$= 1.0 \times 10^{-6}\ mol\ CO_3^{2-}$$

$Ca^{2+}(aq) + CO_3^{2-}(aq) \rightleftharpoons CaCO_3(s)$

$[Ca^{2+}][CO_3^{2-}] = K_{sp}$

$$Q_{sp} = \left[\frac{(0.0250)(0.0018)}{0.0254}\right]\left[\frac{1.0\times10^{-6}}{0.0254}\right] = 7.0\times10^{-8}$$

will precipitate because $Q_{sp}(7.0\times10^{-8}) > K_{sp}(8.7\times10^{-9})$

(b) $2\ Ag^{+}(aq) + CO_3^{2-}(aq) \rightleftharpoons Ag_2CO_3(s)$

$$\frac{1\ mL}{20\ drops}\times 10\ drops = 0.5\ mL = 5\times10^{-4}\ L$$

$(5\times10^{-4}\ L)(0.010\ mol\cdot L^{-1}) = 5\times10^{-6}\ mol\ Na_2CO_3 = 5\times10^{-6}\ mol\ CO_3^{2-}$

$[Ag^{+}]^2[CO_3^{2-}] = K_{sp}$

$$Q_{sp} = \left[\frac{(0.0100)(0.0040)}{0.0105}\right]^2\left[\frac{5\times10^{-6}}{0.0105}\right] = 7\times10^{-9}$$

will precipitate because $Q_{sp}(7\times10^{-9}) > K_{sp}(6.2\times10^{-12})$

11.62 For MOH:

$MOH(s) \rightleftharpoons M^{+}(aq) + OH^{-}(aq)$

$K_{sp} = [M^{+}][OH^{-}]$

$1.0\times10^{-12} = (1.0\times10^{-3})[OH^{-}]$

$[OH^{-}] = 1.0\times10^{-9}\ mol\cdot L^{-1}$

pOH = 9, pH = 5

For $M'(OH)_2$:

$M'(OH)_2(s) \rightleftharpoons M'^{2+}(aq) + 2\ OH^{-}(aq)$

$K_{sp} = [M'^{2+}][OH^{-}]^2$

$1.0\times10^{-12} = (1.0\times10^{-3})[OH^{-}]^2$

$$[OH^{-}] = \sqrt{\frac{1.0\times10^{-12}}{1.0\times10^{-3}}} = 3.2\times10^{-5}\ mol\cdot L^{-1}$$

pOH = 4.5, pH = 9.5

MOH precipitates first at pH = 5, whereas $M'(OH)_2$ does not precipitate until pH = 9.5.

11.64 The K_{sp} values are CaF_2 4.0×10^{-11}

BaF_2 1.7×10^{-6}

$CaCO_3$ 8.7×10^{-9}

$BaCO_3$ 8.1×10^{-9}

The difference in these numbers suggests that there is a greater solubility difference between the fluorides, and therefore this anion should give a better separation. Because different numbers of ions are involved, it is instructive to convert the K_{sp} values into molar solubility. For the fluorides the reaction is

$$MF_2(s) \rightleftharpoons M^{2+}(aq) + 2\ F^-(aq)$$

Change $+x$ $+2x$

$K_{sp} = x(2x)^2$

Solving for x for CaF_2 gives 0.000 22 M and 0.0075 M for BaF_2.

For the carbonates:

$$MCO_3(s) \rightleftharpoons M^{2+}(aq) + CO_3^{2-}(aq)$$

$+x$ $+x$

$K_{sp} = x^2$

Solving this for $CaCO_3$ gives 9.3×10^{-5} M and 9.0×10^{-5} M for $BaCO_3$. Clearly, of the K_{sp} values given, the solubility difference is greatest between the two fluorides, and F^- is the better choice of anion. The solubilities of the two carbonates are almost identical, so this would not be a good choice.

11.66 $MSO_4(s) \rightleftharpoons M^{2+}(aq) + SO_4^{2-}(aq)$

$K_{sp} = [M^{2+}][SO_4^{2-}]$

The K_{sp} values are $BaSO_4$ 1.1×10^{-10}

$$PbSO_4 \quad 1.6 \times 10^{-8}$$

(a) For $BaSO_4$, the maximum concentration of sulfate that allows all the Ba^{2+} to remain in solution is given by

$$1.1 \times 10^{-10} = [Ba^{2+}][SO_4^{\ 2-}] = (0.010)[SO_4^{\ 2-}]$$

$$[SO_4^{\ 2-}] = 1.1 \times 10^{-8}\ mol \cdot L^{-1}$$

Similarly, for $PbSO_4$:

$$1.6 \times 10^{-8} = [Pb^{2+}][SO_4^{\ 2-}] = (0.010)[SO_4^{\ 2-}]$$

$$[SO_4^{\ 2-}] = 1.6 \times 10^{-6}\ mol \cdot L^{-1}$$

(b) The $BaSO_4$ is less soluble and will be essentially all precipitated when the $PbSO_4$ begins to precipitate. The concentration of $SO_4^{\ 2-}$ required for the Pb^{2+} to precipitate will be $1.6 \times 10^{-6}\ mol \cdot L^{-1}$. We can calculate the concentration of Ba^{2+} at this concentration of sulfate ion:

$$1.1 \times 10^{-10} = [Ba^{2+}][SO_4^{\ 2-}] = [Ba^{2+}][1.6 \times 10^{-6}]$$

$$[Ba^{2+}] = 6.9 \times 10^{-5}\ mol \cdot L^{-1}$$

11.68 $Fe^{3+}(aq) + 3\,OH^-(aq) \rightleftharpoons Fe(OH)_3(s)$

$$K_{sp} = [Fe^{3+}][OH^-]^3 = 2.0 \times 10^{-39}$$

S = molar solubility

(a) pH = 11.0; pOH = 3.0; $[OH^-] = 1.0 \times 10^{-3}\ mol \cdot L^{-1}$

$$[Fe^{3+}](1.0 \times 10^{-3})^3 = 2.0 \times 10^{-39}$$

$$S = \frac{2.0 \times 10^{-39}}{1.0 \times 10^{-9}} = 2.0 \times 10^{-30}\ mol \cdot L^{-1} = [Fe^{3+}]$$

= molar solubility of $Fe(OH)_3$ at pH = 11.0

(b) pH = 3.0; pOH = 11.0; $[OH^-] = 1.0 \times 10^{-11}\ mol \cdot L^{-1}$

$$[Fe^{3+}](1.0 \times 10^{-11})^3 = 2.0 \times 10^{-39}$$

$$S = \frac{2.0 \times 10^{-39}}{1.0 \times 10^{-33}} = 2.0 \times 10^{-6}\ mol \cdot L^{-1} = [Fe^{3+}]$$

= molar solubility of $Fe(OH)_3$ at pH = 3.0

(c) $Fe^{2+}(aq) + 2\ OH^-(aq) \rightleftharpoons Fe(OH)_2(s)$

$K_{sp} = [Fe^{2+}][OH^-]^2 = 1.6 \times 10^{-14}$

pH = 8.0; pOH = 6.0; $[OH^-] = 1.0 \times 10^{-6}\ mol \cdot L^{-1}$

$[Fe^{2+}](1.0 \times 10^{-6})^2 = 1.6 \times 10^{-14}$

$$S = \frac{1.6 \times 10^{-14}}{1.0 \times 10^{-12}} = 1.6 \times 10^{-2}\ mol \cdot L^{-1} = [Fe^{2+}]$$

= molar solubility of $Fe(OH)_2$ at pH = 8.0

(d) pH = 6.0; pOH = 8.0; $[OH^-] = 1.0 \times 10^{-8}\ mol \cdot L^{-1}$

$[Fe^{2+}](1.0 \times 10^{-8})^2 = 1.6 \times 10^{-14}$

$$S = \frac{1.60 \times 10^{-14}}{1.0 \times 10^{-16}} = 1.6 \times 10^{2}\ mol \cdot L^{-1} = [Fe^{2+}]$$

= molar solubility of $Fe(OH)_2$ at pH = 6.0

11.70 $BaF_2(s) \rightleftharpoons Ba^{2+}(aq) + 2\ F^-(aq)$ $K_{sp} = 1.7 \times 10^{-6}$

$F^-(aq) + H_2O(l) \rightleftharpoons HF(aq) + OH^-(aq)$ $K_b = 2.9 \times 10^{-11}$

(a) Multiply the second equilibrium equation by 2 and add to the first equilibrium equation:

$BaF_2(s) + 2\ H_2O(l) \rightleftharpoons Ba^{2+}(aq) + 2\ HF(aq) + 2\ OH^-(aq)$

$K = K_{sp}K_b{}^2 = (1.7 \times 10^{-6}) \times (2.9 \times 10^{-11})^2 = 1.4 \times 10^{-27}$

(b) S = molar solubility $K = [Ba^{2+}][HF]^2[OH^-]^2$

$1.4 \times 10^{-27} = (S)(2S)^2(1.0 \times 10^{-7})^2$ (at pH = 7.0, $[OH^-] = 1 \times 10^{-7}$ M)

$1.4 \times 10^{-13} = 4S^3$

$S = [Ba^{2+}] = 3.3 \times 10^{-5}\ mol \cdot L^{-1}$ = molar solubility of BaF_2 at pH = 7.0

(c) $K = [Ba^{2+}][HF]^2[OH^-]^2$

$1.4 \times 10^{-27} = (S)(2S)^2\ (1.0 \times 10^{-10})^2$ (at pH = 4.0)

$1.4 \times 10^{-7} = 4S^3$

$S = [Ba^{2+}] = 3.3 \times 10^{-3}\ mol \cdot L^{-1}$ = molar solubility of BaF_2 at pH = 4.0

11.72 $AgCl(s) \rightleftharpoons Ag^+(aq) + Cl^-(aq)$ $\qquad K_{sp} = 1.6 \times 10^{-10}$

$Ag^+(aq) + 2\ NH_3(aq) \rightleftharpoons Ag(NH_3)_2^+(aq)$ $\qquad K_f = 1.6 \times 10^{+7}$

$AgCl(s) + 2\ NH_3(aq) \rightleftharpoons Ag(NH_3)_2^+(aq) + Cl^-(aq)$ $\qquad K = K_{sp} \times K_f = 2.6 \times 10^{-3}$

$$K = \frac{[Ag(NH_3)_2^+][Cl^-]}{[NH_3]^2}$$

	$AgCl(s) + 2\ NH_3(aq)$	$\rightleftharpoons Ag(NH_3)_2^+(aq)$	$+\ Cl^-(aq)$
Equilibrium concentration —	$1.0 - 2S$	S	S

$$K = \frac{S^2}{(1.0 - 2S)^2} = 2.6 \times 10^{-3}$$

$$\frac{S}{1.0 - 2S} = \sqrt{2.6 \times 10^{-3}} = 5.1 \times 10^{-2}$$

$S = 5.1 \times 10^{-2} - 0.10\ S$

$1.10\ S = 5.1 \times 10^{-2}$

$S = 4.6 \times 10^{-2}\ mol \cdot L^{-1}$

11.74 The calculation for the solubility of the sulfides can be seen to be pH dependent, due to the equilibrium between S^{2-} and HS^- in aqueous solution. The solubilities of metal sulfides can be calculated as a function of pH quantitatively, as shown below.

(a) $Bi_2S_3(s) \rightleftharpoons 2\ Bi^{3+}(aq) + 3\ S^{2-}(aq)$ $\qquad K_{sp} = 1.0 \times 10^{-97}$

$S^{2-}(aq) + H_2O(l) \rightleftharpoons HS^-(aq) + OH^-(aq)$ $\qquad K_{b1} = K_w / K_{a2} = 1.41$

$Bi_2S_3(s) + 3\ H_2O(l) \rightleftharpoons 2\ Bi^{3+}(aq) + 3\ HS^-(aq) + 3\ OH^-(aq)$

$K = K_{sp} \cdot K_{b1}{}^3 = 2.8 \times 10^{-97}$

At $[H_3O^+] = 1.00\ mol \cdot L^{-1}$, $[OH^-] = 1.00 \times 10^{-14}\ mol \cdot l^{-1}$

$[Bi^{3+}]^2[HS^-]^3[OH^-]^3 = 2.8 \times 10^{-97}$

$[Bi^{3+}]^2[HS^-]^3[1.00 \times 10^{-14}]^3 = 2.8 \times 10^{-97}$

$[Bi^{3+}]^2[HS^-]^3 = 2.8 \times 10^{-55}$

If we let x be the number of moles of Bi_2S_3 that dissolve in one L, the concentration of Bi^{3+} will be $2x$ and of HS^- will be $3x$:

$$(2x)^2(3x)^3 = 2.8 \times 10^{-55}$$

$$x = 4.8 \times 10^{-12} \text{ mol} \cdot \text{L}^{-1}$$

Bi_2S_3 is insoluble in 1.00 M HNO_3.

(b) A similar procedure is used for FeS:

$$\text{FeS(s)} \rightleftharpoons \text{Fe}^{2+}\text{(aq)} + \text{S}^{2-}\text{(aq)} \qquad K_{sp} = 6.3 \times 10^{-8}$$

$$\text{S}^{2-}\text{(aq)} + \text{H}_2\text{O(l)} \rightleftharpoons \text{HS}^-\text{(aq)} + \text{OH}^-\text{(aq)} \qquad K_{b1} = K_w / K_{a2} = 1.41$$

$$\text{FeS(s)} + 3\,\text{H}_2\text{O(l)} \rightleftharpoons \text{Fe}^{2+}\text{(aq)} + \text{HS}^-\text{(aq)} + \text{OH}^-\text{(aq)}$$

$$K = K_{sp} \cdot K_{b1} = 8.9 \times 10^{-8}$$

At $[\text{H}_3\text{O}^+] = 1.00 \text{ mol} \cdot \text{L}^{-1}, [\text{OH}^-] = 1.00 \times 10^{-14} \text{ mol} \cdot \text{L}^{-1}$

$$[\text{Fe}^{2+}][\text{HS}^-][\text{OH}^-] = 8.9 \times 10^{-8}$$

$$[\text{Fe}^{2+}][\text{HS}^-][1.00 \times 10^{-14}]^3 = 8.9 \times 10^{-8}$$

$$[\text{Fe}^{2+}][\text{HS}^-] = 8.9 \times 10^{6}$$

If we let x be the number of moles of FeS that dissolve in one L, the concentration of Fe^{2+} will be x and of HS^- will also be x:

$$x^2 = 8.9 \times 10^{6}$$

$$x = 3.0 \times 10^{3} \text{ mol} \cdot \text{L}^{-1}$$

FeS is soluble in 1.00 M HNO_3.

11.76 Both $ZnCl_2$ and $MgCl_2$ are soluble, giving M^{2+} cations and Cl^- ions in solution. If OH^- is added, both will precipitate $M(OH)_2$. However, if you continue to add OH^-, the formation of the complex ion $Zn(OH)_4^{2-}$ will occur; thus $Zn(OH)_2$ will redissolve but the $Mg(OH)_2$ will remain insoluble.

11.78 (a) 1.00 L of solution will have a mass of 8.9×10^2 g and will contain 2.7×10^2 g NH_3 ($17.03 \text{ g} \cdot \text{mol}^{-1}$) or 15.8 moles. The concentration will be approximately 15.8 M. (b) The K_b value for NH_3 is 1.8×10^{-5}. We can set up the typical relationship:

Concentration ($\text{mol} \cdot \text{L}^{-1}$)	$NH_3(aq)$ +	$H_2O(l)$ $\rightleftharpoons$	$NH_4^+(aq)$ +	$OH^-(aq)$
initial	15.8	—	0	0
change	$-x$	—	$+x$	$+x$
equilibrium	$15.8 - x$	—	$+x$	$+x$

$$\frac{x^2}{15.8 - x} = 1.8 \times 10^{-5}$$

assume x is small compared to 15.8

$x^2 = (1.8 \times 10^{-5})(15.8) = 2.8 \times 10^{-4}$

$x = 1.7 \times 10^{-2}$

The assumption was good. $[NH_3] \cong 15.8 \text{ M}$; $[NH_4OH] = 1.7 \times 10^{-2}$ M

(c) 1.00 g $AgNO_3$ ($169.88 \text{ g} \cdot \text{mol}^{-1}$) $= 5.89 \times 10^{-3}$ mol $AgNO_3$

$[AgNO_3] = 5.89 \times 10^{-3}$ M

The K_{sp} of AgOH is 1.5×10^{-8}.

To precipitate the 5.89×10^{-3} moles of Ag^+ present, we will need to have at least 5.89×10^{-3} mol of OH^-. However, from the K_{sp} value of 1.5×10^{-8}, we can calculate that the molar solubility of AgOH will be 1.2×10^{-4} M $\left(\sqrt{1.5 \times 10^{-8}}\right)$, which is about 2.0% of the amount of Ag present. In order to precipitate the Ag^+ so that less than 1% of the Ag^+ is present in solution, we will need to have a slightly higher concentration of OH^-.

$AgOH(s) \rightleftharpoons Ag^+(aq) + OH^-(aq) \qquad K_{sp} = 1.5 \times 10^{-8}$

$1.5 \times 10^{-8} = [Ag^+][OH^-] = [5.89 \times 10^{-5}][OH^-]$

$[OH^-] = 2.5 \times 10^{-4}$ M

We need 5.89×10^{-3} mol + 2.5×10^{-4} mol = 6.14×10^{-3} mol to ensure that at equilibrium at least 99% of the Ag^+ has precipitated.

$(6.14 \times 10^{-3} \text{ mol})(40.00 \text{ g} \cdot \text{mol}^{-1}) = 0.246 \text{ g NaOH}$

(d) The equilibria that we need to satisfy simultaneously include

$NH_3(aq) + H_2O(l) \rightleftharpoons NH_4^+(aq) + OH^-(aq)$ $\quad 1.8 \times 10^{-5}$

$AgOH(s) \rightleftharpoons Ag^+(aq) + OH^-(aq)$ $\quad 1.5 \times 10^{-8}$

$Ag^+(aq) + 2\ NH_3(aq) \rightleftharpoons Ag(NH_3)_2^+(aq)$ $\quad 1.6 \times 10^{7}$

$2\ H_2O(l) \rightleftharpoons H_3O^+(aq) + OH^-(aq)$ $\quad 1.0 \times 10^{-14}$

We begin with the condition that all the silver must be dissolved and that 99% must be in the form of the complex ion.

$$[Ag(NH_3)_2^+] = 0.99 \times (5.89 \times 10^{-3}) = 5.83 \times 10^{-3}$$
$$[Ag^+] = 0.01 \times (5.89 \times 10^{-3}) = 5.89 \times 10^{-5}$$

We can calculate $[OH^-]$ from the K_{sp} relationship:

$$[5.89 \times 10^{-5}][OH^-] = 1.5 \times 10^{-8}$$
$$[OH^-] = 2.5 \times 10^{-4}$$

And $[H_3O^+]$ from K_w:

$$[H_3O^+][2.5 \times 10^{-4}] = 1.0 \times 10^{-14}$$
$$[H_3O^+] = 3.9 \times 10^{-11}$$

We can calculate $[NH_3]$ from the K_f relationship:

$$\frac{[Ag(NH_3)_2^+]}{[Ag^+][NH_3]^2} = 1.6 \times 10^{7}$$
$$[NH_3] = 2.5 \times 10^{-3}$$

And NH_4^+ from the K_b expression for NH_3:

$$\frac{[NH_4^+][OH^-]}{[NH_3]} = 1.8 \times 10^{-5}$$
$$[NH_4^+] = 1.8 \times 10^{-4}$$

These values satisfy all of the equilibrium constants involved reasonably well (within the significant figures).

$[Ag(NH_3)_2^+] = 5.83 \times 10^{-3}$
$[Ag^+] = 5.89 \times 10^{-5}$
$[NH_4^+] = 1.8 \times 10^{-4}$
$[NH_3] = 2.5 \times 10^{-3}$
$[OH^-] = 2.5 \times 10^{-4}$
$[H_3O^+] = 3.9 \times 10^{-11}$

The total amount of ammonia/ammonium that must be added will be $6 \cdot 5.83 \times 10^{-3}\ \text{mol} + 1.8 \times 10^{-4}\ \text{mol} + 2.5 \times 10^{-3}\ \text{mol} = 3.8 \times 10^{-2}\ \text{mol}$. This amount will be supplied by $(3.8 \times 10^{-2}\ \text{mol}) \div 15.8\ \text{mol} \cdot \text{L}^{-1} = 2.4 \times 10^{-3}\ \text{L}$ or 2.4 mL. This will be a negligible volume addition to the 1.00 L of solution.

11.80 (a) The reaction is

$CH_3CO_2^-(aq) + H_2O(l) \rightleftharpoons CH_3COOH(aq) + OH^-(aq)$.

Concentration $(\text{mol} \cdot \text{L}^{-1})$	$CH_3CO_2^-(aq)$	$+ H_2O(l) \rightleftharpoons$	$CH_3COOH(aq)$	$+ OH^-(aq)$
initial	0.026	—	0	0
change	$-x$	—	$+x$	$+x$
equilibrium	$0.026 - x$	—	x	x

$$K_b = \frac{K_w}{K_a} = \frac{[CH_3COOH][OH^-]}{[CH_3CO_2^-]} = \frac{1.0 \times 10^{-14}}{1.8 \times 10^{-5}} = \frac{x^2}{0.026 - x} \approx \frac{x^2}{0.026}$$

$K_b = 5.6 \times 10^{-10}$

$[OH^-] = x = \sqrt{1.44 \times 10^{-11}} = 3.8 \times 10^{-6}\ \text{mol} \cdot \text{L}^{-1}$

and $\text{pOH} = -\log(3.8 \times 10^{-6}) = 5.42$

$\text{pH} = 14.00 - 5.42 = 8.58$

(b) The reaction is

$CH_3COOH(aq) + H_2O(l) \rightleftharpoons CH_3CO_2^-(aq) + H_3O^+(aq)$.

Concentration

$(mol \cdot L^{-1})$	$CH_3COOH(aq)$	$+ H_2O(l) \rightleftharpoons$	$CH_3CO_2^-(aq)$	$+ H_3O^+(aq)$
initial	0.040	—	0	0
change	$-x$	—	$+x$	$+x$
equilibrium	$0.040 - x$	—	x	x

$$K_a = \frac{[CH_3CO_2^-][H_3O^+]}{[CH_3COOH]} = \frac{x^2}{0.040 - x} \approx \frac{x^2}{0.040} = 1.8 \times 10^{-5}$$

$$x = [H_3O^+] = \sqrt{0.040 \times 1.8 \times 10^{-5}} = 8.5 \times 10^{-4}$$

$$pH = -\log[H_3O^+] = -\log(8.5 \times 10^{-4}) = 3.1$$

(c) $(0.026\ mol \cdot L^{-1})(0.150\ L) = 3.9 \times 10^{-3}\ mol\ CH_3CO_2^-$

$$\text{molarity of } CH_3CO_2^- = \frac{3.9 \times 10^{-3}\ mol}{0.150\ L + 0.200\ L} = 1.1 \times 10^{-2}\ mol \cdot L^{-1}$$

moles of $CH_3COOH = (0.020\ mol \cdot L^{-1})(0.200\ L) = 4.0 \times 10^{-3}\ mol$

$$\text{molarity of } CH_3COOH = \frac{4.0 \times 10^{-3}\ mol}{0.150\ L + 0.200\ L} = 1.1 \times 10^{-2}\ mol \cdot L^{-1}$$

and, after mixing:

Concentration

$(mol \cdot L^{-1})$	$CH_3COOH(aq)$	$+ H_2O(l) \rightleftharpoons$	$CH_3CO_2^-(aq)$	$+ H_3O^+(aq)$
initial	1.1×10^{-2}	—	1.1×10^{-2}	0
change	$-x$	—	$+x$	$+x$
equilibrium	$1.1 \times 10^{-2} - x$	—	$1.1 \times 10^{-2} + x$	x

$$K_a = 1.8 \times 10^{-5} = \frac{[CH_3CO_2^-][H_3O^+]}{[CH_3COOH]} = \frac{(1.1 \times 10^{-2} + x)(x)}{1.1 \times 10^{-2} - x}$$

$$1.8 \times 10^{-5} = \frac{(1.1 \times 10^{-2})x}{1.1 \times 10^{-2}} = x$$

$$[H_3O^+] = x = 1.8 \times 10^{-5}\ mol \cdot L^{-1}$$

$$pH = -\log(1.8 \times 10^{-5}) = 4.74$$

11.82 The concentration of acid in the bottle of trichloroacetic acid buffer is found by:

$$\text{pH} = \text{p}K_a - \log\frac{[\text{CCl}_3\text{COOH}]}{[\text{CCl}_3\text{COO}^-]}$$

$$2.95 = 0.52 - \log\frac{[\text{CCl}_3\text{COOH}]}{0.200\ \text{M}}$$

$$[\text{CCl}_3\text{COOH}] = 7.43\times10^{-4}\ \text{M}$$

To increase the pH of the buffer to 3.00, more conjugate base will need to be added. The concentration of conjugate base needed to reach a pH of 3.00 is given by:

$$3.00 = 0.52 - \log\frac{7.43\times10^{-4}\ \text{M}}{[\text{CCl}_3\text{COO}^-]}$$

$$[\text{CCl}_3\text{COO}^-] = 0.224\ \text{M}$$

So the concentration of the conjugate base needs to be increased by 0.024 M by adding:

$$(0.024\ \text{M})(0.150\ \text{L}) = 3.6\times10^{-3}\ \text{mol}$$

or

$$(3.6\times10^{-3}\ \text{mol})(185.37\ \text{g}\cdot\text{mol}^{-1}) = 0.67\ \text{g of sodium trichloroacetate}$$

11.84 (a) $M_a = 0.0329$, $M_b = 0.0257$, $V_a = 20.0$, $V_b = 0.0$ to 50.0

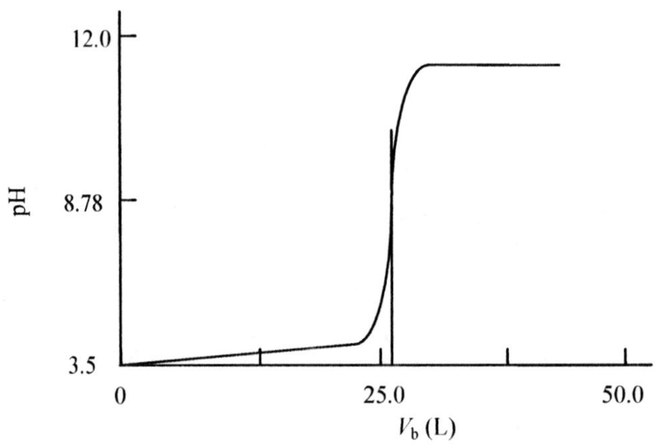

(b) The volume of titrant is calculated from v, which is v_b/v_a where v_a is the volume of the acid and v_b is the volume of the titrant. At the equivalence point $v = 1.28$. $v_a = 20.0$ mL, so $v_b = 1.28 \times 20.0 = 25.6$ mL

(c) 3.50

(d) 8.78

11.86 $[\text{NaOH}] = (0.0118 \text{ L HCl})\left(\frac{2.05 \text{ mol HCl}}{1 \text{ L HCl}}\right)$

$$\times\left(\frac{1 \text{ mol NaOH}}{1 \text{ mol HCl}}\right)\left(\frac{1}{0.0050 \text{ L NaOH}}\right)$$

$$= 4.8 \text{ mol} \cdot \text{L}^{-1}$$

11.88 (a) The acidity constant may be found using:

$$K_a = e^{-\Delta G_r / R\cdot T}$$

$$\Delta G_r = -269.2 \text{ kJ} \cdot \text{mol}^{-1} - (-293.1 \text{ kJ} \cdot \text{mol}^{-1}) = 23.9 \text{ kJ} \cdot \text{mol}^{-1}$$

$$= 23900 \text{ J} \cdot \text{mol}^{-1}$$

$$K_a = e^{-23900 \text{ J}\cdot\text{mol}^{-1} / (8.3145 \text{ J}\cdot\text{K}^{-1}\cdot\text{mol}^{-1})(298 \text{ K})} = 6.47 \times 10^{-5}$$

(b) Given the acidity constant above, the mass of sodium acetate needed is:

$$pK_a = 10^{-K_a} = 4.76$$

$$\text{p}H = \text{p}K_a + \log\frac{[\text{A}^-]}{[\text{HA}]} = 4.0 = 4.19 + \log\frac{[\text{A}^-]}{[\text{HA}]}$$

$$\frac{[\text{A}^-]}{[\text{HA}]} = 10^{4.0-4.19} = 0.646$$

Given $[\text{HA}] = 1.4 \text{ M}$, $[\text{A}^-] = 0.646 \times 1.4 \text{ M} = 0.904 \text{ M}$.

and $(0.904 \text{ M})(2.0 \text{ L})(144.11 \text{ g} \cdot \text{mol}^{-1}) = 261 \text{ g}$

11.90 $C_6H_5COOH(aq) + H_2O(l) \rightleftharpoons H_3O^+(aq) + C_6H_5CO_2^-(aq)$

$$K_\text{a} = \frac{[H_3O^+][C_6H_5CO_2^-]}{[C_6H_5COOH]}$$

$$\text{p}K_\text{a} = \text{pH} - \log\frac{[C_6H_5CO_2^-]}{[C_6H_5COOH]}$$

$$\text{pH} = \text{p}K_\text{a} + \log\frac{[C_6H_5CO_2^-]}{[C_6H_5COOH]}$$

$$\text{(a)}\quad \text{pH} = K_a + \log\left\{\frac{\left[\dfrac{(0.032\ \text{mol}\cdot\text{L}^{-1})(0.0200\ \text{L})}{0.0700\ \text{L}}\right]}{\left[\dfrac{(0.022\ \text{mol}\cdot\text{L}^{-1})(0.0500\ \text{L})}{0.0700\ \text{L}}\right]}\right\}$$

$$\text{pH} = 4.19 + \log\frac{9.1\times10^{-3}}{1.6\times10^{-2}}$$

$\text{pH} = 4.19 - 0.24 = 3.95$ (initial pH)

(b) $0.054\ \text{mmol HCl} = 5.4\times10^{-5}$ mol HCl (a strong acid), producing

5.4×10^{-5} mol C_6H_5COOH from $C_6H_5CO_2^-$ (assume no volume change)

after adding HCl:

$$[C_6H_5COOH] = 1.6\times10^{-2} + \frac{5.4\times10^{-5}\ \text{mol}}{0.0700\ \text{L}} = 1.6\times10^{-2} + 7.7\times10^{-4}$$

$$\approx 1.7\times10^{-2}\ \text{mol}\cdot\text{L}^{-1}$$

$$[C_6H_5CO_2^-] = 9.1\times10^{-3} - \frac{5.4\times10^{-5}\ \text{mol}}{0.0700\ \text{L}} = (9.1\times10^{-3}) - (7.7\times10^{-4})$$

$$= 8.3\times10^{-3}\ \text{mol}\cdot\text{L}^{-1}$$

and $\text{pH} = 4.19 + \log\dfrac{8.3\times10^{-3}}{1.7\times10^{-2}}$

$\text{pH} = 4.19 - 0.31 = 3.88$

$\Delta\ \text{pH} = 3.88 - 3.95 = -0.07$, that is, a decrease of 0.07 pH units

$$\text{(c)}\quad [H_3O^+] = \frac{5.4\times10^{-5}\ \text{mol}}{0.070\ \text{L}} = 7.7\times10^{-4}\ \text{mol}\cdot\text{L}^{-1}$$

$\text{pH} = -\log(7.7\times10^{-4}) = 3.11$

$\Delta\ \text{pH} = 7.00 - 3.11 = -3.89$, that is, a decrease of 3.89 pH units

(d) $(0.0100\ \text{L})(0.054\ \text{mol}\cdot\text{L}^{-1}) = 5.4\times10^{-4}$ mol HCl (a strong acid),

producing 5.4×10^{-4} mol C_6H_5COOH from $C_6H_5CO_2^-$ after adding HCl:

$$\text{molarity of } C_6H_5COOH = \frac{(0.022\ \text{mol}\cdot\text{L}^{-1})(0.0500\ \text{L})}{0.0800\ \text{L}} + \frac{5.4\times10^{-4}\ \text{mol}}{0.0800\ \text{L}}$$

$$= 2.0\times10^{-2}\ \text{mol}\cdot\text{L}^{-1}$$

$$\text{molarity of } C_6H_5CO_2^- = \frac{(0.032\ \text{mol}\cdot\text{L}^{-1})(0.0200\ \text{L})}{0.0800\ \text{L}} - \frac{5.4\times10^{-4}\ \text{mol}}{0.0800\ \text{L}}$$

$$= 1.3\times10^{-2}\ \text{mol}\cdot\text{L}^{-1}$$

$$\text{pH} = 4.19 + \log\frac{1.3\times10^{-3}}{2.0\times10^{-2}}$$

$$\text{pH} = 4.19 - 1.19 = 3.00$$

Δ pH = [3.00 – 3.95], that is, a 0.95 decrease

11.92 $H_2PO_4^-(aq) + H_2O(l) \rightleftharpoons H_3O^+(aq) + HPO_4^{2-}(aq)$

$$K_{a2} = \frac{[H_3O^+][HPO_4^{2-}]}{[H_2PO_4^-]}$$

$$\text{pH} = \text{p}K_{a2} + \log\frac{[HPO_4^{2-}]}{[H_2PO_4^-]}\ \text{(see Exercise 11.12)}$$

$$7.40 = 7.21 + \log x$$

$$\log x = 0.19$$

$$x = 1.5 = \frac{[HPO_4^{2-}]}{[H_2PO_4^-]}$$

$$\text{molarity of } NaH_2PO_4 = 0.10\ \text{mol}\cdot\text{L}^{-1}$$

$$\text{molarity of } Na_2HPO_4 = 1.5\times0.10\ \text{mol}\cdot\text{L}^{-1}$$

$$= 0.15\ \text{mol}\cdot\text{L}^{-1}\text{, or } 0.075\ \text{mol/500 mL}$$

(assuming no volume change),

$$\text{mass of } Na_2HPO_4 = 0.075\ \text{mol } Na_2HPO_4 \times \frac{141.96\ \text{g } Na_2HPO_4}{1\ \text{mol } Na_2HPO_4}$$

$$= 11\ \text{g } Na_2HPO_4$$

11.94 (a) $Ca_5(PO_4)_3(OH)(s) \rightleftharpoons 5\ Ca^{2+}(aq) + 3\ PO_4^{3-}(aq) + OH^-(aq)$

Concentration ($\text{mol}\cdot\text{L}^{-1}$)	Ca^{2+}	PO_4^{3-}	OH^-
initial	0	0	0
change	$+5S$	$+3S$	$+S$
equilibrium	$5S$	$3S$	S

$K_{sp} = [Ca^{2+}]^5[PO_4^{3-}]^3[OH^-]$

$1.0 \times 10^{-36} = (5S)^5 \times (3S)^3 \times (S) = 84\,375\ S^9$

$S = 2.8 \times 10^{-5}\ mol \cdot L^{-1}$ = molar solubility of $Ca_5(PO_4)_3OH$

$Ca_5(PO_4)_3(F)(s)\ \Lambda\ 5\ Ca^{2+}(aq) + 3\ PO_4^{3-}(aq) + F^-(aq)$

Concentration ($mol \cdot L^{-1}$)	Ca^{2+}	PO_4^{3-}	F^-
initial	0	0	0
change	$+5S$	$+3S$	$+S$
equilibrium	$5S$	$3S$	S

$K_{sp} = [Ca^{2+}]^5[PO_4^{3-}]^3[F^-]$

$1.0 \times 10^{-60} = (5S)^5 \times (3S)^3 \times (S) = 84\,375\ S^9$

$S = 6.1 \times 10^{-8}\ mol \cdot L^{-1} = [F^-]$ = molar solubility of $Ca_5(PO_4)_3F$

(b) The overall equation for which we wish to find the free energy change is

$Ca_5(PO_4)_3OH(s) + F^-(aq) \rightleftharpoons Ca_5(PO_4)_3F(s) + OH^-(aq)$ (1)

It can be obtained by combining the two equations given:

$Ca_5(PO_4)_3(OH)(s) \rightleftharpoons 5\ Ca^{2+}(aq) + 3\ PO_4^{3-}(aq) + OH^-(aq)$

$K = 1.0 \times 10^{-36}$ (2)

$Ca_5(PO_4)_3F(s) \rightleftharpoons 5\ Ca^{2+}(aq) + 3\ PO_4^{3-}(aq) + F^-(aq)$

$K = 1.0 \times 10^{-60}$ (3)

Equation (1) = Equation (2) – Equation (3)

$K_1 = K_2/K_3 = (1.0 \times 10^{-36})/(1.0 \times 10^{-60}) = 1.0 \times 10^{24}$

$\Delta G° = -RT \ln K = -(8.314\ J \cdot K^{-1} \cdot mol^{-1})(298\ K) \ln (1.0 \times 10^{24})$

$= -1.4 \times 10^2\ kJ \cdot mol^{-1}$

11.96 (a)

molarity of $ZnCl_2$ = molarity of Zn^{2+}

$$= (1.36\ mg\ ZnCl_2)\left(\frac{10^{-3}\ g}{1\ mg}\right)\left(\frac{1\ mol\ ZnCl_2}{136.27\ g}\right)\left(\frac{1}{0.500\ L}\right)$$

$$= 2.00 \times 10^{-5}\ M\ ZnCl_2$$

pH = 8.00; pOH = 6.00; molarity of $OH^- = 1.00 \times 10^{-6}$ M

$Zn(OH)_2(s) \rightleftharpoons Zn^{2+}(aq) + 2\ OH^-(aq)$

$[Zn^{2+}][OH^-]^2 = Q_{sp}$

$(2.00 \times 10^{-5})(1.0 \times 10^{-6})^2 = 2.0 \times 10^{-17} = Q_{sp}$

Because $Q_{sp}(2.0 \times 10^{-17})$ equals $K_{sp}(2.0 \times 10^{-17})$, no precipitate forms (just barely!)

(b) Because the concentration of hydroxide is higher for (b), we might expect that there would be a precipitate. The corresponding calculation at this pH is

$[OH^-] = 1.0 \times 10^{-2}$ M

$[Zn^{2+}][OH^-]^2 = (2.00 \times 10^{-5})(1.0 \times 10^{-2})^2 = 2.0 \times 10^{-9} = Q_{sp}$

This number exceeds K_{sp} and so we would expect a precipitate to form. The amount of zinc present in solution can be obtained from

$[Zn^{2+}][1.0 \times 10^{-2}]^2 = 2.0 \times 10^{-17}$

$[Zn^{2+}] = 2.0 \times 10^{-13}$ M

However, we must also consider the ensuring formation of the complex ion as additional hydroxide ion is added:

$Zn(OH)_2(aq) + 2\ OH^-(aq) \rightleftharpoons Zn(OH)_4^-(aq)$

The equilibrium constant for this reaction can be obtained by combining the two equilibrium constants for the given reactions:

$Zn(OH)_2(s) \rightleftharpoons Zn^{2+}(aq) + 2\ OH^-(aq) \qquad K_{sp} = 2.0 \times 10^{-17}$

$Zn^{2+}(aq) + 4\ OH^-(aq) \rightleftharpoons Zn(OH)_4^{2-}(aq) \quad K_f = 5 \times 10^{14}$

Adding these two equations gives the desired equilibrium with $K = (2.0 \times 10^{-17})(5 \times 10^{14}) = 1 \times 10^{-2}$.

$$\frac{[Zn(OH)_4^{2-}]}{[OH^-]^2} = 1 \times 10^{-2}$$

$$\frac{[Zn(OH)_4^{2-}]}{(1.0 \times 10^{-2})^2} = 1 \times 10^{-2}$$

$[Zn(OH)_4^{2-}] = 1 \times 10^{-6}$

This represents only about 5% of the zinc present, so the bulk of the zinc will remain as a precipitate.

(c) We can use the relationship from part (b) to determine the concentration of hydroxide ion required to complex with all of the zinc ions ($2.0 \times 10^{-5}\ \text{mol} \cdot \text{L}^{-1}$).

$$\frac{[\text{Zn(OH)}_4{}^{2-}]}{[\text{OH}^-]^2} = 1 \times 10^{-2}$$

$$\frac{2.0 \times 10^{-5}}{[\text{OH}^-]^2} = 1 \times 10^{-2}$$

$$[\text{OH}^-] = 0.04$$

$$\text{pH} = 12.6$$

11.98 In addition to the reaction corresponding to the dissolution of $BaF_2(s)$:

$BaF_2(s) \rightleftharpoons Ba^{2+}(aq) + 2\ F^-(aq) \qquad K = 1.7 \times 10^{-6}$

The buffer will provide a source of $H_3O^+(aq)$ ions that will allow the reaction:

$H_3O^+(aq) + F^- \rightleftharpoons HF(aq) + H_2O(l) \qquad K = 1/(K_a(HF)) = 1/(3.5 \times 10^{-4}) = 2.86 \times 10^3$

These two coupled reactions give two equilibrium expressions, which must be simultaneously satisfied:

$$[\text{F}^-]^2[\text{Ba}^{2+}] = 1.7 \times 10^{-6} \quad \text{and} \quad \frac{[\text{HF}]}{[\text{H}_3\text{O}^+][\text{F}^-]} = 2.86 \times 10^3$$

Given that all fluoride ions come from $BaF_2(s)$ and wind up as either $F^-(aq)$ or $HF(aq)$, and that for every one mole of $Ba^{2+}(aq)$ generated two moles of $F^-(aq)$ are also produced, we can write a third equation which relates the concentration of the fluoride containing species to the concentration of dissolved barium: $[Ba^{2+}] = ½([F^-] + [HF])$.

In the end, the concentration of $Ba^{2+}(aq)$ will be equal to the solubility of $BaF_2(s)$. To determine the equilibrium concentration of $Ba^{2+}(aq)$, we first determine $[H_3O^+]$, which is fixed by the buffer system, and then use the three simultaneous equations above to solve for $[Ba^{2+}]_{eq}$.

The buffer determines the equilibrium concentration of $H_3O^+(aq)$. The initial concentration of $H_3O^+(aq)$ and $NaCH_3CO_2(aq)$ are:

$$[H_3O^+]_i = \frac{(0.055\ L)(0.15\ mol \cdot L^{-1})}{0.10\ L} = 0.0825\ M \quad \text{and}$$

$$[CH_3CO_2^-]_i = \frac{(0.045\ L)(0.65\ mol \cdot L^{-1})}{0.10\ L} = 0.293\ M.$$

To determine their equilibrium concentrations we solve using the familiar method:

Concentration $(mol \cdot L^{-1})$	$CH_3COOH(aq)$ +	$H_2O(l)$ $\rightleftharpoons$	$H_3O^+(aq)$ +	$CH_3CO_2^-(aq)$
initial	0	—	0.0825	0.293
change	$+x$	—	$-x$	$-x$
equilibrium	x	—	$0.0825 - x$	$0.292 - x$

$$K_a = 1.8 \times 10^{-5} = \frac{[H_3O^+][CH_3CO_2^-]}{[CH_3COOH]} = \frac{(0.0825 - x)(0.292 - x)}{(x)}$$

Rearranging this expression we obtain:

$0.024072 - 0.3745\,x + x^2 = 0.$

Using the quadratic formula we find $x = 0.082493$, and $[H_3O^+]$ $= 7.1 \times 10^{-6}$ M

With this equilibrium concentration of $H_3O^+(aq)$ we revisit the three simultaneous equations from above, namely

$[F^-]^2[Ba] = 1.7 \times 10^{-6}$;

$[Ba^{2+}] = \frac{1}{2}([F^-] + [HF])$; and

$$\frac{[HF]}{[H_3O^+][F^-]} = 2.86 \times 10^3.$$

Due to the presence of the buffer, $[H_3O^+] = 7.1 \times 10^{-6}$ and this last equation simplifies to

$$\frac{[HF]}{[F^-]} = 2.03 \times 10^{-2}.$$

Rearranging these three simultaneous equations we find:

$$[\mathrm{Ba}^{2+}] = \frac{1.7\times10^{-6}}{[\mathrm{F}^-]^2}, \quad [\mathrm{HF}] = [\mathrm{F}^-]\times\left(2.03\times10^{-2}\right), \quad \text{and}$$

$$[\mathrm{Ba}^{2+}] = \tfrac{1}{2}\left([\mathrm{F}^-]+[\mathrm{HF}]\right)$$

$$= \frac{1.7\times10^{-6}}{[\mathrm{F}^-]^2} = \tfrac{1}{2}\left[[\mathrm{F}^-]+\left([\mathrm{F}^-]\times\left(2.03\times10^{-2}\right)\right)\right].$$

Solving this expression for $[\mathrm{F}^-]$:

$$[\mathrm{F}^-] = \sqrt[3]{\frac{3.4\times10^{-6}}{1.0203}} = 1.49\times10^{-2}.$$

The equilibrium concentration of Ba^{2+} (aq) is then:

$$[\mathrm{Ba}^{2+}] = \frac{1.7\times10^{-6}}{[\mathrm{F}^-]^2} = \frac{1.7\times10^{-6}}{(1.49\times10^{-2})^2} = 7.62\times10^{-3}\ \mathrm{M}$$

Therefore, the solubility of BaF_2 (s) is 7.6×10^{-3} M

11.100 The K_{sp} values are 1.3×10^{-36} for CuS and 1.3×10^{-15} for MnS. Because the form of the K_{sp} reaction is essentially the same for both compounds, the numbers are comparable directly. MnS is more soluble than CuS and so will remain in solution as the CuS precipitates. CuS will begin to precipitate at $[\mathrm{S}^{2-}] = 6.5\times10^{-36}\ \mathrm{mol\cdot L^{-1}}$ as calculated from

$$K_{sp} = [\mathrm{Cu}^{2+}][\mathrm{S}^{2-}] = 1.3\times10^{-36}$$

$$0.20[\mathrm{S}^{2-}] = 1.3\times10^{-36}$$

$$[\mathrm{S}^{2-}] = 6.5\times10^{-36}\ \mathrm{mol\cdot L^{-1}}$$

11.102 $\mathrm{HgS(s)} \rightleftharpoons \mathrm{Hg}^{2+}\mathrm{(aq)} + \mathrm{S}^{2-}\mathrm{(aq)}$

$$K_{sp} = [\mathrm{Hg}^{2+}][S^{2-}] = S^2 = 1.6\times10^{-52}$$

$$S = \sqrt{1.6\times10^{-52}} = 1.3\times10^{-26}\ \mathrm{mol\cdot L^{-1}} = [\mathrm{Hg}^{2+}]$$

and $$(1.3\times10^{-26}\ \mathrm{mol\cdot L^{-1}})\left(\frac{6.022\times10^{23}\ \text{ions}}{1\ \text{mol}}\right) = \frac{7.8\times10^{-3}\ \text{ions}}{1\ \text{L}}$$

The reciprocal gives the number of liters per ion $= 1.3\times10^{2}$ L/ion

11.104 (a) $ZnS(s) + 2\ H_2O(l) \rightleftharpoons Zn^{2+}(aq) + H_2S(aq) + 2\ OH^-(aq)$

$$K = K_{sp} \cdot K_{b1} \cdot K_{b2} = K_{sp} \cdot \frac{K_w}{K_{a2}} \cdot \frac{K_w}{K_{a1}}$$

$$= 1.6 \times 10^{-24} \cdot \frac{1.00 \times 10^{-14}}{7.1 \times 10^{-15}} \cdot \frac{1.00 \times 10^{-14}}{1.3 \times 10^{-7}} = 1.7 \times 10^{-31}$$

(b) $[Zn^{2+}][H_2S][OH^-]^2 = 1.7 \times 10^{-31}$

at $pH = 7.00, [OH^-] = 1.0 \times 10^{-7}$

$[Zn^{2+}](0.10)(1.0 \times 10^{-7})^2 = 1.7 \times 10^{-31}$

$[Zn^{2+}] = 1.7 \times 10^{-16}\ mol \cdot L^{-1}$

(c) at $pH = 10.00, [OH^-] = 1.0 \times 10^{-4}$

$[Zn^{2+}](0.10)(1.0 \times 10^{-4})^2 = 1.7 \times 10^{-31}$

$[Zn^{2+}] = 1.7 \times 10^{-22}\ mol \cdot L^{-1}$

11.106 (a) Both $Cu(IO_3)_2$ and $Pb(IO_3)_2$ are insoluble compounds, but a small amount of dissolution to give the M^{2+} and IO_3^- ions in solution should occur. The question then becomes one of determining where the equilibrium lies between Cu^{2+} ions, $Cu(IO_3)_2(s)$, Pb^{2+} ions, and $Pb(IO_3)_2(s)$ in this system. The relevant equation is

$Cu(IO_3)_2(s) + Pb^{2+}(aq) \rightleftharpoons Cu^{2+}(aq) + Pb(IO_3)_2(s)$

This information can be obtained from the K_{sp} values:

$Cu(IO_3)_2(s) \rightleftharpoons Cu^{2+}(aq) + 2\ IO_3^-(aq) \qquad K_{sp} = 1.4 \times 10^{-7}$

$Pb(IO_3)_2(s) \rightleftharpoons Pb^{2+}(aq) + 2\ IO_3^-(aq) \qquad K_{sp} = 2.6 \times 10^{-13}$

The first reaction minus the second is the process we want, giving a K value for the overall reaction of

$$K = \frac{1.4 \times 10^{-7}}{2.6 \times 10^{-13}} = 5.4 \times 10^5$$

The form of this equilibrium expression is $K = \frac{[Cu^{2+}]}{[Pb^{2+}]}$. Because the ratio of Cu^{2+} ions to Pb^{2+} ions is 5.4×10^5, this will represent quantitative conversion of the copper iodate to the lead iodate (for 99.99% conversion, the ratio would be $99.99 \div 0.01 = 1 \times 10^4$). Although the thermodynamic analysis indicates that the replacement will take place, it does not tell us how fast it will occur.

(b) The IO_3^- concentration will be determined by the $Pb(IO_3)_2$ dissolution equilibrium because that is the least soluble species:

$$Pb(IO_3)_2(s) \rightleftharpoons Pb^{2+}(aq) + 2\ IO_3^-(aq)$$

$$+x \qquad +2x$$

$$K_{sp} = 2.6 \times 10^{-13} = [Pb^{2+}][IO_3^-]^2 = x(2x)^2$$

$$4x^3 = 2.6 \times 10^{-13}$$

$$x = 4.0 \times 10^{-5}$$

The IO_3^- concentration is $2x = 2(4.0 \times 10^{-5}) = 8.0 \times 10^{-5}\ mol \cdot L^{-1}$.

(c) The free energy of the displacement reaction is calculated from the K of the reaction determined in (a).

$$\Delta G° = -RT \ln K$$

$$= -(8.314\ J \cdot K^{-1} \cdot mol^{-1})(298.2\ K)\ln(5.4 \times 10^5)$$

$$= -32.72\ kJ \cdot mol^{-1}$$

11.108 The dissolution reaction of interest is:

$$AgI(s) \rightleftharpoons Ag^+(aq) + I^-(aq)\ (K_{sp} = 8 \times 10^{-17})$$

The change in the equilibrium constant of this reaction with temperature can be predicted using: $\ln\left(\frac{K_2}{K_1}\right) = \frac{\Delta H_r^o}{R}\left(\frac{1}{T_1} - \frac{1}{T_2}\right)$. The standard enthalpy for the dissolution reaction can be found using the table of thermodynamic values in the back of the text:

$$\Delta H_r^o = 50.38\ kJ \cdot K^{-1} \cdot mol^{-1} - (-61.84\ kJ \cdot K^{-1} \cdot mol^{-1})$$

$$= 112.22\ kJ \cdot K^{-1} \cdot mol^{-1}$$

The solubility constant at 85°C is then found:

$$\ln\left(\frac{K_2}{8\times10^{-17}}\right)=\frac{112,220\ \text{J}\cdot\text{mol}^{-1}}{8.3145\ \text{J}\cdot\text{K}^{-1}\cdot\text{mol}^{-1}}\left(\frac{1}{298\ \text{K}}-\frac{1}{358\ \text{K}}\right)$$

$$K_2=1.6\times10^{-13}$$

11.110 The two equations of interest are:

$H_3O^+(aq) + HCO_3^-(aq) \rightleftharpoons H_2CO_3(aq) + H_2O(l)$ (1)

$H_3O^+(aq) + HCO_3^-(aq) \rightleftharpoons 2\ H_2O(l) + CO_2(g)$ (2)

(a) During exercise, an increase in the amount of carbon dioxide in the muscle will shift reaction (2) above toward reactants, increasing the amount of H_3O^+ (aq) thereby decreasing the pH. (b) and (c) During hyperventilation, CO_2 (g) is expelled from the body at a rapid rate, shifting reaction (2) toward products and increasing the pH of the blood. The resulting condition is known as respiratory alkalosis and can be remedied by breathing into a paper bag, allowing some of the expelled CO_2 (g) to be reabsorbed into the blood.

CHAPTER 12

ELECTROCHEMISTRY

12.2 (a)

$$^{-}O_3S-O-O-SO_3^{-}(aq) \quad + \quad Cr^{3+}(aq) \quad \rightarrow \quad HSO_4^{-}(aq) \quad + \quad Cr_2O_7^{-2}(aq)$$

$^{-}O_3S-O-O-SO_3^{-}(aq)$	$Cr^{3+}(aq)$	$HSO_4^{-}(aq)$	$Cr_2O_7^{-2}(aq)$
↑	↑	↑	↑
+7	+3	+6	+6

(b) $7\,H_2O(l) + 2\,Cr^{3+}(aq) \rightarrow Cr_2O_7^{2-}(aq) + 6\,e^{-} + 14\,H^{+}(aq)$

(c) $2\,H^{+}(aq) + 2\,e^{-} + S_2O_8^{2-}(aq) \rightarrow 2\,HSO_4^{-}(aq)$

(d) $$7\,H_2O(l) + 2\,Cr^{3+}(aq) + 3S_2O_8^{2-}(aq)$$
$$\rightarrow 6\,HSO_4^{-}(aq) + Cr_2O_7^{2-}(aq) + 8\,H^{+}(aq)$$

12.4 (a) $6[Fe^{2+}(aq) \rightarrow Fe^{3+}(aq) + e^{-}]$

$1[Cr_2O_7^{\ 2-}(aq) + 14\ H^{+}\ (aq) + 6\ e^{-} \rightarrow 2\ Cr^{3+}(aq) + 7\ H_2O(l)]$

$$6\ Fe^{2+}(aq) + Cr_2O_7^{\ 2-}(aq) + 14\ H^{+}(aq) + 6\ e^{-} \rightarrow$$
$$6\ Fe^{3+}\ (aq) + 6\ e^{-} + 2\ Cr^{3+}(aq) + 7\ H_2O(l)$$

$$6\ Fe^{2+}(aq) + Cr_2O_7^{\ 2-}(aq) + 14\ H^{+}(aq) \rightarrow$$
$$6\ Fe^{3+}(aq) + 2\ Cr^{3+}(aq) + 7\ H_2O(l)$$

Fe^{2+} is the reducing agent and $Cr_2O_7^{\ 2-}$ is the oxidizing agent.

(b) $5\,[C_2H_5OH(aq) + H_2O(l) \rightarrow CH_3COOH(aq) + 4\ H^{+}(aq) + 4\ e^{-}]$

$4\,[MnO_4^{\ -}(aq) + 8\ H^{+}(aq) + 5\ e^{-} \rightarrow Mn^{2+}(aq) + 4\ H_2O(l)]$

$$5\ C_2H_5OH(aq) + 5\ H_2O(l) + 4\ MnO_4^{\ -}(aq) + 32\ H^{+}\ (aq) + 20\ e^{-} \rightarrow$$
$$5\ CH_3COOH(aq) + 20\ H^{+}(aq) + 20\ e^{-} + 4\ Mn^{2+}(aq) + 16\ H_2O(l)$$

$$5\,C_2H_5OH(aq) + 4\,MnO_4^-(aq) + 12\,H^+(aq) \rightarrow$$
$$5\,CH_3COOH(aq) + 4\,Mn^{2+}(aq) + 11\,H_2O(l)$$

C_2H_5OH is the reducing agent and MnO_4^- is the oxidizing agent.

(c) The reaction is $I^-(aq) + NO_3^-(aq) \rightarrow I_2(aq) + NO(g)$.

$$3[2\,I^-(aq) \rightarrow I_2(aq) + 2\,e^-]$$

$$2\,[NO_3^-(aq) + 4\,H^+(aq) + 3\,e^- \rightarrow NO(g) + 2\,H_2O(l)]$$

$$6\,I^-(aq) + 2\,NO_3^-(aq) + 8\,H^+(aq) + 6\,e^-$$
$$\rightarrow 3\,I_2(aq) + 6\,e^- + 2\,NO(g) + 4\,H_2O(l)$$

$$6\,I^-(aq) + 2\,NO_3^-(aq) + 8\,H^+(aq) \rightarrow 3\,I_2(aq) + 2\,NO(g) + 4\,H_2O(l)$$

I^- is the reducing agent and NO_3^- is the oxidizing agent.

(d) $3[As_2S_3(s) + 8\,H_2O(l) \rightarrow 2\,H_3AsO_4(aq) + 3\,S(s) + 10\,H^+(aq) + 10\,e^-]$

$$10\,[NO_3^-(aq) + 4\,H^+(aq) + 3\,e^- \rightarrow NO(g) + 2\,H_2O(l)]$$

$$3\,As_2S_3(s) + 10\,NO_3^-(aq) + 24\,H_2O(l) + 40\,H^+(aq) + 30\,e^- \rightarrow$$
$$6\,H_3AsO_4(aq) + 9\,S(s) + 10\,NO(g) + 20\,H_2O(l) + 30\,H^+(aq) + 30\,e^-$$

$$3\,As_2S_3(s) + 10\,NO_3^-(aq) + 4\,H_2O(l) + 10\,H^+(aq) \rightarrow$$
$$6\,H_3AsO_4(aq) + 9\,S(s) + 10\,NO(g)$$

As_2S_3 is the reducing agent (both As and S are oxidized) and NO_3^- is the oxidizing agent.

12.6 (a) $Cl_2O_7(g) \rightarrow 2\,ClO_2^-(aq) + 3\,H_2O(l)$ (O's balanced)

$Cl_2O_7(g) + 6\,H_2O(l) \rightarrow 2\,ClO_2^-(aq) + 3\,H_2O(l) + 6\,OH^-(aq)$ (H's balanced)

$Cl_2O_7(g) + 3\,H_2O(l) + 8\,e^- \rightarrow 2\,ClO_2^-(aq) + 6\,OH^-(aq)$ (charge balanced)

$H_2O_2(aq) \rightarrow O_2(g)$ (O's balanced)

$H_2O_2(aq) + 2\,OH^-(aq) \rightarrow O_2(g) + 2\,H_2O(l)$ (H's balanced)

$H_2O_2(aq) + 2\,OH^-(aq) \rightarrow O_2(g) + 2\,H_2O(l) + 2e^-$ (charge balanced)

Combining the two half-reactions:

$1[Cl_2O_7(g) + 3\ H_2O(l) + 8\ e^- \rightarrow 2\ ClO_2^-(aq) + 6\ OH^-(aq)]$

$4[H_2O_2(aq) + 2\ OH^-(aq) \rightarrow O_2(g) + 2\ H_2O(l) + 2\ e^-]$

$Cl_2O_7(g) + 3\ H_2O(l) + 8\ e^- + 4\ H_2O_2(aq) + 8\ OH^-(aq) \rightarrow$
$2\ ClO_2^-(aq) + 6\ OH^-(aq) + 4\ O_2(g) + 8\ H_2O(l) + 8\ e^-$

$Cl_2O_7(g) + 4\ H_2O_2 + 2\ OH^-(aq) \rightarrow 2\ ClO_2^-(aq) + 4\ O_2(g) + 5\ H_2O(l)$

Cl_2O_7 is the oxidizing agent and H_2O_2 is the reducing agent.

(b) $MnO_4^-(aq) \rightarrow MnO_2(s) + 2\ H_2O(l)$ (O's balanced)

$MnO_4^-(aq) + 4\ H_2O(l)$
$\rightarrow MnO_2(s) + 2\ H_2O(l) + 4\ OH^-(aq)$ (H's balanced)

$MnO_4^-(aq) + 2\ H_2O(l) + 3\ e^- \rightarrow$
$MnO_2(s) + 4\ OH^-(aq)$ (charge balanced)

$S^{2-}(aq) \rightarrow S(s) + 2\ e^-$

Combining the two half-reactions:

$2[MnO_4^-(aq) + 2\ H_2O(l) + 3\ e^- \rightarrow MnO_2(s) + 4\ OH^-(aq)]$

$3[S^{2-}(aq) \rightarrow S(s) + 2\ e^-]$

$2\ MnO_4^-(aq) + 4\ H_2O(l) + 6\ e^- + 3\ S^{2-}(aq) \rightarrow$
$2\ MnO_2(s) + 8\ OH^-(aq) + 3\ S(s) + 6\ e^-$

$2\ MnO_4^-(aq) + 4\ H_2O(l) + 3\ S^{2-}(aq) \rightarrow 2\ MnO_2(s) + 8\ OH^-(aq) + 3\ S(s)$

MnO_4^- is the oxidizing agent and S^{2-} is the reducing agent.

(c) $N_2H_4(g) + 2\ H_2O(l) \rightarrow 2\ NO(g)$ (O's balanced)

$N_2H_4(g) + 2\ H_2O(l) + 8\ OH^-(aq) \rightarrow 2\ NO(g) + 8\ H_2O(l)$ (H's balanced)

$N_2H_4(g) + 8\ OH^-(aq) \rightarrow 2\ NO(g) + 6\ H_2O(l) + 8\ e^-$ (charge balanced)

$ClO_3^-(aq) \rightarrow Cl^-(aq) + 3\ H_2O(l)$ (O's balanced)

$ClO_3^-(aq) + 6\ H_2O(l) \rightarrow Cl^-(aq) + 3\ H_2O(l) + 6\ OH^-(aq)$ (H's balanced)

$ClO_3^-(aq) + 3\ H_2O(l) + 6\ e^- \rightarrow Cl^-(aq) + 6\ OH^-(aq)$ (charge balanced)

Combining the two half-reactions:

$$3[N_2H_4(g) + 8\ OH^-(aq) \rightarrow 2\ NO(g) + 6\ H_2O(l) + 8\ e^-]$$

$$4[ClO_3^-(aq) + 3\ H_2O(l) + 6\ e^- \rightarrow Cl^-(aq) + 6\ OH^-(aq)]$$

$$3\ N_2H_4(g) + 24\ OH^-(aq) + 4\ ClO_3^-(aq) + 12\ H_2O(l) + 24\ e^- \rightarrow$$
$$6\ NO(g) + 18\ H_2O(l) + 24\ e^- + 4\ Cl^-(aq) + 24\ OH^-(aq)$$

$$3\ N_2H_4(g) + 4\ ClO_3^-(aq) \rightarrow 6\ NO(g) + 6\ H_2O(l) + 4\ Cl^-(aq)$$

N_2H_4 is the reducing agent and ClO_3^- is the oxidizing agent.

(d) $Pb(OH)_4^{2-}(aq) \rightarrow PbO_2(s) + 2\ H_2O(l) + 2\ e^-$

$ClO^-(aq) + H_2O(l) + 2\ e^- \rightarrow Cl^-(aq) + 2\ OH^-(aq)$

$$Pb(OH)_4^{2-}(aq) + ClO^-(aq) + H_2O(l) + 2\ e^- \rightarrow$$
$$PbO_2(s) + Cl^-(aq) + 2\ H_2O(l) + 2\ OH^-(aq) + 2\ e^-$$

or

$$Pb(OH)_4^{2-}(aq) + ClO^-(aq) \rightarrow PbO_2(s) + Cl^-(aq) + H_2O(l) + 2\ OH^-(aq)$$

$Pb(OH)_4^{2-}$ is the reducing agent and ClO^- is the oxidizing agent.

12.8 We first determine the two half-reactions:

Unbalanced:

$FeHPO_3(aq) \rightarrow Fe(OH)_3(s) + PO_4^{3-}(aq)$

$OCl^-(aq) \rightarrow Cl^-(aq)$

These are then balanced as follows:

The ferrous hydrogen phosphite reaction is complicated because both iron and phosphorus are oxidized. The Fe atom begins with an oxidation number of +2 and ends with an oxidation number of +3, while phosphorus starts as +3 and ends up as +5. There is a total change of 3 electrons.

$FeHPO_3(aq) \rightarrow Fe(OH)_3(s) + PO_4^{3-}(aq) + 3\ e^-$

Balancing the charge by addition of hydroxide for basic solution gives

$FeHPO_3(aq) + 6\ OH^-(aq) \rightarrow Fe(OH)_3(s) + PO_4^{3-}(aq) + 3\ e^-$

Balancing the hydrogen and oxygen atoms by addition of water gives

$FeHPO_3(aq) + 6\ OH^-(aq) \rightarrow Fe(OH)_3(s) + PO_4^{3-}(aq) + 2\ H_2O(l) + 3\ e^-$

The reduction of hypochlorite is simpler to balance. The oxidation number of chlorine changes from +1 in OCl^- to −1 in Cl^-, giving a change of 2 e^-.

$OCl^-(aq) + 2\ e^- \rightarrow Cl^-(aq)$

Balancing the charge by addition of OH^-:

$OCl^-(aq) + 2\ e^- \rightarrow Cl^-(aq) + 2\ OH^-(aq)$

Balancing oxygen and hydrogen atoms by addition of H_2O:

$OCl^-(aq) + H_2O(l) + 2\ e^- \rightarrow Cl^-(aq) + 2\ OH^-(aq)$

Combining the two equations:

$3[OCl^-(aq) + H_2O(l) + 2\ e^- \rightarrow Cl^-(aq) + 2\ OH^-(aq)]$

$2[FeHPO_3(aq) + 6\ OH^-(aq) \rightarrow Fe(OH)_3(s) + PO_4^{3-}(aq) + 2\ H_2O(l) + 3\ e^-]$

$2\ FeHPO_3(aq) + 3\ OCl^-(aq) + 6\ OH^-(aq) \rightarrow 2\ Fe(OH)_3(s) + 2\ PO_4^{3-}(aq) + 3\ Cl^-(aq) + H_2O(l)$

12.10 The anode (oxidation) is written to the left in a cell diagram, the cathode (reduction) to the right.

(a) $Cu^{2+}(aq) + 2\ e^- \rightarrow Cu(s)$ $\quad E° = +0.34$ V (anode)

$2[Cu^+(aq) + e^- \rightarrow Cu(s)]$ $\quad E° = +0.52$ V (cathode)

Therefore, at the anode, after reversal,

$Cu(s) \rightarrow Cu^{2+}(aq) + 2\ e^-$

The cell reaction is

$2\ Cu^+(aq) \rightarrow Cu(s) + Cu^{2+}(aq)$ $\quad E°_{cell} = 0.52\ V - 0.34\ V = 0.18\ V$

(b) $2[Au^{3+}(aq) + 3\ e^- \rightarrow Au(s)]$ $\quad E°(cathode) = +1.40$ V

$3[Cr^{2+}(aq) + 2\ e^- \rightarrow Cr(s)]$ $\quad E°(anode) = -0.90$ V

Therefore, at the anode, after reversal,

$3[Cr(s) \rightarrow Cr^{2+}(aq) + 2\ e^-]$

The cell reaction is, upon addition of the half-reactions,

$3\,Cr(s) + 2\,Au^{3+}(aq) \rightarrow 2\,Au(s) + 3\,Cr^{2+}(aq)$

$E°_{cell} = +1.40\text{ V} - (-0.90\text{ V}) = +2.30\text{ V}$

(c) $AgI(s) + e^- \rightarrow Ag(s) + I^-(aq)$ $E° = -0.15\text{ V}$ (anode)

$AgCl(s) + e^- \rightarrow Ag(s) + Cl^-(aq)$ $E° = +0.22\text{ V}$ (cathode)

Therefore, at the anode, after reversal,

$Ag(s) + I^-(aq) \rightarrow AgI(s) + e^-$

The cell reaction is, upon addition of the half-reactions,

$I^-(aq) + AgCl(s) \rightarrow AgI(s) + Cl^-(aq)$

$E°_{cell} = +0.22\text{ V} - (-0.15\text{ V}) = +0.37\text{ V}$

(d) $2[AgCl(s) + e^- \rightarrow Ag(s) + Cl^-(aq)]$ $E° = +0.22\text{ V}$ (cathode)

$Hg_2Cl_2(s) + 2\,e^- \rightarrow 2\,Hg(l) + 2\,Cl^-(aq)$ $E° = +0.27\text{ V}$ (anode)

Reverse the anode half-reaction: $2\,Hg(l) + 2\,Cl^-(aq) \rightarrow Hg_2Cl_2(s) + 2\,e^-$

and the cell reaction is, upon addition of the half-reactions,

$2\,AgCl(s) + 2\,Hg(l) \rightarrow Hg_2Cl_2(s) + 2\,Ag(s)$

$E°_{cell} = +0.22\text{ V} - 0.27\text{ V} = -0.05\text{ V}$

Note: This balanced equation corresponds to the cell notation as given (the cell is an electrolytic one). The spontaneous process is the reverse of this reaction.

(e) $5[Hg_2^{\ 2+}(aq) + 2\,e^- \rightarrow 2\,Hg(l)]$

$E°(\text{anode}) = +0.79\text{ V}$

$2[MnO_4^{\ -}(aq) + 8\,H^+(aq) + 5\,e^- \rightarrow Mn^{2+}(aq) + 4\,H_2O(l)]$

$E°(\text{cathode}) = +1.51\text{ V}$

Reversing the anode reaction gives

$5[2\,Hg(l) \rightarrow Hg_2^{\ 2+}(aq) + 2\,e^-]$

The cell reaction, upon combination of the half-reactions,

$2\,MnO_4^{\ -}(aq) + 10\,Hg(l) + 16\,H^+(aq)$
$\rightarrow 5\,Hg_2^{\ 2+}(aq) + 2\,Mn^{2+}(aq) + 8\,H_2O(l)$

$E°_{cell} = +1.51 \text{ V} - (+0.79 \text{ V}) = +0.72$

12.12 (a) cathode: $Mn^{2+}(aq) + 2\,e^- \rightarrow Mn(s)$ $E° = -1.18$ V

$Ti^{2+}(aq) + 2\,e^- \rightarrow Ti(s)$ (anode) $E° = -1.63$ V

Reversing the anode reaction:

anode: $Ti(s) \rightarrow Ti^{2+}(aq) + 2\,e^-$

Adding the half-reactions, we have, for the overall reaction:

$Mn^{2+}(aq) + Ti(s) \rightarrow Mn(s) + Ti^{2+}(aq)$

$E°_{cell} = -1.18 \text{ V} - (-1.63 \text{ V}) = +0.45 \text{ V}$

The cell diagram is

$Ti(s)|Ti^{2+}(aq)||Mn^{2+}(aq)|Mn(s)$

(b) cathode: $Fe^{3+}(aq) + e^- \rightarrow Fe^{2+}(aq)$ $E° = +0.77$ V (1)

anode: $H_2(g) \rightarrow 2\,H^+(aq) + 2\,e^-$ $E° = 0.00$ V (2)

Multiplying Eq. 1 by a factor of 2 gives

$2\,Fe^{3+}(aq) + 2\,e^- \rightarrow 2\,Fe^{2+}(aq)$ (3)

Adding Eqs. 2 and 3 gives the overall reaction:

$2\,Fe^{3+}(aq) + H_2(g) \rightarrow 2\,Fe^{2+}(aq) + 2\,H^+(aq)$

$E°_{cell} = +0.77 \text{ V} - 0.00 \text{ V} = +0.77 \text{ V}$

The cell diagram is

$Pt(s)|H_2(g)|H^+(aq)||Fe^{3+}(aq), Fe^{2+}(aq)|Pt(s)$

(c) cathode: $Cu^+(aq) + e^- \rightarrow Cu(s)$ $E° = +0.52$ V (1)

anode: $Cu(s) \rightarrow Cu^{2+}(aq) + 2\,e^-$ $E° = +0.34$ V (2)

Multiplying Eq. 1 by a factor of 2 gives $2\,Cu^+ + 2\,e^- \rightarrow 2\,Cu(s)$ (3)

Upon addition of Eqs. 2 and 3, we have

$2\,Cu^+(aq) \rightarrow Cu(s) + Cu^{2+}(aq)$ $E°_{cell} = 0.52 \text{ V} - 0.34 \text{ V} = 0.18 \text{ V}$

The cell diagram is

$Cu(s)|Cu^{2+}(aq)||Cu^+(aq)|Cu(s)$

(d) $2[MnO_4^-(aq) + 8\ H^+ + 5\ e^- \rightarrow Mn^{2+}(aq) + 4\ H_2O(l)]$ (cathode)

$E° = +1.51$ V

$5[Cl_2(g) + 2\ e^- \rightarrow 2\ Cl^-(aq)]$ (anode)

$E° = +1.36$ V

Reverse the anode reaction $5[2\ Cl^-(aq) \rightarrow Cl_2(g) + 2\ e^-]$, then upon addition

$$2\ MnO_4^-(aq) + 16\ H^+(aq) + 10\ Cl^-(aq) \rightarrow 5\ Cl_2(g) + 2\ Mn^{2+}(aq) + 8\ H_2O(l)$$

$E°_{cell} = +1.51\ V - 1.36\ V = +0.15\ V$

The cell diagram is

$C(gr)|Cl_2(g)\,|\,Cl^-(aq)\,||\,MnO_4^-(aq), H^+(aq), Mn^{2+}(aq)|Pt$

12.14 (a) Eliminating the spectator ions, the reaction is

$Ag^+(aq) + I^-(aq) \rightarrow AgI(s)$

$Ag^+(aq) + e^- \rightarrow Ag(s)$ (cathode) $E° = +0.80$ V

$AgI(s) + e^- \rightarrow Ag(s) + I^-(aq)$ (anode) $E° = -0.15$ V

Reverse the anode reaction $Ag(s) + I^-(aq) \rightarrow AgI(s) + e^-$, then upon addition,

$Ag^+(aq) + I^-(aq) \rightarrow AgI(s)$ $E°_{cell} = +0.80\ V - (-0.15\ V) = +0.95\ V$

The cell diagram is

$Ag(s)|AgI(s)|I^-(aq)\,||\,Ag^+(aq)|Ag(s)$

(b) Rewriting in the notation of this chapter, we have

$H^+(aq, conc) \rightarrow H^+(aq, dil)$

The oxidation reaction at the anode is $\frac{1}{2}H_2(g) \rightarrow H^+(aq) + e^-$ and the reduction at the cathode is $H^+(aq) + e^- \rightarrow \frac{1}{2}H_2(g)$.

The cell diagram is $Pt(s)|H_2(g)\,|\,H^+(aq, dil)\,||\,H^+(aq, conc)\,|\,H_2(g)\,|\,Pt(s)$.

(c) anode: $Zn(s) + 2\ OH^-(aq) \rightarrow ZnO(s) + H_2O(l) + 2\ e^-$

cathode: $Ag_2O(s) + H_2O(l) + 2\ e^- \rightarrow 2\ Ag(s) + 2\ OH^-(aq)$

Adding the two half-reactions gives

$Ag_2O(s) + Zn(s) \rightarrow 2\ Ag(s) + ZnO(s) \qquad E°_{cell} = 1.6\ V$

(See Table 12.1.)

The cell diagram is

$Zn(s)|ZnO(s)|KOH(aq)|Ag_2O(s)\ |\ Ag(s)|steel$

12.16 (a) $Cr_2O_7^{2-}(aq) + 14\ H^+(aq) + 6\ e^- \rightarrow 2\ Cr^{3+}(aq) + 7\ H_2O(l)$

(cathode half-reaction)

$3[Hg_2^{2+}(aq) \rightarrow 2\ Hg^{2+} + 2\ e^-]$ (anode half-reaction × 3)

(b) Adding half-reactions gives

$Cr_2O_7^{2-}(aq) + 3\ Hg_2^{2+}(aq) + 14\ H^+(aq) \rightarrow$
$2\ Cr^{3+}(aq) + 6\ Hg^{2+}(aq) + 7H_2O(l)$

The cell diagram is

$Pt(s)|Hg_2^{2+}(aq), Hg^{2+}(aq)\ ||\ H^+(aq), Cr_2O_7^{2-}(aq), Cr^{3+}(aq)|Pt(s)$

12.18 (a) $Ag^+(aq) + e^- \rightarrow Ag(s) \qquad E°(\text{cathode}) = +0.80\ V$

$Fe^{3+}(aq) + e^- \rightarrow Fe^{2+}(aq) \qquad E°(\text{anode}) = +0.77\ V$

$E°(\text{cell}) = E°(\text{cathode}) - E°(\text{anode}) = +0.80\ V - 0.77\ V = +0.03\ V$

(b) $V^{2+}(aq) + 2\ e^- \rightarrow V(s) \qquad E°(\text{cathode}) = -1.19\ V$

$U^{3+}(aq) + 3\ e^- \rightarrow U(s) \qquad E°(\text{anode}) = -1.79\ V$

$E°(\text{cell}) = -1.19\ V - (-1.79\ V) = +0.60\ V$

(c) $Sn^{4+}(aq) + 2\ e^- \rightarrow Sn^{2+}(aq) \qquad E°(\text{cathode}) = +0.15\ V$

$Sn^{2+}(aq) + 2\ e^- \rightarrow Sn(s) \qquad E°(\text{anode}) = -0.14\ V$

$E°(\text{cell}) = +0.15\ V - (-0.14\ V) = +0.29\ V$

(d) $Au^+(aq) + e^- \rightarrow Au(s) \qquad E°(\text{cathode}) = +1.69\ V$

$Cu^{2+}(aq) + 2\ e^- \rightarrow Cu(s) \qquad E°(\text{anode}) = +0.34\ V$

$E°(\text{cell}) = +1.69\ V - (+0.34\ V) = +1.35\ V$

12.20 (a) $Bi^{3+}(aq) + 3\,e^- \rightarrow Bi(s)$ $E°(\text{cathode}) = +0.20\text{ V}$

$Zn^{2+}(aq) + 2\,e^- \rightarrow Zn(s)$ $E°(\text{anode}) = -0.76\text{ V}$

For the equation, $3\,Zn(s) + 2\,Bi^{3+}(aq) \rightarrow 3\,Zn^{2+}(aq) + 2\,Bi(s)$, $n = 6$

$E°_{\text{cell}} = +0.20\text{ V} - (-0.76\text{ V}) = +0.96\text{ V}$

$$\Delta G°_r = -nFE° = -6(9.6485 \times 10^4\,\text{J}\cdot\text{V}^{-1}\cdot\text{mol}^{-1})(+0.96\text{ V})$$
$$= -5.6 \times 10^2\,\text{kJ}\cdot\text{mol}^{-1}$$

(b) $O_2(g) + 4\,H^+(aq) + 4\,e^- \rightarrow 2\,H_2O(l)$ $E°(\text{cathode}) = +1.23\text{ V}$

$2\,H^+(aq) + 2\,e^- \rightarrow H_2(g)$ $E°(\text{anode}) = 0.00\text{ V}$

Taking $2[2\,H^+(aq) + 2\,e^- \rightarrow H_2(g)]$, reversing it and adding it to the

cathode:

half-reaction gives $2\,H_2(g) + O_2(g) \rightarrow 2\,H_2O(l)$ $n = 4$, $E°_{\text{cell}} = +1.23\text{ V}$

$$\Delta G°_r = -nFE° = -4 \times 9.6485 \times 10^4\text{ C}\cdot\text{mol}^{-1} \times 1.23\text{ V} = -475\text{ kJ}\cdot\text{mol}^{-1}$$

(c) $O_2(g) + 2\,H_2O(l) + 4\,e^- \rightarrow 4\,OH^-(aq)$ $E°(\text{cathode}) = +0.40\text{ V}$

$2\,H_2O(l) + 2\,e^- \rightarrow H_2(g) + 2\,OH^-(aq)$ $E°(\text{anode}) = -0.83\text{ V}$

Taking $2[H_2O(l) + 2\,e^- \rightarrow H_2(g) + 2\,OH^-(aq)]$, reversing it and adding it

to the cathode half-reaction gives $2\,H_2(g) + O_2(g) \rightarrow 2\,H_2O(l)$

$n = 4$, $E°_{\text{cell}} = +1.23\text{ V}$

$\Delta G°_r = nFE° = -475\text{ kJ}\cdot\text{mol}^{-1}$ (same as part b)

(d) $3[Au^+(aq) + e^- \rightarrow Au(s)]$ $E°(\text{cathode}) = +1.69\text{ V}$

$Au^{3+}(aq) + 3\,e^- \rightarrow Au(s)$ $E°(\text{anode}) = +1.40\text{ V}$

Taking $3[Au^+ + e^- \rightarrow Au(s)]$ and adding it to the reverse of the reduction

half-reaction gives $3\,Au^+ \rightarrow Au^{3+} + 2\,Au(s)$ $n = 3$

$E°_{\text{cell}} = +1.69\text{ V} - 1.40\text{ V} = 0.29\text{ V}$

$$\Delta G°_r = -nFE° = -3 \times 9.6485 \times 10^4\text{ C}\cdot\text{mol}^{-1} \times 0.29\text{ J}\cdot\text{C}^{-1}$$
$$= -84\text{ kJ}\cdot\text{mol}^{-1}$$

12.22 The unknown metal ions are reduced, increasing the weight of the unknown metal electrode. The cell can be written as $Fe(s)|Fe^{2+}(aq)||M^{+}(aq)|M(s)$ showing that the M^{+}/M electrode is the cathode, because this is where reduction is occuring.

$Fe^{2+} + 2e \rightarrow Fe(s)$

$E° = -0.44$ V

$E°$ (cathode) = +1.24 V + $E°$ (anode)

$E°$ (cathode) = +1.24 V + (− 0.44 V) = +0.80 V

12.24 (a)

Co^{2+}	$E° = -0.28$ V
Cl_2	$E° = +1.36$ V
Ce^{4+}	$E° = +1.61$ V
In^{3+}	$E° = -0.34$ V

The best oxidizing agent has the highest $E°: Ce^{4+} > Cl_2 > Co^{2+} > In^{3+}$

(b)

NO_3^-	$E° = +0.96$ V
ClO_4^-	$E° = +1.23$ V
HBrO	$E° = +1.60$ V
$Cr_2O_7^{2-}$	$E° = +1.33$ V

The best oxidizing agent has the highest $E°: HBrO > Cr_2O_7^{2-} > ClO_4^- > NO_3^-$

(c)

$O_2 + 4\,H^+ + 4\,e^- \rightarrow 2\,H_2O$	$E° = +1.23$ V
$O_3 + 2\,H^+ + 2\,e^- \rightarrow O_2 + H_2O$	$E° = +2.07$ V
$2\,HClO + 2\,H^+ + 2\,e^- \rightarrow Cl_2 + 2\,H_2O$	$E° = +1.63$ V
$2\,HBrO + 2\,H^+ + 2\,e^- \rightarrow Br_2 + 2\,H_2O$	$E° = +1.60$ V

The order of oxidizing ability in acidic solution (which corresponds to the ease of reduction of the reagent) is $O_3 > HClO > HBrO > O_2$.

(d) $O_2 + H_2O + 4\,e^- \rightarrow 4\,OH^-$ $E° = +0.40$ V

$O_3 + H_2O + 2\,e^- \rightarrow O_2 + 2\,OH^-$ $E° = +1.24$ V

$ClO^- + H_2O + 2\,e^- \rightarrow Cl^- + 2\,OH^-$ $E° = +0.89$ V

$BrO^- + H_2O + 2\,e^- \rightarrow Br^- + 2\,OH^-$ $E° = +0.76$ V

The order of oxidizing ability in basic solution (which corresponds to the ease of reduction of the reagent) is $O_3 > ClO^- > BrO^- > O_2$. Notice that the order in (c) is the same as here, but the relative amounts of difference may be significantly different. For example, HClO and HBrO have very similar oxidizing abilities in acidic solution, but in basic solution the ClO^- ion is substantially more oxidizing than BrO^-. Note also that the oxidizing ability of all of these species is somewhat less in the basic medium.

12.26 (a) Pt^{2+}/Pt $E° = +1.20$ V, Pt^{2+} is oxidizing agent (cathode)

$AgF/Ag, F^-$ $E° = +0.78$ V, Ag is reducing agent (anode)

$Ag(s)|AgF(s)|F^-(aq) \,||\, Pt^{2+}(aq)|Pt(s)$

$E°_{cell} = E°(\text{cathode}) - E°(\text{anode}) = +1.20\text{ V} - 0.78\text{ V} = +0.42\text{ V}$

(b) I_3^-/I^- $E° = +0.53$ V, I_3^- is oxidizing agent (cathode)

Cr^{3+}/Cr^{2+} $E° = -0.41$ V, Cr^{2+} is reducing agent (anode)

$Pt(s)|Cr^{2+}(aq), Cr^{3+}(aq)||I^-(aq),\ I_3^-(aq)\,|\,Pt(s)$

$E°_{cell} = +0.53\text{ V} - (-0.41\text{ V}) = +0.94\text{ V}$

(c) H^+/H_2 $E° = 0.00$ V, H^+ is oxidizing agent (cathode)

Ni^{2+}/Ni $E° = -0.23$ V, Ni is reducing agent (anode)

$Ni(s)|\,Ni^{2+}(aq)||H^+(aq)|H_2(g)|Pt(s)$

$E°_{cell} = 0.00\text{ V} - (-0.23\text{ V}) = +0.23\text{ V}$

(d) $O_3, H^+/O_2$ $E° = +2.07$ V, O_3 is oxidizing agent (cathode)

$O_3/O_2, OH^-$ $E° = +1.24$ V, O_2 is reducing agent (anode)

$Pt(s)|O_3(g), O_2(g)|OH^-(aq)||O_3(g), O_2(g)\,|\,H^+(aq)\,|\,Pt(s)$

$E°_{cell} = +2.07\ V - 1.24\ V = +0.83\ V$

12.28 (a) $Mg^{2+} + Cu \rightarrow$ no reaction; $E°$ for $Cu^{2+}/Cu > E°$ for Mg^{2+}/Mg

(b) $2\ Al + 3\ Pb^{2+} \rightarrow 2\ Al^{3+} + 3\ Pb$

$Pb^{2+} + 2\ e^- \rightarrow Pb$ reduction (cathode) $E° = -0.13\ V$

$Al \rightarrow Al^{3+} + 3\ e^-$ oxidation (anode) $E° = -1.66\ V$

$E°_{cell} = E°(\text{cathode}) - E°(\text{anode}) = -0.13\ V - (-1.66\ V) = +1.53\ V$

$\Delta G°_r = -nFE° = -(6)(9.6485 \times 10^4\ C \cdot mol^{-1})(+1.53\ J \cdot C^{-1})$

$= -886\ kJ \cdot mol^{-1}$; therefore, the reaction is spontaneous

(c) $Hg_2^{2+} + 2\ Ce^{3+} \rightarrow 2\ Ce^{4+} + 2\ Hg$

$E°$ for $Ce^{4+}/Ce^{3+} > E°$ for Hg_2^{2+}/Hg: the reaction is not spontaneous.

(d) $Zn + Sn^{2+} \rightarrow Sn + Zn^{2+}$

$E°$ for $Sn^{2+}/Sn > E°$ for Zn^{2+}/Zn

$Sn^{2+} + 2\ e^- \rightarrow Sn$ $E°(\text{cathode}) = -0.14\ V$

$Zn \rightarrow Zn^{2+} + 2\ e^-$ $E°(\text{anode}) = -0.76\ V$

$\Delta E° = E°(\text{cathode}) - E°(\text{anode}) = -0.14\ V - (-0.76\ V)$

$= +0.62\ V$

$\Delta G°_r = -nFE° = -(2)(9.6485 \times 10^4\ C \cdot mol^{-1})(0.62\ J \cdot C^{-1})$

$= -120\ kJ \cdot mol^{-1}$,

the reaction is spontaneous

(e) $O_2 + H^+ + Hg \rightarrow H_2O + Hg_2^{2+}$ (unbalanced)

$E°$ for $O_2/H_2O > E°$ for Hg_2^{2+}/Hg

$O_2 + 4\ H^+ + 4\ e^- \rightarrow 2\ H_2O$ $E°(\text{cathode}) = +1.23\ V$

$2\ Hg \rightarrow Hg_2^{2+} + 2\ e^-$ $E°(\text{anode}) = +0.79\ V$

Multiplying the second half-reaction by two and adding to the cathode half-reaction gives $4\ Hg + O_2 + 4\ H^+ \rightarrow 2\ Hg_2^{2+} + 2\ H_2O$ (balanced)

$$\Delta E° = E°(\text{cathode}) - E°(\text{anode}) = +1.23\ \text{V} - (+0.79\ \text{V})$$
$$= +0.44\ \text{V}$$
$$\Delta G°_r = -nFE° = -(4)(9.6485 \times 10^4\ \text{C}\cdot\text{mol}^{-1})(0.44\ \text{J}\cdot\text{C}^{-1})$$
$$= -170\ \text{kJ}\cdot\text{mol}^{-1},$$

the reaction is spontaneous

12.30 (a) $2\ Sn^{2+}(aq) \rightarrow Sn(s) + Sn^{4+}(aq)$

(b) $Sn^{2+}(aq) + 2\ e^- \rightarrow Sn(s) \quad E° = -0.14\ \text{V}$

$Sn^{4+}(aq) + 2\ e^- \rightarrow Sn^{2+}(aq) \quad E° = +0.15\ \text{V}$

The desired reaction is obtained by subtracting the second equation from the first.

The $E°$ value is obtained also by subtracting the second value from the first:

$E° = -0.14\ \text{V} - (+0.15\ \text{V}) = -0.29\ \text{V}$

Because the $E°$ value is negative, the process is not spontaneous.

12.32 The appropriate half-reactions are

$Ti^{3+} + e^- \rightarrow Ti^{2+} \qquad E° = -0.37 \qquad (A)$

$Ti^{2+} + 2\ e^- \rightarrow Ti \qquad E° = -1.63 \qquad (B)$

(A) and (B) add to give the desired half-reaction (C):

$Ti^{3+} + 3\ e^- \rightarrow Ti \quad E° = ? \qquad (C)$

In order to calculate the potential of a *half-reaction*, we need to convert the $E°$ values into $\Delta G°$ values:

$$\Delta G°(A) = -nFE°(A) = -1F(-0.37\ \text{V})$$
$$\Delta G°(B) = -nFE°(B) = -2F(-1.63\ \text{V})$$
$$\Delta G°(C) = -nFE°(C) = -3FE°(C)$$
$$\Delta G°(C) = \Delta G°(A) + \Delta G°(B)$$
$$-3FE°(C) = -1F(-0.37\ \text{V}) + [-2F(-1.63\ \text{V})]$$

The constant F will cancel from both sides, leaving:

$-3E°(C) = -1(-0.37 \text{ V}) - 2(-1.63 \text{ V})$

$E°(C) = -[0.37 \text{ V} + 3.26 \text{ V}]/3 = -1.21 \text{ V}$

12.34 (a) The balanced cell reaction is

$2\ Fe^{3+}(aq) + H_2(g) \rightarrow 2\ Fe^{2+}(aq) + 2\ H^+(aq) \quad n = 2$

$E°_{cell} = E°(\text{cathode}) - E°(\text{anode}) = +0.77 \text{ V} - (0.00) \text{ V} = +0.77 \text{ V}$

$$\ln K = \frac{nFE°}{RT} = \frac{2(+0.77 \text{ V})}{0.025\,693 \text{ V}}$$

$$\ln K = \frac{(2)(0.77 \text{ V})}{0.025\,693 \text{ V}} = 59.9$$

$K = 10^{26}$

(b) $Cr(s) + Zn^{2+}(aq) \rightarrow Cr^{2+}(aq) + Zn(s) \qquad n = 2$

$E°_{cell} = E°(\text{cathode}) - E°(\text{anode}) = -0.76 \text{ V} - (-0.91 \text{ V}) = 0.15 \text{ V}$ and

$$\ln K = \frac{nFE°}{RT} = \frac{(2)(+0.15 \text{ V})}{0.025\,693 \text{ V}}$$

$$\ln K = \frac{(2)(0.15 \text{ V})}{0.025\,693 \text{ V}} = +12$$

and $K = 10^5$

12.36 (a) $ClO_4^-(aq) + 2\ H^+(aq) + 2\ e^- \rightarrow ClO_3^-(aq) + H_2O(l)$

$E°(\text{cathode}) = +1.23 \text{ V}$

$Ag^+(aq) + e^- \rightarrow Ag(s) \quad E°(\text{anode}) = +0.80 \text{ V}$

Reversing the latter half-reaction, multiplying by 2, and adding, we have

$ClO_4^-(aq) + 2\ H^+(aq) + 2\ Ag(s) \rightarrow ClO_3^-(aq) + H_2O(l) + Ag^+(aq)$

$n = 2,\ E°_{cell} = +0.43 \text{ V}$

Then, using the Nernst equation, $E = E° - \dfrac{RT}{nF} \ln Q$, we have

$$E = E° - \frac{0.025\ 693}{n} \ln Q$$

$$0.40 \text{ V} = 0.43 \text{ V} - \frac{0.025\ 693 \text{ V}}{2} \ln Q$$

$$\ln Q = -\frac{(2)(-0.03 \text{ V})}{0.025\ 693 \text{ V}} = 2$$

$$\ln Q = 10^1$$

(b) cathode: $Au^{3+}(aq) + 3\ e^- \rightarrow Au(s)$ $\quad E°(\text{cathode}) = +1.40$ V

anode: [reduction potential for

$Cl_2(g) + 2\ e^- \rightarrow 2\ Cl^-(aq)$ is $+\ 1.36$ V $= E°(\text{anode})$]

Multiplying, we have $2[Au^{3+}(aq) + 3\ e^- \rightarrow Au(s)]$

$3[2\ Cl^-(aq) \rightarrow Cl_2(g) + 2\ e^-]$

Adding yields $2\ Au^{3+}(aq) + 6\ Cl^-\ (aq) \rightarrow 2\ Au(s) + 3\ Cl_2(g)$

$E°_{cell} = 0.04$ V

Then, $n = 6$

$$E = E° - \frac{0.025\ 693}{n} \ln Q$$

$$0.00 \text{ V} = 0.04 \text{ V} - \frac{0.025\ 693}{6} \ln Q$$

$$\ln Q = -\frac{(6)(-0.04 \text{ V})}{0.025\ 693 \text{ V}} = 9$$

$$Q = 10^4$$

12.38 (a) Cell reaction: $\quad Pb^{2+}(aq, 0.10 \text{ mol} \cdot L^{-1}) \rightarrow Pb^{2+}(aq, x \text{ mol} \cdot L^{-1})$

$n = 2$, $E_{cell} = 0.050$ V

$$E_{cell} = E°_{cell} - \frac{RT}{nF} \ln Q = -\frac{0.025\ 693 \text{ V}}{2} \ln\left(\frac{x}{0.10 \text{ mol} \cdot L^{-1}}\right) = 0.050 \text{ V}$$

$$\ln\left(\frac{x}{0.10 \text{ M}}\right) = -\frac{2 \times 0.050 \text{ V}}{0.025\ 693 \text{ V}} = -3.9$$

$$\left(\frac{x}{0.10 \text{ M}}\right) = 2 \times 10^{-2};\ x = [Pb^{2+}] = 2 \times 10^{-3} \text{ mol} \cdot L^{-1}$$

(b) Cathode: $Fe^{3+}(aq, x\text{M}) + e^- \rightarrow Fe^{3+}(aq, 0.0010\ \text{M})$

Anode: $Fe^{3+}(aq, 0.10\ \text{M}) + e^- \rightarrow Fe^{2+}(aq, 1.0\ \text{M})$

overall reaction: $Fe^{3+}(aq, x\text{M}) + Fe^{2+}(aq, 1.0\ \text{M}) \rightarrow$

$Fe^{2+}(aq, 0.0010\ \text{M}) + Fe^{3+}\ (aq, 0.10\ \text{M})$

$$E_{\text{cell}} = E°_{\text{cell}} - \frac{RT}{nF}\ln Q$$

$$= -\frac{0.025\ 693\ \text{V}}{1}\ln\left(\frac{0.0010\ \text{M} \times 0.10\ \text{M}}{x\text{M} \times 1.0\ \text{M}}\right) = 0.10\ \text{V}$$

$$\ln\left(\frac{1.0 \times 10^{-4}\ \text{mol} \cdot \text{L}^{-1}}{x}\right) = -\frac{0.10\ \text{V}}{0.025\ 693\ \text{V}} = -3.9$$

$$\frac{1.0 \times 10^{-4}\ \text{mol} \cdot \text{L}^{-1}}{x} = 2 \times 10^{-2}, x = [Fe^{3+}] = 5 \times 10^{-3}\ \text{mol} \cdot \text{L}^{-1}$$

12.40 (a) anode: $Cr(s) \rightarrow Cr^{3+}(aq) + 3\ e^-$ $\quad E°(\text{anode}) = -0.74\ \text{V}$

cathode: $Pb^{2+}(aq) + 2\ e^- \rightarrow Pb(s)$ $\quad E°(\text{cathode}) = -0.13\ \text{V}$

Multiplying half-reactions gives

$3[Pb^{2+}(aq) + 2\ e^- \rightarrow Pb(s)]$

$2[Cr(s) \rightarrow Cr^{3+}(aq) + 3\ e^-]$

overall reaction: $3\ Pb^{2+}(aq) + 2\ Cr(s) \rightarrow 3\ Pb(s) + 2\ Cr^{3+}(aq)$

$E°_{\text{cell}} = +0.61\ \text{V}, n = 6$

$$E = E° - \frac{0.025\ 693\ \text{V}}{n}\ln\frac{[Cr^{3+}]^2}{[Pb^{2+}]^3}$$

$$E = +0.61\ \text{V} - \frac{0.025\ 693\ \text{V}}{6}\ln\frac{(0.37)^2}{(9.5 \times 10^{-3})^3}$$

$$E = +0.61\ \text{V} - 0.0043\ \text{V}\ln 1.6 \times 10^5 = +0.61\ \text{V} - 0.051\ \text{V} = +0.56\ \text{V}$$

(b) $Hg_2Cl_2(s) + 2\ e^- \rightarrow 2\ Hg(l) + 2\ Cl^-(aq)$ $\quad E°(\text{cathode}) = +0.27\ \text{V}$

$2\ H^+(aq) + 2\ e^- \rightarrow H_2(g)$ $\quad E°(\text{anode}) = 0.00\ \text{V}$

Reversing the latter half-reaction and adding, we have

$Hg_2Cl_2(s) + H_2(g) \rightarrow 2\ Hg(l) + 2\ H^+(aq) + 2\ Cl^-(aq)$ $\quad E°_{\text{cell}} = +0.27\ \text{V}$

$$E = E° - \frac{0.025\,693\text{ V}}{n}\ln\frac{[H^+]^2[Cl^-]^2}{P_{H_2}}$$

$$E = +0.27\text{ V} - \frac{0.025\,693\text{ V}}{2}\ln\frac{(10^{-3.5})^2(0.75)^2}{2.0} = +0.27\text{ V} + 0.22\text{ V}$$

$= +0.5$ V

(c) anode: $Sn^{2+}(aq) \rightarrow Sn^{4+}(aq) + 2\,e^-$ $E°(\text{anode}) = +0.15$ V

cathode: $Fe^{3+}(aq) + e^- \rightarrow Fe^{2+}(aq)$ $E°(\text{cathode}) = +0.77$ V

Multiplying half-reactions,

$2[Fe^{3+}(aq) + e^- \rightarrow Fe^{2+}(aq)]$

$Sn^{2+}(aq) \rightarrow Sn^{4+}(aq) + 2\,e^-$

Adding, $2\,Fe^{3+}(aq) + Sn^{2+}(aq) \rightarrow 2\,Fe^{2+}(aq) + Sn^{4+}(aq)$

$E°_{cell} = +0.62$ V

$$E = E° - \frac{0.025\,693\text{ V}}{n}\ln\frac{[Fe^{2+}]^2[Sn^{4+}]}{[Fe^{3+}]^2[Sn^{2+}]}$$

$$E = +0.62\text{ V} - \frac{0.025\,693\text{ V}}{2}\ln\frac{(0.15)^2(0.059)}{(0.15)^2(0.059)}$$

$E = +0.62\text{ V} - 0.0129\text{ V}\ln 1 = +0.62$ V

(d) anode: $Ag(s) + I^-(aq) \rightarrow AgI(s) + e^-$ $E°(\text{anode}) = -0.15$ V

cathode: $AgCl(s) + e^- \rightarrow Ag(s) + Cl^-(aq)$ $E°(\text{cathode}) = +0.22$ V

$Ag(s) + I^-(aq) \rightarrow AgI(s) + e^-$

Adding, $AgCl(s) + I^-(aq) \rightarrow AgI(s) + Cl^-(aq)$ $E°_{cell} = +0.37$ V

$$E = E° - \frac{0.025\,693\text{ V}}{n}\ln\frac{[Cl^-]}{[I^-]}$$

$$E = +0.37\text{ V} - (0.025\,693\text{ V})\ln\frac{0.67}{0.025}$$

$E = +0.37\text{ V} - 0.084\text{ V} = +0.29$ V

12.42 (a) anode: $H_2(g) \rightarrow 2\,H^+(aq) + 2\,e^-$ $E°(\text{anode}) = 0.00$ V

cathode: $AgCl(s) + e^- \rightarrow Ag(s) + Cl^-(aq)$ $E°(\text{cathode}) = 0.22$ V

Multiplying,

$2[AgCl(s) + e^- \rightarrow Ag(s) + Cl^-(aq)]$

$H_2(g) \rightarrow 2\,H^+(aq) + 2\,e^-$

Adding, $2\,AgCl(s) + H_2(g) \rightarrow 2\,H^+(aq) + 2\,Ag(s) + 2\,Cl^-(aq)$

$E°_{cell} = +0.22\text{ V}$

$$E = E° - \frac{0.025\,693\text{ V}}{n}\ln[H^+]^2[Cl^-]^2$$

$$0.30\text{ V} = 0.22\text{ V} - \frac{0.025\,693\text{ V}}{2}\ln\frac{[H^+]^2(1.0)}{(1.0)}$$

$$0.08\text{ V} = -0.0129\text{ V}\ln[H^+]^2$$

$$0.08\text{ V} = -0.025\,693\text{ V}\ln[H^+] = -0.059\text{ V}\times\log[H^+]$$

$$\frac{0.08\text{ V}}{0.059\text{ V}} = -\log[H^+] = \text{pH} = 1.4$$

(b) $Ni^{2+}(aq) + 2\,e^- \rightarrow Ni(s)$ $E°(\text{cathode}) = -0.23\text{ V}$

$Pb(s) \rightarrow Pb^{2+}(aq) + 2\,e^-$ $E°(\text{anode}) = -0.13\text{V}$

Adding half-reactions gives

$Ni^{2+}(aq) + Pb(s) \rightarrow Ni(s) + Pb^{2+}(aq)$

$E°_{cell} = -0.10\text{ V}$

$$E = E° - \frac{0.025\,693\text{ V}}{n}\ln\frac{[Pb^{2+}]}{[Ni^{2+}]}$$

$$0.040\text{ V} = -0.10\text{ V} - \frac{0.025\,693\text{ V}}{2}\ln\frac{[Pb^{2+}]}{0.10}$$

$$0.14\text{ V} = -(0.0129\text{ V})\ln\frac{[Pb^{2+}]}{0.10} = -0.0129\text{ V}(\ln[Pb^{2+}] - \ln[0.10])$$

$$0.14\text{ V} = -0.0129\text{ V}\ln[Pb^{2+}] - 0.030\text{ V}$$

$$0.17\text{ V} = -0.0129\text{ V}\ln[Pb^{2+}]$$

$$\frac{0.17\text{ V}}{-0.0129\text{ V}} = \ln[Pb^{2+}] = -13$$

$$[Pb^{2+}] = 2\times10^{-6}\text{ mol}\cdot\text{L}^{-1}$$

12.44 Cathode: $Pb^{2+}(aq, 0.010M) + 2e^- \rightarrow Pb(s)$ $E° = -0.13\text{ V}$

Anode: $H_2(g, 1.0\text{bar}) \rightarrow 2H^+(aq, \text{pH}) + 2e^-$ $E° = 0\text{ V}$

Total: $Pb^{2+}(aq, 0.010M) + H_2(g, 1.0bar) \rightarrow Pb(s) + 2H^+(aq, pH)$

$E^\circ = -0.13\ V$

$$E = E^\circ - \frac{0.05916\ V}{2}\lg\frac{[H^+]^2}{[Pb^{2+}]}$$

$$0.057\ V = -0.13\ V - \frac{0.05916\ V}{2}\lg\frac{[H^+]^2}{0.010}$$

$$0.187\ V = (-0.05916\ V)\log[H^+] + 0.059\ 16\ V$$

$$pH = \frac{0.187}{0.05916} - 1.000 = 2.2$$

12.46 (a) $Ag_2CrO_4(s) + 2\ e^- \rightarrow 2\ Ag(s) + CrO_4^{2-}(aq)$

(b) To calculate this value we need to determine the $E°$ value for the solubility reaction:

$Ag_2CrO_4(s) \rightarrow 2\ Ag^+(aq) + CrO_4^{2-}(aq) \quad E° = ?$

The relationship $\Delta G° = -RT\ \ln K = -nFE°$ can be used to calculate the value of K_{sp}. The equations that will add to give the net equation we want, are:

$Ag_2CrO_4(s) + 2\ e^- \rightarrow 2\ Ag(s) + CrO_4^{2-}(aq) \quad E° = +0.446\ V$

$Ag(s) \rightarrow Ag^+(aq) + e^- \quad E° = -0.80\ V$

Notice that the second equation is reversed from the reduction reaction given in the Appendix, and consequently the $E°$ value is changed in sign. Adding the first equation to twice the second gives the desired net reaction, and summing the $E°$ values will give the $E°$ value for that process (note that we do not multiply the second equation's $E°$ value by 2).

$E° = (+0.446\ V) + (-0.80\ V) = -0.35\ V$

$$\ln K_{sp} = \frac{nFE°}{RT} = \frac{(2)(9.65 \times 10^4\ C \cdot mol^{-1})(-0.35\ V)}{(8.314\ J \cdot K^{-1} \cdot mol^{-1})(298.2\ K)} = -27$$

$K_{sp} = 10^{-12}$

12.48 (a) $E^\circ = E^\circ(\text{Pb}^{2+}/\text{Pb}) - E^\circ(\text{Ag}^+/\text{Ag}) = -0.13\text{ V} - 0.80\text{ V} = -0.93\text{ V}$

Total cell reaction is:

$2\text{Ag(s)} + \text{Pb}^{2+}(\text{aq}, 0.10\text{ mol L}^{-1}) \rightarrow 2\text{Ag}^+(\text{aq}, 2.1\times10^{-4}\text{ mol L}^{-1}) + \text{Pb(s)}$

$$E = E^\circ - \frac{0.025\ 693\text{ V}}{2}\ln\frac{\left[\text{Ag}^+\right]^2}{\left[\text{Pb}^{2+}\right]}$$

$$= -0.93\text{ V} - \frac{0.025693\text{ V}}{2}\ln\frac{\left(2.1\times10^{-4}\right)^2}{0.10} = -0.74\text{ V}$$

This cell can do work. However, its spontaneous direction is opposite to what we have so far assigned. The maximum work it performs after consuming 1 mol of Pb is

$w = nFE = (2\text{ mol})(96500\text{ C}\cdot\text{mol}^{-1})(0.74\text{ V}) = 1.4\times10^5\text{ J}$

(b) The cell reaction in the spontaneous direction should be:

$2\text{Ag}^+(\text{aq}, 2.1\times10^{-4}\text{ mol L}^{-1}) + \text{Pb(s)} \rightarrow 2\text{Ag(s)} + \text{Pb}^{2+}(\text{aq}, 0.10\text{ mol L}^{-1})$

$$\Delta G_r = -nFE = -2\times96\ 485\text{ C}\cdot\text{mol}^{-1}\times0.74V = -140\text{ kJ}\cdot\text{mol}^{-1}$$

$$\Delta H_r^\circ = \Delta H_f^\circ(\text{Pb}^{2+}) - 2\Delta H_f^\circ(\text{Ag}^+) = -1.7\text{ kJ}\cdot\text{mol}^{-1} - 2\times105.58\text{ kJ}\cdot\text{mol}^{-1}$$

$$= -212.9\text{ kJ}\cdot\text{mol}^{-1}$$

$$\Delta H_r \approx \Delta H_r^\circ$$

$$\Delta S_r = \frac{\Delta H_r - \Delta G_r}{T} = \frac{-212.9\text{ kJ}\cdot\text{mol}^{-1} + 140\text{ kJ}\cdot\text{mol}^{-1}}{298.15\text{ K}} = -240\text{ J}\cdot\text{K}^{-1}\text{mol}^{-1}$$

(negative)

12.50 The change in concentration of Cl^- ions will affect the potential of the calomel electrode. The difference in the potential, relative to the standard potential, will be given by the Nernst equation. First, we calculate the concentration of Cl^- in a saturated KCl solution:

$$[\text{KCl}] = [\text{Cl}^-] = \frac{\left(\dfrac{35\text{ g}}{74.55\text{ g}\cdot\text{mol}^{-1}}\right)}{0.100\text{ L}} = 4.7\text{ mol}\cdot\text{L}^{-1}$$

$Hg_2Cl_2(s) + 2\ e^- \rightarrow 2\ Hg(l) + 2\ Cl^-(aq)$

$$E = E° - \frac{RT}{nF} \ln Q$$

$$= 0.27\ V - \frac{(8.314\ J \cdot K^{-1} \cdot mol^{-1})(298.2\ K)}{2(9.65 \times 10^5\ C \cdot mol^{-1})} \ln Q$$

$$= 0.27\ V - \frac{(8.314\ J \cdot K^{-1} \cdot mol^{-1})(298.2\ K)}{2(9.65 \times 10^5\ C \cdot mol^{-1})} \ln[Cl^-]^2$$

$$= 0.27\ V - \frac{(8.314\ J \cdot K^{-1} \cdot mol^{-1})(298.2\ K)}{2(9.65 \times 10^4\ C \cdot mol^{-1})} \ln(4.7)^2$$

$$= 0.23\ V$$

If this electrode were now set to equal 0, all other potentials would be decreased by 0.23 V (instead of 0.27 V as in Exercise 12.53). The standard hydrogen electrode's potential would be $0.00 - 0.23\ V = -0.23\ V$; the standard reduction potential for Cu^{2+}/Cu would be $0.34\ V - 0.23\ V = +0.11\ V$.

12.52 (a) cathode: $2\ H_2O(l) + 2\ e^- \rightarrow H_2(g) + 2\ OH^-(aq)$

$E°(\text{cathode}) = -0.42\ V\ (pH = 7)$

(rather than $K^+(aq) + e^- \rightarrow K(s)$ $\quad E° = -2.93\ V$)

(b) anode: $2\ H_2O(l) \rightarrow O_2(g) + 4\ H^+(aq) + 4\ e^-$ $\quad E°(\text{anode}) = +0.82\ V$ (pH = 7)

(rather than $2\ Br^-(aq) \rightarrow Br_2(g) + 2\ e^-$ $\quad E°(\text{anode}) = +1.09\ V$)

(c) $E°_{cell} = E°(\text{cathode}) - E°(\text{anode}) = -0.42\ V - (+0.82\ V)$

$E°_{cell} = -1.24\ V$

Therefore, E(minimum supplied) = + 1.24 V

12.54 (a) $Cr(s) \rightarrow Cr^{2+}(aq) + 2\ e^-$ $\quad E° = -0.91\ V$

$Cr(s) \rightarrow Cr^{3+}(aq) + 3\ e^-$ $\quad E° = -0.74\ V$

$2\ H_2O(l) \rightarrow O_2(g) + 4\ H^+(aq) + 4\ e^-$ $\quad E° = +0.81\ V\ (pH = 7)$

Because $E°(\text{Cr}) < E°(\text{H}_2\text{O})$, oxidation of the electrode will occur.

(b) $\text{Pt(s)} \rightarrow \text{Pt}^{2+}\text{(aq)} + 2\ \text{e}^-$ $\qquad E° = +1.20\ \text{V}$

Because $E°(\text{Pt}) > E°(\text{H}_2\text{O})$, oxidation of water will occur.

(c) $\text{Cu(s)} \rightarrow \text{Cu}^{2+}\text{(aq)} + 2\ \text{e}^-$ $\qquad E° = +0.34\ \text{V}$

$\text{Cu(s)} \rightarrow \text{Cu}^{+}\text{(aq)} + \text{e}^-$ $\qquad E° = +0.52\ \text{V}$

Because $E°(\text{Cu}) < E°(\text{H}_2\text{O})$, oxidation of the electrode will occur.

(d) $\text{Ni(s)} \rightarrow \text{Ni}^{2+}\text{(aq)} + 2\ \text{e}^-$ $\qquad E° = -0.23\ \text{V}$

Because $E°(\text{Ni}) < E°(\text{H}_2\text{O})$, oxidation of the electrode will occur.

12.56 $96.5\ \text{kC} = 9.65 \times 10^4\ \text{C} \cong 1.00\ \text{F} = 1.00\ \text{mol e}^-$

(a) $1.00\ \text{mol Ag}^+\text{(aq)} + 1.00\ \text{mol e}^- \rightarrow 1.00\ \text{mol Ag, or } 108\ \text{g}$

(b) $1.00\ \text{mol Cl}^- \rightarrow 0.500\ \text{mol Cl}_2\text{(g)} + 1.00\ \text{mol e}^-$

$0.500\ \text{mol Cl}_2 \text{ at } 298\ \text{K} = 0.500\ \text{mol} \times 24.45\ \text{L}\cdot\text{mol}^{-1} = 12.2\ \text{L}$

(c) $0.500\ \text{mol Cu}^{2+}\text{(aq)} + 1.00\ \text{mol e}^- \rightarrow 0.500\ \text{mol Cu(s)};$

$\text{mass of copper} = 0.500\ \text{mol Cu} \times 63.54\ \text{g}\cdot\text{mol}^{-1} = 31.8\ \text{g}$

12.58 (a) $2\ \text{F}^- \rightarrow \text{F}_2\text{(g)} + 2\ \text{e}^-$

$$\text{moles of e}^- = (7.0\ \text{h})\left(\frac{3600\ \text{s}}{1\ \text{h}}\right)(324\ \text{mA})\left(\frac{10^{-3}\ \text{A}}{1\ \text{mA}}\right)\left(\frac{1\ \text{C}\cdot\text{s}^{-1}}{1\ \text{A}}\right)$$

$$\left(\frac{1\ \text{mol e}^-}{9.6485 \times 10^4\ \text{C}}\right) = 8.5 \times 10^{-2}\ \text{mol e}^-$$

$$\text{volume of F}_2\text{(g)} = (8.5 \times 10^{-2}\ \text{mol e}^-)\left(\frac{1\ \text{mol F}_2\text{(g)}}{2\ \text{mol e}^-}\right)\left(\frac{24.45\ \text{L F}_2\text{(g)}}{1\ \text{mol F}_2\text{(g)}}\right)$$

$$= 1.0\ \text{L F}_2\text{(g)}$$

(b) as in part (a),

$$\text{volume of O}_2\text{(g)} = (8.5 \times 10^{-2}\ \text{mol e}^-)\left(\frac{1\ \text{mol O}_2}{4\ \text{mol e}^-}\right)\left(\frac{24.45\ \text{L}}{1\ \text{mol O}_2}\right) = 0.52\ \text{L}$$

Note: We have assumed that the gases are ideal and so occupy $24.45\ \text{L}\cdot\text{mol}^{-1}$ at 298 K and 1.0 atm.

12.60 (a) $\text{Au}^{3+}(\text{aq}) + 3\ \text{e}^- \rightarrow \text{Au(s)}$

$$\text{current} = (6.66\ \mu\text{g Au})\left(\frac{10^{-6}\ \text{g}}{1\ \mu\text{g}}\right)\left(\frac{1\ \text{mol Au}}{196.97\ \text{g Au}}\right)\left(\frac{3\ \text{mol e}^-}{1\ \text{mol Au}}\right)$$

$$\left(\frac{9.6485\times10^4\ \text{C}}{1\ \text{mol e}^-}\right)\left(\frac{1\ \text{A}}{1\ \text{C}\cdot\text{s}^{-1}}\right)\left(\frac{1}{1800\ \text{s}}\right) = 5.44\times10^{-6}\ \text{A}$$

(b) $\text{Cr}_2\text{O}_7^{\ 2-}(\text{aq}) + 14\ \text{H}^+(\text{aq}) + 12\ \text{e}^- \rightarrow 2\ \text{Cr(s)} + 7\ \text{H}_2\text{O(l)}$

$$\text{time} = (6.66\ \mu\text{g Cr})\left(\frac{10^{-6}}{1\ \mu\text{g}}\right)\left(\frac{1\ \text{mol Cr}}{52.00\ \text{g Cr}}\right)\left(\frac{6\ \text{mol e}^-}{1\ \text{mol Cr}}\right)$$

$$\left(\frac{9.6485\times10^4\ \text{C}}{1\ \text{mol e}^-}\right)\left(\frac{1\ \text{A}}{1\ \text{C}\cdot\text{s}^{-1}}\right)\left(\frac{1}{0.100\ \text{A}}\right) = 0.741\ \text{s}$$

12.62 $\text{Mn}^{n+}(\text{aq}) + n\text{e}^- \rightarrow \text{Mn(s)}$; solve for n

$$\text{moles of Mn} = (4.9\ \text{g Mn})\left(\frac{1\ \text{mol Mn}}{54.94\ \text{g}}\right) = 0.089\ \text{mol Mn}$$

$$\text{total charge used} = (13.7\ \text{h})(3600\ \text{s}\cdot\text{h}^{-1})(350\ \text{mA})\left(\frac{10^{-3}\ \text{A}}{1\ \text{mA}}\right)\left(\frac{1\ \text{C}\cdot\text{s}^{-1}}{1\ \text{A}}\right)$$

$$= 1.73\times10^4\ \text{C}$$

$$\text{moles of e}^- = (1.73\times10^4\ \text{C})\left(\frac{1\ \text{mol e}^-}{9.6485\times10^4}\right) = 0.179\ \text{mol e}^-$$

$n = \dfrac{0.179\ \text{mol e}^-}{0.089\ \text{mol Hg}} = \dfrac{2\ \text{mol charge}}{1\ \text{mol Hg}}$; therefore the species is Mn^{2+} (ox. no. = +2)

12.64 $\text{Ti}^{n+} + n\text{e}^- \rightarrow \text{Ti(s)}$; solve for n.

Charge consumed $= 4.70\ \text{C}\cdot\text{s}^{-1}\times 6.00\ \text{h}\times 3600\ \text{s}\cdot\text{h}^{-1} = 1.02\times10^5\ \text{C}$

$$\text{Moles of charge consumed} = \frac{1.02 \times 10^5\ \text{C}}{9.6485 \times 10^4\ \text{C} \cdot \text{mol}^{-1}} = 1.06\ \text{mol e}^-$$

$$\text{Moles of Ti lost} = \frac{12.57\ \text{g Ti}}{47.88\ \text{g Ti/mol Ti}} = 0.2625\ \text{mol Ti}$$

$$n = \frac{1.06\ \text{mol e}^-}{0.2625\ \text{mol Ti}} = \frac{4.04\ \text{mol e}^-}{1.0\ \text{mol Ti}} = (\text{necessarily})\frac{4\ \text{mol e}^-}{1\ \text{mol Ti}}$$

Therefore, the oxidation number is 4, i.e., the species is Ti^{4+}.

12.66 (a) $Zn^{2+} + 2\,e^- \rightarrow Zn(s)$ $(2\ \text{mol e}^-/1\ \text{mol Zn})$

$$\text{charge used} = (1.0\ \text{mA})\left(\frac{10^{-3}\ \text{A}}{1\ \text{mA}}\right)(31\ \text{d})\left(\frac{24\ \text{h}}{1\ \text{d}}\right)\left(\frac{3600\ \text{s}}{1\ \text{h}}\right)$$

$$= 2.7 \times 10^3\ \text{C}$$

$$\text{moles of e}^-\ \text{used} = (2.7 \times 10^3\ \text{C})\left(\frac{1\ \text{mol e}^-}{96\,500\ \text{C}}\right) = 2.8 \times 10^{-2}\ \text{mol e}^-$$

$$\text{moles of Zn} = (2.8 \times 10^{-2}\ \text{mol e}^-)\left(\frac{1\ \text{mol Zn}}{2\ \text{mol e}^-}\right) = 1.4 \times 10^{-2}\ \text{mol Zn}$$

$$\text{mass of Zn} = (1.4 \times 10^{-2}\ \text{mol Zn})\left(\frac{65.37\ \text{g Zn}}{1\ \text{mol Zn}}\right) = 0.92\ \text{g Zn}$$

(b) Given $2\,H^+ + 2\,e^- \rightarrow H_2(g)$ (hydrogen reduction), corresponding to $Zn^{2+} + 2\,e^- \rightarrow Zn(s)$ (zinc reduction), then 1 mol H_2 = 1 mol Zn.

$$\text{moles of Zn} = \frac{0.92\ \text{g Zn}}{65.37\ \text{g} \cdot \text{mol}^{-1}\ \text{Zn}} = 1.4 \times 10^{-2}\ \text{mol Zn}$$

1.4×10^{-2} mol H_2 is thus formed.

$$PV = nRT \text{ and } V = \frac{nRT}{P}$$

$$= \frac{(1.4 \times 10^{-2}\ \text{mol H}_2)(0.083\,14\ \text{L} \cdot \text{bar} \cdot \text{K}^{-1} \cdot \text{mol}^{-1})(298\ \text{K})}{1\ \text{bar}}$$

$V = 0.35$ L

(c) Neither method is ideal for measuring the amount of current compared to today's ammeters. The Zn method has the advantage that the solid zinc is more robust and masses tend to be easier to measure than volumes. Hydrogen, as a gas, is more difficult to keep for long periods of time. It can be readily lost, leading to inaccurate readings. H_2 gas is also potentially hazardous and explosions could result.

12.68 cathode: $2H^+(aq) + 2e^- \rightarrow H_2(g)$ $E^\circ(H^+/H_2) = 0$ V

anode: $Si(s) + 2\ H_2O(l) \rightarrow SiO_2(s) + 4H^+(aq) + 4e^-$ $E^\circ(SiO_2/Si) = -0.84$ V

Total: $Si(s) + 2\ H_2O(l) \rightarrow SiO_2(s) + 2\ H_2(g)$

$E^\circ = E^\circ(H^+/H_2) - E^\circ(H^+, SiO_2/Si) = 0\ V - (-0.84\ V) = 0.84\ V$

12.70 See Box 12.1 and Table 12.2.

(a) $H_2SO_4(aq)$

(b) $PbO_2(s)$

(c) $PbSO_4(s) + H^+(aq) + 2\ e^- \rightarrow Pb(s) + HSO_4^-(aq)$

12.72 The overall reaction in discharge is the sum of the anode and cathode reactions in Table 12.2.

$PbO_2(s) + Pb(s) + 2\ H_2SO_4(aq) \rightarrow 2\ PbSO_4(s) + 2\ H_2O(l)$

So, as the reaction progresses, H_2SO_4 is converted to H_2O. Because the density of H_2SO_4 is greater than that of H_2O, the density of the electrolyte in the battery decreases as the battery is discharged.

12.74 (a) Lead-antimony grids have a larger surface area than smooth plates, so the battery can generate large currents (briefly) for starting a car; also, precipitate can adhere to grids better than to a smooth surface.

(b) $PbSO_4(s)(Pb^{2+})$; see the anode reaction in Table 12.2.

(c) Generally, 6 separate cells are linked in series: $6 \times 2\ V = 12\ V$.

12.76 In order to answer this exercise, we need to identify the reduction potentials of the metal ions.

There are

$Ag^+ + e^- \rightarrow Ag$ $\quad E° = +0.80$ V

$Cu^{2+} + 2\,e^- \rightarrow Cu$ $\quad E° = +0.34$ V

$Ni^{2+} + 2\,e^- \rightarrow Ni$ $\quad E° = -0.23$ V

(a) Metals that have reduction potentials in the range -0.23 V to $+\,0.34$ V will satisfy this requirement. The metals from Appendix 2B that are appropriate include Sn, In, Pb, Fe, and Bi.

(b) Metals that have reduction potentials in the range $+0.34$ to $+\,0.80$ would be appropriate. The only metal in this range that is appropriate is Hg.

(c) Any metal with a reduction potential higher than 0.80 V will leave all three metal ions in solution. These include Au and Pt. Notice that Hg has a potential in this range for oxidation to Hg^{2+}; however, the reduction potential of the Hg_2^{2+}/Hg couple is lower, so that Hg can reduce Ag^+ with the formation of the Hg_2^{2+} ion.

(d) It is not possible to find a metal that will reduce Ag^+ and Ni^{2+} but not Cu^{2+}, as the Cu^{2+} potential lies between those of Ag^+ and Ni^{2+}. Any metal that would not reduce Cu^{2+} would also not reduce Ni^{2+}.

12.78 (a) $Cu^+(aq) + e^- \rightarrow Cu\,(s)$

$$Q = (3.05\,\text{g Cu})\left(\frac{1\,\text{mol Cu}}{63.546\,\text{g Cu}}\right)\left(\frac{1\,\text{mol e}^-}{1\,\text{mol Cu}}\right)\left(\frac{96485\text{ C}}{1\,\text{mol e}^-}\right) = 4631\text{ C}$$

$Sn^{2+}\,(aq) + 2\,e^- \rightarrow Sn\,(s)$

$$m\,(\text{Sn deposited}) = (4631\text{ C})\left(\frac{1\,\text{mol e}^-}{96485\text{C}}\right)\left(\frac{1\,\text{mol Sn}}{2\,\text{mol e}^-}\right)\left(\frac{118.7\text{g Sn}}{1\,\text{mol Sn}}\right)$$

$$= 2.85\text{ g Sn}$$

$Fe^{3+}\,(aq) + 3\,e^- \rightarrow Fe\,(s)$

$$m\,(\text{Fe deposited}) = (4631\ \text{C})\left(\frac{1\,\text{mol}\,e^-}{96485\,\text{C}}\right)\left(\frac{1\,\text{mol Fe}}{3\,\text{mol}\,e^-}\right)\left(\frac{55.847\,\text{g Fe}}{1\,\text{mol Fe}}\right)$$

$$= 0.893\ \text{g Fe}$$

(b) $Q = It;\quad t = Q/I = \dfrac{4631\,\text{C}}{3.5\,\text{A}} = 1.3 \times 10^3$ seconds

12.80 (a) Water containing dissolved ions.

(b) Galvanizing steel (coating it with a film of zinc) is a form of rust protection. Zinc lies below iron in the electrochemical series, so if a scratch exposes the iron, the more strongly reducing zinc can release electrons to the iron. The zinc, not the iron, is oxidized. A sacrificial anode is one that is more easily oxidized than the metal to be preserved from oxidation. Therefore, it is preferentially oxidized, affording the desired protection.

(c) Elements below iron in the electrochemical series, for example Mg, are used as sacrificial anodes to protect the steel hulls of ships. Pb, below iron in the series, would work but would pose environmental hazards.

12.82 (a) Perhaps. Technically speaking, its potential is below the iron couple. From a practical point of view, however, owing to its passivation, it may not readily interact with its surrounding aqueous medium.

(b) Zn: yes, $E°$ below iron couple.

Ag: no, $E°$ above iron couple.

Cu: no, $E°$ above iron couple.

Mg: yes, $E°$ below iron couple.

(c) the moisture (and its dissolved ions) in the surrounding soil

12.84 The appropriate half-reactions are:

$Ti^{4+} + e^- \rightarrow Ti^{3+}$ $\quad E° = 0.00$ $\quad$ (A)

$Ti^{3+} + e^- \rightarrow Ti^{2+}$ $\quad E° = -0.37$ $\quad$ (B)

$Ti^{2+} + e^- \rightarrow Ti$ $\quad E° = -1.63$ $\quad$ (C)

(A), (B), and (C) add to give the desired half-reaction (D):

$Ti^{4+} + 4\,e^- \rightarrow Ti \qquad E° = ? \qquad (D)$

In order to calculate the potential of a half-reaction we need to convert the $E°$ values into $\Delta G°$ values:

$$\Delta G°(A) = -nFE°(A) = -1\,F(0.00\text{ V})$$
$$\Delta G°(B) = -nFE°(B) = -1\,F(-0.37\text{ V})$$
$$\Delta G°(C) = -nFE°(C) = -2\,F(-1.63\text{ V})$$
$$\Delta G°(D) = \Delta G°(A) + \Delta G°(B) + \Delta G°(C)$$
$$-4\,FE°(C) = [-1\,F(0.00\text{ V})] + [-1\,F(-0.37\text{ V})] + [-2\,F(-1.63\text{ V})]$$

The constant $-F$ will cancel from both sides leaving:

$$4\,E°(C) = 1(0.00\text{ V}) + [1(-0.37\text{ V})] + [2(-1.63\text{ V})]$$
$$E°(C) = -3.63\text{ V}/4 = -0.91\text{ V}$$

12.86 The strategy is to find the $E°$ value for the solubility reaction and then find appropriate half-reactions that add to give that solubility reaction. One of these half-reactions (A) is our unknown, the other (B) is obtained from Appendix 2B. The potential for the combination of (A) and (B) is obtained using the K_{sp} value.

$Ni(OH)_2(s) + 2\,e^- \rightarrow Ni(s) + 2\,OH^-(aq) \qquad E° = ? \qquad (A)$

$Ni(s) \rightarrow Ni^{2+}(aq) + 2\,e^- \qquad E° = +0.23\text{ V} \qquad (B)$

$Ni(OH)_2(s) \rightarrow Ni^{2+}(aq) + 2\,OH^-(aq) \qquad E° = -0.51\text{ V} \qquad (C)$

$$E°(C) = \frac{RT \ln K_{sp}}{nF}$$
$$= \frac{(8.314\text{ J}\cdot\text{K}^{-1}\cdot\text{mol}^{-1})(298.2\text{ K})\ln(6.5\times10^{-18})}{2(9.65\times10^{4}\text{ C}\cdot\text{mol}^{-18})}$$
$$= -0.51\text{ V}$$

$$-0.51\text{ V} = E°(A) + (+0.23\text{ V})$$
$$E°(A) = -0.74\text{ V}$$

12.88 See Appendix 2B or Table 12.1.

(a) $O_2 + 4\,H^+ + 4\,e^- \rightarrow 2\,H_2O \quad E°(\text{cathode}) = +1.23\text{ V}$

(b) $O_2 + 2\,H_2O + 4\,e^- \rightarrow 4\,OH^-$ $E°(\text{cathode}) = +0.40$ V

(1) $MnO_4^- + e^- \rightarrow MnO_4^{2-}$ $E° = +0.56$ V

(2) $MnO_4^- + 8\,H^+ + 5\,e^- \rightarrow Mn^{2+} + 4\,H_2O$ $E° = +1.51$ V

The possible combinations of the half-reactions in (a) and (b) with (1) and (2) are

(1) and (a) $4\,MnO_4^- + 2\,H_2O \rightarrow 4\,MnO_4^{2-} + O_2 + 4\,H^+$ $E° = -0.67$

(1) and (b) $4\,MnO_4^- + 4\,OH^- \rightarrow 4\,MnO_4^{2-} + O_2 + 2\,H_2O$ $E° = +0.16$

(2) and (a) $4\,MnO_4^- + 12\,H^+ \rightarrow 4\,Mn^{2+} + 5\,O_2 + 6\,H_2O$ $E° = +0.28$

((2) and (b) give the same result as (2) and (a)).

From these potentials we see that the reduction of MnO_4^- to MnO_4^{2-} is spontaneous in basic solution, whereas reduction to Mn^{2+} is spontaneous in acidic solution. The potential for the reduction in basic solution is less positive, so MnO_4^- is more stable in basic solution.

12.90 Consider $Al^{3+}(aq) + 3\,e^- \rightarrow Al(s)$ $(E° = -1.66)$. With this half-reaction as the anode reaction and one or both of the given reduction reactions, a cell with a positive potential can be constructed. Two adjacent filled teeth, simultaneously in contact with the aluminum, could behave as two independent cells at different potentials, corresponding to the two possible reduction half-reactions. Current will then flow between them, stimulating the pain sensors.

The two possible cell reactions are

$$3\,Hg_2^{2+}(aq) + 4\,Ag(s) + 2\,Al(s) \rightarrow 2\,Ag_2Hg_3(s) + 2\,Al^{3+}(aq) \qquad E°_{cell} = +2.51\text{ V}$$

$$3\,Sn^{2+}(aq) + 9\,Ag(s) + 2\,Al(s) \rightarrow 3\,Ag_3Sn(s) + 2\,Al^{3+}(aq) \qquad E°_{cell} = +1.61\text{ V}$$

12.92 $n\,(\text{KBr needed}) = n\,(Cu^+ \text{ initial}) = 0.035\text{ L} \times 0.012\text{ M} = 4.2 \times 10^{-4}$ mol

$$V\,(\text{KBr added}) = \frac{4.2 \text{ x } 10^{-4} \text{ mol}}{0.010 \text{ M}} \times \frac{1000 \text{ mL}}{1 \text{ L}} = 42.0 \text{ mL}$$

$Cu^{+}(aq) + e^{-} \rightarrow Cu(s)$ $\qquad E° = +0.521$ V

$$E = E° - \frac{0.025693 \text{ V}}{\text{n}} \ln \frac{1}{[Cu^{+}]}$$

Need $[Cu^{+}]$ at equivalence point.

$CuBr(s) \rightarrow Cu^{+}(aq) + Br^{-}(aq)$ $\quad K_{sp} = 5.2 \times 10^{-9}$

$K_{sp} = [Cu^{+}][Br^{-}] = s^{2} = 5.2 \times 10^{-9}$

$s = 7.2 \times 10^{-5}$ M $= [Cu^{+}]$ at the equivalence point

$$E = +0.521 \text{ V} - \frac{0.025693 \text{ V}}{1 \text{mol}} \ln \frac{1}{(7.2 \times 10^{-5})}$$

$E = +0.276$ V is the potential at the equivalence point.

12.94 (a) The chrome-plated piece of metal is the anode. It is immersed in 200 mL of 0.0080 M $CrCl_3(aq)$ in one of the 250-mL beakers. The copper wire is the cathode in 200 mL of 0.12 M $CuSO_4(aq)$ in the other beaker. These two electrodes are connected to each other through the voltmeter by clips and wires. Each end of the salt bridge is submerged in one of the two solutions in the beakers to complete the circuit.

(b) at cathode: $Cu^{2+}(aq) + 2\,e^{-} \rightarrow Cu(s)$ $\quad E°(\text{cathode}) = +0.34$ V

at anode: $Cr(s) \rightarrow Cr^{3+}(aq) + 3\,e^{-}$ $\quad E°(\text{anode}) = -0.74$ V

(c) $3\,Cu^{2+}(aq) + 2\,Cr(s) \rightarrow 3\,Cu(s) + 2\,Cr^{3+}(aq)$ $\quad E°_{cell} = +1.08$ V

(d) $Cr(s)\,|\,0.0080 \text{ M } CrCl_3(aq)||0.12 \text{ M } CuSO_4(aq)|Cu(s)$

(e) $$E_{cell} = E°_{cell} - \frac{0.025\,693 \text{ V}}{n} \ln \frac{[Cr^{3+}]^2}{[Cu^{2+}]^3}$$

$$E_{cell} = 1.08 \text{ V} - \frac{0.025\,693 \text{ V}}{6} \ln \frac{(0.0080)^2}{(0.12)^3}$$

$E_{cell} = 1.08 \text{ V} - 0.004\,28 \ln 0.037$

$E_{cell} = 1.08 \text{ V} + 0.014 \text{ V} = 1.09 \text{ V}$

12.96 (a) $Pb^{2+}(aq) + 2\,e^- \rightarrow Pb(s)$ $E°(\text{cathode}) = -0.13\text{ V}$

$Zn(s) \rightarrow Zn^{2+}(aq) + 2\,e^-$ $E°(\text{anode}) = -0.76\text{ V}$

$Pb^{2+}(aq) + Zn(s) \rightarrow Zn^{2+}(aq) + Pb(s)$ $E°_{\text{cell}} = +0.63\text{ V}$

$$E_{\text{cell}} = E°_{\text{cell}} - \left(\frac{0.0257\text{ V}}{n}\right)\ln\left(\frac{[Zn^{2+}]}{[Pb^{2+}]}\right)$$

$$0.66\text{ V} = 0.63\text{ V} - \left(\frac{0.0257\text{ V}}{2}\right)\ln\left(\frac{[Zn^{2+}]}{0.10}\right)$$

$$-2.33 = \ln\left(\frac{[Zn^{2+}]}{0.10}\right)$$

$$[Zn^{2+}] = 9.7\times10^{-3}\text{ mol}\cdot\text{L}^{-1}$$

(b) We begin with the Nernst equation from part (a)

$$E_{\text{cell}} = 0.63\text{ V} - \left(\frac{0.0257\text{ V}}{2}\right)\ln\frac{[Zn^{2+}]}{(0.10)}$$

$$E_{\text{cell}} = 0.63\text{ V} - (0.0128\text{ V})[\ln[Zn^{2+}] - \ln(0.10)]$$

$$E_{\text{cell}} = 0.63\text{ V} - (0.0128\text{ V})[\ln[Zn^{2+}] + 2.30]$$

$$E_{\text{cell}} = 0.60\text{ V} - 0.0128\ln[Zn^{2+}]$$

12.98 Cell reaction:

$$14\,H^+(aq) + Cr_2O_7^{2-}(aq) + 6\,Fe(CN)_6^{4-}(aq) \rightarrow 2\,Cr^{3+}(aq) + 6\,Fe(CN)_6^{3-}(aq) + 7\,H_2O(l)$$

The equilibrium constant is related to $E°$ for the cell,

$$E° = \frac{0.025693\text{ V}}{n}\ln K$$

We can use the Nernst Equation to get $E°$.

$$E = E° - \frac{0.025693\text{ V}}{6}\ln Q$$

$$1.18 = E° - \frac{0.025693\text{ V}}{6}\ln\frac{(0.01)^2(1\times10^{-4})^6}{(0.1)(1.0)^6}$$

$$1.18 = E° + 0.27$$

$$E° = 0.91\text{ V} = \frac{0.025693\text{ V}}{6}\ln K$$

$\ln K = 2.125 \times 10^2$

$K = 2.0 \times 10^{92}$

12.100 (a) $Cu^{2+} + 2\,e^- \rightarrow Cu \quad E° = +0.34$ V

$Cu^{2+} + e^- \rightarrow Cu^+ \quad E° = +0.15$ V

Here, we do not combine the half-reactions to obtain an overall cell reaction; instead, we combine them to obtain a new half-reaction. Therefore, the procedure $E° = E°(\text{cathode}) - E°(\text{anode})$ **cannot** be used here to obtain $E°_{Cu^+/Cu}$. Reduction potentials are not extensive physical properties and therefore cannot be directly added or subtracted to obtain a third reduction potential. However, the related $\Delta G°_r$ value for the half-reactions is an extensive property, so we first calculate $\Delta G°_{r,Cu^+/Cu}$ and from it obtain $E°_{Cu^+/Cu°}$. Then,

$Cu^{2+} + 2\,e^- \rightarrow Cu$

$Cu^+ \rightarrow Cu^{2+} + e^-$

$Cu^+ + e^- \rightarrow Cu$ (new half-reaction)

$$Cu^{2+}/Cu^+ \quad \Delta G°_r = -nFE° = -(2)(9.65 \times 10^4\ C \cdot mol^{-1})(+0.34\ J \cdot C^{-1})$$
$$= -66\ kJ \cdot mol^{-1}$$

$$Cu^+/Cu \quad \Delta G°_r = -nFE° = -(1)(9.65 \times 10^4\ C \cdot mol^{-1})(-0.15\ J \cdot C^{-1})$$
$$= +15\ kJ \cdot mol^{-1}$$

$$\Delta G°_{r,Cu^+/Cu} = \Delta G°_{r,Cu^{2+}/Cu^+} + \Delta G°_{r,Cu^+/Cu} = -66\ kJ + 15\ kJ$$
$$= -51\ kJ \cdot mol^{-1}$$

$$\text{Then, } E°_{Cu^+/Cu} = \frac{\Delta G°_{r,Cu^+/Cu}}{-nF} = \frac{-51 \times 10^3\ J}{-(1)(9.65 \times 10^4\ C \cdot mol^{-1})} = +0.53\ V$$

12.102 Set up a cell in which one electrode is the silver-silver chloride electrode and the other is the hydrogen electrode. The E of this cell will be sensitive to $[H^+]$ and can be used to obtain pH.

$$2\,H^+(aq) + 2\,e^- \rightarrow H_2(g) \qquad E°(\text{cathode}) = 0.00\text{ V}$$

$$2\,AgCl(s) + 2\,e^- \rightarrow 2\,Ag(s) + 2\,Cl^-(aq) \qquad E°(\text{anode}) = -0.22\text{ V}$$

$$2\,AgCl(s) + H_2(g, 1\text{ atm}) \rightarrow 2\,Ag(s) + 2\,Cl^-(aq) + 2\,H^+(aq)$$

$$E°_{\text{cell}} = 0.22\text{ V}$$

(a) If $[Cl^-] = 1.0\text{ mol}\cdot L^{-1}$:

$$E = E° - \left(\frac{0.0257\text{ V}}{2}\right)\ln([H^+]^2) = 0.22\text{ V} - (0.0257)\ln[H^+]$$

$\ln[H^+] = 2.303\log[H^+] = -2.303(\text{pH})$,
so $E = 0.22\text{ V} + 0.0592\text{ V} \times \text{pH}$, and

$$\text{pH} = \frac{E - 0.22\text{ V}}{0.0592\text{ V}}$$

By measuring E of this cell, pH can be obtained.

(b) $\text{pOH} = 14.00 - \text{pH}$

12.104 (a) The variation of potential for

$$Ag^+ + e^- \rightarrow Ag \qquad E° = +0.80\text{ V}$$

with Ag^+ concentration is derived from the Nernst equation:

$$E = E° - \frac{RT}{nF}\ln Q$$

$$E = +0.80\text{ V} - \frac{0.059\,16}{1}\log\frac{1}{[Ag^+]}$$

$$E = +0.80\text{ V} + 0.059\,16\log[Ag^+]$$

(b) The potentials for $AgX + e^- \rightarrow Ag + X^-$ and K_{sp} values for AgX are

Compound	$E°$	K_{sp}
AgCl	+0.22 V	1.6×10^{-10}
AgBr	+0.07 V	7.7×10^{-13}
AgI	−0.15 V	1.5×10^{-16}

From the K_{sp} values, we can determine $[Ag^+]$ at 1 M X^- for each: $[Ag^+] = K_{sp}$ (because $K_{sp} = [Ag^+][X^-]$). If we use these values in the relationship $E = +0.80\text{ V} + 0.059\ 16 \log[Ag^+]$, we calculate the following potentials:

AgX	$[Ag^+]$, $mol \cdot L^{-1}$	$E = +0.80\text{ V} + 0.059\ 16 \log[Ag^+]$, V
AgCl	1.6×10^{-10}	+0.22
AgBr	7.7×10^{-13}	+0.08
AgI	1.5×10^{-16}	−0.14

As can be seen, these values are essentially, within the limitations of the data, the values obtained as the reduction potentials of the silver halides. This tells us that the solubility determines the concentration of Ag^+ ion in solution, which is the reason the potentials differ.

12.106 The potential for electrolyzing a metal cation will change as the concentration of the metal decreases during an electrolysis reaction. This will require increasing the potential across the cell. We do not need to know the actual beginning concentration in order to calculate this answer. The Nernst equation will allow us to calculate the potential difference between $E1$ and $E2$:

$$E1 = E° - \frac{0.059\,16}{1} \log \frac{1}{[M^+(1)]}$$

$$E2 = E° - \frac{0.059\,16}{1} \log \frac{1}{[M^+(2)]}$$

Subtracting the first equation from the second gives

$$E2 - E1 = -\frac{0.059\,16}{1}\left\{\log \frac{1}{[M^+(2)]} - \log \frac{1}{[M^+(1)]}\right\}$$

$$E2 - E1 = -\frac{0.059\,16}{1} \log \frac{[M^+(1)]}{[M^+(2)]}$$

Because we want 99.99% of the metal to be plated out, the final concentration $[M^+(2)]$ will be $0.0001 \times [M^+(1)]$:

$$E2 - E1 = -\frac{0.059\,16}{1} \log \frac{[M^+(1)]}{0.0001 \times [M^+(1)]} = -0.24\text{ V}$$

The potential will need to be increased by about 0.24 V in order to plate out 99.99% of the metal.

12.108 (a) Galvanic from half cell (1) to half cell (2); therefore, half cell (1) is anode (M is oxidized by X^{2+}). Hence $E°(M^+/M) < E°(X^{+2}/X)$.

Galvanic from half cell (1) to SHE; therefore, half cell (1) is anode (M is oxidized by H_3O^+). Hence $E°(M^+/M) < E°(\text{SHE}) = 0$.

$E°(M^+/M) < 0$; $E°(M^+/M) = -0.25$ V

$E°(X^{+2}/X) > 0$; $E°(X^{+2}/X) = +0.65$ V

(b) Anode: $M \rightarrow M^+ + e^-$ $\quad E° = 0.25$ V

Cathode: $X^{2+} + 2\,e^- \rightarrow X$ $\quad E° = 0.65$ V

Net: $X^{2+} + 2\,M \rightarrow 2\,M^+ + X$ $\quad E° = 0.90$ V

12.110 $P_{Cl_2} = P_{\text{total}} - P_{H_2O}$

$$= 770.\text{ Torr} - 17.54\text{ Torr} = 752\text{ Torr}\left(\frac{1\text{ atm}}{760\text{ Torr}}\right)$$

$$= 0.990\text{ atm}$$

$$n_{Cl_2} = \frac{P_{Cl_2}V}{RT} = \frac{(0.990\text{ atm})(20.0\text{ L})}{(0.08206\text{ atm}\cdot\text{L}\cdot\text{K}^{-1}\text{mol}^{-1})(293.15\text{ K})} = 0.823\text{ mol}$$

$$Q = (0.823\text{ mol Cl}_2)\left(\frac{2\text{ mol e}^-}{1\text{ mol Cl}_2}\right)\left(\frac{96\ 485\text{ C}}{\text{mol e}^-}\right) = 1.59\times10^5\text{ C}$$

$$t = \frac{1.59\times10^5\text{ C}}{2.0\text{ A}} = 7.94\times10^4\text{ s}\left(\frac{1\text{ hour}}{3600\text{ s}}\right) = 22\text{ hours}$$

12.112 The reaction of interest is

(1) $HF(aq) \leftrightarrow H^+(aq) + F^-(aq)$ $\qquad K_a = \frac{[H^+][F^-]}{[HF]} = ?$

We have

(2) $F_2(g) + 2\ H^+(aq) + 2e^- \rightarrow 2\ HF(aq)$ $\qquad E^\circ = +3.03\text{ V}$

and from Appendix 2B,

(3) $F_2(g) + 2e^- \rightarrow 2\ F^-(aq)$ $\qquad E^\circ = +2.87\text{ V}$

Since combining reactions (1/2)(3) – (1/2)(2) = (1),

$$E_{(1)}^\circ = +2.87\text{ V} - 3.03\text{ V} = -0.16\text{ V}$$

$$\Delta G^\circ = -nFE^\circ = -RT\ln K$$

$$K = e^{\frac{nFE^\circ}{RT}} = e^{\frac{(96\ 485)(-0.16)}{(8.314)(298.15)}} = e^{-6.228} = 2.0\times10^{-3}$$

(We note that this value for K_a does not match the one given in Table 10.1 on page 376, 3.5×10^{-4}. There may be a mistake in the data given here or the thermodynamic values may have been determined from different experiments.)

CHAPTER 13
CHEMICAL KINETICS

13.2 (a) $\text{rate}(N_2O_5) = -\text{rate}(O_2) \times \left(\dfrac{2 \text{ mol } N_2O_5}{1 \text{ mol } O_2}\right) = -2 \text{ rate}(O_2)$

(b) $\text{rate}(NO_2) = -\text{rate}(N_2O_5) \times \left(\dfrac{4 \text{ mol } NO_2}{2 \text{ mol } N_2O_5}\right) = -2 \text{ rate}(N_2O_5)$

(c) $\text{rate}(NO_2) = \text{rate}(O_2) \times \left(\dfrac{4 \text{ mol } NO_2}{1 \text{ mol } O_2}\right) = 4 \text{ rate}(O_2)$

13.4 (a) $\text{rate of change of } [ClO^-] = -\text{rate}(O_2) \times \left(\dfrac{3.6 \text{ mol } Cl^-}{L \cdot \text{min}}\right)\left(\dfrac{3 \text{ mol } ClO^-}{2 \text{ mol } Cl^-}\right)$

$$= 5.4 \text{ mol} \cdot L^{-1} \cdot \text{min}^{-1}$$

(b) $3.6 \text{ mol} \cdot L^{-1} \cdot \text{min}^{-1} \div 2 = 1.8 \text{ mol} \cdot L^{-1} \cdot \text{min}^{-1}$

13.6 (a) $\text{rate of formation of } MnO_4^- = \left(2.0 \text{ mol} \dfrac{MnO_4^{2-}}{L \cdot \text{min}}\right)\left(\dfrac{2 \text{ mol } MnO_4^-}{3 \text{ mol } MnO_4^{2-}}\right)$

$=5.4 \text{ mol} \cdot L^{-1} \cdot \text{min}^{-1}$

(b) $\text{rate of reaction of } H^+(aq) = \left(2.0 \text{ mol} \dfrac{MnO_4^{2-}}{L \cdot \text{min}}\right)\left(\dfrac{4 \text{ mol } H^+}{3 \text{ mol } MnO_4^{2-}}\right)$

$= 2.7 \text{ mol } H^+ \cdot L^{-1} \cdot \text{min}^{-1}$

(c) $\text{rate of reaction of } H^+(aq) = \left(2.0 \text{ mol} \dfrac{MnO_4^{2-}}{L \cdot \text{min}}\right)\left(\dfrac{4 \text{ mol } H^+}{3 \text{ mol } MnO_4^{2-}}\right)$

$2.7 \text{ mol } \cdot L^{-1} \cdot \text{min}^{-1} \div 3 = 0.67 \text{ mol } \cdot L^{-1} \cdot \text{min}^{-1}$

13.8 (a) and (c)

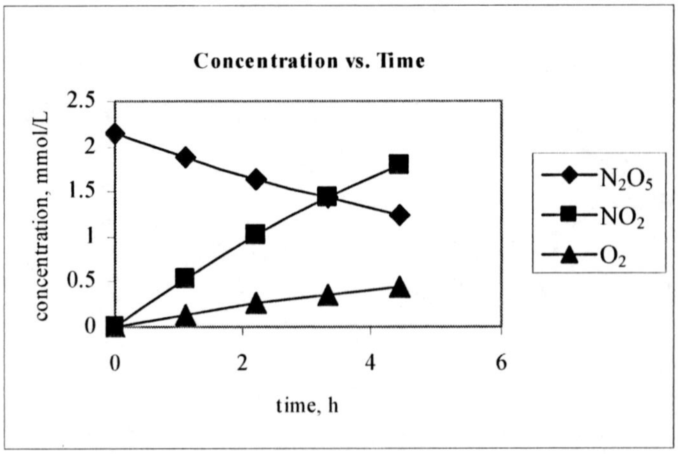

(b) The rates are calculated from the slopes of the lines drawn tangent to the curve at each point. Doing this by hand on the graph leads to a fair amount of error in determination of the slopes. One can also determine the values by taking the derivative of the curve fitted to the data points followed by insertion of the specific time values into the derivative equation. Fitting the data to the function concentration = A·exp(-B·time) the constants A and B are found to be: 2.1514 and 0.1224, respectively. The instantaneous rates are, therefore:

time	rate, $mmol \cdot L^{-1} \cdot h^{-1}$
0	−0.2024
1.11	−0.2092
2.22	−0.2154
3.33	−0.2210
4.44	−0.2260

13.10 (a) rate ($Torr \cdot s^{-1}$) = $k_o P^o = k_o$

$k_o = Torr \cdot s^{-1}$

(b) rate ($Torr \cdot s^{-1}$) = $k_1 P$

$k_1 = (Torr \cdot s^{-1})/Torr = s^{-1}$

(c) rate ($Torr \cdot s^{-1}$) = $k_2 P^2$

$k_2 = (Torr \cdot s^{-1})/(Torr^2) = Torr^{-1} \cdot s\ s^{-1}$

13.12 $\text{rate} = k[C_2H_6] = 5.55\times10^{-4}\,\text{s}^{-1}\left(\dfrac{0.250\text{ g }C_2H_6}{0.500\text{ L}}\right)\left(\dfrac{1\text{ mol }C_2H_6}{30.07\text{ g }C_2H_6}\right)$

$\text{rate} = 9.1\times10^{-6}\,(\text{mol }C_2H_6)\,\text{L}^{-1}\cdot\text{s}^{-1}$

13.14 (a) The units of the rate constant indicate that the reaction is second order, thus

$$\text{rate} = k[NO_2]_0^2 = \left(\frac{0.54\text{ L}}{\text{mol}\cdot\text{s}}\right)\left[\left(\frac{0.420\text{ g}}{0.15\text{ L}}\right)\left(\frac{1\text{ mol }NO_2}{46.01\text{ g }NO_2}\right)\right]^2$$

$$= 2.0\times10^{-3}\,\text{mol}\cdot\text{L}^{-1}\cdot\text{s}^{-1}$$

(b) The rate increases by a factor of $\left(\dfrac{750}{420}\right)^2 = 3.19$

13.16 rate = $k[NO]^a[O_2]^b$, where the orders a and b are to be determined. When the concentration of NO is doubled the rate increases by a factor of 4, that is

$$2^a = 4,\text{ so } a = 2$$

When both the concentration of NO and O_2 are doubled, the rate increases by a factor of 8, that is

$$2^2 \times 2^b = 8,\text{ so } b = 1$$

Therefore, the rate law is

$$\text{rate} = k[NO]^2[O_2]$$

13.18 (a) and (b) rate = $k[NO_2]_0[O_3]_0$, because increasing the concentration of either reactant by a factor while holding the other concentration constant increases the rate by that same factor; the reaction is first order in each reactant and second order overall.

$$k = \frac{\text{rate}}{[NO_2]_0[O_3]_0}$$

(c) $$= \left(\frac{11.4\text{ mmol}}{\text{L}\cdot\text{s}}\right)\left(\frac{\text{L}}{0.38\times10^{-3}\text{ mmol}}\right)\left(\frac{\text{L}}{0.70\times10^{-3}\text{ mmol}}\right)$$

$$= 0.38\times10^{-3}\ \text{L}\cdot\text{mmol}^{-1}\cdot\text{s}^{-1}$$

(d) $\text{rate} = \left(\dfrac{4.3\times10^{-7}\ \text{L}}{\text{mmol}\cdot\text{s}}\right)\left(\dfrac{0.66\times10^{-3}\ \text{mmol}}{\text{L}}\right)\left(\dfrac{0.18\times10^{-3}\ \text{mmol}}{\text{L}}\right)$

$= 5.1\ \text{mmol}\cdot\text{L}^{-1}\cdot\text{s}^{-1}$

13.20 (a) rate = $k[\text{A}]_0^{a}[\text{B}]_0^{b}[\text{C}]_0^{c}$

$$\frac{\text{rate (exp 1)}}{\text{rate (exp 2)}} = \frac{[A]_0^a\ \text{(exp 1)}}{[A]_0^a\ \text{(exp 2)}}$$

$$\left(\frac{3.7}{0.66}\right) = \left(\frac{2.06}{0.87}\right)^a$$

$5.6 = 2.37^a$

by inspection, $a = 2$.

$$\frac{\text{rate (exp 4)}}{\text{rate (exp 3)}} = \frac{[\text{A}]_0^2[\text{C}]_0^c\ \text{(exp 4)}}{[\text{A}]_0^2[\text{C}]_0^c\ \text{(exp 3)}}$$

$$\left(\frac{0.072}{0.013}\right) = \left(\frac{1.00}{0.50}\right)^2\left(\frac{1.00}{0.50}\right)^c$$

$5.5 = 4\times 2.0^c$ or $2.0^c = 1.4$

$c\log 2 = 0.15$

$c = 0.48 \cong 0.5$

$$\frac{\text{rate (exp 2)}}{\text{rate (exp 3)}} = \frac{[\text{A}]_0^2[\text{B}]_0^b[\text{C}]_0^{0.5}\ \text{(exp 2)}}{[\text{A}]_0^2[\text{B}]_0^b[\text{C}]_0^{0.5}\ \text{(exp 3)}}$$

$$\left(\frac{0.66}{0.013}\right) = \left(\frac{0.87}{0.50}\right)^2\left(\frac{3.05}{0.50}\right)^b\left(\frac{4.00}{0.50}\right)^{0.5}$$

$51 = 8.6\times 6.1^b$ or $6.1^b = 5.9$

$b\log 6.1 = 0.77$

$b = 0.99 \cong 1$

Therefore, rate = $k[\text{A}]_0^2[\text{B}]_0[\text{C}]_0^{0.5}$

(b) order = $2 + 1 + \frac{1}{2} = 3.5$

(c) $k = \dfrac{\text{rate}}{[A]_0^2[B]_0[C]_0^{0.5}}$

Using data from experiment 1

$$k = \frac{3.7\times10^{-3}\ \text{mol}\cdot\text{L}^{-1}\cdot\text{s}^{-1}}{\left(2.06\times10^{-3}\ \text{mol}\cdot\text{L}^{-1}\right)^2\left(3.05\times10^{-3}\ \text{mol}\cdot\text{L}^{-1}\right)\left(4.00\times10^{-3}\ \text{mol}\cdot\text{L}^{-1}\right)^{0.5}}$$

$$k = 4.5\times10^{6}\ \left(\text{L}\cdot\text{mol}^{-1}\right)^{5/2}\cdot\text{s}^{-1}$$

13.22 (a) $k = \dfrac{\ln\left(\dfrac{[A]_0}{[A]_t}\right)}{t} = \dfrac{\ln\left(\dfrac{[A]_0}{\frac{1}{4}[A]_0}\right)}{t} = \dfrac{\ln 4}{38\ \text{min}} = 0.036\ \text{min}^{-1}$

(b) $[A]_t = 0.039(\text{mol A})\cdot\text{L}^{-1} - \left(\dfrac{0.0095\ \text{mol B}}{1\ \text{L}}\right)\left(\dfrac{2\ \text{mol A}}{1\ \text{mol B}}\right)$

$= 0.020\ \text{mol}\cdot\text{L}^{-1}$

$$k = \frac{\ln\left(\dfrac{0.039\ \text{mol}\cdot\text{L}^{-1}}{0.020\ \text{mol}\cdot\text{L}^{-1}}\right)}{75\ \text{s}} = 8.9\times10^{-3}\text{s}^{-1}$$

(c) $[A]_t = 0.040(\text{mol A})\cdot\text{L}^{-1} - \left(\dfrac{0.030\ \text{mol B}}{1\ \text{L}}\right)\left(\dfrac{2\ \text{mol A}}{3\ \text{mol B}}\right)$

$= 0.020\ \text{mol}\cdot\text{L}^{-1}$

$$k = \frac{\ln\left(\dfrac{0.040\ \text{mol}\cdot\text{L}^{-1}}{0.020\ \text{mol}\cdot\text{L}^{-1}}\right)}{8.8\ \text{min}} = 7.9\times10^{-2}\text{min}^{-1}$$

13.24 (a) $t_{1/2} = \dfrac{0.693}{k} = \dfrac{0.693}{0.15\ \text{s}^{-1}} = 4.6\ \text{s}$

(b) $[N_2O_5]_t = [N_2O_5]_0 e^{-kt} = 0.0567\ \text{mol}\cdot\text{L}^{-1}\times e^{-\left(0.15\text{s}^{-1}\times 2.0\ \text{s}\right)}$

(c) $t = \dfrac{\ln\left(\dfrac{[A]_0}{[A]_t}\right)}{k} = \dfrac{\ln\left(\dfrac{[N_2O_5]_0}{[N_2O_5]_t}\right)}{k} = \dfrac{\ln\left(\dfrac{0.0567}{0.0135}\right)}{0.15\ \text{s}^{-1}} = 9.6\ \text{s}$

$= (9.6\ \text{s})\left(\dfrac{1\ \text{min}}{60\ \text{s}}\right) = 0.16\ \text{min}$

13.26 $t_{1/2} = \dfrac{0.693}{k} = \dfrac{0.693}{5.74\ \text{h}^{-1}} = 0.121\ \text{h}$

(a) $\dfrac{[\text{A}]}{[\text{A}]_0} = \dfrac{1}{8} = \left(\dfrac{1}{2}\right)^3$; therefore, 3 half-lives elapse

$t = 3 \times 0.121\ \text{h} = 0.363\ \text{h}$

(b) $t = \dfrac{\ln\left(\dfrac{1}{0.05}\right)}{5.74\ \text{h}^{-1}} = 0.52\ \text{h}$

(c) $t = \dfrac{\ln\left(\dfrac{[\text{A}]_0}{[\text{A}]_t}\right)}{k} = \dfrac{\ln 10}{5.74\ \text{h}^{-1}} = 0.40\ \text{h}$

13.28 $C_2H_6(g) \rightarrow 2\ CH_3(g)$

(a) $t_{1/2} = \dfrac{0.693}{k} = \dfrac{0.693}{1.98\ \text{h}^{-1}} = 0.350\ \text{h}$

(b) $t = \dfrac{\ln\left(\dfrac{[N_2O_5]_0}{[N_2O_5]}\right)}{k} = \dfrac{\ln\left(\dfrac{1.15\times10^{-3}\ \text{mol/0.5 L}}{2.35\times10^{-4}\ \text{mol/0.5 L}}\right)}{1.98\ \text{h}^{-1}} = 0.802\ \text{h}$

(c) Because the volume is fixed,

$$\ln\left(\frac{\text{mass}_0}{\text{mass left}}\right) = kt = \left(\frac{1.98}{\text{h}}\right)\left(\frac{1\ \text{h}}{60\ \text{min}}\right)(45\ \text{min}) = 1.48$$

$$\text{mass left} = \frac{\text{mass}_0}{e^{1.48}} = \frac{6.88\ \text{mg}}{4.39} = 1.57\ \text{mg}$$

13.30 $[\text{A}]_{95\text{s}} = 0.035\ \text{mol}\cdot\text{L}^{-1} - \dfrac{1\ \text{mol A}}{2\ \text{mol B}} \times 0.0120\ (\text{mol B})\cdot\text{L}^{-1} = 0.0275\ \text{mol}\cdot\text{L}^{-1}$

$$k = \frac{\ln\left(\dfrac{[\text{A}]_0}{[\text{A}]_t}\right)}{t} = \frac{\ln\left(\dfrac{0.0335}{0.0275}\right)}{95\ \text{s}} = 2.1\times10^{-3}\ \text{s}^{-1}$$

$[\text{A}]_t = 0.0335\ \text{mol}\cdot\text{L}^{-1} - \dfrac{1\ \text{mol A}}{2\ \text{mol B}} \times 0.0190\ (\text{mol B})\cdot\text{L}^{-1} = 0.024\ \text{mol}\cdot\text{L}^{-1}$

$$t = \frac{\ln\left(\dfrac{0.0335}{0.024}\right)}{2.1\times10^{-3}\text{s}^{-1}} = 1.6\times10^{2}\text{s}$$

additional time $= 160\ \text{s} - 95\ \text{s} = 6\times10^{1}\text{s}$

13.32 (a)

time, s	$[I_2](mol{\cdot}L^{-1})$	$1/[I_2](mol{\cdot}L^{-1})$
0	0.00100	1000
1	0.00043	2300
2	0.00027	3700
3	0.00020	5000
4	0.00016	6300

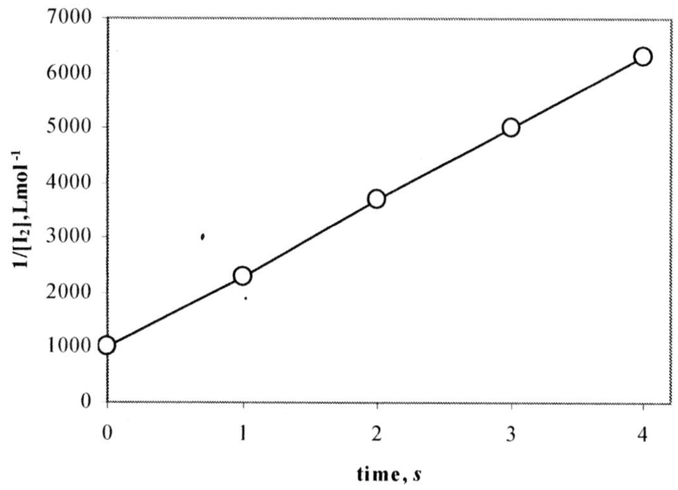

(b) It can be seen that a plot of $1/[I_2]$ versus time is a straight line with an intercept of 1000 L mol^{-1} and a slope of 1300 L mol^{-1} s^{-1}. Therefore, the reaction is second order with a rate constant k = 1.3103 L mol^{-1} s^{-1}.

13.34 Start by finding an expression for the half-life of a second-order reaction:

$$[A]_t = \frac{[A]_0}{1+[A]_0 kt}$$

$$\frac{[A]_{t_{1/2}}}{[A]_0} = \frac{1}{1+[A]_0 kt_{1/2}} = \frac{1}{2}$$

Therefore, $1+[A]_0 kt_{1/2} = 2, \quad kt_{1/2} = 1$, and

$$t_{1/2} = \frac{1}{k[\text{A}]_0}$$

Also, for a second order reaction,

$$k = \frac{\dfrac{1}{[\text{A}]_t} - \dfrac{1}{[\text{A}]_0}}{t}$$

(a) $k = \dfrac{1}{t_{1/2}[\text{A}]_0} = \dfrac{1}{(100\ \text{s})(2.5\times10^{-3}\ \text{mol}\cdot\text{L}^{-1})} = 4.0\ \text{L}\cdot\text{mol}^{-1}\cdot\text{s}^{-1}$

(b) $[\text{A}]_t = \dfrac{[\text{A}]_0}{1+[\text{A}]_0 kt}$

$$[\text{A}] = \frac{0.30\ \text{mol A}}{\text{L}} - \left(\frac{0.010\ \text{mol C}}{\text{L}}\right)\left(\frac{3\ \text{mol A}}{1\ \text{mol C}}\right) = 0.27\ (\text{mol A})\cdot\text{L}^{-1}$$

$$k = \frac{\dfrac{1\ \text{L}}{0.27\ \text{mol}} - \dfrac{1\ \text{L}}{0.30\ \text{mol}}}{200\ \text{s}} = 1.9\times10^{-3}\ \text{L}\cdot\text{mol}^{-1}\cdot\text{s}^{-1}$$

13.36 Using equations introduced in the previous problem:

(a) $t_{1/2} = \dfrac{1}{k[\text{A}]_0} = \dfrac{1}{(0.54\ \text{L}\cdot\text{mol}^{-1}\cdot\text{s}^{-1})(0.20\ \text{mol}\cdot\text{L}^{-1})} = 9.3\ \text{s}$

(b) $\frac{1}{16}[\text{A}]_0 = (0.0625)(0.20\ \text{mol}\cdot\text{L}^{-1}) = 0.0125\ \text{mol}\cdot\text{L}^{-1} = [\text{A}]$

$$t = \frac{\dfrac{1}{[\text{A}]_t} - \dfrac{1}{[\text{A}]_0}}{k} = \frac{\dfrac{1}{0.0125\ \text{mol}\cdot\text{L}^{-1}} - \dfrac{1\ \text{L}}{0.200\ \text{mol}\cdot\text{L}^{-1}}}{0.54\ \text{L}\cdot\text{mol}^{-1}\cdot\text{s}^{-1}} = 1.4\times10^{2}\ \text{s}$$

(c) $\frac{1}{9}[\text{A}]_0 = 0.111...\times 0.20\ \text{mol}\cdot\text{L}^{-1} = 0.022\ \text{mol}\cdot\text{L}^{-1} = [\text{A}]$

13.38 $rate = -\dfrac{1}{a}\dfrac{\text{d}[\text{A}]}{\text{d}t} = k[\text{A}]^2$

$$\frac{\text{d}[\text{A}]}{[\text{A}]^2} = -ak\ \text{d}t$$

integrate from $[\text{A}]_0$ at $t = 0$ to $[\text{A}]_t$ at t:

$$\int_{[\text{A}]_0}^{[\text{A}]_t} \frac{\text{d}[\text{A}]}{[\text{A}]^2} = -ak\int_0^t \text{d}t$$

$[\text{A}]_0^{-1} - [\text{A}]_t^{-1} = -akt$, and $[\text{A}]_t^{-1} = akt + [\text{A}]_0^{-1}$

13.40 Given: $\frac{d[A]}{dt} = -k[A]^n$, we can derive an expression for the amount of time needed for the inital concentration of A, $[A]_0$, to decrease by 1/2. Begin by obtaining the integrated rate law for a third-order reaction by separation of variables:

$$\int_{[A]_0}^{[A]_t} [A]^{-n} d[A] = \int_0^t -k dt = -\frac{1}{n-1}\left[[A]_t^{-(n-1)} - [A]_0^{-(n-1)}\right] = -kt$$

To obtain an expression for the half-life, let $[A]_t = \frac{1}{2}[A]_0$ and $t = t_{1/2}$:

$$-\frac{1}{n-1}\left[(\tfrac{1}{2}[A]_0)^{-(n-1)} - [A]_0^{-(n-1)}\right] = -kt_{1/2}$$

solving for the half-life:

$$t_{1/2} = \frac{-\frac{[A]_0^{-(n-1)}}{n-1}\left(\frac{1}{2}^{-(n-1)} - 1^{-(n-1)}\right)}{k}$$

13.42 The overall reaction is:

2,3-dihydroxy-2,3-dimethylbutane(aq) + $Pb(O_2CCH_3)_4$(aq) →

$2CH_3C(=O)CH_3$(aq) + $Pb(O_2CCH_3)_2$(aq) + 2 CH_3COOH(aq).

The intermediate is:

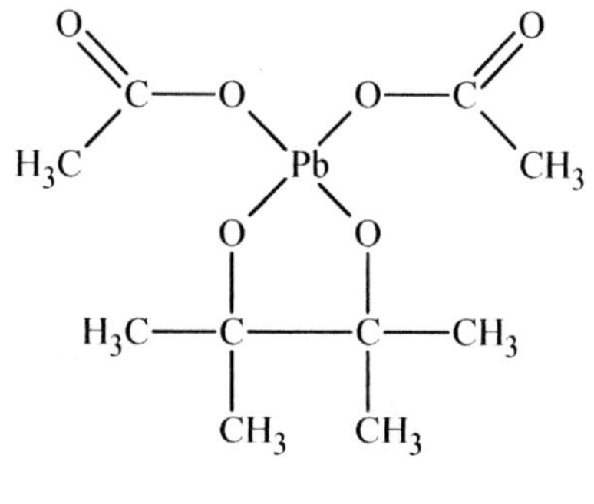

13.44 The second step is rate-determining; therefore,

rate = $k[I^-][HClO]$, but HClO is an intermediate and must be eliminated.

$k_1[ClO^-][H_2O] = k_1'[HClO][OH^-]$

giving

$$[\text{HClO}] = \frac{k_1}{k_1'}\frac{[\text{ClO}^-][\text{H}_2\text{O}]}{[\text{OH}^-]}$$

Substituting, we write: $\text{rate} = k_2\left(\frac{k_1}{k_1'}\right)\frac{[\text{ClO}^-][\text{H}_2\text{O}][\text{I}^-]}{[\text{OH}^-]}$

Because H_2O is the solvent, $[H_2O]$ = constant; therefore

$$\text{rate} = k\frac{[\text{ClO}^-]\,[\text{I}^-]}{[\text{OH}^-]}$$

13.46 The rate increases in proportion to the first power of $[O_2]$ and to the second power of [NO]. Therefore, for $2\ NO(g) + O_2(g) \rightarrow 2\ NO_2(g)$, we can write this rate expression: rate = $k[NO]^2[O_2]$

First, consider mechanism (a):

Step 1:

$$\text{NO} + \text{O}_2 \underset{k_r}{\overset{k_f}{\rightleftharpoons}} \text{NO}_3;\ K_{eq} = \frac{k_f}{k_r} = \frac{[\text{NO}_3]}{[\text{NO}][\text{O}_2]};\ [\text{NO}_3] = K_{eq}[\text{NO}][\text{O}_2]$$

Step 2:

$NO + NO_3 \rightarrow 2\ NO_2$ (slow and, therefore, rate-controlling)

$\text{rate} = k_2[\text{NO}][\text{NO}_3] = k_2K_{eq}[\text{NO}]^2[\text{O}_2]$, or

$\text{rate} = k[\text{NO}]^2[\text{O}_2]$, with $k = k_2K_{eq}$, which agrees with the experimental result.

In mechanism (b), step 1 is rate-controlling and is second order with respect to [NO], but does not involve $[O_2]$. Because the other reactions are fast, the expected rate equation would be: rate = $k[NO]_2$, but this does not agree with the observed rate law.

13.48 (a) True (b) True (c) True

13.50 The overall rate of formation of A is $\text{rate} = -k[\text{A}]^2 + k'[\text{B}]^2$ because the first reaction reduces [A] while the second increases [A]. Given the 1:1 stoichiometry of the reaction, if no B was present at the beginning of the reaction, [A] and [B] at any time are related by the equation:

$[A]+[B]=[A]_o$ where $[A]_o$ is the initial concentration of A. Therefore,

$$\frac{d[A]}{dt}=-k[A]^2+k'([A]_o-[A])^2=(k+k')[A]^2+k'\left([A]_0^2-2[A]_o[A]\right)$$

The solution of this first-order differential equation is:

$$[A]=\frac{k'+ke^{-(k'+k)t}}{k'+k}[A]_o$$

As $t\to\infty$ the concentrations reach their equilibrium values given by:

$$[A]_{eq}=\frac{k'[A]_o}{k'+k}\quad\text{and}\quad[B]_{eq}=[A]_o-[A]_\infty=\frac{k[A]_o}{k+k'}$$

taking the ratio of products over reactants we see that:

$$\frac{[B]_{eq}}{[A]_{eq}}=\frac{k}{k'}=K$$ where K is the equilibrium constant for the reaction.

13.52 (a) Prepare the following table and graph:

T, K	$1/T$, K^{-1}	k, s^{-1}	ln k
660	1.52×10^{-3}	7.2×10^{-4}	−7.2
680	1.47×10^{-3}	2.2×10^{-3}	−6.1
720	1.39×10^{-3}	1.7×10^{-2}	−4.1
760	1.32×10^{-3}	0.11	−2.2

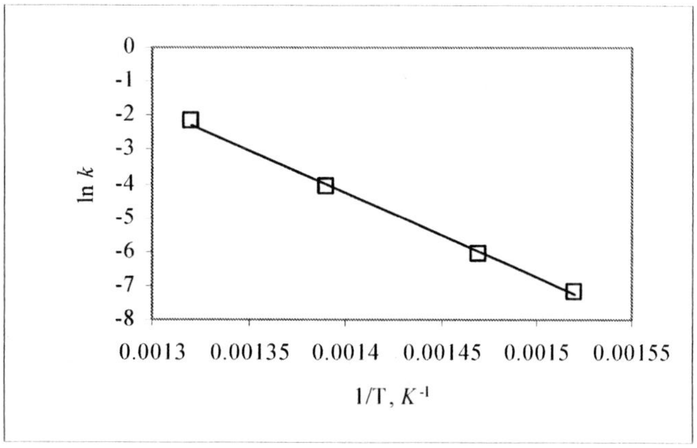

$$\ln k = \ln A - \frac{E_a}{R} \times \frac{1}{T}$$

The slope of the plot can be determined from a graphing program using a least squares fitting routine or from two points on the graph. The slope from the graphing program held here gave a value for the slope of -2.52×10^4. From two points:

$$\text{slope of plot} = -\frac{E_a}{R} = \frac{[-2.2-(-7.2)]K}{(1.32-1.52)\times 10^{-3}} = -2.5 \times 10^4 \text{ K}$$

$E_a = -\text{slope} \times R = 2.5 \times 10^4 \text{ K} \times 8.314 \times 10^{-3} \text{ kJ}\cdot\text{K}^{-1}\cdot\text{mol}^{-1} = 2.1 \times 10^2$ $\text{kJ}\cdot\text{mol}^{-1}$

(b) By inspecting the graph ($T = 400°\text{C} = 673$ K; $\frac{1}{T} = 1.49 \times 10^{-3}$), we can adequately estimate

$\ln k = -6.5$, $k = 2 \times 10^{-3}\ \text{s}^{-1}$

The value of k can also be calculated as follows:

$$\ln\left(\frac{k'}{k}\right) = \frac{E_a}{R}\left[\frac{1}{T} - \frac{1}{T'}\right] = 2.5\times 10^4\,\text{K}\left[\frac{1}{673\text{ K}} - \frac{1}{760\text{ K}}\right] = 4.3$$

$$\frac{k'}{k} = 70 \text{ (one significant figure)},\ k = \frac{0.11}{70} = 2 \times 10^{-3}$$

13.54 k' = rate constant at $T' = 630$ K

$$E_a = \frac{R\ \ln\left(\frac{k'}{k}\right)}{\left(\frac{1}{T} - \frac{1}{T'}\right)} = \frac{\left(0.08314\ \text{kJ}\cdot\text{K}^{-1}\cdot\text{mol}^{-1}\right)\ln\left(\frac{6.0\times 10^{-5}}{2.4\times 10^{-6}}\right)}{\left(\frac{1}{575\text{ K}} - \frac{1}{630\text{ K}}\right)}$$

$$E_a = 1.8\times 10^2\ \text{kJ}\cdot\text{mol}^{-1}$$

13.56 k' = rate constant at $T' = 37°\text{C} = 3130$ K

$$\ln\left(\frac{k'}{k}\right) = \frac{38\ \text{kJ}\cdot\text{mol}^{-1}}{(0.08314\ \text{kJ}\cdot\text{K}^{-1}\cdot\text{mol}^{-1})}\left(\frac{1}{298\ \text{K}} - \frac{1}{310\ \text{K}}\right) - 0.59$$

$$\frac{k'}{k} = 1.8,\ k' = 1.8\times 1.5\times 10^{10}\ \text{L}\cdot\text{mol}^{-1}\cdot\text{s}^{-1} = 2.7\times 10^{10}\ \text{L}\cdot\text{mol}^{-1}\cdot\text{s}^{-1}$$

13.58 $$\ln\left(\frac{k'}{k}\right) = \frac{E_\text{a}}{\text{R}}\left(\frac{1}{T} - \frac{1}{T'}\right) = \frac{E_\text{a}}{\text{R}}\left(\frac{T - T'}{TT'}\right)$$

$$\ln\left(\frac{k'}{k}\right) = \left(\frac{384\ \text{kJ}\cdot\text{mol}^{-1}}{8.314\ \times 10^{-3}\ \text{kJ}\cdot\text{K}^{-1}\cdot\text{mol}^{-1}}\right)\left(\frac{1073\ \text{K}\ -\ 973\ \text{K}}{973\ \text{K}\ \times\ 1073\ \text{K}}\right)$$

$$\ln\left(\frac{k'}{k}\right) = 81$$

$$k' = 81 \times 5.5 \times 10^{-4}\ \text{s}^{-1} = 4.5 \times 10^{-2}\ \text{s}^{-1}$$

13.60 (a) The equilibrium constant will be given by the ratio of the rate constant of the forward reaction to the rate constant of the reverse reaction:

$$K = \frac{k'}{k} = \frac{36.4\ \text{L}\cdot\text{mol}^{-1}\cdot\text{h}^{-1}}{24.3\ \text{L}\cdot\text{mol}^{-1}\cdot\text{h}^{-1}} = 1.50$$

(b) The reaction profile corresponds to a plot similar to that in Fig. 13.31b. The reaction is exothermic—the forward reaction has a lower activation barrier than the reverse reaction.

(c) Raising the temperature will increase the rate constant of the reaction with the higher activation barrier relatively more than that of the reaction with the lower energy barrier. We expect the rate of the forward reaction to go up substantially more than for the reverse reaction in this case. k will increase less than k', and consequently the equilibrium constant K will decrease. This is consistent with Le Chatelier's principle.

13.62 (a) cat = catalyzed, uncat = uncatalyzed

$$E_{\text{a,cat}} = \frac{62}{88}E_\text{a} = 0.705\ E_\text{a}$$

$$\frac{\text{rate(cat)}}{\text{rate(uncat)}} = \frac{k_{\text{cat}}}{k_{\text{uncat}}} = \frac{\text{A}e^{-E_{\text{a,cat}}/RT}}{\text{A}e^{-E_{\text{a}}/RT}} = \frac{e^{-0.705\,E_{\text{a}}/RT}}{e^{-E_{\text{a}}/RT}} = e^{0.295\,E_{\text{a}}/RT}$$

$$= e^{0.295 \times 88\ \text{kJ}\cdot\text{mol}^{-1}/RT} = e^{\frac{26\ \text{kJ}\cdot\text{mol}^{-1}}{RT}} = e^{\frac{26\ \text{kJ}\cdot\text{mol}^{-1}}{(8.314\times10^{-3}\ \text{kJ}\cdot\text{K}^{-1}\cdot\text{mol}^{-1})(300\ \text{K})}}$$

$$= 3 \times 10^4$$

(b) The last step of the calculation is repeated with $T = 500$ K:

$$\frac{\text{rate (cat)}}{\text{rate (uncat)}} = e^{\frac{26\ \text{kJ}\cdot\text{mol}^{-1}}{RT}} = e^{\frac{26\ \text{kJ}\cdot\text{mol}^{-1}}{(8.314\times10^{-3}\ \text{kJ}\cdot\text{K}^{-1}\cdot\text{mol}^{-1})(500\ \text{K})}} = 5 \times 10^2$$

The rate enhancement is lower at higher temperatures.

13.64 cat = catalyzed, uncat = uncatalyzed

$$\frac{\text{rate (cat)}}{\text{rate (uncat)}} = \frac{k_{\text{cat}}}{k_{\text{uncat}}} = 500 = \frac{\text{A}e^{-E_{\text{a,cat}}/RT}}{\text{A}e^{-E_{\text{a}}/RT}}$$

$$\ln 500 = \frac{-E_{\text{a,cat}}}{RT} + \frac{E_{\text{a}}}{RT}$$

$$E_{\text{a,cat}} = E_{\text{a}} - RT \ln 500$$

$$E_{\text{a,cat}} = 106\ \text{kJ}\cdot\text{mol}^{-1} - (8.314 \times 10^{-3}\ \text{kJ}\cdot\text{K}^{-1}\cdot\text{mol}^{-1})(310\ \text{K})(\ln 500)$$

$$= 90\ \text{kJ}\cdot\text{mol}^{-1}$$

13.66 Overall reaction:

$CH_3COOCH_2CH_3(aq) + H_2O(aq) \rightarrow CH_3COOH(aq) \rightarrow CH_3CH_2OH(aq)$

The intermediate is ethoxide ion, $CH_3CH_2O^-$. The catalyst is OH^-.

13.68 (a) True. (b) True. (c) False. The equilibrium constant for a reaction is unaffected by the presence of a catalyst. (d) False. The pathway of a reaction is altered by a catalyst but the beginning and ending points are the same, so the amount of energy released (or gained) is not affected by the presence of a catalyst.

13.70 The Michaelis-Menten mechanism of enzyme reaction:

$$E + S \underset{k_1'}{\overset{k_1}{\rightleftharpoons}} ES$$

$$ES \xrightarrow{k_2} E + \text{product}$$

Where E, S, and ES represent the enzyme, substrate, and enzyme-substrate complex, respectively. The Michaelis constant is:

$$K_M = \frac{k_1' + k_2}{k_1}$$

If this constant increases, the rates of the two reactions that lead to decomposition of the enzyme-substrate complex increase compared to the rate of the enzyme-substrate formation reaction. Therefore, a high Michaelis constant indicates an unstable enzyme-substrate complex.

13.72 (a) The best way to determine reaction order is to prepare plots of [A] versus time, ln[A] versus time, and 1/[A] versus time. The first will give a linear relationship when the reaction is zero order with respect to [A], the second will be linear when the reaction is first order with respect to [A], and the last will be linear when the reaction is second order with respect to [A]. These plots are shown below.

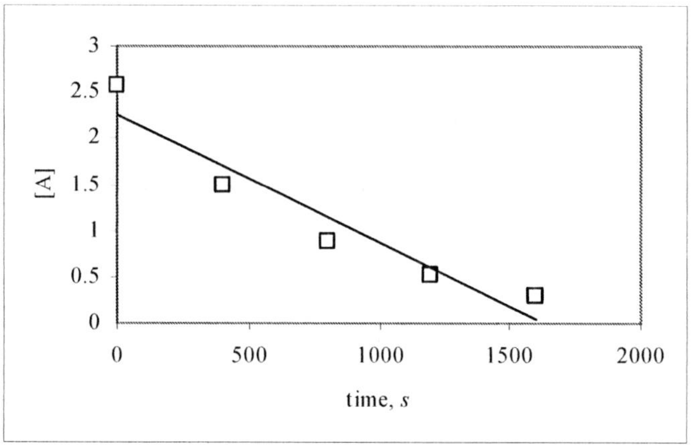

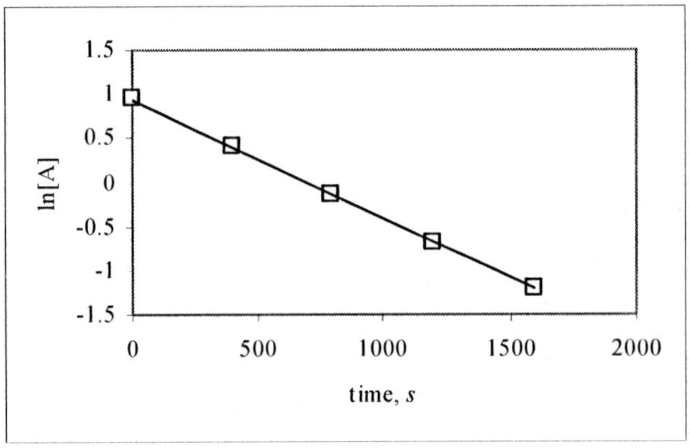

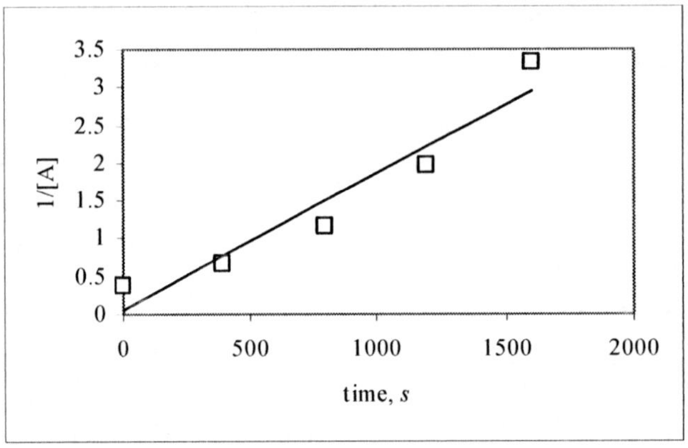

As can be readily seen from the graphs, the only one that gives a straight line relationship is the plot of ln[A] versus time, indicating that the reaction is first order with respect to [A].

(b) The rate constant for the reaction can be determined from the slope of the plot of ln[A] versus time. Because $\ln[A]_t = -kt + \ln[A]_0$, the slope of the line will be $-k$ and the intercept should correspond to $\ln[A]_0$. The equation determined from the graph is $\ln[A] = -0.0013\ \tau + 0.9415$. Therefore, $k = 0.0013\ s^{-1}$ and $[A]_0$ is calculated to be 2.56, which is in good agreement with the known starting concentration.

13.74 (a) $[A]_t = [A]_0 e^{-kt}$

$[H_2O_2]_t = [H_2O_2]_0 e^{-kt}$

$[H_2O_2]_t = 0.35\ mol{\cdot}L^{-1} \times e^{(-0.0410\ min^{-1} \times 10\ min)}$

$= 0.23\ mol{\cdot}L^{-1}$

(b) $\ln\left(\frac{[A]_t}{[A]_0}\right) = kt$

$$\ln\left(\frac{0.50}{0.10}\right) = 0.0410\ \text{min}^{-1} \times t$$

$$t = \frac{\ln 5}{0.0410\ \text{min}^{-1}} = 39\ \text{min}$$

(c) A reduction of $\frac{1}{4}$, means $\frac{3}{4}$ remains.

$$t = \frac{\ln\left(\frac{[H_2O_2]_0}{[H_2O_2]}\right)}{k} = \frac{\ln\left(\frac{4}{3}\right)}{0.0410\ \text{min}^{-1}} = 7.0\ \text{min}$$

(d) A reduction of 75%, means 25% or $\frac{1}{4}$ remains.

$$t = \frac{\ln\left(\frac{[H_2O_2]_0}{[H_2O_2]}\right)}{k} = \frac{\ln\left(\frac{4}{1}\right)}{k} = \frac{1.39}{0.0410\ \text{min}^{-1}} = 34\ \text{min}$$

13.76 Refer to Figure 13.25b. For an exothermic reaction, the activation energy for the reverse reaction is greater than that for the forward reaction.

$E_{a,\text{reverse}} = E_{a,\text{forward}} - \Delta H = 100\ \text{kJ·mol}^{-1} - (-200\ \text{kJ·mol}^{-1}) = 300\ \text{kJ·mol}^{-1}$

13.78 The given mechanism is:

$$A + B \underset{k_1'}{\overset{k_1}{\rightleftharpoons}} C + D$$

$$C + E \xrightarrow{k_2} X + Y$$

(a) the rate of formation of X = k_2[C][E]

The dependence of this rate on the concentration of the intermediate C can be can be removed by applying the steady-state approximation. Setting the rate of formation of C equal to zero and solving for [C]:

rate of formation of C = 0 = k_1 [A][B] – k_1'[C][D] – k_1 [C][E]

$$[\mathrm{C}] = \frac{k_1[\mathrm{A}][\mathrm{B}]}{k_1'[\mathrm{D}] + k_2[\mathrm{E}]}$$

Substituting this expression for [C] into the rate equation for the appearance of X above:

$$\text{rate of formation of X} = \frac{k_2 k_1[\mathrm{A}][\mathrm{B}][\mathrm{E}]}{k_1'[\mathrm{D}] + k_2[\mathrm{E}]}$$

(b) In the limit of a large excess of E, the denominator of the expression above simplifies to k_2[E] (k_2[E] being much larger than k_1'[D]) and the expression for the rate of formation of X becomes:

$$\text{rate of formation of X} = \frac{k_2 k_1[\mathrm{A}][\mathrm{B}][\mathrm{E}]}{k_2[\mathrm{E}]} = k_1[\mathrm{A}][\mathrm{B}]$$

(c) This expression makes physical sense in that if there is a large excess of E, any small amount of C produced will quickly react to form X and Y and stand little chance of reacting with D to reform the reactants. The overall rate of the mechanism would, therefore, depend only on the rate-limiting step, A + B → C + D, making the rate second order overall.

13.80 The overall reaction is 2 $CH_3C(=O)CH_3 \rightarrow CH_3C(=O)CH_2C(OH)(CH_3)_2$. The intermediates are

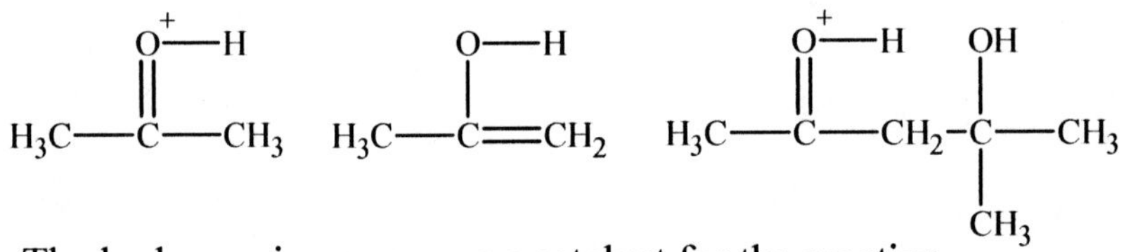

The hydrogen ion serves as a catalyst for the reaction.

13.82 (a)

T, K	$1/T$, K^{-1}	k, 10^{-3} $L\cdot mol^{-1}\cdot s^{-1}$	$\ln(k \times 10^3)$	$\ln k$
297	0.003 37	4.8	1.57	–5.34
301	0.003 32	7.8	2.05	–4.85
305	0.003 28	13	2.56	–4.34
309	0.003 24	20	3.00	–3.91
313	0.003 19	32	3.47	–3.44

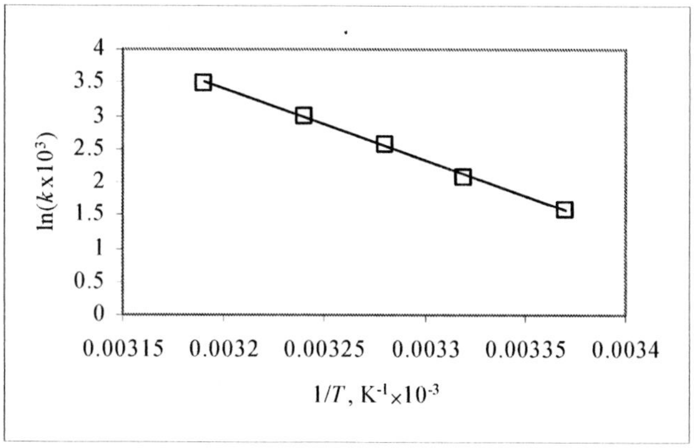

$$\ln k = \ln A - \frac{E_a}{RT}$$

slope of straight line above (the ordinates are given as ln k)

$$\text{slope of line} = \frac{(3.47 - 1.57)\text{ K}}{0.003\ 19 - 0.003\ 37} = -1.1 \times 10^4 \text{ K} = -\frac{E_a}{R}$$

$E_a = (1.1 \times 10^4 \text{ K})(8.314\ 51 \times 10^{-3} \text{ kJ}\cdot\text{K}^{-1}\cdot\text{mol}^{-1}) = 91 \text{ kJ}\cdot\text{mol}^{-1}$

(b) $T = 310$ K, $1/T = 3.23 \times 10^{-3}$ K^{-1}

From the plot, $\ln(k \times 10^3) \cong 3.10$, $k \times 10^3 \cong 22$, $k \cong 2.2 \times 10^{-2}$ $L\cdot mol^{-1}\cdot s^{-1}$

(c) The balanced equation is

sucrose(aq) $\rightarrow$ glucose(aq) + fructose(aq)

We will use the enthalpies of formation for the solid based upon the assumption that the solvation of the sugars is negligible. In fact, this is not unreasonable, because the solvation energy of the sucrose should largely compensate for the solvation of the glucose and fructose in the Hess' Law calculation.

$\Delta H°_{rxn} = -1268 \text{ kJ}\cdot\text{mol}^{-1} + -1266 \text{ kJ}\cdot\text{mol}^{-1} - (-2222 \text{ kJ}\cdot\text{mol}^{-1})$

$= -312 \text{ kJ·mol}^{-1}$

The overall reaction profile will be

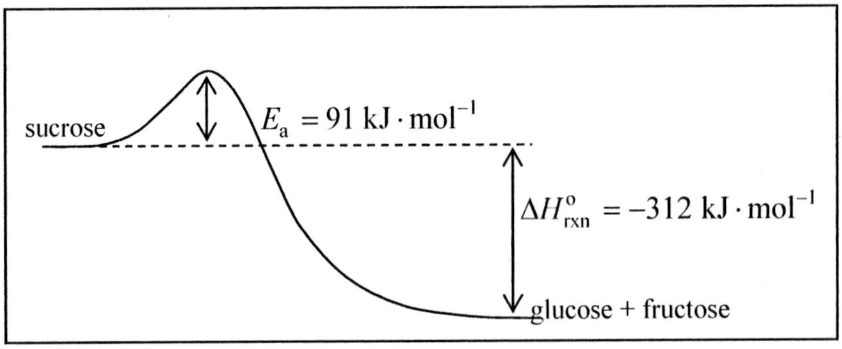

13.84 (a) The mechanism assumed is

$$E + S \underset{k_1'}{\overset{k_1}{\rightleftharpoons}} ES$$

$$ES \xrightarrow{k_2} E + P$$

Where E is the free enzyme, S is the substrate, ES is the enzyme-substrate complex, and P is the product.

The rate-determining step is the formation of products from the activated complex ES, so we can write

rate = k_2 [ES]

But because ES is an intermediate, we cannot leave it in the final rate expression. We will make a steady-state approximation in order to determine the concentration of ES present in solution. We also need to realize that $[E]_0$ is the starting concentration of enzyme. Because a significant portion of the enzyme will be bound to the substrate, the actual concentration of E at any given time will be equal to $[E]_0 - [ES]$.

In the steady-state approach, the rate of formation of the activated complex ES added to its rate of disappearance will be set equal to 0.

Rate of formation of ES = $k_1[E][S] = k_1([E]_0 - [ES])[S]$

Rate of disappearance of ES = $-k_2 [ES] + -k_1'[ES]$

$k_1([E]_0 - [ES])[S] + (-k_2 [ES] - k_1'[ES]) = 0$

$k_1([E]_0 - [ES])[S] - (k_2 + k_1'[ES]) = 0$

$k_1([E]_0 - [ES])[S] = (k_2 + k_1'[ES])$

$k_1[E]_0[S] - k_1[ES][S] = (k_2 + k_1')[ES]$

$(k_2 + k_1')[ES] + k_1[ES][S] = k_1[E]_0[S]$

$[ES](k_2 + k_1' + k_1[S]) = k_1[E]_0[S]$

$$[ES] = \frac{k_1[E]_0[S]}{k_2 + k_1' + k_1[S]}$$

The rate expression then becomes

$$\text{rate} = k_2[ES] = \frac{k_2 k_1[E]_0[S]}{k_2 + k_1' + k_1[S]}$$

Dividing both numerator and denominator by k_1, we obtain

$$\text{rate} = \frac{k_2[E]_0[S]}{\frac{k_2 + k_1'}{k_1} + [S]}$$

$$= \frac{k_2[E]_0[S]}{K_M + [S]}$$

q.e.d.

[S]

(b) It is easy to confirm that the rate is independent of substrate concentration at high concentrations of substrate by noting that under these conditions $[S] >> K_M$ so that $K_M + [S] \approx [S]$. The [S] in numerator and denominator will cancel, leaving rate = $k_2[E]_0$.

13.86 According to collision theory, the rate of a reaction is dictated by the frequency of collisions and the fraction of collisions that supply sufficient energy to initiate the reaction. As stated in the "How Do We Do That" box on page 513:

Rate of reaction = (collision frequency) × (fraction with sufficient energy)

The collision frequency is calculated using:

collision frequency = $\sigma v_{rel} N_A^2[A][B]$

where the relative velocity, v_{rel} is:

$$v_{rel} = \left(\frac{8RT}{\pi\mu}\right)^{1/2}$$

and the fraction of collisions providing at least enough energy to overcome the activation energy for the reaction is given by:

fraction with sufficient energy $= e^{-E_a/RT}$

If we take the reactants to be A and B, the ratio of reaction rates: rate (320 K) / rate (300 K) is:

$$\frac{\text{rate (320 K)}}{\text{rate (300 K)}} = \frac{\sigma\left(v_{rel,\,320\text{ K}}\right)N_A^2[\text{A}][\text{B}]}{\sigma\left(v_{rel,\,300\text{ K}}\right)N_A^2[\text{A}][\text{B}]}\cdot\frac{e^{-E_a/R\cdot 320\text{ K}}}{e^{-E_a/R\cdot 300\text{ K}}}$$

$$= \frac{\sigma\left(\left(\frac{8R\cdot 320\text{ K}}{\pi\mu}\right)^{1/2}\right)N_A^2[\text{A}][\text{B}]}{\sigma\left(\left(\frac{8R\cdot 300\text{ K}}{\pi\mu}\right)^{1/2}\right)N_A^2[\text{A}][\text{B}]}\cdot\frac{e^{-E_a/R\cdot 320\text{ K}}}{e^{-E_a/R\cdot 300\text{ K}}}$$

Canceling those terms which are not dependent on the temperature, this expression simplifies to:

$$\frac{\text{rate (320 K)}}{\text{rate (300 K)}} = \frac{(320\text{ K})^{1/2}}{(300\text{ K})^{1/2}}\cdot\frac{e^{-E_a/R\cdot 320\text{ K}}}{e^{-E_a/R\cdot 300\text{ K}}}$$

$$= 1.033\cdot e^{(-E_a/R\cdot 320\text{ K})+(E_a/R\cdot 300\text{ K})}$$

$$= 1.033\cdot e^{\left(25\text{ kJ}\cdot\text{mol}^{-1}/0.0083145\text{ kJ}\cdot\text{K}^{-1}\cdot\text{mol}^{-1}\right)\left(\frac{1}{300\text{ K}}-\frac{1}{320\text{ K}}\right)}$$

$$= 1.9$$

13.88 The reaction between A and B has two possible pathways that lead to different products:

$A + B \rightarrow C + D$ (1)

$A + B \rightarrow E + F$ (2)

Reaction (1) proceeds by a mechanism in which A first fragments into C and G. Mechanism for reaction (1):

$A \rightarrow C + G$ slow

$B + G \rightarrow D$ fast

Reaction (2) proceeds by a direct combination of A and B having a second-order, bimolecular rate law.

(a) Reaction (1) will have a rate law dependent only upon the concentration of A, rate = k_1[A], whereas reaction (2) will have a rate law that depends upon both A and B, rate = k_2[A][B]. If we wish the rate of reaction (2) to proceed faster, then we can increase the concentration of [B]. This should not affect the rate of reaction (1).

(b) The rate at which A disappears will be determined by summing the rates of consumption of A for both pathways: $-\dfrac{d[\text{A}]}{t} = k_1[\text{A}] + k_2[\text{A}][\text{B}]$.

(c) We can use the equation from (b) to determine the rate constants because we know the rates at two sets of concentrations. The rate of disappearance of A for Experiment 1 is going to be 0.0045 $\text{mol·L}^{-1}\text{·h}^{-1}$ + 0.0015 $\text{mol·L}^{-1}\text{·h}^{-1}$ (because 1 mole of A is consumed for each mole of C produced and for each mole of E produced). For Experiment 2, the value is 0.0045 $\text{mol·L}^{-1}\text{·h}^{-1}$ + 0.000 50 $\text{mol·L}^{-1}\text{·h}^{-1}$. This data allows us to construct two simultaneous equations:

$$0.0060\ \text{mol·L}^{-1}\text{·h}^{-1} = k_1(0.100\ \text{mol·L}^{-1}) + k_2(0.100\ \text{mol·L}^{-1})(0.100\ \text{mol·L}^{-1})$$

$$0.0050\ \text{mol·L}^{-1}\text{·h}^{-1} = k_1(0.100\ \text{mol·L}^{-1}) + k_2(0.100\ \text{mol·L}^{-1})(0.050\ \text{mol·L}^{-1})$$

or

$$0.0060\ \text{mol·L}^{-1}\text{·h}^{-1} = k_1(0.100\ \text{mol·L}^{-1}) + k_2(0.0100\ \text{mol}^2\text{·L}^{-2})$$

$$0.0050\ \text{mol·L}^{-1}\text{·h}^{-1} = k_1(0.100\ \text{mol·L}^{-1}) + k_2(0.0050\ \text{mol}^2\text{·L}^{-2})$$

Solving these simultaneously, we obtain

$k_2 = 0.020\ \text{L·mol}^{-1}\text{·h}^{-1}$ and $k_1 = 0.040\ \text{h}^{-1}$

13.90 Let P = pollutant

(a) rate = R – $k[\text{P}]_{eq}$ = 0 at equilibrium

$$[\text{P}]_{eq} = +\left(\frac{\text{R}}{k}\right)$$

(b) Because $[P]_{eq}$ = constant, there is no overall half-life, but on average, for an individual molecule, there is a 50% probability that it will have decayed after $t_{1/2}$ where

$$t_{1/2} = \frac{0.693}{k}$$

13.92 (a) $K = \frac{k_1}{k_1'}$ (1)

(b) From Section 13.7

$$\ln\frac{k_1}{k_2} = -\frac{E_{\text{a(forward)}}}{R}\left(\frac{1}{T_1} - \frac{1}{T_2}\right) \quad (2)$$

$$\ln\frac{k_1'}{k_2'} = -\frac{E_{\text{a(reverse)}}}{R}\left(\frac{1}{T_1} - \frac{1}{T_2}\right) \quad (3)$$

(c) and (d) To show the relationship between the kinetic and thermodynamic treatment, we need to find an expression for K in terms of the rate constants. This can be done by subtracting equation (3) from equation (2):

$$\ln\frac{k_1}{k_2} - \ln\frac{k_1'}{k_2'} = -\frac{E_{\text{a(forward)}}}{R}\left(\frac{1}{T_1} - \frac{1}{T_2}\right) - \left[-\frac{E_{\text{a(reverse)}}}{\text{r}}\left(\frac{1}{T_1} - \frac{1}{T_2}\right)\right]$$

$$\ln\left(\frac{k_1}{k_2}\cdot\frac{k_2'}{k_1'}\right) = -\left(\frac{E_{\text{a(forward)}}}{R} - \frac{E_{\text{a(reverse)}}}{R}\right)\left(\frac{1}{T_1} - \frac{1}{T_2}\right)$$

$$\ln\frac{K_1}{K_2} = -\frac{(E_{\text{a(forward)}} - E_{\text{a(reverse)}})}{R}\left(\frac{1}{T_1} - \frac{1}{T_2}\right)$$

This is now in a form similar to the van't Hoff equation derived in Chapter 9:

$$\ln\frac{K_1}{K_2} = -\frac{\Delta H^\circ_{\text{r}}}{R}\left(\frac{1}{T_1} - \frac{1}{T_2}\right)$$

in which $\Delta H^\circ_{\text{r}}$ corresponds to $E_{\text{a(forward)}} - E_{\text{a(reverse)}}$.

13.94 (a) Table 13.1 gives the rate constant for the reaction. Using the data from Appendix 2A, the equilibrium constant for the reaction cyclopropane → propene can be calculated at 773 K. From that value and the value for the rate of conversion of cyclopropane to propene, the rate constant for the conversion of propene to cyclopropane can be estimated:

$\Delta H°_r = 20.42\ \text{kJ·mol}^{-1} - 53.30\ \text{kJ·mol}^{-1} = -32.88\ \text{kJ·mol}^{-1}$

$\Delta S°_r = 266.6\ \text{J·K}^{-1}\text{·mol}^{-1} - 237.4\ \text{J·K}^{-1}\text{·mol}^{-1} = 29.2\ \text{J·K}^{-1}\text{·mol}^{-1}$

$\Delta G°_r = -32.88\ \text{kJ·mol}^{-1} - (773\ \text{K})(29.2\ \text{J·K}^{-1}\text{·mol}^{-1})/(1000\ \text{J·kJ}^{-1})$

$= -55.5\ \text{kJ·mol}^{-1}$

$\Delta G°_r = -RT \ln K$

$(-55.5\ \text{kJ·mol}^{-1})(1000\ \text{J·kJ}^{1}) = (8.314\ \text{J·K}^{1}\text{·mol}^{1})(773\ \text{K}) \ln K$

$\ln K = 8.64$

$K = 5.7 \times 10^3$

$$K = \frac{k}{k'}$$

$$5.7 \times 10^3 = \frac{6.7 \times 10^{-4}\ \text{s}^{-1}}{k'}$$

$k' = 1.2 \times 10^{-7}\ \text{s}^{-1}$

(b) The reaction is exothermic, which means that the activation energy for the reverse reaction will be larger than the activation energy for the forward reaction. This means that the rate constant for the conversion of propene to cyclopropane will be decreased more than the rate constant of the conversion of cyclopropane to propene upon decreasing the temperature. The result is that K will become larger as T is lowered.

13.96 $CH_3OH + H^+ \underset{k_1'}{\overset{k_1}{\rightleftharpoons}} CH_3OH_2^+$

$CH_3OH_2^+ + Br^- \xrightarrow{k_1} CH_3Br + H_2O$

(a) The rate law will be based upon the second step, which is the slow step of the reaction.

$\text{rate} = k_2[CH_3OH_2^+][Br^-]$

The rate law cannot be left in this form because $CH_3OH_2^+$ is a reactive intermediate. We can express the concentration of $CH_3OH_2^+$ in terms of the reactants and products, using the fast equilibrium established in the first step.

$$K = \frac{k_1}{k_1'} = \frac{[CH_3OH_2^+]}{[CH_3OH][H^+]}$$

$$[CH_3OH_2^+] = \frac{k_1}{k_1'}[CH_3OH][H^+]$$

The rate then becomes

$$\text{rate} = \frac{k_2 k_1}{k_1'}[CH_3OH][H^+][Br^-]$$

The steady state approximation is similar to the fast equilibrium approach, except that k_2 may not be very much smaller than k_1'. This means that the rate of the reaction to final products is fast enough to disturb the equilibrium established in the pre-equilibrium step. To use this approximation, we set the rate of formation of the intermediate equal to its rate of disappearance:

rate of formation of $CH_3OH_2^+ = k_1[CH_3OH][H^+]$

rate of disappearance of $CH_3OH_2^+ = k_1'[CH_3OH_2^+] + k_2[CH_3OH_2^+][Br^-]$

$k_1'[CH_3OH_2^+] + k_2[CH_3OH_2^+][Br^-] = k_1[CH_3OH][H^+]$

$[CH_3OH_2^+](k_1' + k_2[Br^-]) = k_1[CH_3OH][H^+]$

$$[CH_3OH_2^+] = \frac{k_1[CH_3OH][H^+]}{k_1' + k_2[Br^-]}$$

We then place this expression in the same rate law as in part (a).

$$\text{rate} = \frac{k_2 k_1[CH_3OH][H^+][Br^-]}{k_1' + k_2[Br^-]}$$

(b) When k_2 is very much less than k_1', the expressions will be the same. They will also be the same at low concentrations of Br^-.

(c) At high concentrations of Br^-, the steady-state approximation rate will have $k_2[Br^-] >> k_1'$. Under those conditions, the rate law becomes rate = $k_1[CH_3OH][H^+]$. The fast equilibrium rate law would not change at high concentrations of Br^-.

13.98 The reaction profile as given:

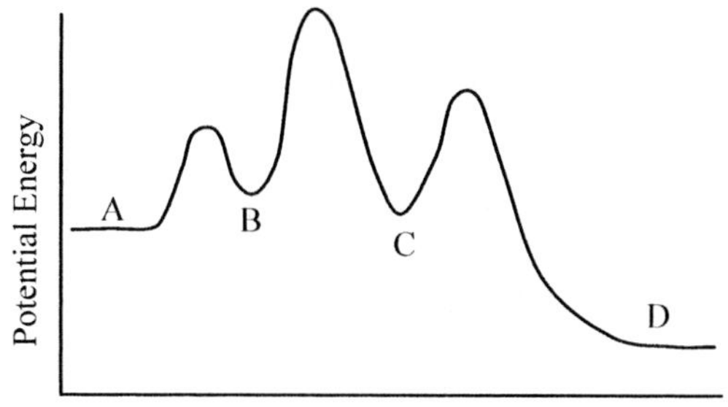

(a) Overall, the reaction is exothermic. The product possesses less potential energy than the reactant. (b) There are two intermediates, B and C. (c) There are three activated complexes, one complex at the maximum in the curve between the reactant A and the intermediate B, an activated complex between the intermediates B and C, and a third activated complex between the intermediate C and the final product D. (d) The reaction to form C from B would be rate limiting as this step has the largest activation barrier. (e) The reaction to form D from C would be the fastest as the energy released on going from the activated complex between the intermediates B and C to the intermediate C is sufficient to send C over the final reaction barrier to D.

13.100 (a) To convert from one isomer to the other, a single C-C bond is broken and a single C-N bond is formed, giving an approximate enthalpy of reaction of:

$305\ \text{kJ}\cdot\text{mol}^{-1} - 348\ \text{kJ}\cdot\text{mol}^{-1} = -43\ \ \text{kJ}\cdot\text{mol}^{-1}$. (b) The rate constant at 300 K is found using:

$\ln\frac{k'}{k}=\frac{E_a}{R}\left(\frac{1}{T}-\frac{1}{T'}\right)$. Substituting into this equation we find:

$$\ln\frac{k'}{6.6\times10^{-4}\ \text{s}^{-1}}=\frac{161000\ \text{J}\cdot\text{mol}^{-1}}{8.3145\ \text{J}\cdot\text{K}^{-1}\cdot\text{mol}^{-1}}\left(\frac{1}{500\ \text{K}}-\frac{1}{300\ \text{K}}\right)$$

Solving for k':

$k'=4.04\times10^{-15}\ \text{s}^{-1}$

At this rate, the time required to reduce the initial concentration of the isomer by 75% is found by:

$$\ln\frac{[\text{CH}_3\text{NC}]_t}{[\text{CH}_3\text{NC}]_0}=\ln(0.75)=-k\cdot t=-4.04\times10^{-15}\ \text{s}^{-1}\cdot t$$

$t=7.12\times10^{13}\ \text{s}$

(c) We begin by finding the rate constant at which the concentration of the isomer will reduce by 75% in one hour:

$$\ln\frac{[\text{CH}_3\text{NC}]_t}{[\text{CH}_3\text{NC}]_0}=\ln(0.75)=-k\cdot t=-k\cdot 1\ \text{h}\left(\frac{60\ \text{min}}{\text{h}}\right)\left(\frac{60\ \text{s}}{\text{min}}\right)$$

Solving for k:

$k=7.99\times10^{-5}\ \text{s}^{-1}$

We can now determine at what temperature this rate constant is reached using the equation from part (b) above:

$$\ln\frac{k'}{k}=\frac{E_a}{R}\left(\frac{1}{T}-\frac{1}{T'}\right).$$

$$\ln\frac{7.99\times10^{-5}\ \text{s}^{-1}}{6.6\times10^{-4}\ \text{s}^{-1}}=\frac{161000\ \text{J}\cdot\text{mol}^{-1}}{8.3145\ \text{J}\cdot\text{K}^{-1}\cdot\text{mol}^{-1}}\left(\frac{1}{500\ \text{K}}-\frac{1}{T}\right)$$

Solving for T:

$T=474\ \text{K}$

(d) Argon serves as a collision partner. Collisions between the CH_3NC (g) and argon atoms provide the energy needed to overcome the activation energy and allow the isomerization reaction to proceed toward products. The argon atoms also serve as an energy sink, accepting the energy released during the isomerization reaction. (e) At high argon concentrations, collisions resulting in reaction are between the CH_3NC (g) and argon atoms:

CH_3NC (g) + Ar (g) → CH_3CN (g) +Ar (g).

Given this elementary reaction one would predict that the reaction was first order with respect to CH_3NC (g) concentration and would appear first order overall if the concentration of Ar (g) was large and unchanging. If, however, the concentration of Ar (g) was greatly reduced, the isomerization reaction would also proceed through collisions between CH_3NC (g) molecules, making the reaction second order with respect to CH_3NC (g).

CHAPTER 14

THE ELEMENTS: THE FIRST FOUR MAIN GROUPS

14.2 (a) thallium (b) bismuth (c) tin (d) aluminum

14.4 (a) rubidium (b) germanium (c) selenium (d) aluminum

14.6 bismuth < arsenic < nitrogen

14.8 fluorine

14.10 tin

14.12 (a) $SnCl_2$. The radius of a metal atom with a higher oxidation number is smaller than that for the same metal ion with a lower oxidation number. (b) Here we should consider the Lewis structures:

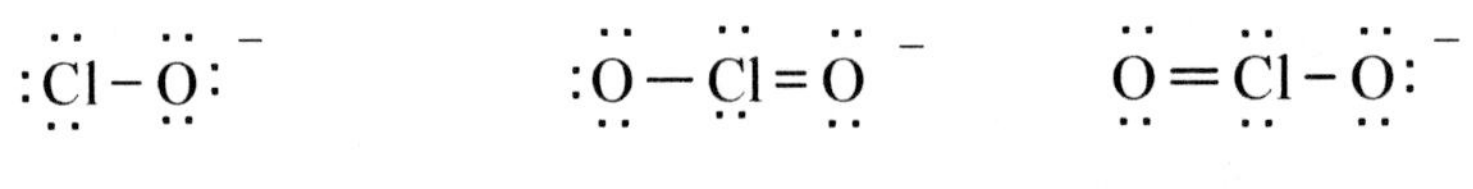

single bond resonance structures give a bond order of 1.5

We expect the Cl–O bond in $Cl-O^-$ to be longer because it is a single bond, whereas the Cl–O bond order in ClO_2^- is 1.5 and the bond should be shorter. One could also predict the Cl–O bond distance in ClO_2^- to be shorter because the Cl atom is in a higher oxidation state. (c) PH_3. The radius of phosphorus is larger than the radius of nitrogen.

14.14 $Ca(s) + H_2(g) \rightarrow CaH_2(s)$

14.16 (a) molecular (b) molecular (c) saline (d) metallic

14.18 (a) basic (b) basic (c) amphoteric (d) acidic

14.20 (a) HNO_3 (c) H_3PO_4 (d) H_2SeO_4

14.22 (a) $NaH(s) + H_2O(l) \rightarrow NaOH(aq) + H_2(g)$

(b) $CH_4(g) + H_2O(g) \xrightarrow{800^\circ C,\ Ni} CO(g) + 3\ H_2(g)$

(c) $H_2C{=}CH_2(g) + H_2(g) \xrightarrow{\Delta, \text{pressure},\ Ni} CH_3CH_3(g)$

The oxidation number of C in $H_2C{=}CH_2$ is -2, and -3 in $H_3C{-}CH_3$. C has been reduced.

(d) $Mg(s) + 2\ HCl(aq) \rightarrow MgCl_2(aq) + H_2(g)$

14.24 (a) $CO(g) + H_2O(g) \rightarrow CO_2(g) + H_2(g)$

$$\Delta H^\circ_r = \Delta H^\circ_f(CO_2, g) - [\Delta H^\circ_f(CO, g) + \Delta H^\circ_f(H_2O, g)]$$

$$= (-393.51\ kJ\cdot mol^{-1}) - [(-110.53\ kJ\cdot mol^{-1}) + (-241.82\ kJ\cdot mol^{-1})]$$
$$= -41.16\ kJ\cdot mol^{-1}$$

(b) $\Delta S^\circ_r = S^\circ(CO_2, g) + S^\circ(H_2, g)$

$$- [S^\circ(CO, g) + S^\circ(H_2O, g)]$$
$$= 213.74\ J\cdot K^{-1}\cdot mol^{-1} + 130.68\ J\cdot K^{-1}\cdot mol^{-1} - [197.67\ J\cdot K^{-1}\cdot mol^{-1}$$
$$+ 188.83\ J\cdot K^{-1}\cdot mol^{-1}] = -42.08\ J\cdot K^{-1}\cdot mol^{-1}$$

(c) $CO(g) + H_2O(g) \rightarrow CO_2(g) + H_2(g)$

$$\Delta G^\circ_r = \Delta G^\circ_f(CO_2, g) - [\Delta G^\circ_f(CO, g) + \Delta G^\circ_f(H_2O, g)]$$
$$= (-394.36\ kJ\cdot mol^{-1})$$
$$- [(-137\cdot 17\ kJ\cdot mol^{-1}) + (-228.57\ kJ\cdot mol^{-1})]$$
$$= -28.62\ kJ\cdot mol^{-1}$$

ΔG°_r can also be calculated from ΔH°_r and ΔS°_r:

$$\Delta G^\circ_r = \Delta H^\circ_r - T\Delta S^\circ_r$$
$$= -41.16\ \text{kJ}\cdot\text{mol}^{-1} - (298\ \text{K})(-42.08\ \text{J}\cdot\text{K}^{-1}\cdot\text{mol}^{-1})/(1000\ \text{J}\cdot\text{kJ}^{-1})$$
$$= -28.62\ \text{kJ}\cdot\text{mol}^{-1}$$

14.26 (a) $N_2(g) + 3\ H_2(g) \xrightarrow{400\text{-}600^\circ\text{C, Fe, 150-160 atm}} 2\ NH_3(g)$

(b) $H_2(g) + F_2(g) \rightarrow 2\ HF(g)$

(c) $2\ Cs(s) + H_2(g) \xrightarrow{\Delta} 2\ CsH(s)$

(d) $Cu^{2+}(aq) + H_2(g) \rightarrow Cu(s) + 2\ H^+(aq)$

14.28 (a) $Li \rightarrow Li^+ + e^-$ $\qquad E^\circ = +3.05\ \text{V}$

$Cu^{2+} + 2\ e^- \rightarrow Cu$ $\qquad E^\circ = +0.34\ \text{V}$

$2\ Li + Cu^{2+} \rightarrow Cu + 2\ Li^+$ $\qquad E^\circ = +3.39\ \text{V}$

The maximum potential possible is + 3.39 V.

(b) The difficulty is isolating the two half cells but still maintaining electrical contact. Ions need to flow through the system to maintain charge balance in the reaction. In this case, a material that allows appropriate counter anions but not Cu^{2+} ions or water to pass through would be necessary. It is necessary to keep the Li compartment free of water because lithium and water react vigorously. Not only would this destroy the cell but it would circumvent the desired electrochemical process.

14.30 (a) ns^1

(b) It is relatively easy to remove the one valence electron of the alkali metals. All of them have a low first ionization energy.

14.32 (a) $K(s) + O_2(g) \rightarrow KO_2(s)$

(b) $Na_2O(s) + H_2O(l) \rightarrow 2\ NaOH(aq)$

(c) $2\ Li(s) + 2\ HCl(aq) \rightarrow 2\ LiCl(aq) + H_2(g)$

(d) $2\ Cs(l) + I_2(s) \rightarrow 2\ CsI(s)$

14.34 (a) $\Delta G°_r = -2\Delta G°_f(NaCl, s) = -2(-384.14\ kJ \cdot mol^{-1}) = 768.28\ kJ \cdot mol^{-1}$

(b) The reduction of sodium ions is a one-electron process,

$Na^+ + 1\ e^- \rightarrow Na(s);$

therefore, $1\ mol\ Na^+ = 1\ mol\ e^-$.

$$\text{current} = \left(\frac{400.\ g\ Na}{4.00\ h}\right)\left(\frac{1\ mol\ Na}{22.99\ g\ Na}\right)\left(\frac{96\ 485\ C}{1\ mol\ Na}\right)\left(\frac{1\ h}{3600\ s}\right)$$

$$= 117\ C \cdot s^{-1} = 117\ A$$

14.36 $Ba(s) + H_2(g) \rightarrow BaH_2(s)$

14.38 Due to the greater polarizing ability of the smaller Mg^{2+} ion on the carbonate ion, we expect $MgCO_3$ to be less stable than $CaCO_3$. Therefore, we expect that the oxide formed is MgO and that the carbonate is $CaCO_3$. Furthermore, for $MgCO_3(s) \rightarrow MgO(s) + CO_2(g)$, $\Delta G°_r$ is roughly +50 kJ, and for $CaCO_3(s) \rightarrow CaO(s) + CO_2(g)$, $\Delta G°_r$ is roughly +130 kJ (as calculated from the data in Appendix 2A). Thus, the formation of MgO versus that of CaO is thermodynamically favored.

14.40 The trend of decreasing lattice enthalpies down the group is related to the size of the ions (which increases down the group). Smaller ions have a greater concentration of charge than larger ions and can more strongly attract ions of opposite charge. Thus, the ions at the top of the group have larger lattice enthalpies than those at the bottom.

14.42 (a) $Mg^{2+}(aq, seawater) + Ca(OH)_2(aq) \rightarrow Mg(OH)_2(s) + Ca^{2+}(aq)$ then $Mg(OH)_2(s) + 2\ HCl(aq) \rightarrow MgCl_2(aq) + 2\ H_2O(l)$, and after drying and melting $MgCl_2(l) \xrightarrow{\text{electrolysis}} Mg(s) + Cl_2(g)$

(b) $Ca(s) + 2\ H_2O(l) \rightarrow Ca(OH)_2(aq) + H_2(g)$

14.44 (a) $Mg(s) + Br_2(l) \rightarrow MgBr_2(s)$

(b) $3\ BaO(s) + 2\ Al(s) \xrightarrow{\Delta} Al_2O_3(s) + 3\ Ba(s)$

(c) $CaO(s) + SiO_2(s) \xrightarrow{\Delta} CaSiO_3(l)$

14.46

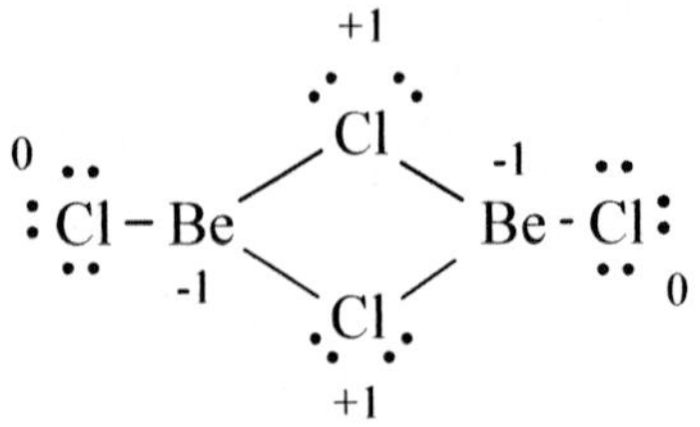

14.48 (a) $CaO(s) + H_2O(l) \rightarrow Ca(OH)_2(aq)$

$$\Delta H° = -1002\cdot 82\ kJ \cdot mol^{-1} - (-635.09\ kJ \cdot mol^{-1} - 285.83\ kJ \cdot mol^{-1})$$
$$= -81.9\ kJ \cdot mol^{-1}$$

$$q = 25\ kg\ CaO\left(\frac{1\ mol\ CaO}{56.08 \times 10^{-3}\ kg\ CaO}\right)\left(\frac{-81.9\ kJ}{1\ mole}\right)$$
$$= -37 \times 10^3\ kJ$$

(b) The $Ca(OH)_2(aq)$ is first treated with CO_2 to precipitate insoluble $CaCO_3$ from the solution. The $CaCO_3$ is then heated to produce CaO and CO_2.

14.50 $B_2O_3(s) + 3\ Mg(s) \xrightarrow{\Delta} 2\ B(s) + 3\ MgO(s)$

14.52 (a) $Al_2O_3(s) + 2\ OH^-(aq) + 3\ H_2O(l) \rightarrow 2[Al(OH)_4]^-(aq)$

(b) $Al_2O_3(s) + 6\ H_3O^+(aq) + 3\ H_2O(l) \rightarrow 2[Al(H_2O)_6]^{3+}(aq)$

(c) $2\ B(s) + 2\ NH_3(g) \xrightarrow{\Delta} 2\ BN(s) + 3\ H_2(g)$

14.54 (a) BF_3: industrial catalyst used for its Lewis acidity

(b) $NaBH_4$: industrial reducing agent

(c) $Al_2(SO_4)_3$ is used in papermaking as a cellulose coagulant.

14.56 (a) Assuming that BF will display the same molecular orbital energy level order as N_2, the ground-state electron configuration will be

$\sigma_{1s}{}^2\sigma_{1s}^{*\,2}\sigma_{2s}{}^2\sigma_{2s}^{*\,2}\pi_{2p}{}^4\sigma_{2p}{}^2$;

$\text{BO} = \frac{1}{2}(10 - 4) = 3$

(b) The molecular orbitals in CO and BF differ in shape and energy since the atoms involved in bonding have different atomic orbital energies and sizes. The orbitals of BF will more closely resemble those of HF while those of CO will more closely resemble those of N_2 (see Chapter 3).

14.58 $2\ Al_2O_3(s) + 3\ C(gr) \rightarrow 4\ Al(s) + 3\ CO_2(g)$

$$3.6 \times 10^6 \text{ tonnes} = 3.6 \times 10^9 \text{ kg} = 3.6 \times 10^{12} \text{ g}$$

$$\text{mass of carbon lost} = (3.6 \times 10^{12} \text{ g}) \times \left(\frac{1 \text{ mol Al}}{26.98 \text{ g Al}}\right)\left(\frac{3 \text{ mol C}}{4 \text{ mol Al}}\right)\left(\frac{12.01 \text{ g C}}{1 \text{ mol C}}\right) = 1.2 \times 10^{12} \text{ g C}$$

14.60 $\Delta G^\circ_r = -nFE^\circ$, for $Al^{3+}(aq) + 3\ e^- \rightarrow Al(s)$

$$n = 3 = -3 \times (9.65 \times 10^4 \text{ C}\cdot\text{mol}^{-1})(-1.66 \text{ V})$$
$$= 4.80 \times 10^5 \text{ J}\cdot\text{mol}^{-1} = 480 \text{ kJ}\cdot\text{mol}^{-1}$$

The formation reaction for Al^{3+} (aq) is the reverse of the half-reaction above. Therefore, $\Delta G^\circ_f(Al^{3+}, aq) = -480 \text{ kJ}\cdot\text{mol}^{-1}$. The standard free energy of formation of the ions in solution includes contributions from a number of sources, chief among them being the ionization energy of the metal and the hydration energy of the ion. The third ionization energy of Al is less than that of Tl, and its energy of hydration is much greater due to its much smaller size. So Al^{3+} (aq) has a much greater tendency to form than Ti^{3+}; that is, its $\Delta G°_f$ would tend to be more negative.

14.62 Carbon occurs widely in the earth. Coal is largely carbon, though many other substances are present. Coal is converted into coke by distillation of its volatile components. Pure graphite is obtained from the coke by passing a large electric current through rods of the coke for several days. Carbon black, finely divided graphite, can be produced by the destructive distillation of gaseous hydrocarbons, which can be obtained from petroleum.

14.64 Carbon atoms have a special ability to form π-bonds; silicon atoms do not form them. This can be partly understood in terms of atomic size. In the larger silicon atoms, the sideways overlap of *p*-orbitals that results in π-bonds becomes much less effective. Because the distance between nuclei is greater in silicon than in carbon, the *p*-orbitals cannot overlap as strongly. Silicon occurs only in a diamond-like crystalline form; it does not form a graphite-like structure because of the weakness of π-bonding in silicon.

14.66 (a) $Sn(s) + 2\ OH^-(aq) + 2\ H_2O \rightarrow [Sn(OH)_4]^{2-} + H_2(g)$

(b) $C(s) + H_2O(g) \xrightarrow{\Delta} CO(g) + H_2(g)$

(c) $CH_4(g) + H_2O(g) \xrightarrow{800°C,\ Ni} CO(g) + 3\ H_2(g)$

(d) $Al_4C_3(s) + 12\ H_2O(l) \rightarrow 4\ Al(OH)_3(s) + 3\ CH_4(g)$

14.68 Refer to Fig. 14.43. Each Si atom is at the center of a tetrahedron formed from four O atoms at the corners. Each corner O atom is joined to another Si atom in a neighboring tetrahedron. The Si—O—Si bond angle should be about 109.5°.

14.70 $2\,CO(g) + O_2(g) \rightarrow 2\,CO_2(g)$

$$\Delta H^\circ_r = \Delta H^\circ_r(\text{products}) - \Delta H^\circ_r(\text{reactants})$$

$$\Delta H^\circ_r = (2)(-393.51\ \text{kJ}\cdot\text{mol}^{-1}) - [(2)(-110.53\ \text{kJ}\cdot\text{mol}^{-1})]$$

$$= -565.96\ \text{kJ}\cdot\text{mol}^{-1}$$

$$\Delta S^\circ_r = S^\circ(\text{products}) - S^\circ(\text{reactants})$$

$$\Delta S^\circ_r = (2)(213.74\ \text{J}\cdot\text{K}^{-1}\cdot\text{mol}^{-1}) - [(2)(+197.67\ \text{J}\cdot\text{K}^{-1}\cdot\text{mol}^{-1}) + 205.14\ \text{J}\cdot\text{K}^{-1}\cdot\text{mol}^{-1}]$$

$$\Delta S^\circ_r = -173.00\ \text{J}\cdot\text{K}^{-1}\cdot\text{mol}^{-1}$$

$$\Delta G^\circ_r = \Delta H^\circ_r - T\Delta S^\circ_r$$

$$\Delta G^\circ_r = -565.96\ \text{kJ}\cdot\text{mol}^{-1} - (298\ \text{K})(-173.00\ \text{J}\cdot\text{K}^{-1}\cdot\text{mol}^{-1})/(1000\ \text{J}\cdot\text{kJ}^{-1})$$

$$\Delta G^\circ_r = -5.144\times 10^2\ \text{kJ}\cdot\text{mol}^{-1}$$

$$\Delta G^\circ_r = -RT\ln K = \Delta H^\circ_r - T\Delta S^\circ_r$$

when $K = 1$, $\ln K = 0$, $\Delta G^\circ_r = 0$, and $T\Delta S^\circ_r = \Delta H^\circ_r$

$$T = \frac{\Delta H^\circ_r}{\Delta S^\circ_r} = \frac{-565.96\times 10^3\ \text{J}\cdot\text{mol}^{-1}}{-173.00\ \text{J}\cdot\text{K}^{-1}\cdot\text{mol}^{-1}} = 3271\ \text{K}\ (\text{for } T < 3271\ \text{K},\ K < 1)$$

14.72 $SiCl_4$ can act as a Lewis acid, whereas CCl_4 cannot. The silicon atom is much bigger than the carbon atom and can expand its octet by using its *d*-orbitals, so it can accommodate the lone pair of an attacking Lewis base. A carbon atom is much smaller and has no low-lying *d*-orbitals, so it cannot act as a Lewis acid.

14.74 (a) If each silicon tetrahedron shares 2 O atoms, the empirical formula of the anion will be SiO_3^{2-}. This may exist either as rings (as in $Si_3O_9^{6-}$) or in chains. The formula is K_2SiO_3.

(b) If the tetrahedra share 3 O atoms, they will tend to form sheet-like layers of silicate anions. The simplest formula for such an arrangement is $[Si_4O_{10}]^{4-}$ giving a formula of $K_4Si_4O_{10}$ or $K_2Si_2O_4$.

14.76 Fluorite is CaF_2. It reacts with H_2SO_4 as follows:

$$CaF_2(s) + H_2SO_4(aq) \rightarrow CaSO_4(s) + 2\ HF(aq)$$

The HF produced in the reaction then etches the glass, forming SiF_6^{2-}.

14.78
$$SiO_2 + 2\ OH^- \rightarrow SiO_2(OH)_2^{2-}$$
$$SiO_2 + 3\ OH^- \rightarrow SiO_3(OH)^{3-} + H_2O$$
$$SiO_2 + 4\ OH^- \rightarrow SiO_4^{4-} + 2\ H_2O$$
$$2\ SiO_2 + 6\ OH^- \rightarrow Si_2O_7^{6-} + 3\ H_2O$$
$$SiO_2 + 2\ OH^- \rightarrow SiO_3^{2-} + H_2O$$

14.80 In the aluminosilicate before heating, the particles are held together by relatively weak forces, such as dipole-dipole interactions and hydrogen bonds. The particles have O—H groups covalently bonded to their surfaces. Heating causes these O—H groups to eliminate water in a procedure analogous to forming an anhydride. When the water is eliminated, the bonds holding the particles together are covalent and the rigidity of the structure is increased.

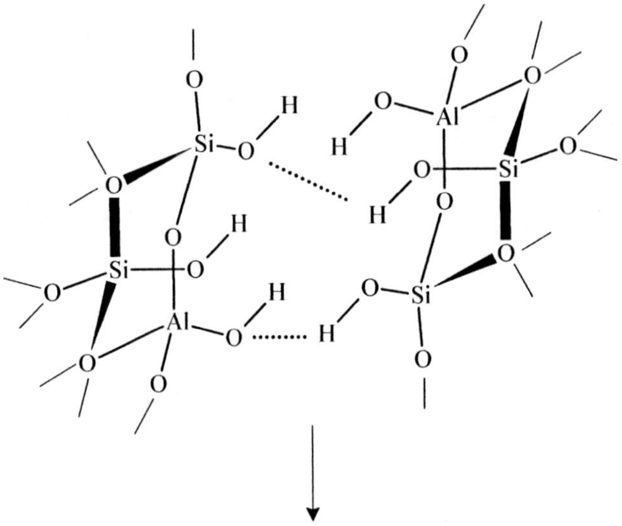

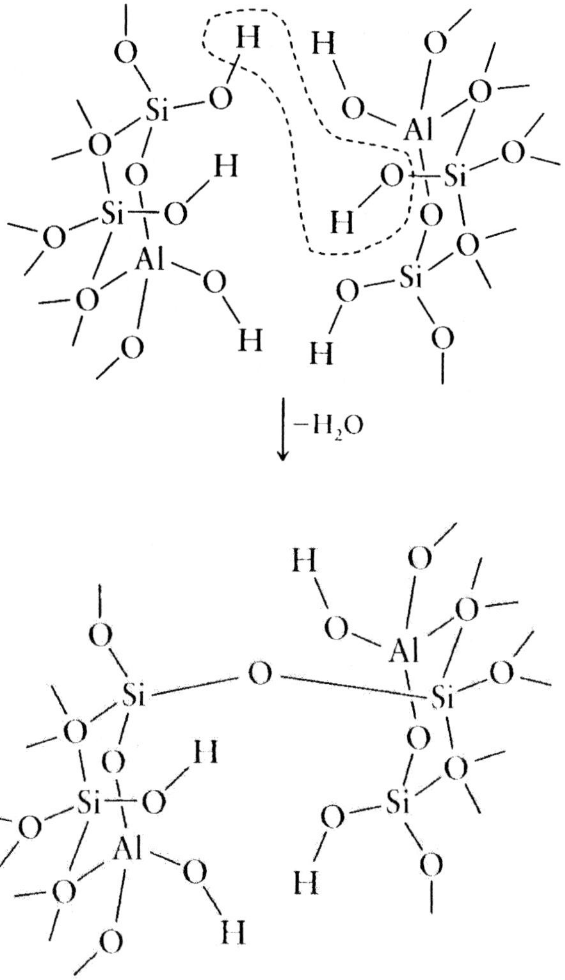

14.82 $Li^+ + e^- \rightarrow Li$ $\qquad E° = -3.05$ V

$S + 2\,e^- \rightarrow S^{2-}$ $\qquad E° = -0.48$ V

The potential for the overall reaction $2\,Li + S \rightarrow Li_2S$ will be

$E° = -0.48 \text{ V} - (-3.05 \text{ V}) = +2.57 \text{ V}$.

14.84

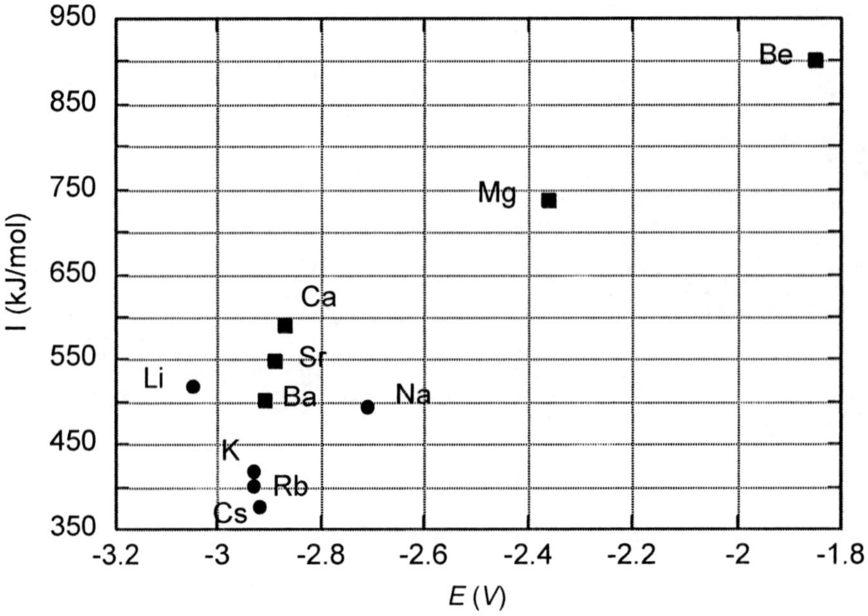

In general, as the atomic number increases down a Group (Group 1 Li to Na, Group 2 Be to Ba), both the first ionization energy and the standard reduction potential decrease (algebraically). This trend is seen more clearly in Group 2 than in Group 1. This correlation between ionization energy and standard reduction potential is not surprising, because both are measures of the ease of removal of electrons from atoms, though under different conditions. Discussion of the anomalous behavior of Li, due to the very small size of Li^+, is given in the solution to Exercise 14.83; this graph confirms the anomaly.

14.86 In aqueous solution, the Be^{2+} ion exists as $[Be(H_2O)_4]^{2+}$. It can react as an acid, with water acting as a base:

$$[Be(H_2O)_4]^{2+}(aq) + H_2O(l) \rightleftharpoons [Be(H_2O)_3OH]^{+}(aq) + H_3O^{+}(aq)$$

$$[Be(H_2O)_3OH]^{+}(aq) + H_2O(l) \rightleftharpoons Be(H_2O)_2(OH)_2 + H_3O^{+}(aq)$$

$$Be(H_2O)_2(OH)_2(aq) + H_2O(l) \rightleftharpoons [Be(H_2O)(OH)_3]^{-} + H_3O^{+}(aq)$$

$$[Be(H_2O)(OH)_3]^{-} + H_2O(l) \rightleftharpoons [Be(OH)_4]^{2-} + H_3O^{+}(aq)$$

14.88

Element	Electronegativity
Ge	2.0
Sn	2.0
Tl	2.0
In	1.8
Ga	1.6
Al	1.6

	Element	$E°$(V)
Increasing ↑	Al	−1.66
reducing	Ga	−0.53
strength	Tl	−0.34
	Sn	−0.14
	In	−0.14

NOTE: The $E°$ value for Ge is not available in Appendix 2B.

14.90 (a) SiH_4, as opposed to CH_4, can function as a Lewis acid and can consequently accept the lone pair of an attacking Lewis base. This results from the larger size of the Si atom and the presence of low-lying empty *d*-orbitals, which are not present in carbon.

(b) $SiH_4(g) + 2\,H_2O(l) \xrightarrow{OH^-} SiO_2(s) + 4\,H_2(g)$

14.92 All ammonium salts are soluble in water, as are all Group 1 salts; all these ions have a + 1 charge; the ionic radii of the Group 1 cations range from 58 pm for Li^+ to 170 pm for Cs^+, and the ammonium ion is in this range, ~149 pm. The ammonium ion forms ionic compounds analogous to the compounds of the Group 1 cations; for example, NH_4Cl is analogous to NaCl, $(NH_4)_2CO_3$ is analogous to Na_2CO_3, etc.

14.94 Mg [Ne] $3s^2$

Ca [Ar] $4s^2$

Sr [Ar] $3d^{10}\,4s^2\,4p^6\,5s^2$

Ba [Kr] $4d^{10}\,5s^2\,5p^6\,6s^2$

Ra [Kr] $4d^{10}\,4f^{14}\,5s^2\,5p^6\,5d^{10}\,6s^2\,6p^6\,7s^2$

Al [Ne] $3s^2\,3p^1$

Ga [Ar] $3d^{10}\,4s^2\,4p^1$

In $[Ar]\ 3d^{10}\ 4s^2\ 4p^6\ 4d^{10}\ 5s^2\ 5p^1$

Tl $[Kr]\ 4d^{10}\ 4f^{14}\ 5s^2\ 5p^6\ 5d^{10}\ 6s^2\ 6p^1$

In passing from Mg to Ca down Group 2, the outer electron configuration remains the same, ns^2, so the ionization energy is expected to decrease as electrons in the $4s^2$ sublevel are farther from the nucleus than those in the $3s^2$ sublevel and are well shielded. In passing from Al to Ga down Group 13, the nuclear charge increases by 18, but the shielding does not increase proportionately, because the electron configuration has changed from $ns^2\ np^1$ to $(n-1)\ d^{10}\ ns^2\ np^1$, and *d*-electrons are not as effective at shielding outer electrons as *s*- or *p*-electrons. Therefore, the outer electron in Ga "sees" a higher effective nuclear charge than the outer electron in aluminum, and its ionization energy increases rather than decreases. Similar arguments apply to the increase in ionization energy in passing from In to Tl down Group 13. The electron configuration changes by the addition of $(n-2)f^{14}$, and *f*-electrons do not shield outer electrons very efficiently.

14.96

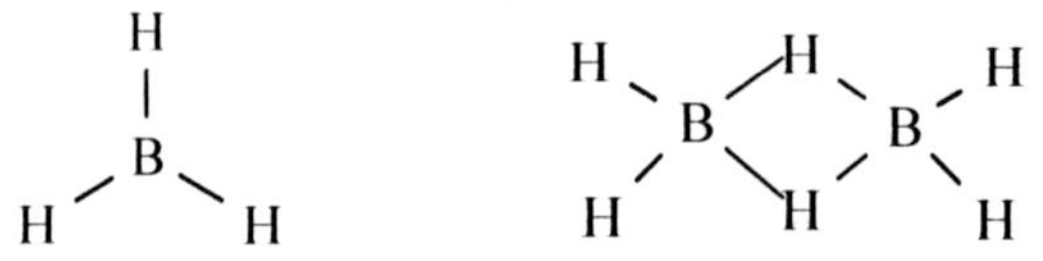

(1) $B(s) + \frac{3}{2}H_2(g) \rightarrow BH_3(g)$ $\Delta H°_f = +100\ kJ \cdot mol^{-1}$

(2) $B(s) \rightarrow B(g)$ $\Delta H°_f = +563\ kJ \cdot mol^{-1}$

(3) $\frac{1}{2}H_2(g) \rightarrow H(g)$ $\Delta H°_f = +218\ kJ \cdot mol^{-1}$

(4) $2\ B(s) + 3\ H_2(g) \rightarrow B_2H_6(g)$ $\Delta H°_f = +36\ kJ \cdot mol^{-1}$

(a) Reverse equation 1, multiply equation 3 by 3, and add to equation 2:

$BH_3(g) \rightarrow B(s) + \frac{3}{2} H_2(g)$ $\Delta H° = -100 \text{ kJ} \cdot \text{mol}^{-1}$

$B(s) \rightarrow B(g)$ $\Delta H° = +563 \text{ kJ} \cdot \text{mol}^{-1}$

$\frac{3}{2} H_2(g) \rightarrow 3 H(g)$ $\Delta H° = 3(218) \text{ kJ} \cdot \text{mol}^{-1}$

$BH_3(g) \rightarrow B(g) + 3 H(g)$ $\Delta H° = +1117 \text{ kJ} \cdot \text{mol}^{-1}$
$= 3\Delta H(B - H)$

$\Delta H(B - H) = +372 \text{ kJ} \cdot \text{mol}^{-1}$

Assume terminal B—H bonds have the same bond enthalpy as B—H bonds in BH_3, that is, $372 \text{ kJ} \cdot \text{mol}^{-1}$.

Reverse equation 4, multiply equation 2 by 2, and equation 3 by 6. Then add.

$B_2H_6(g) \rightarrow 2 B(s) + 3 H_2(g)$ $\Delta H° = -36 \text{ kJ} \cdot \text{mol}^{-1}$

$2 B(s) \rightarrow 2 B(g)$ $\Delta H° = 2 \times 563 \text{ kJ} \cdot \text{mol}^{-1}$

$3 H_2(g) \rightarrow 6 H(g)$ $\Delta H° = 6 \times 218 \text{ kJ} \cdot \text{mol}^{-1}$

$B_2H_6(g) \rightarrow 2 B(g) + 6 H(g)$ $\Delta H° = 2398 \text{ kJ} \cdot \text{mol}^{-1}$

(b) $\Delta H° = 2398 \text{ kJ} \cdot \text{mol}^{-1} = 4 \times \Delta H(\text{B—H}) + 4 \times \Delta H(\text{B—H—B})$

$$\Delta H(\text{B—H—B}) = \frac{2398 \text{ kJ} \cdot \text{mol}^{-1} - 4 \times 372 \text{ kJ} \cdot \text{mol}^{-1}}{4}$$
$$= 228 \text{ kJ} \cdot \text{mol}^{-1}$$

As bond length and bond enthalpy are (very roughly) inversely related, the stronger terminal B—H bonds are expected to be shorter.

14.98 (a) sp^2

(b) A tube with a diameter of about 1.3 nm will have a circumference of $2\pi r$ or πd. Thus, the circumference of a 1.3×10^{-9} m diameter nanotube will be $(1.3 \times 10^{-9} \text{ m})(\pi) = 4.1 \times 10^{-9}$ m.

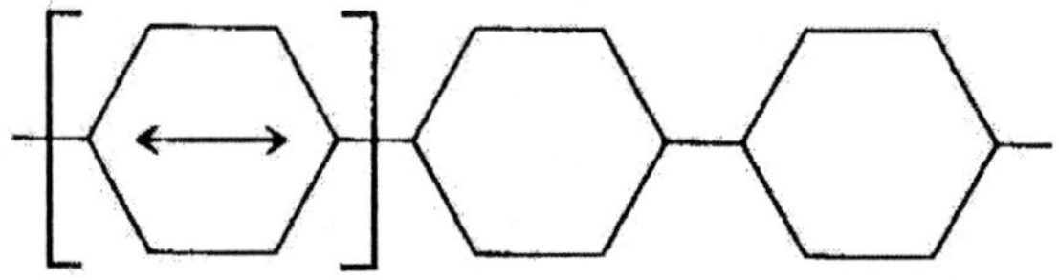

To calculate the number of C_6 rings that will be strung together, we need to calculate the distance across the C_6 ring as shown by the arrow.

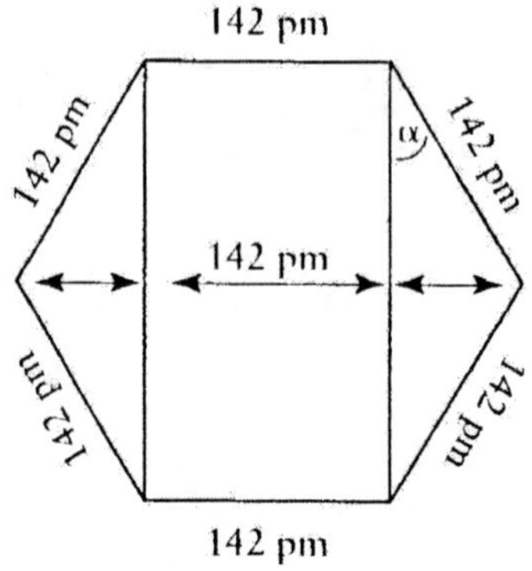

The total distance from one carbon to the opposite carbon on the ring will be given by

$d = 142\text{ pm} + 2\,(142\text{ pm} \times \sin 30°) = 284\text{ pm}$

One repeat unit will be one benzene ring plus one C—C bond,

$284\text{ pm} + 142\text{ pm} = 426\text{ pm}$ or $4.26 \times 10^{-10}\text{ m}$. There are thus about

$4.1 \times 10^{-9} \div 4.26 \times 10^{-10} = 10$ of units strung together around the smallest nanotube known.

14.100 The relevant numbers from the Appendices are the atomic radii of Ca and Sr, and their molar masses. The values are

Element	Atomic radius	Molar mass
Ca	197 pm	$40.08\text{ g}\cdot\text{mol}^{-1}$
Sr	215 pm	$87.62\text{ g}\cdot\text{mol}^{-1}$

For a metal that packs in a face-centered cubic lattice, there are a total of four metal atoms in the unit cell. From Chapter 5 we learned that the edge length, a, of the unit cell is related to the radius of the atom by the relationship $4\,r = a\sqrt{2}$. Using the atomic radii given above, the unit cell edge lengths for Ca and Sr will be 5.57×10^{-8} cm and 6.08×10^{-8} cm, respectively. The densities will be given by

$$d_{\mathrm{Ca}} = \frac{4\left(\dfrac{40.08\ \mathrm{g \cdot mol^{-1}}}{6.022 \times 10^{23}\ \mathrm{mol^{-1}}}\right)}{(5.57 \times 10^{-8}\ \mathrm{cm})^3} = 1.54\ \mathrm{g \cdot cm^{-3}}$$

$$d_{\mathrm{Sr}} = \frac{4\left(\dfrac{87.62\ \mathrm{g \cdot mol^{-1}}}{6.022 \times 10^{23}\ \mathrm{mol^{-1}}}\right)}{(6.08 \times 10^{-8}\ \mathrm{cm})^3} = 2.59\ \mathrm{g \cdot cm^{-3}}$$

Strontium is more dense than calcium.

14.102 In C_{60} the carbon atoms are sp^2 hybridized and are nearly planar. However, the curvature of the molecule introduces some strain at the carbon atoms so that there is some tendency for some of the carbon atoms to undergo conversion to sp^3 hybridization. However, to make every carbon sp^3 hybridized would introduce much more strain on the carbon cage and, after a certain point, further addition of hydrogen becomes unfavorable.

14.104 Both the antisymmetrical stretch (ν_3) and the doubly degenerate bend (ν_2) are vibrational modes that deform the molecule such that it is lop-sided in terms of its charge distribution and overall dipole moment, so these vibrations can absorb infrared radiation.

14.106 The equation for the process is

$$\mathrm{SiCl_4 + 2\ H_2(g) \rightarrow Si(s) + 4\ HCl(g)}$$

To determine the temperature range for the spontaneity of the reaction, we need to know $\Delta H°$, and $\Delta S°$, for the process. These are easily calculated from the data given and from data in Appendix 2A.

$$\Delta H°_r = 4(\Delta H°_{f,HCl}) - (\Delta H°_{f,SiCl_4})$$
$$= 4(-92.31\ kJ \cdot mol^{-1}) - (-662.75\ kJ \cdot mol^{-1})$$
$$= +293.51\ kJ \cdot mol^{-1}$$
$$\Delta S°_r = 4(S°_{m,HCl}) + S°_{m,Si} - [S°_{m,SiCl_4} + 2S°_{m,H_2}]$$
$$= 4(186.91\ J \cdot K^{-1} \cdot mol^{-1}) + 18.83\ J \cdot K^{-1} \cdot mol^{-1}$$
$$- [330.86\ J \cdot K^{-1} \cdot mol^{-1} + 2(130.68\ J \cdot K^{-1} \cdot mol^{-1})]$$
$$= +174.25\ J \cdot K^{-1} \cdot mol^{-1}$$

Because $\Delta H°_r$ and $\Delta S°_r$ are both positive, the reaction will be spontaneous at elevated temperatures. The temperature at which the process becomes spontaneous can be calculated by determining the temperature at which $\Delta G°_r$ is equal to 0.

$$\Delta G°_r = (293.51 \times 10^3\ J \cdot mol^{-1}) - T(+174.25\ J \cdot K^{-1} \cdot mol^{-1}) = 0$$
$$T = 1684\ K \text{ or } 1411°C$$

The reaction should be spontaneous approximately at temperatures above 1411°C.

14.108 (a) A reasonable, simple unit cell can be obtained by considering the nitrogen atoms to be at the corners of the unit cell.

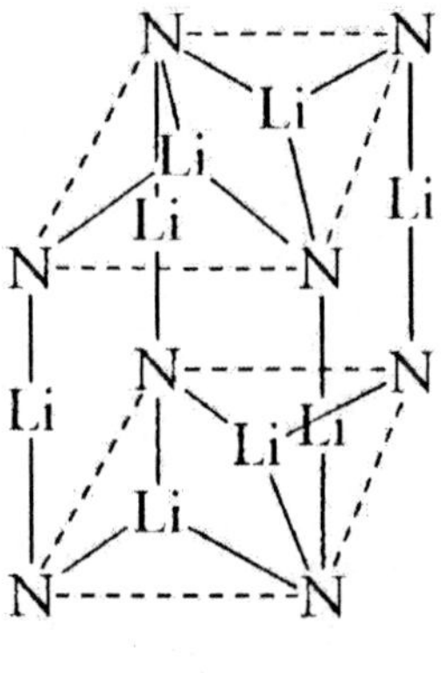

(b) The unit cell appears to be hexagonal.

(c) There is one nitrogen atom ($\frac{1}{8}$ of the eight corners) in the unit cell and three lithium atoms ($\frac{1}{4} \times 4$ atoms along edges plus $\frac{1}{2} \times 4$ atoms in the faces) in this unit cell, giving overall one formula unit per unit cell.

(d) The density is determined from the volume of the unit cell and the mass of the atoms that comprise it. The distances between planes containing the nitrogen atoms will be $2 \times 194 \text{ pm} = 338 \text{ pm}$. The dimensions of the unit cell that lie within the N-containing plane can be calculated using simple geometry, given that all the Li—N distances are 213 pm and all the angles are $60°$.

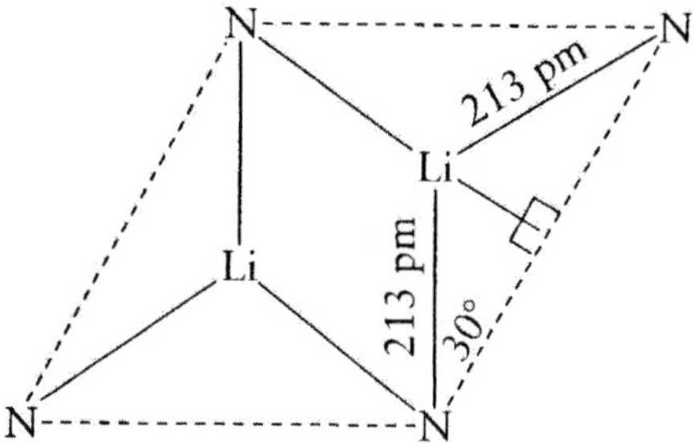

$$\tfrac{1}{2} d_{\text{N-N}} = 231 \text{ pm} \times \cos 30°$$
$$d_{\text{N-N}} = 2 \times 231 \text{ pm} \times \cos 30°$$
$$= 369 \text{ pm}$$

The area of the rhombohedron is given by $b \times h$. We know b, which we have just calculated to be 369 pm. The quantity h can be calculated from $\sin 60° = h/369 \text{ pm}$ or $h = 320 \text{ pm}$. The area is $369 \text{ pm} \times 320 \text{ pm} = 1.18 \times 10^5 \text{ pm}^2$. The volume is $1.18 \times 10^5 \text{ pm}^2 \times 388 \text{ pm} = 4.58 \times 10^7 \text{ pm}^3$ per unit cell. Because one pm is equal to 1×10^{-10} cm, the volume of the unit cell is $4.58 \times 10^{-23} \text{ cm}^3$. There is only one Li_3N formula unit in the unit cell, which will give a mass of $34.83 \text{ g} \cdot \text{mol}^{-1} \div 6.02 \times 10^{23}$ formula units $\cdot \text{mol}^{-1} = 5.78 \times 10^{-23} \text{ g} \cdot \text{unit cell}^{-1}$.

The density calculated is $5.78 \times 10^{-23} \text{ g} \div 4.58 \times 10^{-23} \text{ cm}^3 = 1.26 \text{ g} \cdot \text{cm}^{-3}$, which agrees with the experimental value.

14.110 (a) The dolomite and calcite structures are essentially identical. The cations and anions are located in the same positions in both structures.

(b) The structures differ in that in one the cations are all Ca^{2+} ions (calcite), and in the other the cations are a mix of Ca^{2+} and Mg^{2+}. (c) The Ca^{2+} ions and Mg^{2+} ions are randomly distributed over the cation positions in the unit cell.

14.112 $2\ CsO_2(s) + CO_2(g) \rightarrow Cs_2CO_3(s) + \frac{3}{2}\ O_2(g)$

$$n_{O_2} = (30.0\ g)\left(\frac{1\ mol\ CsO_2}{164.9\ g\ CsO_2}\right)\left(\frac{\frac{3}{2}\ mol\ O_2}{2\ mol\ CsO_2}\right) = 0.1364\ mol$$

$$V = \frac{nRT}{P} = (0.1364\ mol)\frac{(0.08206\ L\cdot atm\cdot K^{-1}\cdot mol^{-1})(298\ K)}{(1.00\ atm)}$$

$$= 3.34\ L$$

14.114

CHAPTER 15

THE ELEMENTS: THE LAST FOUR MAIN GROUPS

15.2 (a) 0; (b) 0; (c) -3; (d) -2; (e) $+3$; (f) $+5$; (g) $+5$; (h) $+3$

15.4 (a) $HNO_2 + H_2NNH_2 \longrightarrow 2\ H_2O + HN_3$

$$(15.0\text{ g }H_2NNH_2)\left(\frac{1\text{ mol }H_2NNH_2}{32.05\text{ g }H_2NNH_2}\right)\left(\frac{1\text{ mol }HN_3}{1\text{ mol }H_2NNH_2}\right)$$

$$\left(\frac{43.04\text{ g }HN_3}{1\text{ mol }HN_3}\right) = 20.1\text{ g }HN_3$$

(b) $N_2O(g) + 2\ NaNH_2(l) \xrightarrow{175°C} NaN_3(l) + NaOH(l) + NH_3(g)$

(c) The oxidation number of nitrogen in hydrazine is -2, while in hydrazoic acid it is nominally $-1/3$. This process corresponds to an oxidation of the hydrazine.

15.6 $2\ NaN_3 \longrightarrow 2\ Na + 3\ N_2$

$$\text{number of moles of } N_2(g) = n\frac{PV}{RT} = \frac{(1.5\text{ atm})(100\text{ L})}{(0.0821\text{ L}\cdot\text{atm}\cdot\text{mol}^{-1}\cdot\text{K}^{-1})(293\text{ K})}$$

$$= 6.24\text{ mol}$$

$$\text{mass of } NaN_3 \text{ needed} = (6.24\text{ mol }N_2)\left(\frac{2\text{ mol }NaN_3}{3\text{ mol }N_2}\right)\left(\frac{65.02\text{ g }NaN_3}{1\text{ mol }NaN_3}\right)$$

$$= 270\text{ g }NaN_3$$

15.8 $P_4O_6 - H_3PO_3$; $P_4O_6(s) + 6\ H_2O(l) \longrightarrow 4\ H_3PO_3(aq)$

$P_4O_{10} - H_3PO_4$; $P_4O_{10}(s) + 6\ H_2O(l) \longrightarrow 4\ H_3PO_4(aq)$

15.10 (a)

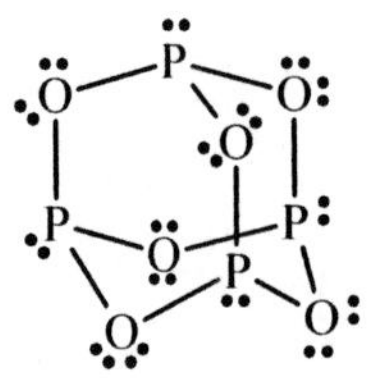

(b) Yes. Each oxygen atom has two bonds to phosphorus atoms and two lone pairs. Each phosphorus atom has three bonds to oxygen atoms and one lone pair.

(c) The phosphorus atoms can be replaced by CH fragments and the oxygen atoms by CH_2 groups.

(d) The molecule would be $C_{10}H_{16}$. This compound is known as adamantane.

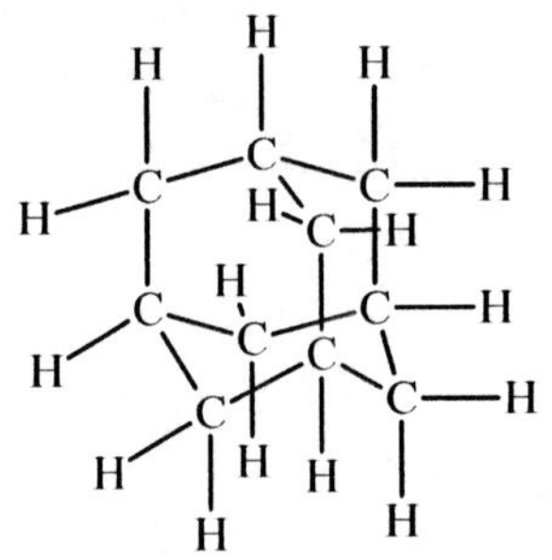

15.12 First we balance the equation:

$$PCl_5(g) + 4\,H_2O(l) \longrightarrow 5\,HCl(aq) + H_3PO_4(aq)$$

$$\begin{aligned}\Delta G°_r &= 5(-131.23\text{ kJ}\cdot\text{mol}^{-1}) + (-1142.54\text{ kJ}\cdot\text{mol}^{-1}) \\ &\quad - [(-305.0\text{ kJ}\cdot\text{mol}^{-1}) + 4(-237.13\text{ kJ}\cdot\text{mol}^{-1})] \\ &= -545.17\text{ kJ}\cdot\text{mol}^{-1}\end{aligned}$$

The reaction is spontaneous.

15.14 (a) $Na_2O(s) + H_2O(l) \longrightarrow 2\,NaOH(aq)$

(b) $Na_2O_2(s) + 2\,H_2O(l) \longrightarrow 2\,NaOH(aq) + H_2O_2(aq)$

(c) $SO_2(g) + H_2O(l) \longrightarrow H_2SO_3(aq)$

(d) $2\,SO_2(g) + O_2(g) \xrightarrow{500°C,\ V_2O_5} 2\,SO_3(g)$

15.16 (a) $FeS(s) + 2\,HCl(aq) \longrightarrow FeCl_2(aq) + H_2S(g)$

(b) $H_2(g) + S(s) \xrightarrow{\Delta} H_2S(g)$

(c) $S_2Cl_2(l) + Cl_2(g) \xrightarrow{FeCl_3} 2\ SCl_2(l)$

15.18 (a) The Lewis structure of oleum is

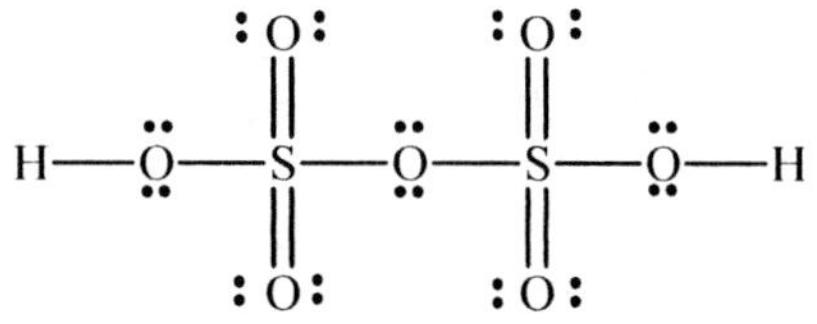

(a) The formal charges are 0 on all S and O atoms.

(b) +6

15.20 The system is $HS^-(aq) + H_2O(l) \rightleftharpoons H_2S(aq) + OH^-(aq)$

Concentration ($mol \cdot L^{-1}$)	HS^- +	$H_2O \rightleftharpoons$	H_2S +	OH^-
initial	0.050	—	0	0
change	$-x$	—	$+x$	$+x$
equilibrium	$0.050 - x$	—	x	x

$$K_b = \frac{K_w}{K_a} = \frac{1.0 \times 10^{-14}}{1.3 \times 10^{-7}} = 7.7 \times 10^{-8}$$

$$K_b = \frac{x^2}{0.050 - x} \approx \frac{x^2}{0.050} = 7.7 \times 10^{-8}$$

$x = 6.2 \times 10^{-5} = [OH^-]$

$pOH = 4.21$

$pH = 14.00 - 4.21 = 9.79$

15.22 (a) $PbS(s) + 2\ O_2(g) \longrightarrow PbSO_4(s)$ (1)

$PbS(s) + 4\ O_2(g) \longrightarrow PbO_2(s) + SO_2(g)$ (2)

(b) We do not have available the free energy of formation of PbS(s), so we need to discover a different method of determining this instead of simply comparing the signs and relative magnitudes of the free energies of the two reactions. The appropriate reaction is

$PbSO_4(s) \longrightarrow PbO_2(s) + SO_2(g)$

Calculate $\Delta G°_r$

$$\Delta G°_r = \Delta G°_f(PbO_2, s) + \Delta G°_f(SO_2, g) - \Delta G°_f(PbSO_4, s)$$
$$= [-217.33 + (-300.19) - (-813.14)]\ kJ \cdot mol^{-1} = +295.62\ kJ \cdot mol^{-1}$$

The positive $\Delta G°_r$ is unfavorable to the formation of PbO_2. Thus, $PbSO_4$ is favored as a product over $PbO_2 + SO_2$.

15.24 (a) $H_2O_2 + 2\,H^+ + 2\,e^- \longrightarrow 2\,H_2O$ $\qquad E° = +1.78$ V

$2(Fe^{2+} \longrightarrow Fe^{3+} + e^-)$ $\qquad E° = -0.77$ V

$H_2O_2 + 2\,Fe^{2+} + 2\,H^+ \longrightarrow 2\,Fe^{3+} + 2\,H_2O$ $\qquad E° = +1.01$ V

The potential is positive so this process will be spontaneous.

(b) The potential must be obtained from the Nernst equation. From part (a), the standard potential that will be appropriate for 1.00 M $H^+(aq)$ is 1.01 V. The Nernst equation expression for this reaction will be

$$E = +1.101\ V - \left(\frac{0.025\ 693}{2}\right) \ln \frac{[Fe^{3+}]^2}{[H_2O_2][Fe^{2+}]^2[H^+]^2}$$

To find the standard potential in neutral solution, the only parameter that changes is the concentration of H^+. All other concentrations in the quotient will remain equal to 1.

$$E = +1.01\ V - \left(\frac{0.025\ 693}{2}\right) \ln \frac{1}{[H^+]^2}$$

$$E = +1.01\ V - \left(\frac{0.025\ 693}{2}\right) \ln \frac{1}{(1.0 \times 10^{-7})^2}$$

$E = +0.60$ V

The reaction is still spontaneous in neutral solution; however, the reaction becomes less favorable as the pH increases.

(c) Raising the pH above 7 would be complicated by the precipitation of insoluble iron hydroxide complexes.

15.26 (a) CaOCl; (b) It is an effective but mild oxidizing agent.

15.28 (a) IF_7: F = −1; therefore I = +7

(b) $NaIO_4$: Na = +1, O = −2; therefore I = +7

(c) HBrO(aq): H = +1, O = −2; therefore Br = +1

(d) $NaClO_2$: Na = +1, O = −2; therefore Cl = +3

15.30 (a) $Cl_2(g) + H_2O(l) \longrightarrow HClO(aq) + HCl(aq)$

(b) $Cl_2(g) + 2\,OH^-(aq) \longrightarrow ClO^-(aq) + Cl^-(aq) + H_2O(l)$

(c) $3\,Cl_2(g) + 6\,OH^-(aq) \xrightarrow{\Delta} ClO_3^-(aq) + 5\,Cl^-(aq) + 3\,H_2O(l)$

(d) In part (a), Cl goes from $N_{ox} = 0$ to $N_{ox} = +1$ (in HClO) and from $N_{ox} = 0$ to $N_{ox} = -1$ (in HCl). In part (b), Cl goes from $N_{ox} = 0$ to $N_{ox} = +1$ (in ClO^-) and from $N_{ox} = 0$ to $N_{ox} = -1$ (in Cl^-). In part (c), Cl goes from $N_{ox} = 0$ to $N_{ox} = +5$ (in ClO_3^-) and from $N_{ox} = 0$ to $N_{ox} = -1$ (in Cl^-). So in each case, Cl is both oxidized and reduced.

15.32 (a) HIO < HBrO < HClO (Fluorine does not form oxoacids.)

(b) The electronegativity of the halogens increases in the order I to Cl; therefore, there is a greater electron-withdrawing effect upon the HO bond in this order, resulting in a greater tendency to release H and thus greater acid strength.

15.34

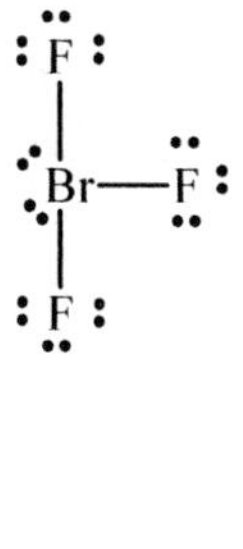

AX_3E_2, T-shaped, *dsp*^{3}

15.36

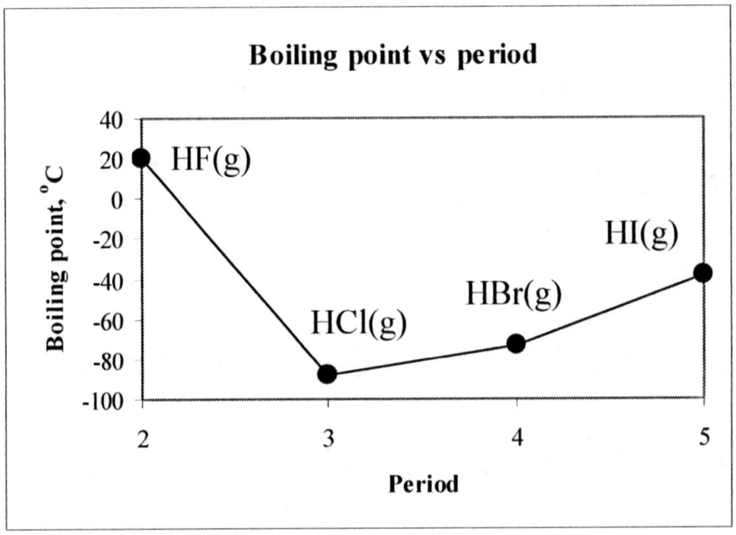

HF is anomalous, due to the hydrogen bonding that exists between HF molecules in the liquid state. The upward trend from HCl to HI is normal and parallels the increase in molecular weight and London forces.

15.38 (a) The following two half-cell reactions involving O_3 can occur:

$$O_3 + 2\,H^+ + 2\,e^- \longrightarrow O_2 + H_2O \qquad E° = +2.07\ V$$

$$O_3 + H_2O + 2\,e^- \longrightarrow O_2 + 2\,OH^- \qquad E° = +1.24\ V$$

These may be compared to

$$F_2 + 2\,e^- \longrightarrow 2\,F^- \qquad E° = +2.87\ V$$

Therefore, fluorine is a stronger oxidizing agent than ozone.

(b) No. While the reduction potential of ozone is considerably different in acidic (+2.07 V) versus basic (+1.24 V) solution, fluorine is still a much stronger oxidant (+2.87 V).

15.40 $$[I_2] = (0.028\,45\ L)\left(\frac{0.025\ mol\ S_2O_3^{2-}}{1\ L}\right)\left(\frac{1\ mol\ I_2}{2\ mol\ S_2O_3^{2-}}\right)\left(\frac{1}{0.025\,00\ L}\right)$$

$$= 0.014\ mol \cdot L^{-1}$$

15.42 Because there is no data available for the free energy of formation of HClO, we must turn to the electrochemical data in Appendix 2B. The appropriate half-reactions are

$$Cl_2 + 2\,e^- \longrightarrow 2\,Cl^- \qquad E° = +1.36\ V$$

$$Cl_2 + 2\,H_2O \longrightarrow 2\,HClO + 2\,H^+ + 2\,e^- \qquad E° = -1.63\ V$$

$$2\,Cl_2 + 2\,H_2O \longrightarrow 2\,HClO + 2\,H^+ + 2\,Cl^- \qquad E° = -0.27\ V$$

The $E°$ will be the same for the reaction balanced as

$$Cl_2 + H_2O \longrightarrow HClO + H^+ + Cl^- \qquad E° = -0.27\ V$$

but n will be 1 e^- rather than 2 e^- as in the previous equation.

$\Delta G°_r$ can now be calculated from $\Delta G°_r = -nFE°$

$$\Delta G°_r = -(1)(96\,485\ J\cdot V^{-1}\cdot mol^{-1})(-0.27\ V) = +26\ kJ\cdot mol^{-1}$$

(b) The equilibrium constant can be calculated from $\Delta G°_r = -RT \ln K$

$$K = e^{-\frac{\Delta G°_r}{RT}} = e^{-\frac{26\,000\ J\cdot mol^{-1}}{(8.314\ J\cdot K^{-1}\cdot mol^{-1})(298\ K)}} \cong 2.8 \times 10^{-5}$$

15.44 Krypton and xenon are obtained by the distillation of liquid air.

15.46 (a) XeO_3:O = −2; therefore Xe = +6

(b) XeO_6^{4-}:O = − 2; therefore Xe = +8

(c) XeF_2 : F = −1; therefore Xe = +2

(d) $HXeO_4^-$:H = +1, O = −2; therefore Xe = +6

15.48 (a) $XeF_6(s) + 3\,H_2O(l) \longrightarrow XeO_3(s) + 6\,HF(aq)$

(b) $Pt(s) + XeF_4(s) \longrightarrow Xe(g) + PtF_4(s)$

(c) $Kr(g) + F_2(g) \xrightarrow{\text{electric discharge}} KrF_2(s)$

15.50 H_4XeO_6, in which Xe has an oxidation number of +8, is expected to be a stronger oxidizing agent than H_2XeO_4, in which the oxidation number is lower, +6.

15.52 Both these names refer to types of colloids, but the nature of what is suspended in what changes. An emulsion is a suspension of small particles of one liquid in another liquid. An example of this is a mixture of oil and vinegar. A gel is a type of solid emulsion in which a liquid or solid phase is suspended in a solid. Gels are characterized by being soft but holding their shape. Gelatin desserts are a type of gel. Gels are also formed when some water-sensitive metal complexes are hydrolyzed. These gels are industrially important as precursors to metal oxide materials.

15.54 (a) hydrogen bonds (b) hydrogen bonds (c) hydrogen bonds, dipole-dipole forces, and London forces

15.56 The energy of the emitted radiation is always lower than that of the exciting radiation.

15.58 In a neon light, the emission of the radiation, which is seen as light, is occurring directly from the gas when it is excited by a voltage. In a fluorescent light, the gas that is excited by the voltage emits its energy as radiation, which then excites a fluorescent material that coats the walls of the fluorescent light bulb. The radiation observed is therefore not the radiation of the gas being excited.

15.60 As discussed in Chapter 6, the enthalpy of the reaction:

$P_4(g) \rightarrow 2\ P_2(g)$

May be estimated by summing the energy needed to break the bonds of $P_4(g)$ and subtracting the energy released upon formation of the 2 P–P bonds. This sum is given to us as 217 kJ mol^{-1}. Therefore, the bond enthalpy of P_2 is:

$$\text{Bond enthalpy of } P_2 = \frac{6(200\ \text{kJ}\cdot\text{mol}^{-1}) - 217\ \text{kJ}\cdot\text{mol}^{-1}}{2} = 492\ \text{kJ}\cdot\text{mol}^{-1}$$

This bond enthalpy is significantly less than the bond enthalpy of N_2, an isoelectronic compound, because in P_2 bonding involves electrons in a valence shell which is further from the nuclei, resulting in a weaker bond.

15.62 $S + 2\ e^- \longrightarrow S^{2-}(aq) \qquad E° = -0.48\ \text{V}$

$O_2 + 4\ H^+ + 4\ e^- \longrightarrow 2\ H_2O \qquad E° = +1.23\ \text{V}$

$Cl_2 + 2e^- \longrightarrow 2\ Cl^- \qquad E° = +1.36\ \text{V}$

$H_2O_2 + 2\ H^+ + 2\ e^- \longrightarrow 2\ H_2O \qquad E° = +1.78\ \text{V}$

Because a larger value of $E°$ corresponds to greater oxidizing power, the order is

$S < O_2 < Cl_2 < H_2O_2$

15.64 Stage 1: $4\ NH_3(g) + 5\ O_2(g) \longrightarrow 4\ NO(g) + 6\ H_2O(g)$

$\Delta H°_r = \Delta H°_f(\text{products}) - \Delta H°_f(\text{reactants})$

$$\Delta H°_r = [(4)(90.25\ \text{kJ}\cdot\text{mol}^{-1}) + (6)(-241.82\ \text{kJ}\cdot\text{mol}^{-1})] - [(4)(-46.11\ \text{kJ}\cdot\text{mol}^{-1}]$$

$\Delta H°_r = -905.48\ \text{kJ}\cdot\text{mol}^{-1}$

$\Delta S°_r = S°(\text{products}) - S°(\text{reactants})$

$H_2O_2 + OH^- \longrightarrow O_2 + H_2O + H^+ + 2\ e^-$

$\Delta S°_r = +180.52\ \text{J}\cdot\text{K}^{-1}\cdot\text{mol}^{-1}$

$\Delta G°_r = \Delta H°_r - T\Delta S°_r$

The reaction is spontaneous at all temperatures because $\Delta H°$ is negative and $\Delta S°$ is positive.

Stage 2: $2\ NO(g) + O_2(g) \longrightarrow 2\ NO_2(g)$

$\Delta H°_r = \Delta H°_f(\text{products}) - \Delta H°_f(\text{reactants})$

$\Delta H°_r = [(2)(33.18 \text{ kJ} \cdot \text{mol}^{-1})] - [(2)(90.25 \text{ kJ} \cdot \text{mol}^{-1})]$

$\Delta H°_r = -114.14 \text{ kJ} \cdot \text{mol}^{-1}$

$H_2O_2 \longrightarrow H^+ + HO_2^-$

$\Delta S°_r = \{[(2)(240.06)] - [(2)(210.76) + (205.14)]\} \text{ J} \cdot \text{K}^{-1} \cdot \text{mol}^{-1}$

$\Delta S°_r = -146.54 \text{ J} \cdot \text{K}^{-1} \cdot \text{mol}^{-1}$

Because both $\Delta H°_r$ and $\Delta S°_r$ are negative, the reaction is spontaneous at relatively low temperatures (up to 779 K), but above this temperature it is nonspontaneous.

Stage 3: $3\ NO_2(g) + H_2O(l) \longrightarrow 2\ HNO_3(aq) + NO(g)$

$\Delta H°_r = \Delta H°_f(\text{products}) - \Delta H°_f(\text{reactants})$

$$\Delta H°_r = [(2)(-207.36 \text{ kJ} \cdot \text{mol}^{-1}) + (90.25 \text{ kJ} \cdot \text{mol}^{-1})] - [(3)(33.18 \text{ kJ} \cdot \text{mol}^{-1}) + (-285.83 \text{ kJ} \cdot \text{mol}^{-1})]$$

$\Delta H°_r = -138.18 \text{ kJ} \cdot \text{mol}^{-1}$

$H_2O_2 + 2\ H^+ + 2\ e^- \longrightarrow 2\ H_2O$

$$\Delta S°_r = [(2)(146.4) + (210.76)] \text{ J} \cdot \text{K}^{-1} \cdot \text{mol}^{-1} - [(3)(240.06) + (69.91)] \text{ J} \cdot \text{K}^{-1} \cdot \text{mol}^{-1}$$

$\Delta S°_r = -286.5 \text{ J} \cdot \text{K}^{-1} \cdot \text{mol}^{-1}$

The reaction is spontaneous up to 482 K; above this temperature it is nonspontaneous.

15.66 (a) $2\ KClO_3(s) \xrightarrow{MnO_2,\ 70°-100°C} 2\ KCl(s) + 3\ O_2(g)$

This is a redox reaction that is used for the laboratory preparation of $O_2(g)$. It illustrates the mild instability of chlorates at higher temperatures. Cl is reduced and O is oxidized.

(b) $CaF_2(s) + 2\ H_2SO_4(\text{conc}) \longrightarrow Ca(HSO_4)_2(aq) + 2\ HF(g)$

This is a Brønsted acid-base reaction. H^+ is the acid and F^- is the base.

(c) $OF_2(g) + 2\ OH^-(aq) \longrightarrow O_2(g) + 2\ F^-(aq) + H_2O(l)$

This is a redox reaction. O in $OF_2\ (N_{ox} = +2)$ is reduced, O in $OH^-\ (N_{ox} = -2)$ is oxidized.

(d) $2\ H_2S(g) + 3\ O_2(g) \longrightarrow 2\ SO_2(g) + 2\ H_2O(l)$

This is a redox reaction that illustrates the reducing ability of H_2S. This reaction is one of the principal sources of SO_2 in the atmosphere. S(– 2) is oxidized and O (zero) is reduced.

15.68 (a) $SCl_2(l) + 2\ C_2H_4 \rightarrow S(C_2H_4Cl)_2$

(b) The S atom in mustard gas is sp^3 hybridized.

(c) Step 1: $SCl_2 + C_2H_4 \rightarrow Cl\text{–}S\text{–}CH_2CH_2Cl$

Step 2: $Cl\text{–}S\text{–}CH_2CH_2Cl + C_2H_4 \rightarrow ClCH_2CH_2\text{–}S\text{–}CH_2CH_2Cl$

One plausible scenario is that a lone pair of electrons on the sulfur atom adds into the π^* orbital on ethene. This will leave a vacant *p*-orbital on the adjacent carbon atom, which can then act as an acceptor toward a chlorine atom. As the C—Cl bond is formed, the S—Cl bond would be broken. This process could then be repeated for the other lone pair on sulfur with another molecule of ethene.

15.70 (a) Cl_2O

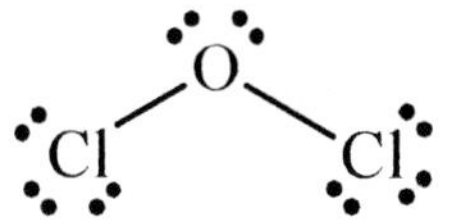

(b) I_2O_5

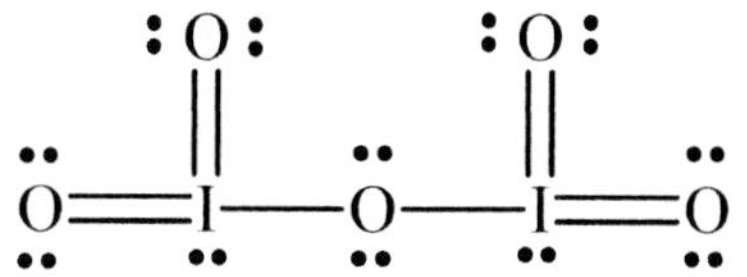

(c) Cl_2O_7

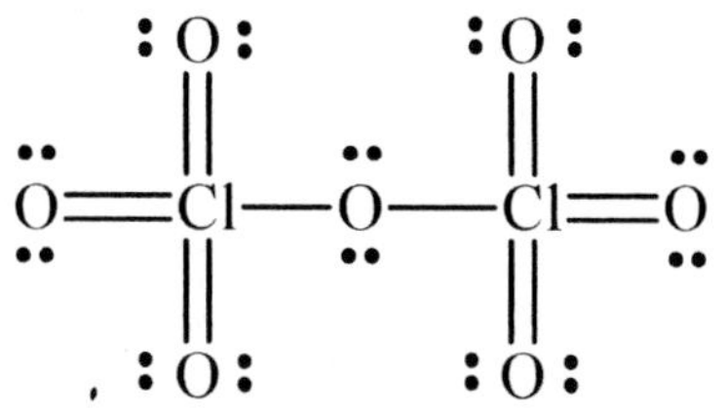

15.72 (a) $$[Cl^-] = \left(\frac{2.798\text{ g AgCl}}{0.05000\text{ L}}\right)\left(\frac{1\text{ mol AgCl}}{143.32\text{ g AgCl}}\right)\left(\frac{1\text{ mol Cl}^-}{1\text{ mol AgCl}}\right)$$

$$= 0.3904\text{ mol}\cdot\text{L}^{-1}$$

(b) AgF is soluble; therefore this method cannot be used for the determination of F^- ions in solution.

15.74 (a)

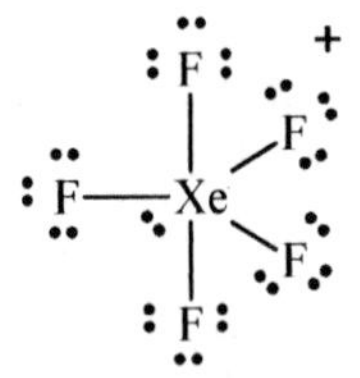

AX_5E, square pyramidal

(b)

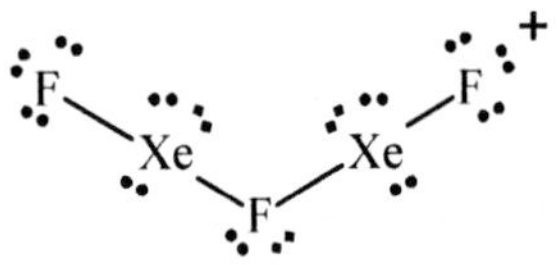

15.76 The method of organization is determined by whether one wishes to emphasize similarities or differences. It seems easier and more logical to emphasize similarities rather than differences. Trends within groups from top to bottom tend to be quantitative (small numerical differences), rather than qualitative (large numerical differences leading to distinctly different behavior), as are observed from left to right within a period. Trends from metallic to nonmetallic behavior across a period would be more apparent in the organization by period, as would the related trends in ionization energy and electron affinity. One could also more readily see changes in valence as represented, say, by the formulas of common compounds, such as the oxides and chlorides. Trends in melting points and boiling points of the elements and their compounds would also be more apparent, as well as many other physical properties. So, there are advantages and disadvantages to both methods of organization. Organization by group still seems preferable, as organization by period would not permit the generalizations and summaries that are useful features of the present arrangement.

15.78 (a) (1) $4\ Zn(s) + 6\ H_2O(l) + NO_3^-(aq) \rightarrow$

$$NH_3(aq) + 9\ OH^-(aq) + 4\ Zn^{2+}(aq)$$

(2) $NH_3(aq) + HCl(aq) \rightarrow NH_4Cl(aq) + H_2O(l)$

(3) $HCl(aq) + NaOH(aq) \rightarrow NaCl(aq) + H_2O(l)$

(b) number of moles of unreacted $HCl = 0.028\ 22\ L \times 0.00150\ mol \cdot L^{-1}$

$$= 4.23 \times 10^{-5}\ mol$$

total number of moles of

$HCl = 0.050\ 00\ L \times 0.00250\ mol \cdot L^{-1} = 1.25 \times 10^{-4}\ mol$

moles of HCl reacted = moles of NH_3

$$= (1.25 \times 10^{-4}\ mol - 4.23 \times 10^{-5}\ mol)$$

$$= 8.3 \times 10^{-5}\ mol\ NH_3$$

$$[NO_3^-] = \frac{(8.3 \times 10^{-5}\ mol\ NH_3)\left(\dfrac{1\ mol\ NO_3^-}{1\ mol\ NH_3}\right)}{0.025\ 00\ L} = 0.33 \times 10^{-3}\ mol \cdot L^{-1}$$

15.80 (a) The system is:

Concentration ($mol \cdot L^{-1}$)	$I_2(s)$ +	$I^-(aq)$ $\rightleftharpoons$	$I_3^-(aq)$
initial	—	0.50	0
change	—	$-x$	$+x$
equilibrium	—	$0.50 - x$	x

The equilibrium constant is given by:

$$K = 698 = \frac{x}{0.50 - x} \quad \text{and}$$

$x = 0.50$

The concentration of $I_3^-(aq)$ at equilibrium is, therefore, 0.50 M

(b) The mass of $I_2(s)$ consumed during the reaction is:

$(0.50\text{ M})(0.250\text{ L})\left(126.90\text{ g}\cdot\text{mol}^{-1}\right) = 16\text{ g}$

The amount of $I_2(s)$ remaining after equilibrium is reached is:

$35.0\text{ g} - 16\text{ g} = 19\text{ g}$

15.82 The ionic radii of the halide ions increase down the group. The smaller the ion, the greater the lattice enthalpy of the compounds of the ion due to the greater concentration of the charge in the ion. Increased lattice enthalpy results in higher melting and boiling points; the ions cannot as easily break free from each other. Thus, melting and boiling points decrease from fluoride to iodide for ionic halides. The predominant forces between covalent halogen compounds are London forces, which are greater for larger atoms. Thus, the melting and boiling points increase from fluoride to iodide for molecular halides.

15.84 The balanced chemical equation is:

$2\ K_2MnF_6 + 4\ SbF_5 \rightarrow 4\ KSbF_6 + 2\ MnF_3 + F_2$

The amount, in moles, of F_2 produced is found by:

$$\frac{125\text{ g } K_2MnF_6}{247.14\text{ g}\cdot\text{mol}^{-1}}\left(\frac{1\text{ mol } F_2}{2\text{ mol } K_2MnF_6}\right) = 0.253\text{ mol } F_2$$

The volume of this amount of fluorine gas is:

$$V = \frac{nRT}{P} = \frac{(0.253\text{ mol})\left(0.08206\text{ L}\cdot\text{atm}\cdot\text{K}^{-1}\cdot\text{mol}^{-1}\right)(298\text{ K})}{1\text{ atm}}$$

$$= 6.18\text{ L}$$

15.86 (a)

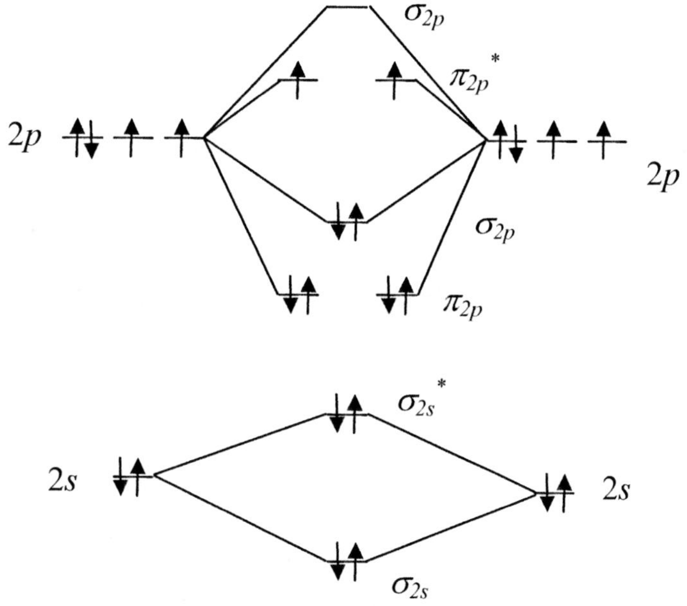

The O—O bond order is 2 (eight electrons in bonding orbitals and four electrons in antibonding orbitals). O_2 is paramagnetic with two unpaired electrons.

(b) The paramagnetism of O_2 is not explained by its Lewis structure, although the Lewis structure does predict the correct bond order.

(c) The highest occupied molecular orbital is antibonding.

(d) The superoxide ion has one more electron than O_2. This electron will populate an antibonding orbital, reducing the bond order to 1.5. The ion will remain paramagnetic but will have only one unpaired electron. The peroxide ion places another electron in an antibonding orbital. The bond order will be 1 and the ion will diamagnetic.

15.88 (a) The Lewis structure of N_2O_3 has two resonance forms:

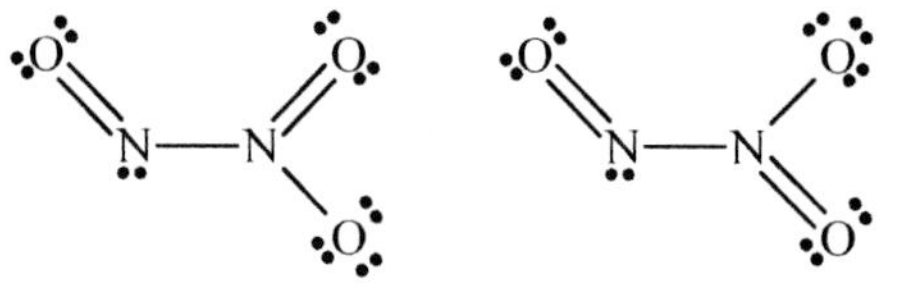

(b) The N—N bond order is 1.

(c) Because of its small size, N can readily form multiple bonds to other atoms. This is not as easy for P and consequently it prefers a structure with only single bonds.

15.90 (a) The unit cell is hexagonal.

(b) There are ten Ca^{2+} ions and six PO_4^{3-} ions in the unit cell.

(c) Two Ca^{2+} ions lie completely within the unit cell plus 16 Ca^{2+} ions lying on the cell faces (1/2 is in the cell for a total of eight ions in the cell). The six phosphate ions lie completely within the unit cell.

(d) The X^- ions are not shown.

15.92 Ion-ion forces are among the strongest intermolecular interactions. Therefore, the interactions in (a) are the strongest shown. Hydrogen bonding is stronger than a dipole-dipole interaction indicating that the interactions in (c) are stronger than those shown in (d).

Chapter 16

The Elements: The *d* Block

16.2 Five *d*-block elements can be found in Appendix 2B with positive standard potentials. They are

Ag, $E° = +0.80$ V

Cu, $E° = +0.34$ V

Au, $E° = +1.40$ V

Pt, $E° = +1.20$ V

Hg, $E° = +0.79$ V

16.4 (a) Mn (b) Zn (c) One might expect gold to be larger because it is a third row transition metal and silver is in row two, but because of the lanthanide contraction, they are essentially the same size (both have atomic radii equal to 144 pm). (d) Mo

16.6 (a) Co is very slightly higher.

(b) Fe

(c) Cr is slightly higher.

(d) Because silver is larger, one expects it to have a lower first ionization potential, which is the case (731 $kJ \cdot mol^{-1}$).

(e) One might expect the third row transition metal to have a lower first ionization energy; however, due to the lanthanide contraction, the ionization potential for silver is less than for gold (731 $kJ \cdot mol^{-1}$ vs. 890 $kJ \cdot mol^{-1}$).

16.8 Chromium has a slightly smaller atomic radius (129 pm vs. 135 pm for V) and a somewhat larger atomic mass (52.00 g·mol^{-1} vs. 50.94 g·mol^{-1} for V). Based on these two facts alone, you would expect the density of Cr to be larger by a factor of

$$\left(\frac{135}{129}\right)^2 \times \left(\frac{52.00}{50.94}\right) = 1.16$$

In actuality, the ratio is

$$\left(\frac{7.19}{6.11}\right) = 1.18$$

16.10 A compound in which an element has a high oxidation number tends to be a good oxidizing agent. The trend toward higher oxidation state stability increases from top to bottom of the group; conversely, lower oxidation state stability increases from bottom to top. Therefore, Mn would (and does) have a more stable (low) oxidation state (+2); therefore MnO_4^- would be the best oxidizing agent.

16.12 In MCl_4, M has an oxidation number of +4. Referring to Figure 16.6, we see that Zr has the most stable +4 oxidation state.

16.14 (a) $FeCr_2O_4(s) + 4\,C(s) \xrightarrow{\Delta} Fe(s) + 2\,Cr(s) + 4\,CO(g)$

(b) $2\,CrO_4^{2-}(aq) + 2\,H_3O^+(aq) \rightarrow Cr_2O_7^{2-}(aq) + 3\,H_2O(l)$

(c) $3\,MnO_2(s) + 4\,Al(s) \xrightarrow{\Delta} 3\,Mn(l) + 2\,Al_2O_3(s)$

16.16 (a) Fe_3O_4, which is actually a mixture of iron(II) oxide, FeO, and iron(III) oxide, Fe_2O_3

(b) iron(II) disulfide, FeS_2

(c) iron(II) titanate, $FeTiO_3$

(d) iron(II) chromite, $FeCr_2O_4$

16.18 (a) $E°(Ni^{2+}/Ni) = -0.23$ V

$E°(H^+/H_2) = 0.00$ V

The products are Ni^{2+}, H_2, and Cl^-.

(b) $E°(Ti^{2+}/Ti) = -1.63$ V

$E°(Ti^{3+}/Ti^{2+}) = -0.37$ V

$E°(Ti^{4+}/Ti^{3+}) = 0.00$ V

The products are Ti^{4+}, H_2, and Cl^-.

(c) $E°(Pt^{2+}/Pt) = +1.20$ V

$E°(MnO^{4-}/Mn^{2+}) = +1.51$ V

The products are Pt^{2+}, Mn^{2+}, Cl^-, K^+, and H_2O.

16.20 (a) $Cr_2O_3(s) + 2\ Al(s) \rightarrow Al_2O_3(s) + 2\ Cr(l)$

(b) $2\ Cu(s) + H_2O(l) + O_2(g) + CO_2(g) \rightarrow Cu_2(OH)_2CO_3(s)$

(c) $Ni(s, impure) + 4\ CO(g) \xrightarrow{50\,°C} Ni(CO)_4(g)$ followed by

$Ni(CO)_4(g) \xrightarrow{200\,°C} Ni(s, pure) + 4\ CO(g)$

16.22

Fe^{2+}	$[Ar]3d^6$	Fe^{3+}	$[Ar]3d^5$
Co^{2+}	$[Ar]3d^7$	Co^{3+}	$[Ar]3d^6$
Ni^{2+}	$[Ar]3d^8$	Ni^{3+}	$[Ar]3d^7$

Oxidation of Fe^{2+} to Fe^{3+} readily occurs because a half-filled *d*-subshell is obtained, which is not the case with either Co^{2+} or Ni^{2+}. Half-filled subshells are low-energy electron arrangements; therefore, they have a strong tendency to form.

16.24 Chalcopyrite, which contains $CuFeS_2$, is crushed and ground, and the sulfide is extracted in enriched form by froth flotation. "Roasting," or heating in air, follows:

$2\ CuFeS_2(s) + 3\ O_2(g) \xrightarrow{\Delta} 2\ CuS(s) + 2\ FeO(s) + 2\ SO_2(g)$

Compressed air is then forced through the resulting CuS:

$CuS(s) + O_2(g) \rightarrow Cu(l) + SO_2(g)$

16.26 (a) hexacyanoferrate(III) ion

Let x = oxidation number to be determined

$x(Fe) + 6 \times (-1) = -3$

$x(Fe) = -3 - (-6) = +3$

(b) pentaquahydroxoiron(III) ion

$x(Fe) + 1 \times (-1) + 5 \times (0) = +2$

$x(Fe) = +2 - (-1) = +3$

(c) tetraammineaquachlorocobalt(III) ion

$x(Co) + 1 \times (-1) + 4 \times (0) + 1 \times (0) = +2$

$x(Co) = + 2 - (-1) = + 3$

(d) tris(ethylenediamine)iridium(III) ion

$x(Ir) + 3 \times (0) = + 3$

$x(Ir) = + 3$

16.28 (a) $[Cr(OH)_2(NH_3)_3(H_2O)]Cl$

(b) $K_2[PtCl_4]$

(c) $[NiCl_2(H_2O)_4]SO_4$

(d) $K_3[Rh(C_2O_4)_3]$

(e) $[RhCl(OH)(C_2O_4)_2]^{3-} \cdot 8\ H_2O$

16.30 The chloride and cyanide ions can only be monodentate with respect to one metal ion. Ethylenediamine tetraacetate may bond through four oxygen atoms and two nitrogen atoms, making it a hexadentate ligand if all the arms are simultaneously bound to one metal center. The molecule $N(CH_2CH_2NH_2)_3$ has four nitrogen atoms, each having a lone pair that can bind to a metal center. It can function as a tetradentate ligand.

(a) M—Cl (b) M—C≡N

(c) 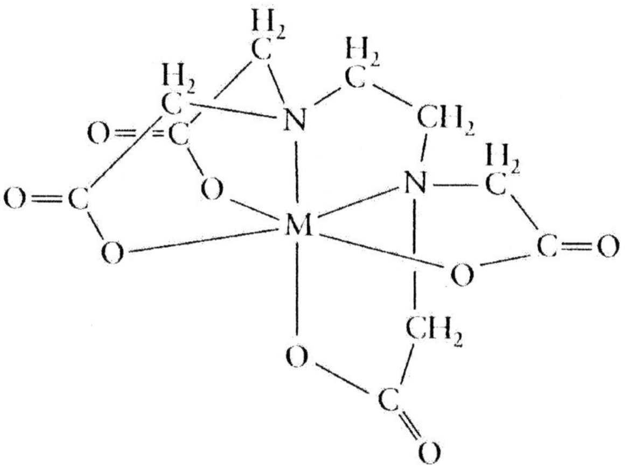

(d) 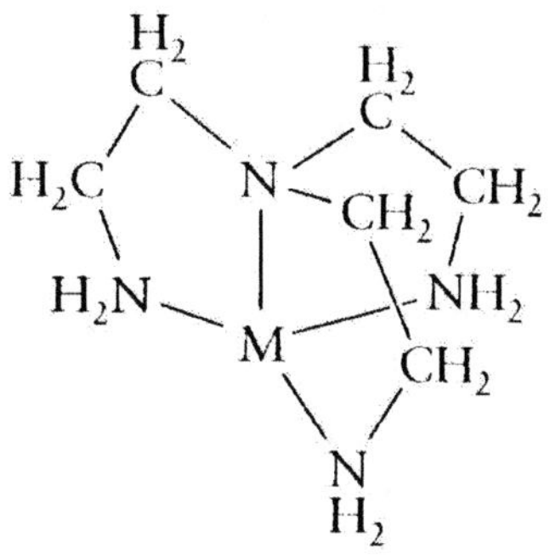

16.32 As shown below, the molecules (a) and (b) can function as chelating ligands. In (a) the two phenyl rings are free to rotate around in the single bond that connects them, making it possible for them to place the two nitrogen atoms in a proper geometrical arrangement to chelate a metal ion. The two amine groups in (c) are arranged so that they would not be able to coordinate simultaneously to the same metal center. It is possible for each of the amine groups in (c) to coordinate to two different metal centers, however. This is not classified as chelating. When a single ligand binds to two different metal centers, it is known as a *bridging* ligand.

(a) 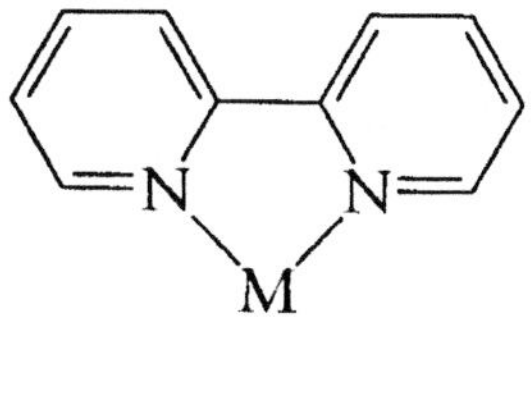
(b) 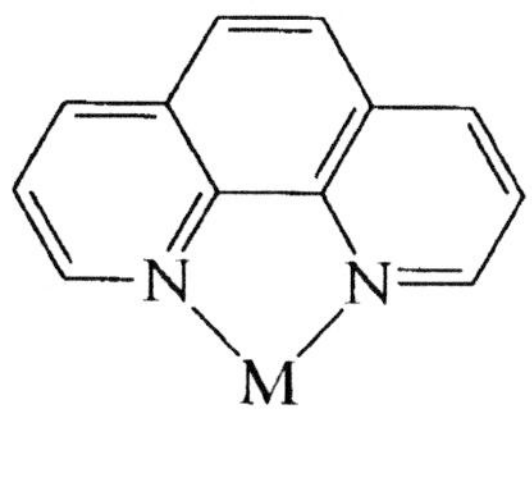
(c) 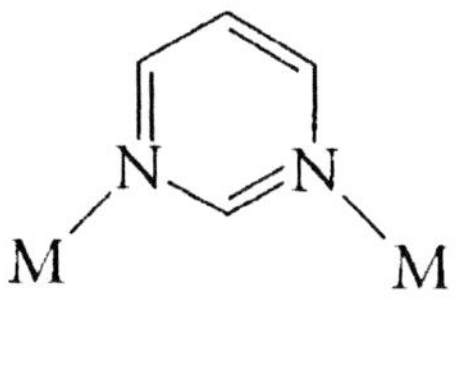

16.34 (a) 6 (en is bidentate) (b) 6 (ox is bidentate) (c) 4 (d) 5

16.36 (a) coordination isomerism
(b) hydrate isomerism
(c) linkage isomerism
(d) coordination isomerism

16.38

(a) no; only 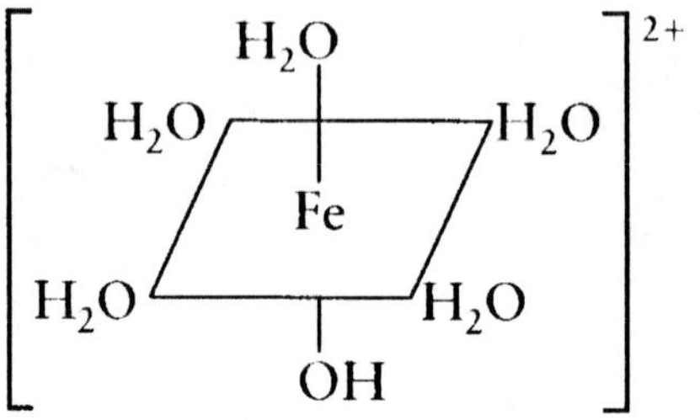

(b) yes: 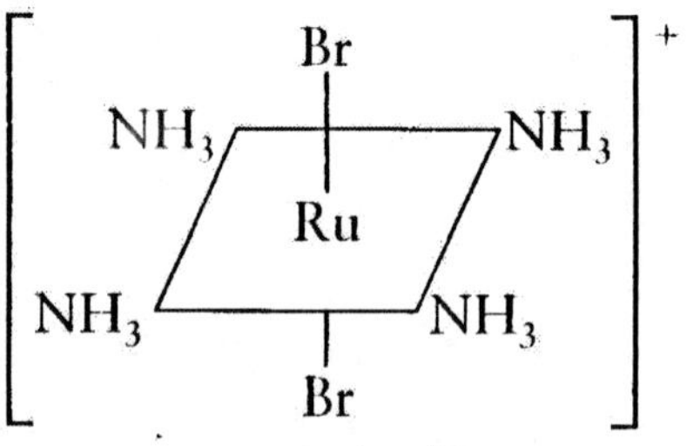

trans-tetraamminedibromo-ruthenium(II) ion

and 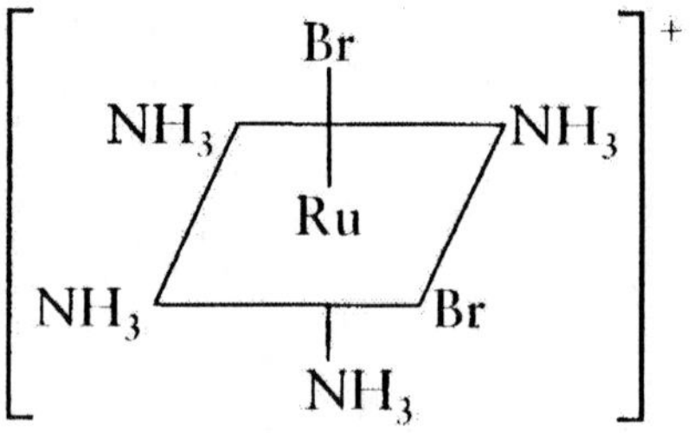

cis-tetraamminedibromo-ruthenium(II) ion

(c) yes: 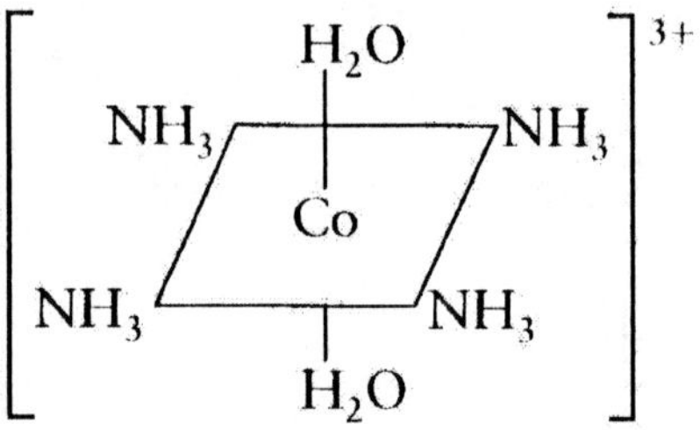

trans-diaquatetraammine cobalt(III) ion

and 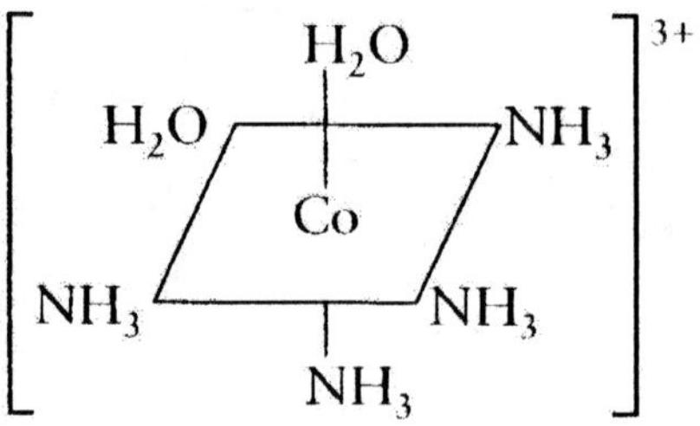

cis-diaquatetraammine cobalt(III) ion

16.40

(a)

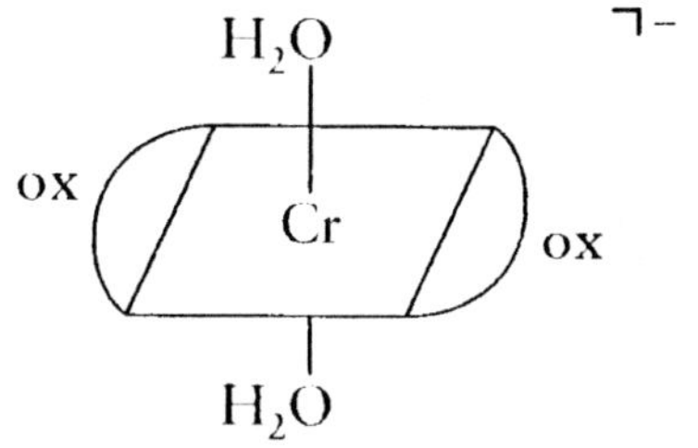

trans-diaquabis(oxalato)chromate(III) ion

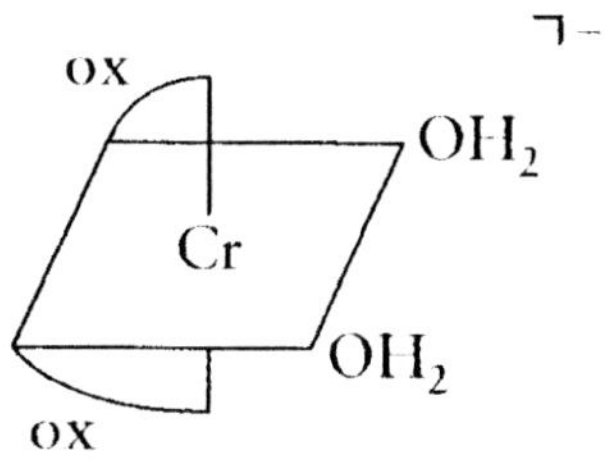

cis-diaquabis(oxalato)chromate(III) ion

A second cis isomer is the mirror image of the first cis complex (same name).

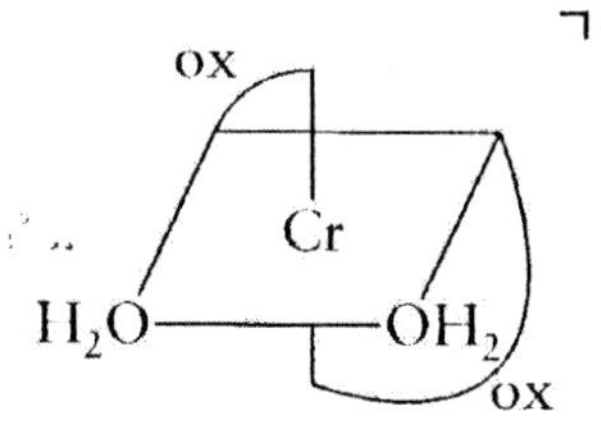

(b) The cis isomers are optically active.

16.42 Complex (a) is chiral; the enantiomeric pair is illustrated.

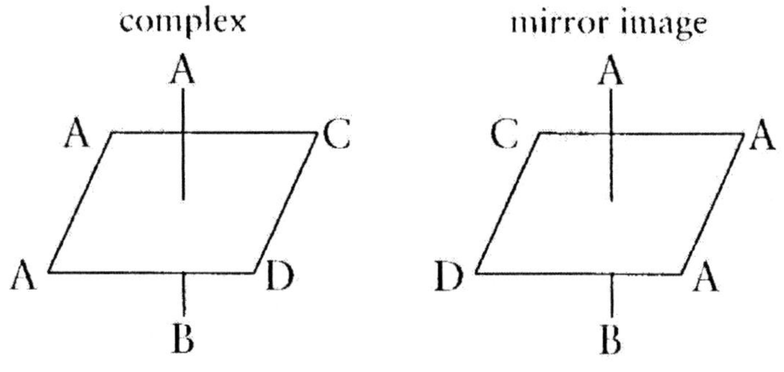

Complex (b) is not chiral.

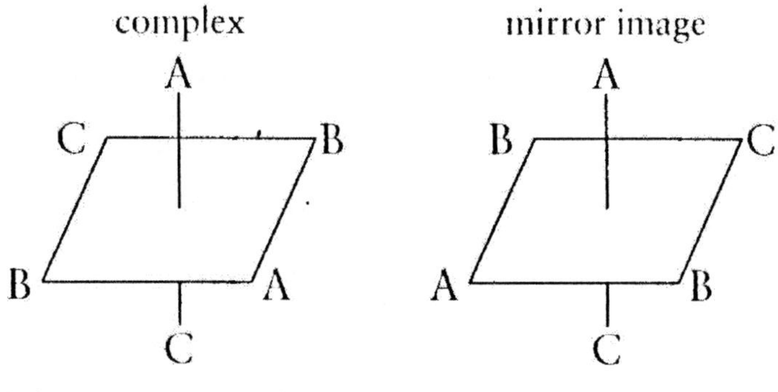

A 90° counterclockwise rotation of the mirror image about the vertical A-C axis will make the complex and mirror image match. The two complexes, (a) and (b) are not even isomers, so they cannot be enantiomers.

16.44 (a) 3; (b) 6; (c) 6; (d) 9; (e) 4; (f) 6

16.46 (a) 2; (b) 3; (c) 8; (d) 8; (e) 3; (f) 8

16.48 (a) octahedral: weak-field ligand, 10 e^-

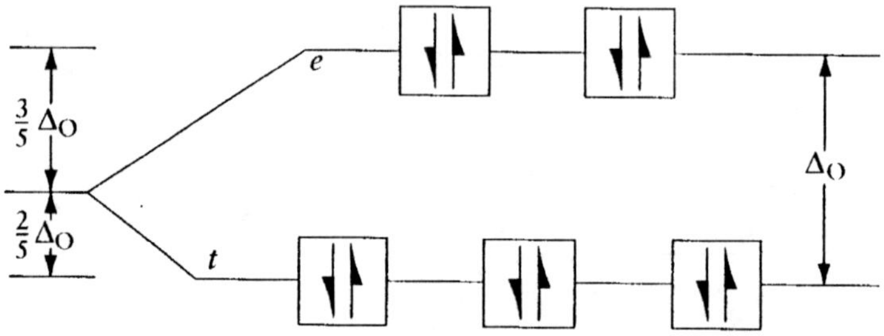

no unpaired electrons

(b) tetrahedral: weak-field ligand, 7 e^-

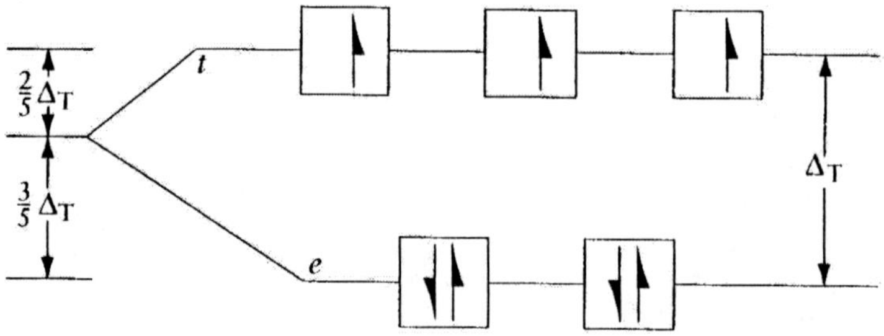

3 unpaired electrons

(c) octahedral: strong-field ligand, 6 e⁻

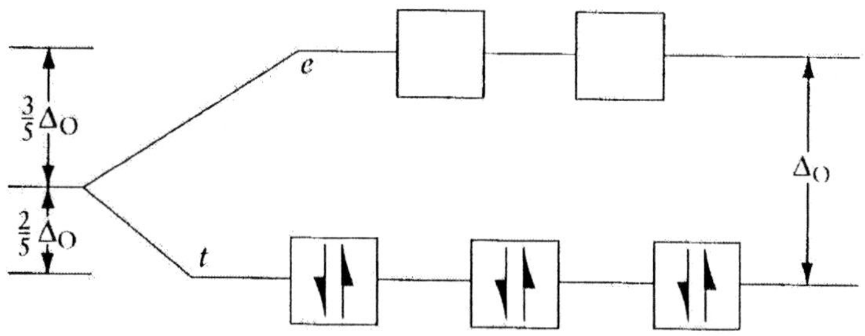

no unpaired electrons

(d) octahedral: weak-field ligand, 6 e⁻

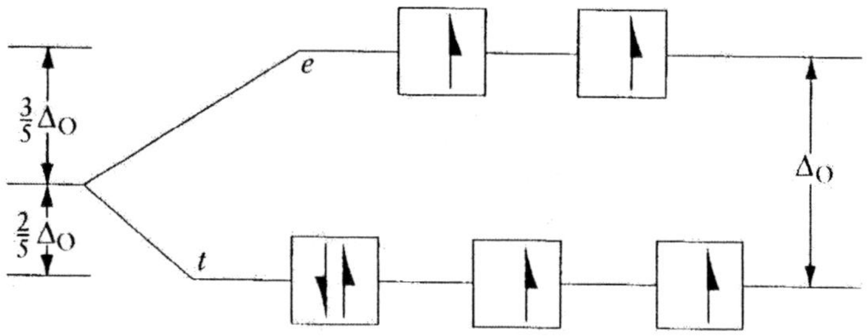

4 unpaired electrons

16.50 (a) $[FeF_6]^{3-}$

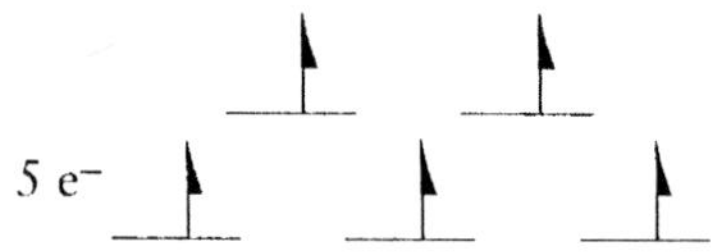

5 unpaired electrons

(b) $[Co(ox)_3]^{4-}$

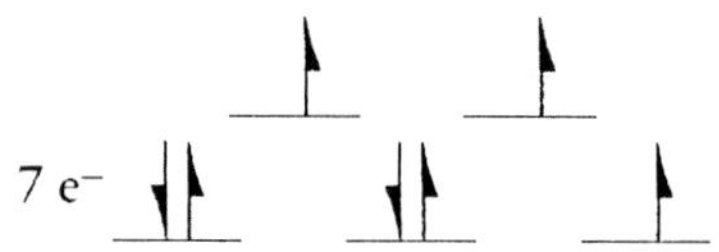

3 unpaired electrons

16.52 Because the splitting between the energies of the atomic orbitals increases when a weak-field ligand is replaced with a strong-field ligand, light absorption will occur at shorter wavelengths and the color of the complex will change, for example, from green to orange-yellow. Furthermore, the number of unpaired electrons may decrease (due to pairing of the electrons caused by the strong-field ligands) and the compound may become less paramagnetic (less total spin).

16.54 NH_3 is a strong-field ligand; therefore, the splitting is large between the t_{2g} and e_g orbitals. Co^{3+} has six *d* electrons; therefore, the configuration is $t_{2g}^{\ 6}$. The splitting is large and therefore the transition is a high-energy transition. This leads to a short wavelength. The metal in $[CoCl_6]^{3-}$ is a Co^{3+} and therefore again has six *d* electrons. Cl^- is a weak-field ligand; therefore, the splitting is small and the configuration is $t_{2g}^{\ 4}e_g^{\ 2}$. Because the splitting is small, the transition is a low-energy transition. This leads to a longer wavelength.

16.56 (a) $[CuBr_4]^{2-} \rightarrow [Cu(H_2O)_6]^{2+}$

(b) violet → blue is expected

Br^- is a weak-field ligand. H_2O is a stronger weak-field ligand. The color shift is from shorter to longer wavelength. The weaker-field complex absorbs the longer wavelengths and transmits the shorter (in this case, violet) color. The opposite is true for the stronger-field complex. Also, the $[CuBr_4]^{2-}$ ion is tetrahedral, and tetrahedral complexes in general have smaller values for Δ than octahedral complexes.

16.58 Cu(II) compounds contain one unpaired electron ($3d^9$ configuration); Cu(I) compounds have no unpaired electrons ($3d^{10}$). Therefore, Cu(II) compounds may be colored and paramagnetic, but Cu(I) compounds are not.

16.60 (a) $\Delta_O = \dfrac{hc}{\lambda} = \dfrac{(6.63 \times 10^{-34}\ \text{J}\cdot\text{s}^{-1})(3.00 \times 10^{8}\ \text{m}\cdot\text{s}^{-1})}{295 \times 10^{-3}\ \text{m}} = 6.75 \times 10^{-19}\text{J}$

(b) $\Delta_O = \dfrac{hc}{\lambda} = \dfrac{(6.63 \times 10^{-34}\ \text{J}\cdot\text{s}^{-1})(3.00 \times 10^{8}\ \text{m}\cdot\text{s}^{-1})}{435 \times 10^{-9}\ \text{m}} = 4.75 \times 10^{-19}\text{J}$

(c) $\Delta_O = \dfrac{hc}{\lambda} = \dfrac{(6.63 \times 10^{-34}\ \text{J}\cdot\text{s}^{-1})(3.00 \times 10^{8}\ \text{m}\cdot\text{s}^{-1})}{540 \times 10^{-9}\ \text{m}} = 3.68 \times 10^{-19}\text{J}$

The above numbers can be multiplied by 6.022×10^{23} to obtain the energies in $\text{kJ}\cdot\text{mol}^{-1}$.

(a) $6.74 \times 10^{-19}\ \text{J} \times 6.022 \times 10^{23}\ \text{mol}^{-1} = 4.06 \times 10^{5}\ \text{J}\cdot\text{mol}^{-1}$

$= 406\ \text{kJ}\cdot\text{mol}^{-1}$

(b) 4.57×10^{-19} J $\times$ 6.022×10^{23} mol^{-1} = 2.75×10^5 J·mol^{-1}

= 275 kJ·mol^{-1}

(c) 3.68×10^{-19} J $\times$ 6.022×10^{23} mol^{-1} = 2.22×10^5 J·mol^{-1}

= 222 kJ·mol^{-1}

$H_2O < NH_3 < CN^-$ (spectrochemical series)

16.62 The t_{2g} set, which is comprised of the d_{xy}, d_{xz} and d_{yz} orbitals

16.64 (a) NH_3 is neither a π-acid nor a π-base because it does not have any empty π-type antibonding orbitals nor does it have any extra lone pairs of electrons to donate. (b) ox has extra lone pairs in addition to the one that is used to form the σ-bond to the metal and so it can act as a π-base, donating electrons in a *p*-orbital to an empty *d*-orbital on the metal. (c) The F^- ion has extra lone pairs in addition to the one that is used to form the σ-bond to the metal and so it can act as a π-base, donating electrons in a *p*-orbital to an empty *d*-orbital on the metal. (d) CO is a π-acid ligand accepting electrons into the empty π^*-orbital created by the C—O multiple bond. $F^- <$ ox $< NH_3 <$ CO. Note that the spectrochemical series orders the ligands as π-bases < σ-bond only ligands < π-acceptors.

16.66 Bonding. In a complex that forms only σ-bonds, the t_{2g} set of orbitals is non-bonding. If the ligands can function as π-acceptors, the t_{2g} set becomes bonding by interaction with the π^*-orbitals on the ligands.

16.68 Antibonding. In a complex that forms only σ-bonds, the t_{2g} set of orbitals is non-bonding. If the ligands can function as π-donors, the t_{2g} set becomes antibonding by interacting with the filled *p*-orbitals on the ligands.

16.70 Ethylenediamine is a σ-bond-forming ligand that cannot function as either a π-acceptor or a π-donor. CO, on the other hand, can function as a π-acceptor ligand. When it does so, it lowers the energy of the t_{2g} set of orbitals, making Δ larger.

16.72 (a) Limestone is converted to CaO, which helps remove acidic (nonmetal oxide) anhydride and amphoteric impurities from the ore.

(b) $CaO(s) + SiO_2(s) \xrightarrow{\Delta} CaSiO_3(l)$

$CaO(s) + Al_2O_3(s) \xrightarrow{\Delta} Ca(AlO_2)_2(l)$

$6\ CaO(s) + P_4O_{10}(s) \xrightarrow{\Delta} 2\ Ca_3(PO_4)_2(l)$

16.74 Chromium

16.76 Up to 40% zinc in copper.

16.78 +3

16.80 A diamagnetic substance has no unpaired electrons and is weakly pushed out of a magnetic field. Paramagnetism refers to the presence of unpaired electrons in a substance. A paramagnetic compound is pulled toward a magnetic field. Ferromagnetism is an extensive property that occurs when the unpaired electrons on a number of metal ions within a sample align with each other. Paramagnetism is a property of any substance with unpaired electrons, whereas ferromagnetism is a property of certain substances that can become permanently magnetized. Their spins become aligned, and this alignment can be retained even in the absence of a magnetic field. In a paramagnetic substance, the alignment is lost when the magnetic field is removed. Antiferromagnetism is the opposite of ferromagnetism—it occurs when the unpaired electrons on a number of metal ions within a sample pair between the metal ions, so that the overall magnetism cancels.

16.82 $n\,(H_2O) = 2.387\text{ g } H_2O \times \dfrac{1\text{ mol } H_2O}{18.016\text{ g } H_2O} = 0.1324\text{ mol}$

$n\,(Cl_2) = 1.57\text{ g } Cl_2 \times \dfrac{1\text{ mol } Cl_2}{70.90\text{ g } Cl_2} = 0.0221\text{ mol } Cl_2$

$$n(H_2O) = \frac{0.1324\ \text{mol}}{0.0221\ \text{mol}} = 5.99$$

Six water molecules are present in the complex.

16.84

Concentration (mol/L)	$Ni^{2+}(aq)$	+ $6\ NH_3(aq)$	→ $[Ni(NH_3)_6]^{2+}(aq)$
Start	0.10	0.75	0.0
Assume complete rxn.			
New initial	0.0	0.15	0.10
Change	$+x$	$+6x$	$-x$
Equilibrium	x	$0.15+6x$	$0.10-x$

$$K = 1.0 \times 10^9 = \frac{[Ni(NH_3)_6^{2+}]}{[Ni^{2+}][NH_3]^6} = \frac{(0.10 - x)}{(x)(0.15 + 6x)^6} \approx \frac{(0.10)}{(x)(0.15)^6}$$

$x = 8.8 \times 10^{-6}$ M

$[Ni^{2+}] = 8.8 \times 10^{-6}$ M

$[NH_3] = 0.15$ M

$[Ni(NH_3)_6]^{2+} = 0.10$ M

16.86

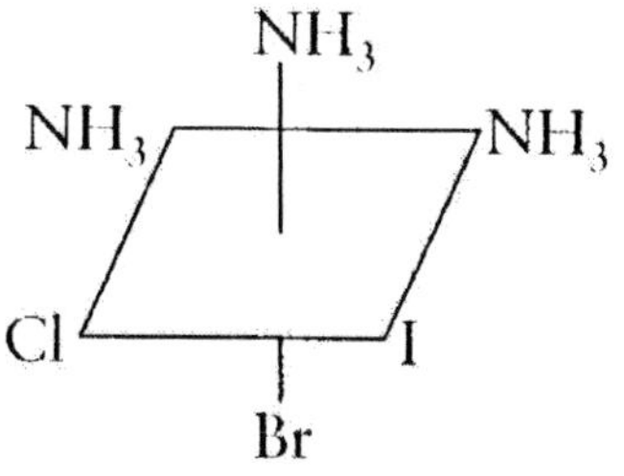

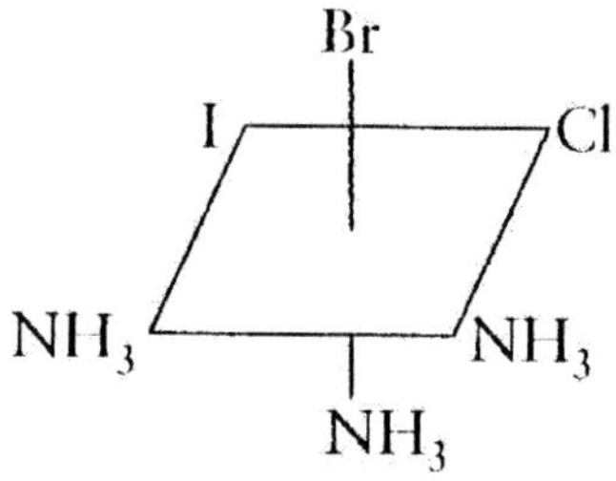

The above isomers are chiral and enantiomers of each other.

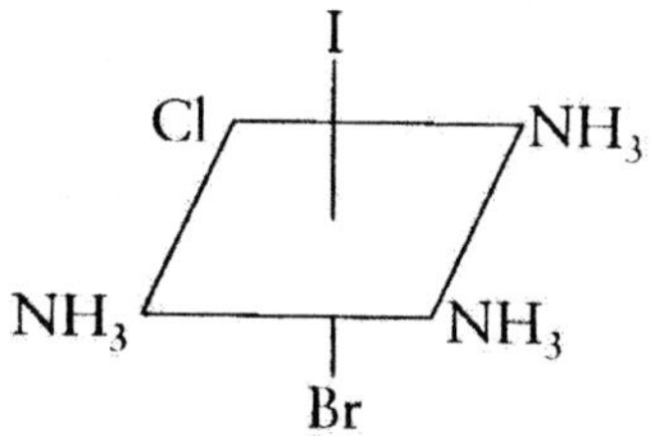

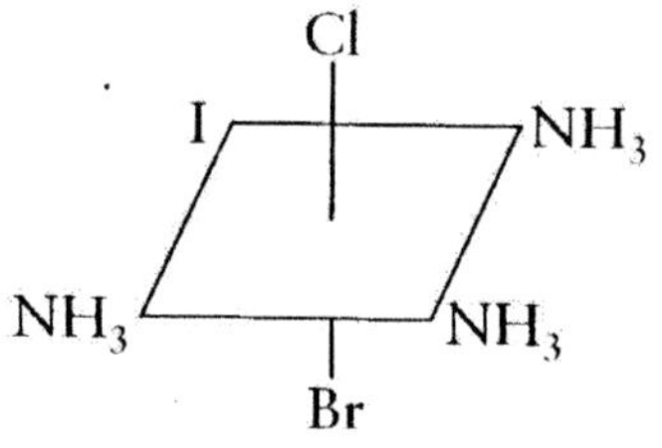

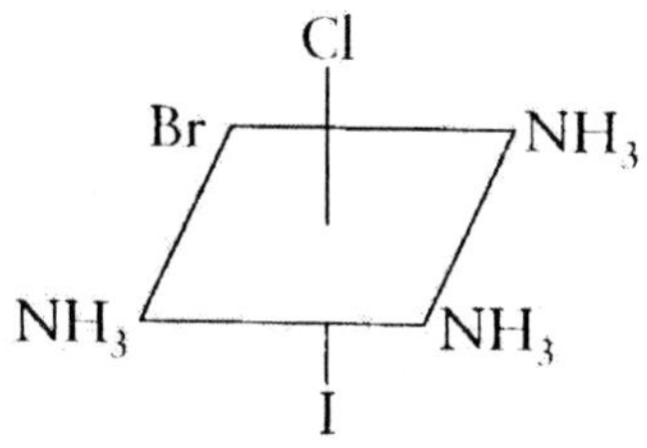

These isomers are not chiral.

16.88 Isomerism is possible. The following are the different isomers:

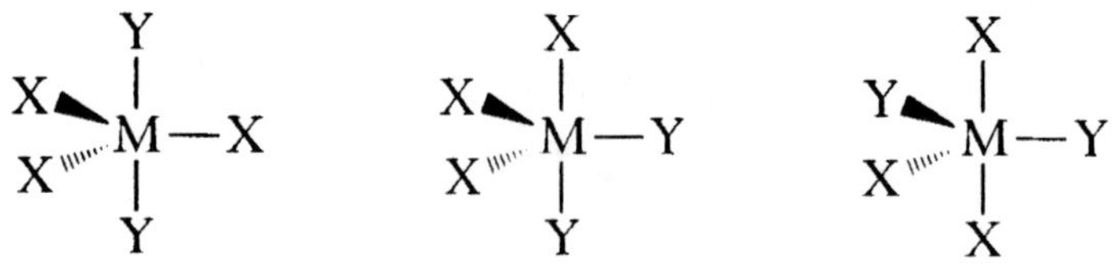

16.90 (a) strong field (b) weak field

16.92 Although the hybride ion H^- is similar to halide ions because it has a −1 charge, it is different because it has no extra lone pairs of electrons. H^- would therefore be a stronger-field ligand than the halide ions, which have lone pairs that will tend to make Δ smaller.

16.94 All of these species might be able to bind to the iron by donating a lone pair to the empty d_{z^2} orbital. However, CO can bind more strongly than O_2 because it has the ability to act as a π-acceptor through an empty π^* orbital (see Figure 14.36, π-backbonding). Nitrite anion may be able to act as a π-acceptor, too, so it will probably bind fairly tightly to the iron center. Chloride and boron trifluoride would be expected to bind more weakly, if at all, since they only have lone pairs to donate and have no empty π^* orbitals. The relative energy of the σ-donating orbital on each ligand compared to that of the σ-accepting orbital on iron would also be important in determining the stability of the bond. If these energies do not match closely enough, a bond will not form.

16.96 See Figure 16.28. Remove (mentally) the ligands along the ±*z* axes. From the shape of the atomic orbitals and their orientation with respect to the *x*, *y*, and *z* axes, it is clear that the $d_{x^2-y^2}$-orbital will have the greatest overlap, therefore repulsion, from the ligands in the *xy* plane. The d_{xy}-orbital will have the next strongest repulsion from these ligands, d_{z^2} next, and finally, the weakest repulsion will be with d_{xz} and d_{yz}. Therefore, the energy-level diagram will be (schematically):

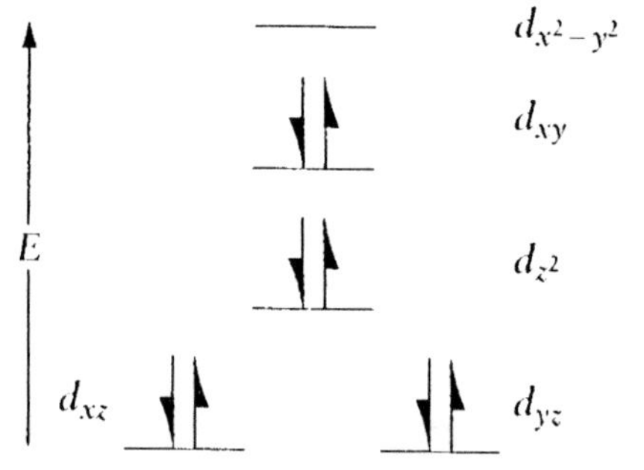

The building-up principle is illustrated in the diagram with a d^8 ion (Pd^{2+}). The separation in energy between the $d_{x^2-y^2}$ and d_{xy} orbitals determines whether or not there will be unpaired electrons.

16.98 If the sample were 100% pure, its optical rotation would be 48.0 degrees(mol·L^{-1})$^{-1}$·cm^{-1}. Because the rotation is only 46.5 degrees (mol·L^{-1})$^{-1}$·cm^{-1}, we know that the percentage of the sample of A that is giving rise to the rotation is (46.5 ÷ 48.0) × 100 = 96.9%. But because the impurity is A*, this is not all of the A in the sample.

Consider the case where [A] = [A*], which gives rise to no rotation of light because the degree of rotation of A* will exactly cancel the rotation of A. If we have a mixture of 10% A* and 90% A, then the rotation os 10% of A* will cancel an equal amount of rotation by A. Thus, the observed rotation will be 80% of the value of pure A or 0.80 × 48.0 degrees·(mol·L^{-1})$^{-1}$·cm^{-1} = 38 degrees·(mol·L^{-1})$^{-1}$·cm^{-1}. For the specific case in hand, the 3.1% rotation that is lost must be due to a 1:1 mixture of A and A*. So the total amount of A in the sample will be 96.9% + 1/2(3.1%) = 98.5%.

16.100 (a) $CuSO_4(s) + 5\ H_2O(l) \rightarrow CuSO_4 \cdot 5\ H_2O(s)$

The pentahydrate has a complicated structure. It consists of the $Cu(H_2O)_4{}^{2+}$ ion bonded to two sulfate ions. The four water molecules of the tetrahydrate ion form an approximate square planar structure around the copper atom, with the sulfate oxygen atoms in the remaining two coordination sites. The fifth water is hydrogen-bonded to the sulfate and not directly associated with the Cu^{2+} ion.

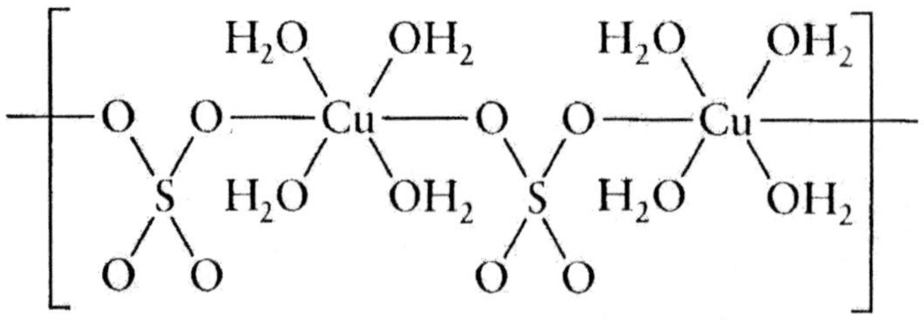

(b) $CuSO_4(s) + 6\ H_2O(l) \rightarrow [Cu(H_2O)_6]^{2+}\ (aq) + SO_4^{2-}\ (aq)$

(c) $[Cu(H_2O)_6]^{2+}\ (aq) + 4\ NH_3(aq) \rightarrow [Cu(NH_3)_4]^{2+}\ (aq) + 6\ H_2O(l)$

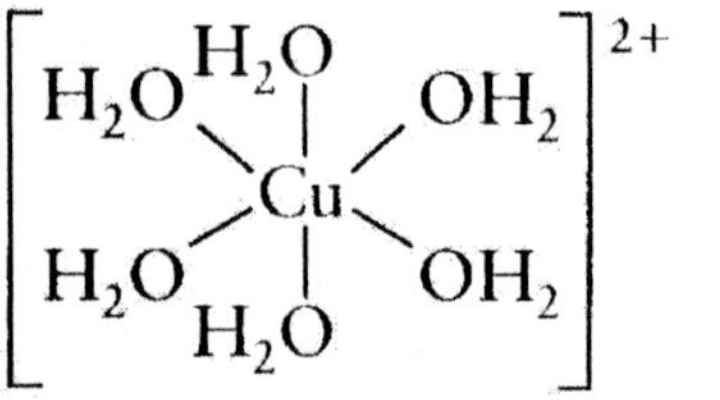

$[Cu(H_2O)_6]^{2+}$

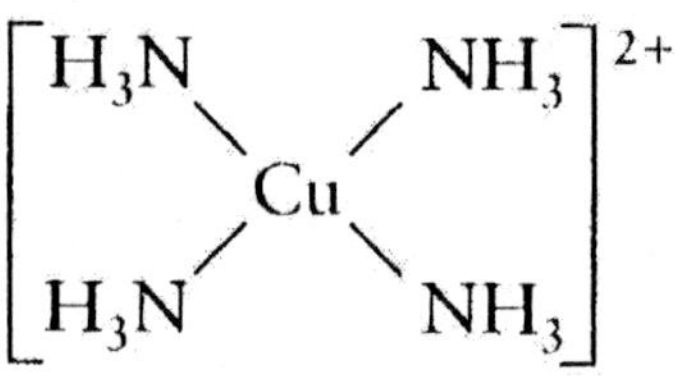

$[Cu(NH_3)_4]^{2+}$

16.102 (a) Iron is four and six coordinate, oxygen is four coordinate.

(b) Iron is approximately octahedral or tetrahedral, and oxygen is tetrahedral.

(c) Iron atoms:

1/4 × 12 octahedral iron atoms on the unit cell edges	3 atoms
1/2 × 12 atoms on the unit cell faces	6 atoms
7 octahedral atoms completely inside the unit cell	7 atoms
8 tetrahedral atoms completely inside the unit cell	8 atoms
Total	24 atoms

Oxygen atoms:

1/4 × 24 atoms on unit cell edges	6 atoms
1/2 × 24 atoms on unit cell faces	12 atoms
14 atoms completely within the unit cell	14 atoms
Total	32 atoms

The overall formula is $Fe_{24}O_{32}$ or Fe_3O_4.

(d) The unit cell is face centered cubic.

(e) The edge length is 839 pm and there are eight formula units per unit cell. The mass in the unit cell is

$8 \times 231.55\ \text{g·mol}^{-1} \div 6.02 \times 10^{23}$ formula units·mol^{-1} $= 3.08 \times 10^{-21}\ \text{g·mol}^{-1}$

The volume of the unit cell is

$(839\ \text{pm} \times 10^{-10}\ \text{cm·pm}^{-1})^3 = 5.90 \times 10^{-22}\ \text{cm}^3$

density $= 3.08 \times 10^{-21}\ \text{g·mol}^{-1} \div 5.90 \times 10^{-22}\ \text{cm}^3 = 5.22\ \text{g·cm}^3$

16.104 (a) trigonal bipyramidal

(b) +1

(c) There are six possible isomers (including the one shown on the Web site).

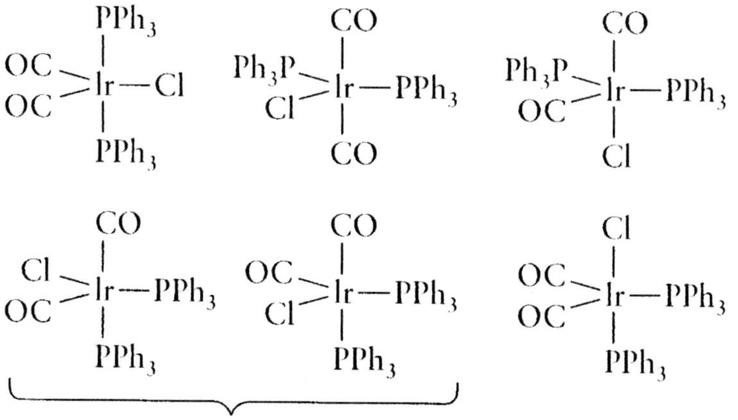

nonsuperimposable mirror images

nonsuperimposable mirror images

16.106 $Hg_2^{2+} + 2\ e^- \rightarrow 2\ Hg$ (1) $E° = +0.79$ V

$2\ Hg^{2+} + 2\ e^- \rightarrow Hg_2^{2+}$ (2) $E° = +0.92$

Reversing Eq. 2 gives

$Hg_2^{2+} \rightarrow 2\ Hg^{2+} + 2\ e^-$ (3)

The cathode process is given by Eq. 1; the anode process by Eq. 3. Combining them yields the overall (disproportionation) reaction:

$2\ Hg_2^{2+} \rightarrow 2\ Hg^{2+} + 2\ Hg$

The potential of this reaction is

$E°_{cell} = E°_{cathode} - E°_{anode}$

$E°_{cell} = +0.79\ V - (+0.92\ V) = -0.13\ V$

$$\ln K = \frac{nFE}{RT} = \frac{-0.13\ V}{0.025\ 693\ V} = -5.1$$

$K = 6 \times 10^{-3}$

Chapter 17

Nuclear Chemistry

17.2 $\Delta E = h\nu$, $\nu = \dfrac{\Delta E}{h}$, $\lambda = \dfrac{c}{\nu}$

(a) $\nu = \dfrac{9.6 \times 10^{-13}\ \text{J}}{6.63 \times 10^{-34}\ \text{J} \cdot \text{s}} = 1.4 \times 10^{21}\ \text{Hz}$

$\lambda = \dfrac{3.00 \times 10^{8}\ \text{m} \cdot \text{s}^{-1}}{1.4 \times 10^{21}\text{s}^{-1}} = 2.1 \times 10^{-13}\ \text{m}$

(b) $\nu = \dfrac{2.2 \times 10^{-14}\ \text{J}}{6.63 \times 10^{-34}\ \text{J} \cdot \text{s}} = 3.3 \times 10^{19}\ \text{Hz}$

$\lambda = \dfrac{3.00 \times 10^{8}\ \text{m} \cdot \text{s}^{-1}}{3.3 \times 10^{19}\text{s}^{-1}} = 9.1 \times 10^{-12}\ \text{m}$

(c) $\nu = \dfrac{4.7 \times 10^{-16}\ \text{kJ} \times \left(\dfrac{10^{3}\ \text{J}}{1\ \text{kJ}}\right)}{6.63 \times 10^{-34}\ \text{J} \cdot \text{s}} = 7.1 \times 10^{20}\ \text{Hz}$

$\lambda = \dfrac{3.00 \times 10^{8}\ \text{m} \cdot \text{s}^{-1}}{7.1 \times 10^{20}\text{s}^{-1}} = 4.2 \times 10^{-13}\ \text{m}$

(d) $\nu = \dfrac{3.8 \times 10^{-16}\ \text{kJ} \times \left(\dfrac{10^{3}\ \text{J}}{1\ \text{kJ}}\right)}{6.63 \times 10^{-34}\ \text{J} \cdot \text{s}} = 5.7 \times 10^{20}\ \text{Hz}$

$\lambda = \dfrac{3.00 \times 10^{8}\ \text{m} \cdot \text{s}^{-1}}{5.7 \times 10^{20}\text{s}^{-1}} = 5.3 \times 10^{-13}\ \text{m}$

17.4 energy of 1 MeV $= \left(\dfrac{10^{6}\ \text{eV}}{1\ \text{MeV}}\right)\left(\dfrac{1.602 \times 10^{-19}}{1\ \text{eV}}\right) = 1.602 \times 10^{-13}\ \text{J} \cdot \text{MeV}^{-1}$

(a) $\Delta E = (5.30\ \text{MeV})\left(\dfrac{1.602 \times 10^{-13}}{1\ \text{MeV}}\right) = 8.49 \times 10^{-13}\ \text{J}$

$$v = \frac{\Delta E}{h} = \frac{8.49 \times 10^{-13}\ \text{J}}{6.63 \times 10^{-34}\ \text{J} \cdot \text{s}} = 1.28 \times 10^{21}\ \text{s}^{-1} = 1.28 \times 10^{21}\ \text{Hz}$$

$$\lambda = \frac{c}{v} = \frac{3.00 \times 10^{8}\ \text{m} \cdot \text{s}^{-1}}{1.28 \times 10^{21}\ \text{s}^{-1}} = 2.34 \times 10^{-13}\ \text{m}$$

(b) $\Delta E = (0.26\ \text{MeV}) \left(\frac{1.602 \times 10^{-13}\ \text{J}}{1\ \text{MeV}} \right) = 4.2 \times 10^{-14}\ \text{J}$

$$v = \frac{\Delta E}{h} = \frac{4.2 \times 10^{-14}\ \text{J}}{6.63 \times 10^{-34}\ \text{J} \cdot \text{s}} = 6.3 \times 10^{19}\ \text{s}^{-1} = 6.3 \times 10^{19}\ \text{Hz}$$

$$\lambda = \frac{c}{v} = \frac{3.00 \times 10^{8}\ \text{m} \cdot \text{s}^{-1}}{6.3 \times 10^{19}\ \text{s}^{-1}} = 4.8 \times 10^{-12}\ \text{m}$$

(c) $\Delta E = (5.4\ \text{MeV}) \left(\frac{1.602 \times 10^{-13}\ \text{J}}{1\ \text{MeV}} \right) = 8.7 \times 10^{-13}\ \text{J}$

$$v = \frac{\Delta E}{h} = \frac{8.7 \times 10^{-13}\ \text{J}}{6.63 \times 10^{-34}\ \text{J} \cdot \text{s}} = 1.3 \times 10^{21}\ \text{Hz}$$

$$\lambda = \frac{c}{v} = \frac{3.00 \times 10^{8}\ \text{m} \cdot \text{s}^{-1}}{1.3 \times 10^{21}\ \text{s}^{-1}} = 2.3 \times 10^{-13}\ \text{m}$$

17.6 (a) ${}^{228}_{89}\text{Ac} \rightarrow {}^{0}_{-1}\text{e} + {}^{228}_{90}\text{Th}$

(b) ${}^{212}_{86}\text{Rn} \rightarrow {}^{4}_{2}\alpha + {}^{208}_{84}\text{Po}$

(c) ${}^{221}_{87}\text{Fr} \rightarrow {}^{4}_{2}\alpha + {}^{217}_{85}\text{At}$

(d) ${}^{230}_{91}\text{Pa} + {}^{0}_{-1}\text{e} \rightarrow {}^{230}_{90}\text{Th}$

17.8 (a) ${}^{233}_{92}\text{U} \rightarrow {}^{0}_{-1}\text{e} + {}^{233}_{93}\text{Np}$

(b) ${}^{56}_{27}\text{Co} \rightarrow {}^{1}_{1}\text{p} + {}^{55}_{26}\text{Fe}$

(c) ${}^{158}_{67}\text{Ho} \rightarrow {}^{0}_{-1}\text{e} + {}^{158}_{66}\text{Dy}$

(d) ${}^{212}_{84}\text{Po} \rightarrow {}^{4}_{2}\alpha + {}^{208}_{82}\text{Pb}$

17.10 (a) ${}^{14}_{6}\text{C} \rightarrow {}^{14}_{7}\text{N} + {}^{0}_{-1}\text{e}$ (β emission)

(b) ${}^{19}_{10}\text{Ne} \rightarrow {}^{19}_{9}\text{F} + {}^{0}_{+1}\text{e}$ (positron emission)

(c) $^{188}_{79}Au \rightarrow {}^{188}_{78}Pt + {}^{0}_{+1}e$ (positron emission)

(d) $^{229}_{92}U \rightarrow {}^{225}_{90}Th + {}^{4}_{2}\alpha$ (α emission)

17.12 (a) $^{74}_{36}Kr \rightarrow ? + {}^{0}_{+1}e$

$^{74}_{36}Kr \rightarrow {}^{74}_{35}Br + {}^{0}_{+1}e$

(b) $^{174}_{72}Hf \rightarrow ? + {}^{4}_{2}\alpha$

$^{174}_{72}Hf \rightarrow {}^{170}_{70}Yb + {}^{4}_{2}\alpha$

(c) $^{98}_{43}Tc \rightarrow ? + {}^{0}_{-1}e$

$^{98}_{43}Tc \rightarrow {}^{98}_{44}Ru + {}^{0}_{-1}e$

(d) $^{41}_{20}Ca + {}^{0}_{-1}e \rightarrow ?$

$^{41}_{20}Ca + {}^{0}_{-1}e \rightarrow {}^{41}_{19}K$

17.14 (a) $^{60}_{29}Cu$ is proton rich (*A* is below the band of stability); β^{+} decay is most likely:

$^{60}_{29}Cu \rightarrow + {}^{0}_{+1}e + {}^{60}_{28}Ni$

(b) $^{140}_{54}Xe$ is neutron rich (*A* is above the band of stability); β decay is most likely:

$^{140}_{54}Xe \rightarrow + {}^{0}_{-1}e + {}^{140}_{55}Cs$

(c) $^{246}_{95}Am$ has Z>83 and therefore, although neutron rich, most likely undergoes α decay:

$^{246}_{95}Am \rightarrow {}^{4}_{2}\alpha + {}^{242}_{93}Np$

(d) $^{240}_{93}Np$ has Z>83, and therefore, although slightly neutron rich, most likely undergoes α decay: $^{240}_{93}Np \rightarrow + {}^{4}_{2}\alpha + {}^{236}_{91}Pa$

17.16 α $^{237}_{93}Np \rightarrow {}^{4}_{2}\alpha + {}^{233}_{91}Pa$

β $^{233}_{91}Pa \rightarrow {}^{0}_{-1}e + {}^{233}_{92}U$

α ${}^{233}_{92}U \rightarrow {}^{4}_{2}\alpha + {}^{229}_{90}Th$

α ${}^{229}_{90}Th \rightarrow {}^{4}_{2}\alpha + {}^{225}_{88}Ra$

β ${}^{225}_{88}Ra \rightarrow {}^{0}_{-1}e + {}^{225}_{89}Ac$

α ${}^{225}_{89}Ac \rightarrow {}^{4}_{2}\alpha + {}^{221}_{87}Fr$

α ${}^{221}_{87}Fr \rightarrow {}^{4}_{2}\alpha + {}^{217}_{85}At$

α ${}^{217}_{85}At \rightarrow {}^{4}_{2}\alpha + {}^{213}_{83}Bi$

β ${}^{213}_{83}Bi \rightarrow {}^{0}_{-1}e + {}^{213}_{84}Po$

α ${}^{213}_{84}Po \rightarrow {}^{4}_{2}\alpha + {}^{209}_{82}Pb$

β ${}^{209}_{82}Pb \rightarrow {}^{0}_{-1}e + {}^{209}_{83}Bi$

17.18 (a) ${}^{20}_{10}Ne + {}^{1}_{1}p \rightarrow {}^{21}_{11}Na + \gamma$

(b) ${}^{1}_{1}H + {}^{1}_{1}p \rightarrow {}^{2}_{1}H + {}^{0}_{+1}e$

(c) ${}^{15}_{7}N + {}^{1}_{1}p \rightarrow {}^{12}_{6}C + {}^{4}_{2}\alpha$

(d) ${}^{20}_{10}Ne + {}^{4}_{2}\alpha \rightarrow {}^{24}_{12}Mg + \gamma$

17.20 (a) ${}^{20}_{9}F + \gamma \rightarrow {}^{0}_{-1}e + {}^{20}_{10}Ne$

(b) ${}^{44}_{22}Ti + {}^{0}_{-1}e \rightarrow {}^{0}_{+1}e + {}^{44}_{20}Ca$

(c) ${}^{241}_{95}Am + {}^{11}_{5}B \rightarrow 4\,{}^{1}_{0}n + {}^{248}_{100}Fm$

(d) ${}^{243}_{95}Am + {}^{1}_{0}n \rightarrow 4\,{}^{0}_{-1}e + {}^{244}_{96}Cm$

17.22 (a) ${}^{245}_{98}Cf + {}^{12}_{6}C \rightarrow {}^{257}_{104}Rf$

(b) ${}^{209}_{83}Bi + {}^{58}_{26}Fe \rightarrow {}^{266}_{109}Mt + {}^{1}_{0}n$

${}^{266}_{109}Mt \rightarrow {}^{4}_{2}\alpha + {}^{262}_{107}Bh$

${}^{262}_{107}Bh$ is the daughter nucleus.

17.24 (a) ununsepium, Uus (b) unbiennium, Ube (c) Bibibiium, Bbb

17.26 activity = $(10.7\text{ Bq})\left(\dfrac{1\text{ Ci}}{3.7\times10^{10}\text{ Bq}}\right)\left(\dfrac{1\ \mu\text{Ci}}{10^{-6}\text{ Ci}}\right) = 2.9\times10^{-4}\ \mu\text{Ci}$

17.28 dis = disintegration(s); dis·s^{-1} = disintegrations per second = dps = Bq

(a) activity = $\left(\dfrac{590\text{ clicks}}{100\text{ s}}\right)\left(\dfrac{1000\text{ dis}}{1\text{ click}}\right)\left(\dfrac{1\text{ Ci}}{3.7\times10^{10}\text{ dps}}\right) = 1.6\times10^{-7}\text{ Ci}$

(b) activity = $\left(\dfrac{2.7\times10^{4}\text{ clicks}}{1.5\text{ h}}\right)\left(\dfrac{1\text{ h}}{3600\text{ s}}\right)\left(\dfrac{1000\text{ dis}}{1\text{ click}}\right)\left(\dfrac{1\text{ Ci}}{3.7\times10^{10}\text{ dps}}\right)$

$= 1.4\times10^{-7}\text{ Ci}$

(c) activity = $\left(\dfrac{159\text{ clicks}}{1.0\text{ min}}\right)\left(\dfrac{1\text{ min}}{60\text{ s}}\right)\left(\dfrac{1000\text{ dis}}{1\text{ click}}\right)\left(\dfrac{1\text{ Ci}}{3.7\times10^{10}\text{ dps}}\right)$

$= 7.2\times10^{-8}\text{ Ci}$

17.30 dose in rads = $\left(\dfrac{2.0\text{ J}}{5.0\text{ g}}\right)\left(\dfrac{10^{3}\text{ g}}{1\text{ kg}}\right)\left(\dfrac{1\text{ rad}}{10^{-2}\text{ J}\cdot\text{kg}^{-1}}\right) = 4.0\times10^{4}\text{ rad}$

dose equivalent in rems = Q × dose in rads (for α radiation, Q is about 20 rem/rad)

$= \dfrac{20\text{ rem}}{1\text{ rad}}\times4.0\times10^{4}\text{ rad} = 8.0\times10^{5}\text{ rem}$

$8.0\times10^{5}\text{ rem}\div100\text{ rem/Sv} = 8.0\times10^{3}\text{ Sv}$

17.32 $2.0\text{ mrad}\cdot\text{day}^{-1} = 2.0\times10^{-3}\text{ rad}\cdot\text{day}^{-1}$

dose equivalent in rem = Q × dose in rad

dose rate in rem = Q × dose rate in rad (Q = about 20 for α radiation)

dose rate in rem = $(20\text{ rem}\cdot\text{rad}^{-1})(2.0\times10^{-3}\text{ rad}\cdot\text{day}^{-1})$

$= 4.0\times10^{-2}\text{ rem}\cdot\text{day}^{-1}$

Then

$100\text{ rem} = 4.0\times10^{-2}\text{ rem}\cdot\text{day}^{-1}\times\text{time}$

time = $\dfrac{100\text{ rem}}{4.0\times10^{-2}\text{ rem}\cdot\text{day}^{-1}} = 2.5\times10^{3}$ day (not far from 7 years)

17.34 $t_{1/2} = \frac{0.693}{k}$ (or use the ln key on your calculator to obtain a more precise value for ln 2)

(a) $t_{1/2} = \frac{0.693}{5.3 \times 10^{-10}\ y^{-1}} = 1.3 \times 10^{9}$ y

(b) $t_{1/2} = \frac{0.693}{0.132\ y^{-1}} = 5.25$ y

(c) $t_{1/2} = \frac{0.693}{3.85 \times 10^{-3}\ s^{-1}} = 180$ s

17.36 Activity = disintegrations per second

Initial activity $\propto N_0$, final activity $\propto N$

$N = N_0 e^{-kt}$

$$\frac{\text{final activity}}{\text{initial activity}} = \frac{N}{N_0} = e^{-kt}$$

final activity = initial activity × e^{-kt}

$$k = \frac{0.693}{28.1\ y} = 2.47 \times 10^{-2}\ y^{-1}$$

final activity = 3.0×10^{4} Bq × $e^{-(2.47\times10^{-2}\ y^{-1}\times 50\ y)} = 8.7 \times 10^{3}$ Bq

17.38 $k = \frac{0.693}{t_{1/2}}$, percentage remaining = 100% × $\frac{N}{N_0}$ = 100% × e^{-kt}

(a) $k = \frac{0.693}{28.1\ y} = 0.0247\ y^{-1}$

percentage remaining = 100% × $e^{-(0.0247\ y^{-1}\times 7.5\ y)}$ = 83%

For this type of problem, an alternative relationship may be used:

$\left(\frac{1}{2}\right)^{h}$ = fraction of original sample remaining (h = number of half-lives)

Thus, $\left(\frac{1}{2}\right)^{\frac{7.5}{28.1}} = 0.83 = 83\%$

(b) $k = \dfrac{0.693}{8.05 \text{ d}} = 0.0861 \text{ d}^{-1}$

percentage remaining = $100\% \times e^{-(0.0861 \text{ d}^{-1} \times 7.0 \text{ d})} = 55\%$

17.40 (a) $k = \dfrac{0.693}{t_{1/2}} = \dfrac{0.693}{10.8 \text{ y}} = 0.0642 \text{ y}^{-1}$

fraction remaining = $\dfrac{N}{N_0} = e^{-kt} = e^{-(0.0642 \text{ y}^{-1} \times 50.0 \text{ y})} = 0.0404$

(Alternatively, $\left(\dfrac{1}{2}\right)^{50.0/10.8} = 0.0404$)

(b) t = elapsed time

$$t = -\frac{t_{1/2}}{\ln 2} \ln\left(\frac{N}{N_0}\right)$$

$$t = -\frac{5.73 \times 10^3 \text{ y}}{\ln 2} \ln\left(\frac{0.75 \times N_0}{N_0}\right) = 2.4 \times 10^3 \text{ y}$$

17.42 activity $\propto N$

t = elapsed time

5500 disintegrations/24.0 h = 229 disintegrations/h

$$t = -\frac{t_{1/2}}{\ln 2} \ln\left(\frac{N}{N_0}\right)$$

$$t = -\frac{5.73 \times 10^3}{\ln 2} \ln\left(\frac{229}{920}\right) = 11\,500 \text{ y}$$

17.44 In each case: $k = \dfrac{0.693}{t_{1/2} \text{ (in s)}}$

activity in Bq = $k \times N$, activity in Ci = $\dfrac{\text{activity in Bq}}{3.7 \times 10^{10} \text{ Bq/Ci}}$

Note: Bq (= disintegrating nuclei per second) has the units of nuclei·s^{-1}.

(a) $k = \left(\dfrac{0.693}{7.1\times10^{8}\ \text{y}}\right)\left(\dfrac{1\ \text{y}}{3.17\times10^{7}\ \text{s}}\right) = 3.1 \times 10^{-7}\ \text{s}^{-1}$

$N = (1.0\ \text{g})\left(\dfrac{1\ \text{mol}}{267\ \text{g}}\right)\left(\dfrac{6.022\times10^{23}\ {}^{235}\text{U nuclei}}{1\ \text{mol}}\right) = 2.3 \times 10^{21}$ nuclei

activity $= (3.1 \times 10^{-17}\ \text{s}^{-1})(2.3 \times 10^{21}\ \text{nuclei})\left(\dfrac{1\ \text{Ci}}{3.7\times10^{10}\ \text{Bq}}\right)$

$= 1.9 \times 10^{-6}$ Ci

(b) $k = \left(\dfrac{0.693}{5.26\ \text{y}}\right)\left(\dfrac{1\ \text{y}}{3.17\times10^{7}\ \text{s}}\right) = 4.16 \times 10^{-9}\ \text{s}^{-1}$

$N = (0.010)(1.0\ \text{g})\left(\dfrac{1\ \text{mol}}{60\ \text{g}}\right)\left(\dfrac{6.022\times10^{23}\ \text{nuclei}}{1\ \text{mol}}\right)$

$= 1.0 \times 10^{20}$ nuclei

activity $= (4.16 \times 10^{-9}\ \text{s}^{-1})(1.0 \times 10^{20}\ \text{nuclei})\left(\dfrac{1\ \text{Ci}}{3.7\times10^{10}\ \text{Bq}}\right) = 11$ Ci

(c) $k = \left(\dfrac{0.693}{26.1\ \text{h}}\right)\left(\dfrac{1\ \text{h}}{3600\ \text{s}}\right) = 7.38 \times 10^{-6}\ \text{s}^{-1}$

$N = (5.0 \times 10^{-3}\ \text{g})\left(\dfrac{1\ \text{mol}}{200\ \text{g}}\right)\left(\dfrac{6.022\times10^{23}\ \text{nuclei}}{1\ \text{mol}}\right) = 1.5 \times 10^{19}$ nuclei

activity $= (7.38 \times 10^{-6}\ \text{s}^{-1})(1.5 \times 10^{19}\ \text{nuclei})\left(\dfrac{1\ \text{Ci}}{3.7\times10^{10}\ \text{Bq}}\right)$

$= 3.0 \times 10^{3}$ Ci

17.46 $t_{1/2} = 0.0033$ s

$k = \dfrac{0.693}{t_{1/2}} = \dfrac{0.693}{0.0033\ \text{s}} = 2.1 \times 10^{2}\ \text{s}^{-1}$

activity $\propto N$; and, because $N = N_0 e^{-kt}$

final activity = initial activity $\times\ e^{-kt} = 0.10\ \mu\text{Ci}\ e^{-(2.1\times10^{2}\ \text{s}^{-1}\times1.0\ \text{s})}$

$= 6.3 \times 10^{-93}\ \mu$Ci ($\approx$no activity)

Alternatively:

$$0.10\ \mu\text{Ci} \times \left(\frac{1}{2}\right)^{1.0/0.0033} = 6.0 \times 10^{-93}\ \mu\text{Ci (i.e., no activity)}$$

17.48 $1.20\ \text{Ci} \times \left(\frac{1}{2}\right)^{5.0/5.26} = 0.62\ \text{Ci}$

17.50 Activity = rate of decay = $\frac{\Delta N}{\Delta t} = k \times N$

$$k = \frac{\Delta N}{\Delta t \cdot N} = \frac{(3.4 \times 10^{13}\ \text{disintegrations})}{(15\ \text{min})(3.25 \times 10^{18}\ \text{atoms})}\left(\frac{60\ \text{min}}{1\ \text{hour}}\right)\left(\frac{24\ \text{hours}}{1\ \text{day}}\right)$$
$$= 0.0010\ \text{day}^{-1}$$

(a) $\frac{\Delta N}{N}100\% = \left(1 - e^{-k\Delta t}\right) \times 100\% = \left(1 - e^{-0.0010 \times 150}\right) \times 100\% = 78\%$

(b) $N_{remain} = N_0 \times (1 - 0.78) = 3.25 \times 10^{18} \times 0.22 = 7.2 \times 10^{17}$ atoms

(c) $t_{\frac{1}{2}} = \frac{\ln 2}{k} = \frac{\ln 2}{0.0010\ \text{day}^{-1}} = 690$ day

17.52 See full solution in Exercise 17.51. Plot below shows Y_{max}= 32 days.

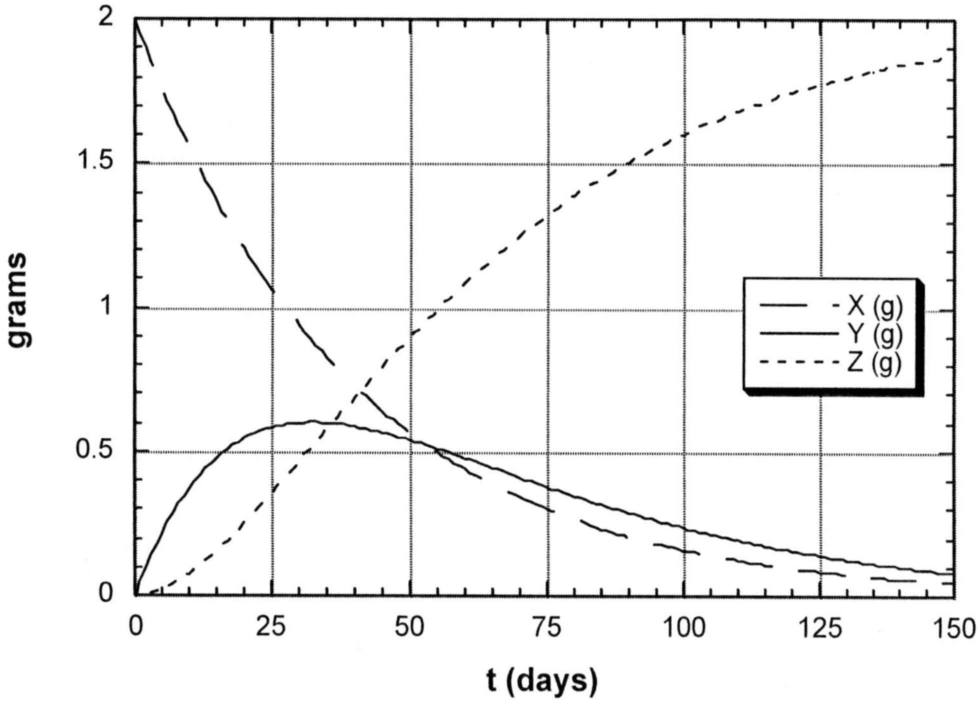

17.54 A suitable radiolabeled compound can be dissolved in a solution at a known concentration and injected into the animal's bloodstream. After a few minutes the blood stream will be equilibrated. At this point, a blood sample can be taken so that the activity of the radioisotope in the blood can be determined. From this, one can calculate the dilution factor (the amount by which the original concentration of the injected material was diluted) and subsequently the total blood volume of the animal can be determined.

17.56 The O—D bond does not dissociate as readily as the O—H bond, as seen by the higher pK_a value for D_2O. This added reluctance to dissociate makes many reactions involving O—D bonds slower than those involving O—H bonds. Because many reactions occurring in the body involve the breaking of O—H bonds, these reactions will be significantly slowed if the H atoms are replaced by D atoms. So although D and H are chemically very similar, they do not behave exactly the same in subtle ways.

17.58 (a) $E = mc^2 = (1.00\ \text{kg})(3.00 \times 10^8\ \text{m}\cdot\text{s}^{-1})^2$

$= 9.00 \times 10^{16}\ \text{kg}\cdot\text{m}^2\cdot\text{s}^{-2} = 9.00 \times 10^{16}\ \text{J}$

(b) $E = mc^2 = (0.454\ \text{kg})(3.00 \times 10^8\ \text{m}\cdot\text{s}^{-1})^2$

$= 4.09 \times 10^{16}\ \text{kg}\cdot\text{m}^2\cdot\text{s}^{-2} = 4.09 \times 10^{16}\ \text{J}$

(c) $E = mc^2 = (1.674\,93 \times 10^{-27}\ \text{kg})(3.00 \times 10^8\ \text{m}\cdot\text{s}^{-1})^2$

$= 1.51 \times 10^{-10}\ \text{kg}\cdot\text{m}^2\cdot\text{s}^{-2} = 1.51 \times 10^{-10}\ \text{J}$

(d) $E = mc^2 = (0.001\,0079\ \text{kg}\cdot\text{mol}^{-1} \div 6.02 \times 10^{23}\ \text{atoms}\cdot\text{mol}^{-1})$

$\times (3.00 \times 10^8\ \text{m}\cdot\text{s}^{-1})^2$

$= 1.51 \times 10^{-10}\ \text{kg}\cdot\text{m}^2\cdot\text{s}^{-2} = 1.51 \times 10^{-10}\ \text{J}$

17.60 (a) $\Delta m = \dfrac{\Delta E}{c^2} = \dfrac{3\times10^{11}\ \text{J}}{(3.00\times10^8\ \text{m}\cdot\text{s}^{-1})^2} = -3 \times 10^{-6}\ \text{kg} = -3 \times 10^{-3}\ \text{g}$

(b) The mass of a He atom is

$4.00 \div 6.02 \times 10^{23}$ g·mol^{-1} = 6.64×10^{-24} g or 6.64×10^{-27} kg.

The mass of an H atom is

$1.0079 \div 6.02 \times 10^{23}$ g·mol^{-1} = 1.67×10^{-24} g or 1.67×10^{-27} kg.

The mass of a neutron is $1.674\ 93 \times 10^{-27}$ kg.

$6\ \text{D} \rightarrow 2\ {}^{4}\text{He} + 2\ {}^{1}\text{H} + 2\ \text{n}$

The mass of the products side of the equation is

$(2 \times 6.64 \times 10^{-27}\ \text{kg})$

$+ (2 \times 1.67 \times 10^{-27}\ \text{kg}) + (2 \times 1.67 \times 10^{-27}\ \text{kg}) = 2.00 \times 10^{-26}\ \text{kg}.$

The mass on the reactants side of the equation is given by

$6 \times (0.002\ 014\ \text{g·mol}^{-1} \div 6.02 \times 10^{23}\ \text{atoms·mol}^{-1}) = 2.01 \times 10^{-23}\ \text{kg}.$

The difference in mass is 1×10^{-25} kg. The total loss of mass from (a) is 3×10^{-6} kg, so we need 3×10^{19} times the equation as written. This corresponds to $(3 \times 10^{19}) \times (6\ \text{atoms D}) \div 6.02 \times 10^{23}\ \text{atoms·mol}^{-1}$ $= 3 \times 10^{-4}$ mol D, or 6×10^{-4} g D.

17.62 (a) ${}^{98}_{42}\text{Mo}$

$42\ {}^{1}\text{H} + 56\ \text{n} \rightarrow {}^{98}_{42}\text{Mo}$

$\Delta m = 97.9055\ \text{u} - (42 \times 1.0078\ \text{u} + 56 \times 1.0087\ \text{u}) = -0.9093\ \text{u}$

$\Delta m = 0.9093\ \text{u} \times 1.6605 \times 10^{-27}\ \text{kg·u}^{-1} = -1.510 \times 10^{-27}\ \text{kg}$

$E_{\text{bind}} = -1.510 \times 10^{-27}\ \text{kg} \times (2.998 \times 10^{8}\ \text{m·s}^{-1})^2 = -1.357 \times 10^{-10}\ \text{J}$

$$E_{\text{bind}}/\text{nucleon} = \frac{-1.357 \times 10^{-10}\ \text{J}}{98\ \text{nucleons}} = -1.385 \times 10^{-12}\ \text{J·nucleon}^{-1}$$

(b) ${}^{151}_{63}\text{Eu}$

$63\ {}^{1}\text{H} + 88\ \text{n} \rightarrow {}^{151}_{63}\text{Eu}$

$\Delta m = 150.9196\ \text{u} - (63 \times 1.0078\ \text{u} + 88 \times 1.0087\ \text{u}) = -1.3374\ \text{u}$

$\Delta m = -1.3374\ \text{u} \times 1.6605 \times 10^{-27}\ \text{kg·u}^{-1} = -2.2208 \times 10^{-27}\ \text{kg}$

$E_{\text{bind}} = -2.2208 \times 10^{-27}\ \text{kg} \times (2.9979 \times 10^{8}\ \text{m·s})^2 = -1.9959 \times 10^{-10}\ \text{J}$

$$E_{\text{bind}}/\text{nucleon} = \frac{-1.9959 \times 10^{-10}\ \text{J}}{151\ \text{nucleons}} = -1.3218 \times 10^{-12}\ \text{J·nucleon}^{-1}$$

(c) $^{56}_{26}\text{Fe}$:

$26\ {}^{1}\text{H} + 30\ \text{n} \rightarrow {}^{56}_{26}\text{Fe}$

$\Delta m = 55.9349\ \text{u} - (26 \times 1.0078\ \text{u} + 30 \times 1.0087\ \text{u}) = -0.5289\ \text{u}$

$$\Delta m = -0.5289\ \text{u} \times \left(\frac{1.661 \times 10^{-27}\ \text{kg}}{1\ \text{u}}\right) = -8.785 \times 10^{-28}\ \text{kg}$$

$E_{\text{bind}} = -8.785 \times 10^{-28}\ \text{kg} \times (2.998 \times 10^{8}\ \text{m·s}^{-1})^{-2} = -7.896 \times 10^{-11}\ \text{J}$

$$E_{\text{bind}}/\text{nucleon} = \frac{-7.896 \times 10^{-11}\ \text{J}}{56\ \text{nucleons}} = -1.410 \times 10^{-12}\ \text{J·nucleon}^{-1}$$

(d) $^{232}_{90}\text{Th}$

$90\ {}^{1}\text{H} + 142\ \text{n} \rightarrow {}^{232}_{90}\text{Th}$

$\Delta m = 232.0382\ \text{u} - (90 \times 1.0078\ \text{u} + 142 \times 1.0087\ \text{u}) = -1.8992\ \text{u}$

$\Delta m = -1.8992\ \text{u} \times 1.6605 \times 10^{-27}\ \text{kg·u}^{-1} = -3.154 \times 10^{-27}\ \text{kg}$

$E_{\text{bind}} = -3.154 \times 10^{-27}\ \text{kg} \times (2.998 \times 10^{8}\ \text{m·s}^{-1})^{2} = -2.835 \times 10^{-10}\ \text{J}$

$$E_{\text{bind}}/\text{nucleon} = \frac{-2.835 \times 10^{-10}\ \text{J}}{232\ \text{nucleons}} = -1.222 \times 10^{-12}\ \text{J·nucleon}^{-1}$$

(e) Because ^{56}Fe has the largest binding energy per nucleon, it is the most stable.

17.64 (a) $^{7}_{3}\text{Li} + {}^{1}_{1}\text{H} \rightarrow {}^{7}_{4}\text{Be} + {}^{1}_{0}\text{n}$

$7.0160\ \text{u} + 1.0078\ \text{u} \rightarrow 7.0169\ \text{u} + 1.0087\ \text{u}$

$8.0238\ \text{u} \rightarrow 8.0256\ \text{u}$

$\Delta m = 0.0018\ \text{u}$

$$\Delta m = (0.0018\ \text{u})\left(\frac{1.661 \times 10^{-27}\ \text{kg}}{1\ \text{u}}\right) = 3.0 \times 10^{-30}\ \text{kg}$$

$\Delta E = \Delta mc^2 = (3.0 \times 10^{-30}\ \text{kg})(3.00 \times 10^{8}\ \text{m·s}^{-1})^{2} = 2.7 \times 10^{-13}\ \text{J}$

$$\left(\frac{2.7 \times 10^{-13}\ \text{J}}{8.0238\ \text{u}}\right)\left(\frac{1\ \text{u}}{1.661 \times 10^{-24}\ \text{g}}\right) = 2.0 \times 10^{10}\ \text{J·g}^{-1}$$

(b) $^{59}_{27}\text{Co} + {}^{2}_{1}\text{D} \rightarrow {}^{60}_{27}\text{Co} + {}^{1}_{1}\text{H}$

$58.9332\ \text{u} + 2.0141\ \text{u} \rightarrow 59.9529\ \text{u} + 1.0078\ \text{u}$

$60.9473 \text{ u} \rightarrow 60.9607 \text{ u}$

$\Delta m = 0.0134 \text{ u}$

$$\Delta m = (0.0134 \text{ u})\left(\frac{1.6605 \times 10^{-27} \text{ kg}}{1 \text{ u}}\right) = 2.23 \times 10^{-29} \text{ kg}$$

$$\Delta E = \Delta mc^2 = (2.23 \times 10^{-29} \text{ kg})(3.00 \times 10^8 \text{ m·s}^{-1})^2 = 2.01 \times 10^{-12} \text{ J}$$

$$\left(\frac{2.01 \times 10^{-12} \text{ J}}{60.9473 \text{ u}}\right)\left(\frac{1 \text{ u}}{1.661 \times 10^{-24} \text{ g}}\right) = 1.99 \times 10^{10} \text{ J·g}^{-1}$$

(c) ${}^{40}_{19}\text{K} + {}^{0}_{-1}\text{e} \rightarrow {}^{40}_{18}\text{Ar}$

$39.9640 \text{ u} + 0.0005 \text{ u} \rightarrow 39.9624 \text{ u}$

$39.9645 \text{ u} \rightarrow 39.9624 \text{ u}$

$\Delta m = -0.0021 \text{ u}$

$$\Delta m = (-0.0021 \text{ u})\left(\frac{1.661 \times 10^{-27} \text{ kg}}{1 \text{ u}}\right) = -3.5 \times 10^{-30} \text{ kg}$$

$$\Delta E = \Delta mc^2 = (-3.5 \times 10^{-30} \text{ kg})(3.00 \times 10^8 \text{ m·s}^{-1})^2 = -3.2 \times 10^{-13} \text{ J}$$

$$\left(\frac{-3.2 \times 10^{-13} \text{ J}}{39.9645 \text{ u}}\right)\left(\frac{1 \text{ u}}{1.661 \times 10^{-24} \text{ g}}\right) = -4.8 \times 10^{9} \text{ J·g}^{-1}$$

(d) ${}^{10}_{5}\text{B} + {}^{1}_{0}\text{n} \rightarrow {}^{4}_{2}\text{He} + {}^{7}_{3}\text{Li}$

$10.0129 \text{ u} + 1.0087 \text{ u} \rightarrow 4.0026 \text{ u} + 7.0160 \text{ u}$

$11.0216 \text{ u} \rightarrow 11.0186 \text{ u}$

$\Delta m = -0.0030 \text{ u}$

$$\Delta E = \Delta mc^2 = (-0.0030 \text{ u})\left(\frac{1.661 \times 10^{-27} \text{ kg}}{1 \text{ u}}\right) = -5.0 \times 10^{-30} \text{ J}$$

$$\left(\frac{-5.0 \times 10^{-30} \text{ J}}{11.0216 \text{ u}}\right)\left(\frac{1 \text{ u}}{1.661 \times 10^{-24} \text{ g}}\right) = -2.7 \times 10^{-7} \text{ J·g}^{-1}$$

17.66 (a) ${}^{234}_{94}\text{Pu} \rightarrow {}^{230}_{92}\text{U} + {}^{4}_{2}\alpha$

$\Delta m = 230.0339 \text{ u} + 4.0026 \text{ u} - 234.0433 \text{ u} = -0.0076 \text{ u}$

$$\Delta E = \Delta mc^2 = (-0.0076 \text{ u})\left(\frac{1.661\times10^{-27}\text{ kg}}{1\text{ u}}\right)(3.00\times10^8\text{ m·s}^{-1})^2$$

$= -1.1 \times 10^{-12}$ J

(b) $\dfrac{1.00\times10^{-6}\text{ g}}{234.0433\text{ g}\cdot\text{mol}^{-1}} = 4.27\times10^{-9}$ mol

$$k = \frac{0.693}{t_{1/2}} = \frac{0.693}{8.8\text{ h}} = 0.079\text{ h}^{-1}$$

$$\frac{N}{N_0} = e^{-(0.070\text{ h}^{-1})(24\text{ h})} = 0.15$$

If N is 0.15 N_0, then 85% of the sample decayed in the 24 h period.

4.27×10^{-9} mol $\times$ 0.85 $\times$ 6.02×10^{23} atoms·mol^{-1} = 2.2×10^{15} atoms

total energy released = 2.2×10^{15} atoms $\times$ 1.1×10^{-12} J·atom^{-1}

= 2.4×10^3 J, or 2.4 kJ

17.68 (a) ${}^{239}_{94}\text{Pu} + {}^{1}_{0}\text{n} \rightarrow {}^{98}_{42}\text{Mo} + {}^{138}_{52}\text{Te} + 4\,{}^{1}_{0}\text{n}$

(b) ${}^{239}_{94}\text{Pu} + {}^{1}_{0}\text{n} \rightarrow {}^{100}_{43}\text{Tc} + {}^{137}_{51}\text{Sb} + 4\,{}^{1}_{0}\text{n}$

(c) ${}^{239}_{94}\text{Pu} + {}^{1}_{0}\text{n} \rightarrow {}^{104}_{45}\text{Rh} + {}^{133}_{49}\text{In} + 3\,{}^{1}_{0}\text{n}$

17.70 (a) 99.0% removal corresponds to $N = 0.010\ N_0$

$$k = \frac{\ln 2}{t_{1/2}} = \frac{0.693}{3.82\text{ d}} = 0.181\text{ d}^{-1}$$

$\ln\left(\dfrac{N}{N_0}\right) = -kt$, or $\ln\left(\dfrac{N_0}{N}\right) = kt$, therefore

$$t = \frac{1}{k}\ln\left(\frac{N_0}{N}\right) = \frac{1}{0.181\text{ d}^{-1}}\ln\left(\frac{N_0}{0.010\,N_0}\right)$$

$t = 25$ d

(b) Because it takes only 25 days for 99.0% of the radon to be removed by disintegration, it does not seem likely that radon formed deep in the earth's crust could leak to the surface in such a relatively short time. So, most of the radon observed must have been formed near the Earth's surface.

(c) Radon enters homes from the soil into the basements. It is naturally given off by concrete, cinder block, and stone building materials. An absorbing material could be used to absorb the emitted radon gas before it entered the basement. Most of the radon would disintegrate before it freed itself from the absorbing material, because the half-life is short. One such absorbing material is polyester cloth.

17.72 First, determine the decay constant from the half-life of ${}^{3}_{1}T$. Then calculate the number of T nuclei in 1.0 mg of T. The activity is then given by

rate = $k \times N$

which is the number of disintegrations per second (Bq). Then convert to absorbed dose in rad and dose equivalent in rem, using the information in Table 17.4

$$k = \frac{0.693}{t_{1/2}} = \frac{0.693}{12.3\ \text{y} \times 3.17 \times 10^{7}\ \text{s}\cdot\text{y}^{-1}} = 1.78 \times 10^{-9}\ \text{s}^{-1}$$

$$N = 1.0 \times 10^{-3}\ \text{g} \times \frac{1\ \text{mol}}{3.0\ \text{g}} \times 6.022 \times 10^{23}\ \text{mol}^{-1} = 2.0 \times 10^{20}\ \text{nuclei}$$

activity = $1.78 \times 10^{-9}\ \text{s}^{-1} \times 2.0 \times 10^{20} = 3.6 \times 10^{11}$ dps = 3.6×10^{11} Bq

If 100% of the energy produced by these disintegrations were absorbed by the 1.0 g of tissue, then the energy absorbed per second would be

$E = 0.0186\ \text{MeV} \times 1.602 \times 10^{-13}\ \text{J}\cdot\text{MeV}^{-1} \times 3.6 \times 10^{11}\ \text{s}^{-1}$

$= 1.1 \times 10^{-3}\ \text{J}\cdot\text{s}^{-1}$

$$\text{absorbed dose} = 1.1 \times 10^{-3}\ \text{J}\cdot\text{s}^{-1}\cdot\text{g}^{-1} \times 10^{3}\ \text{g}\cdot\text{kg}^{-1} \times \left(\frac{1\ \text{rad}}{1 \times 10^{-2}\ \text{J}\cdot\text{kg}^{-1}}\right)$$

$= 1.1 \times 10^{2}\ \text{rad}\cdot\text{s}^{-1}$

For 10% of the energy absorbed:

dose equivalent = $Q \times$ absorbed dose

$= 1\ \text{rem/rad} \times 1.1 \times 10^{2}\ \text{rad}\cdot\text{s}^{-1} \times 0.10$

$= 11\ \text{rem}\cdot\text{s}^{-1}$

17.74 The decay process can generate much heat, which would speed up the corrosion rate. The decay process can also result in the production of new, possibly corrosive chemicals as a result of nuclear fission and nuclear transmutation. Chemical breakdown can occur as a result of nuclear bombardment, resulting in new, highly corrosive gases and other substances.

17.76 $t_{1/2} = 12.3$ y; $k = \dfrac{0.693}{12.3 \text{ y}} = 0.0563 \text{ y}^{-1}$; activity $\propto$ N

$$t\,(= \text{age}) = -\frac{1}{k}\ln\left(\frac{N}{N_0}\right)$$

$$N = 0.091\ N_0;\ \frac{N}{N_0} = \frac{0.091\ N_0}{N_0} = 0.091$$

$$t\,(= \text{age}) = -\frac{1}{0.0563 \text{ y}^{-1}}\ln(0.091) = 42.6 \text{ y}$$

17.78 If $m = m_0 \times 2^{-\frac{t}{t_{1/2}}}$ then $m_0 = m \times 2^{\frac{t}{t_{1/2}}}$.

$$m_{\text{P}}^0 = 0.0254\text{g} \times 2^{\frac{90}{14.28}} = 2.00\text{g}$$

$$m_{\text{S}}^0 = 1.466\text{g} \times 2^{\frac{90}{87.2}} = 3.00\text{g}$$

$$m_{\text{Fe}}^0 = 0.744\text{g} \times 2^{\frac{90}{44.6}} = 3.00\text{g}$$

Therefore

$$\text{P\%} = \frac{2.00}{8.00} \times 100\% = 25.0\%$$

$$\text{S\%} = \text{Fe\%} = \frac{3.00}{8.00} \times 100\% = 37.5\%$$

17.80 There are several important properties for any effective ligand. First, it must bond very strongly to the metal ion or else the metal ion will be liberated from the ligand and find its way into other body tissues.

Secondly, the ligand itself should not be toxic. Additionally, the ligand must have a very high specific binding attraction to the tissue in question.

17.82 (a) Pu^{3+}, Pu^{4+}; (b) Pu^{3+}, 108 pm; Pu^{4+}, 93 pm; Fe^{2+}, 82 pm; Fe^{3+}, 67 pm; (c) While the radii of the Pu ions are larger than those of iron, the charge to radius ratios are similar: Pu^{3+}, 0.028; Pu^{4+}, 0.043; Fe^{2+}, 0.024; Fe^{3+}, 0.045. Because of this, the Pu^{3+} binds similarly to Fe^{2+} and Pu^{4+} binds similarly to Fe^{3+}. The redox potentials for the reduction of the higher oxidation state species to the next lower oxidation number are also similar ($E°_{\text{red}}Pu^{4+} = 0.98$ V versus $E°_{\text{red}}Fe^{3+} = 0.77$ V) so that the plutonium also mimics the iron reasonably well in its redox chemistry.

17.84 $^{1}_{0}n + {}^{0}_{1}e \rightarrow {}^{1}_{1}p + 2\gamma$

$$\Delta E = \Delta m \cdot c^2 = 2m_e c^2 = 2 \times 9.10\ 939 \times 10^{-31}\,\text{kg} \times \left(2.99\ 792 \times 10^{8}\,\text{m} \cdot \text{s}^{-1}\right)^2$$

$$\Delta E = (1.63\ 742 \times 10^{-13}\,\text{J})\left(\frac{1\ \text{MeV}}{1.60\ 218 \times 10^{-13}\,\text{J}}\right) = 1.02\ 199\ \text{MeV}$$

17.86 (a) The boiling temperature can be calculated as:

$$T = \frac{\Delta H}{\Delta S}$$

Since the amount of energy required to vaporize each isotopomer varies with the zero-point energy of the two hydrogen bonds for each water molecule, ΔH can be represented as

$$\Delta H = \Delta H_0 - 2E_0$$

The zero point energy should be calculated as

$$E_0 = \frac{1}{2}h\nu, \quad \nu = \frac{1}{2\pi}\sqrt{\frac{k}{m}}$$

where m is supposed to be a reduced mass. As hydrogen is much lighter than the surrounding O network, the reduced mass is approximately just the mass of H (or D, or T). Combining these relationships, and assuming all quantities to be independent of temperature, we get following model for the normal boiling point:

$$T = a - \frac{b}{\sqrt{m}}$$

The unknown parameters a and b are

$$a = \frac{\Delta H_0}{\Delta S},\ b = \frac{\hbar\sqrt{k}}{\Delta S}$$

These unknown parameters can be determined from the normal boiling points of H_2O (100.00 °C) and D_2O (101.42 °C):

$$373.15\ \text{K} = a - \frac{b}{\sqrt{1.008\ \text{u}}} = a - (0.99602\ \text{u}^{-\frac{1}{2}})b$$

$$374.57\ \text{K} = a - \frac{b}{\sqrt{2.0140\ \text{u}}} = a - (0.70464\ \text{u}^{-\frac{1}{2}})b$$

Subtracting the first expression from the second gives

$$1.42\ \text{K} = (0.29138\ \text{u}^{-\frac{1}{2}})b$$

$$b = 4.873\ \text{K}\cdot\text{u}^{\frac{1}{2}}$$

(Note that the subtraction yields 1.42 K on the left-hand side and limits b to three significant figures. We will keep one extra digit for use in subsequent calculations.)

Substituting b back into the first expression gives

$$373.15\ \text{K} = a - \frac{4.873\ \text{K}\cdot\text{u}^{\frac{1}{2}}}{\sqrt{1.00797\ \text{u}}}$$

$$a = 378.00\ \text{K}$$

So our model for the boiling point of water, including different zero-point energies for different masses of hydrogen isotopes, is

$$T = 378.0 - \frac{4.873}{\sqrt{m}}$$

(b) The normal boiling point of T_2O can be predicted from our model.

$$T = 378.00\ \text{K} - \frac{4.873\ \text{K}\cdot\text{u}^{\frac{1}{2}}}{\sqrt{3.01605\ \text{u}}} = 375.19\ \text{K}$$

$$= 102.04^{\circ}\text{C}$$

Given the crude nature of our model, this prediction is in good agreement with the experimental value of 101.51 °C. (See W.M. Jones, *J. Chem. Phys.* **48**, 207 (1968).)

17.88 Nuclei that are positron emitters are proton-rich and lie below the band of stability.

(a) O stable at 16, so ^{18}O is neutron rich, will emit an electron rather than a positron; not good for PET. $^{18}_{8}O \rightarrow ^{18}_{9}F + ^{0}_{-1}e$

(b) N stable at 14, so ^{13}N is proton-rich, but it is likely to emit a proton to reach stable $^{12}_{6}C$; not good for PET. $^{13}_{7}N \rightarrow ^{12}_{6}C + ^{1}_{1}p$

(c) C stable at 12, so ^{11}C is proton rich, will emit a positron; good for PET. $^{11}_{6}C \rightarrow ^{11}_{5}B + ^{0}_{1}e$

(d) F stable at 19, so ^{20}F is neutron rich, will emit an electron rather than a positron; not good for PET. $^{20}_{9}F \rightarrow ^{20}_{10}Ne + ^{0}_{-1}e$

(e) O stable at 16, so ^{15}O is proton-rich, but it will emit a proton to reach stable $^{14}_{7}N$; not good for PET. $^{15}_{8}O \rightarrow ^{14}_{7}N + ^{1}_{1}p$

CHAPTER 18

ORGANIC CHEMISTRY I: THE HYDROCARBONS

18.2 (a) C_6H_{12}, alkane; (b) C_5H_8, alkene; (d) C_3H_4, alkene; (d) C_8H_{18}, alkane

18.4 (a) $CH_3CH_2CH(CH_3)_2$ or C_5H_{12}, alkane; (b) $C_{10}H_8$, aromatic hydrocarbon (c) C_8H_{18}, alkene; (d) C_4H_4, alkene and alkyne

18.6 (a) C_9H_{16}, alkene; (b) C_6H_{10}, alkyne; (c) C_9H_{16}, alkane; (d) $C_{10}H_8$, aromatic hydrocarbon

18.8 (a) nonane; (b) pentane; (c) hexane; (d) octane

18.10 (a) decyl; (b) ethyl; (c) butyl; (d) octyl

18.12 (a) hexane; (b) 2, 2, 3-trimethylpetane; (c) 2, 2, 4-trimethylpentane; (d) 2, 2-dimethylbutane

18.14 (a) 3-phenyl-1-octene; (b) 3-chloro-4, 5-dimethyl-1-hexyne

18.16 (a) $CH_3CH(CH_3)CH_2CH(CH_2CH_3)CH_2CH_3$

or $(CH_3)_2CHCH_2CH(CH_2CH_3)_2$

(b) CH_2=$CHC(CH_3)_2CH_2CH_3$

(c) *cis*-CH_3CH=$CHCH(CH_3)CH_2CH_3$

(d) *trans*-$CH_3CH{=}CHCH_3$

18.18 (a)

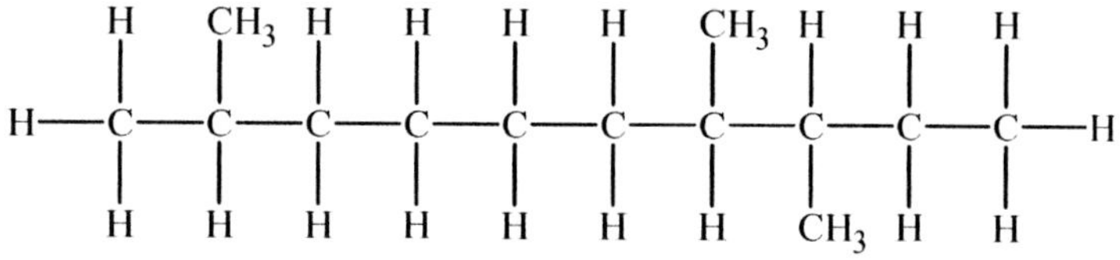

(b)

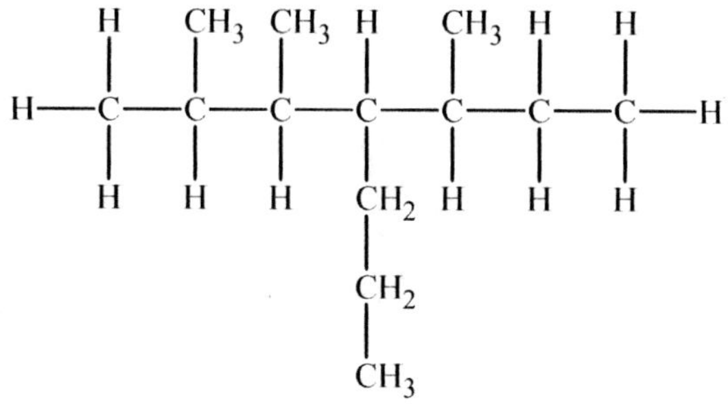

(c)

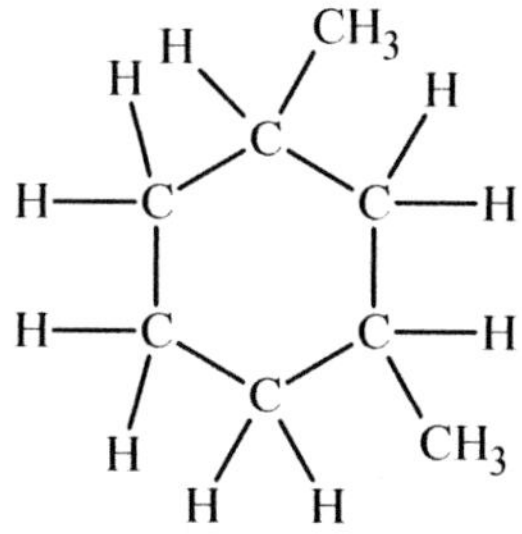

(d)

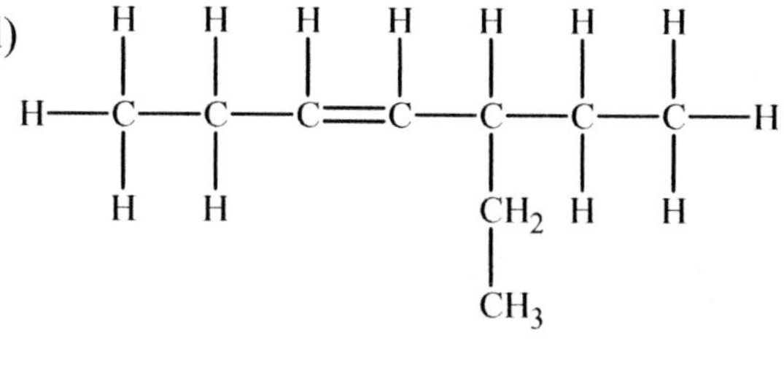

18.20 (a)

(b)

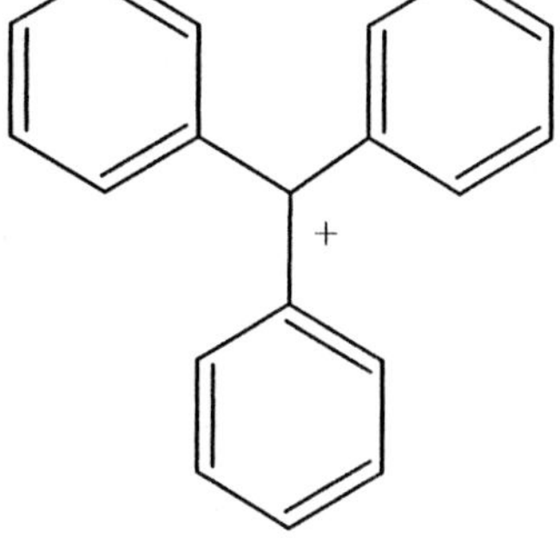

(c)

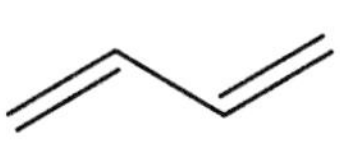

18.22 (a) C_4H_8

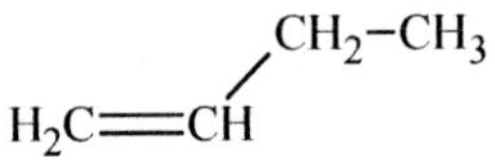

1-Butene

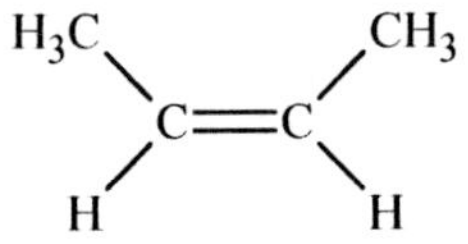

cis-2-Butene

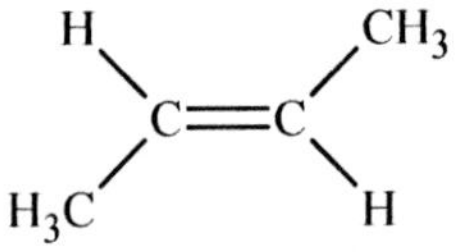

trans-2-Butene

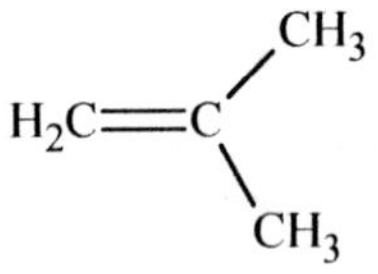

2-Methyl-1-propene

(b) C_5H_{10}

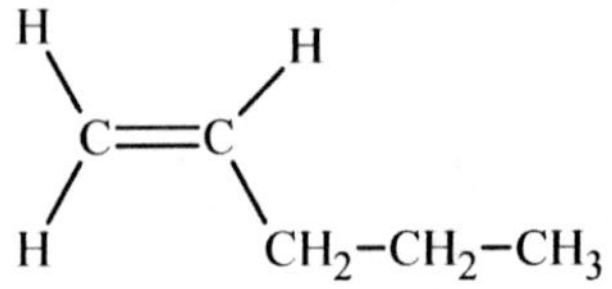

1-Pentene

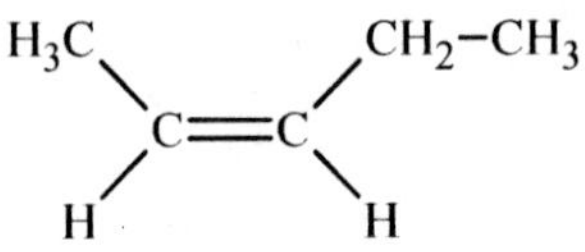

cis-2-Pentene

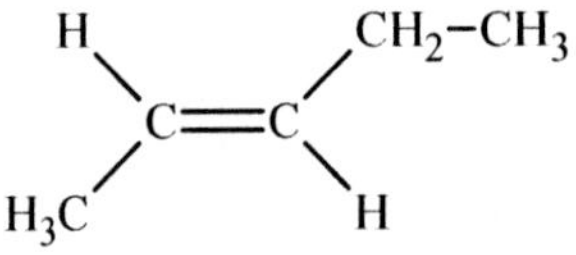

trans-2-Pentene

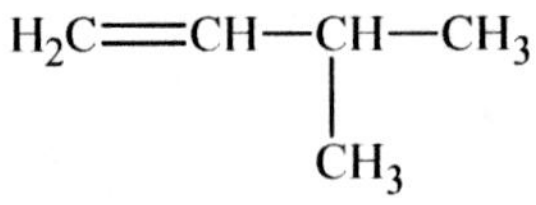

3-Methyl-1-butene

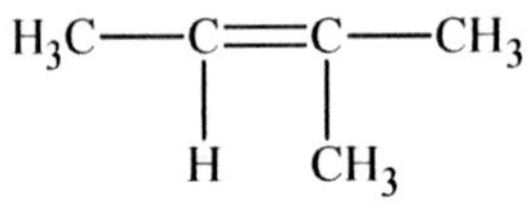

2-Methyl-2-butene

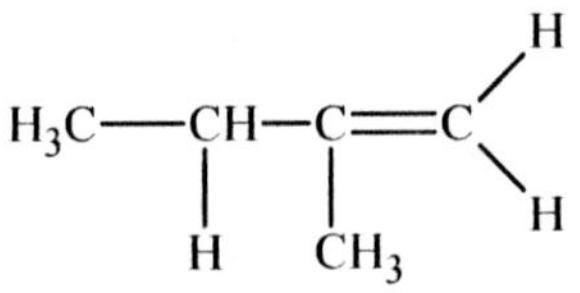

2-Methyl-1-butene

18.24 (a)

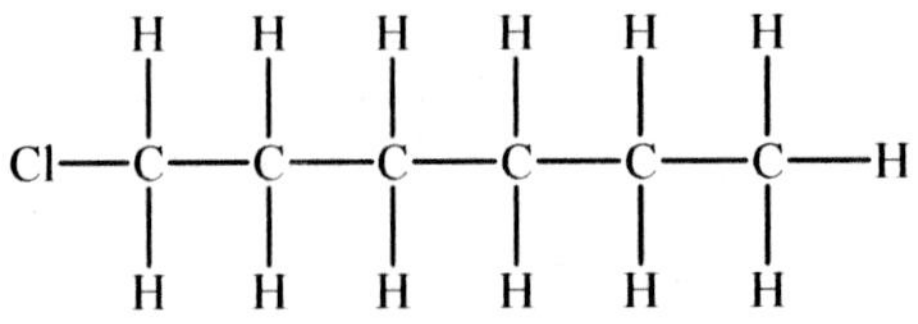

1-Chlorohexane

$C_6H_{13}Cl$

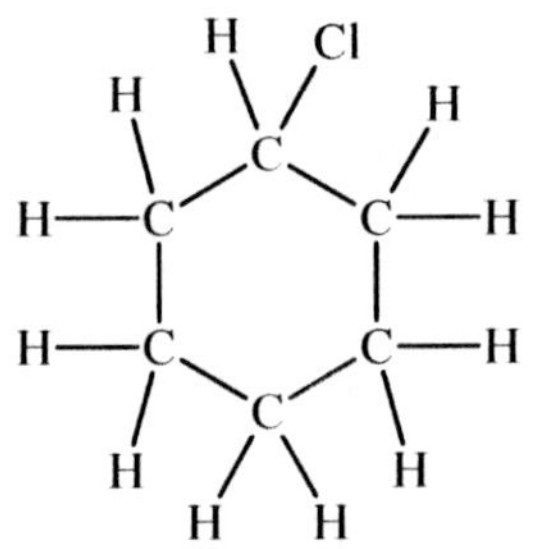

Chlorocyclohexane

$C_6H_{11}Cl$

Different formula, not isomers.

(b) Same formula, C_4H_8, but different bonding arrangements: structural isomers.

(c) Same formula, $C_2H_2Cl_2$, same bonding arrangement, but different geometry: geometrical isomers.

18.26 (a)

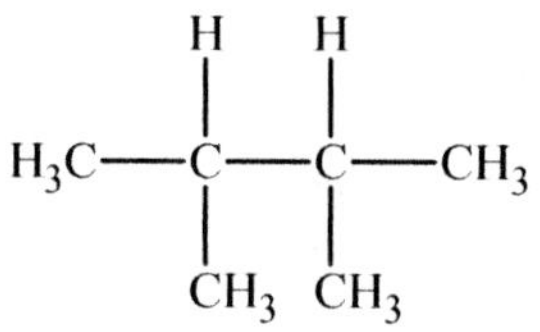

(b)

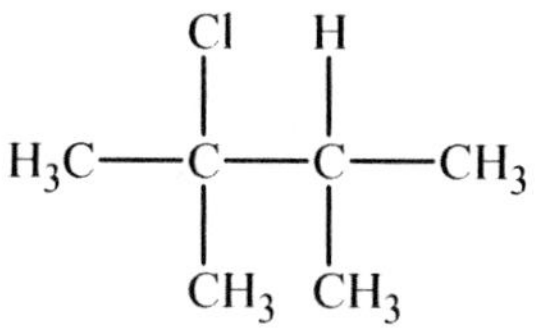

and

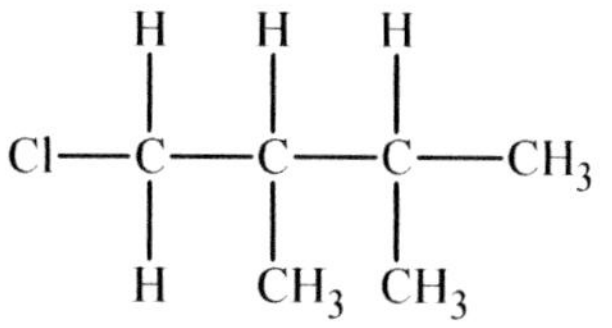

Note: Both H's and all four methyl groups are equivalent in the hydrocarbon.

18.28 (a), (b), and (d) display optical activity.

(a)

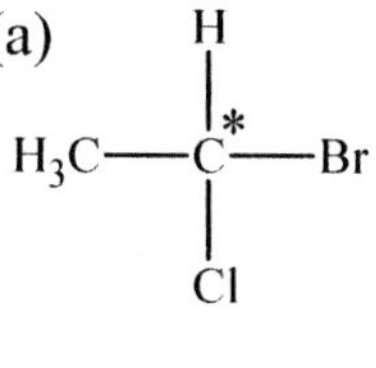

(b)

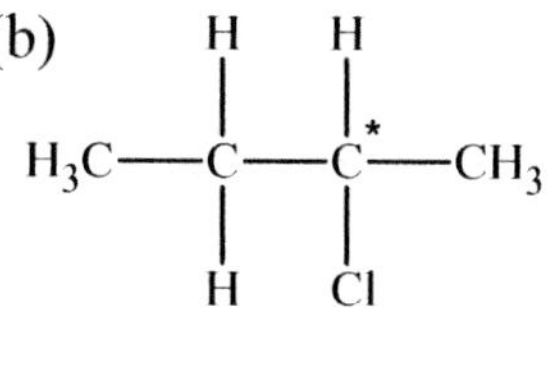

(d)

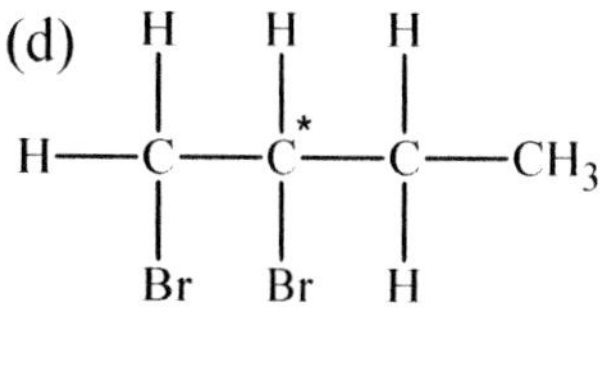

18.30 $C(CH_3)_4$ (9.5°C) < $CH_3CH_2CH(CH_3)_2$ (28°C)

< $CH_3CH_2CH_2CH_2CH_3$ (36°C)

Because the molecular formulas are all identical, the compounds will have the same molar mass. The difference in boiling points can then be attributed to the degree of branching; the more highly branched compounds have the lower boiling points.

* Indicates the chiral carbon atoms. (c) is optically inactive.

18.32 $CH_4(g) + 4\,F_2(g) \longrightarrow CF_4(g) + 4\,HF(g)$

break 4 C–H bonds	$4(+412\ kJ \cdot mol^{-1})$
break 4 F–F bonds	$4(+158\ kJ \cdot mol^{-1})$
form 4 C–F bonds	$4(-484\ kJ \cdot mol^{-1})$
form 4 H–F bonds	$4(-565\ kJ \cdot mol^{-1})$
Total	$-1916\ kJ \cdot mol^{-1}$

The corresponding reaction with chlorine is much less exothermic for two reasons. First, it requires much less energy to break the F-F bond, and second, the bonds formed between F and C or H are also stronger than the corresponding bonds to Cl. For this reason, direct fluorinations of hydrocarbons are extremely hazardous and may lead to fires and explosions. Chlorinations are also hazardous but are more easily done than fluorinations.

18.34 One monochlorocylopropane:

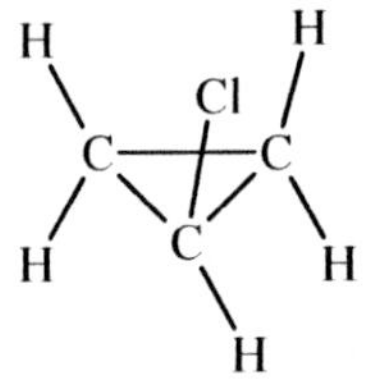

Three dichlorocyclopropanes:

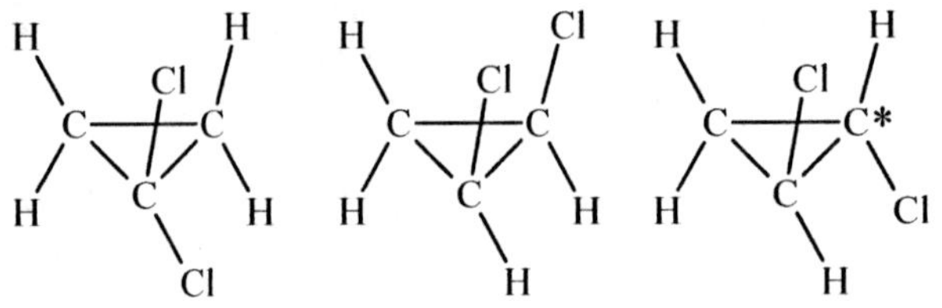

Three trichlorocyclopropanes:

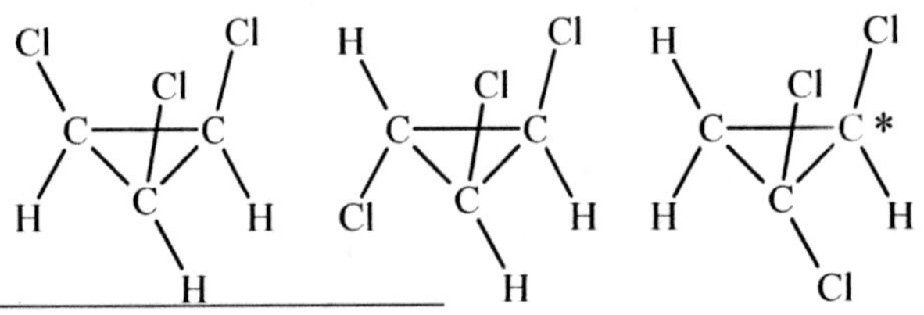

*Indicates a structure with a nonsuperimposable mirror image.

Three tetrachlorocyclopropanes:

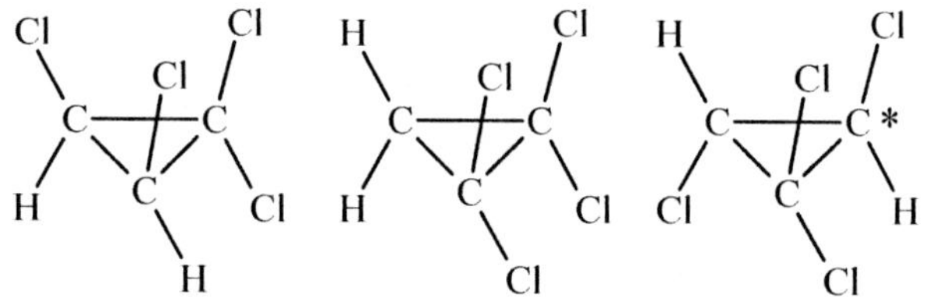

One pentachlorocyclopropane:

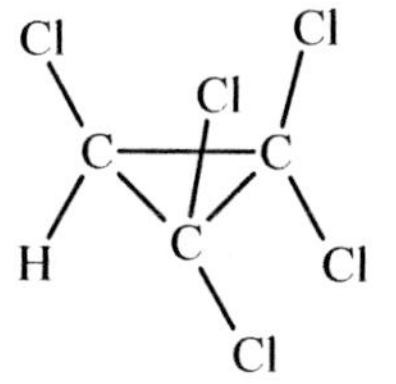

One hexachlorocyclopropane:

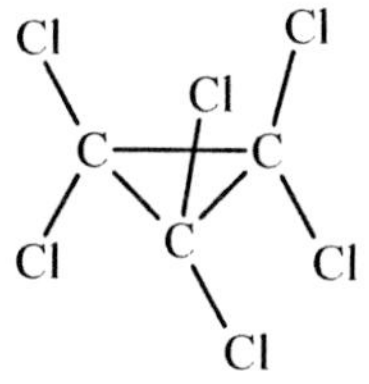

18.36

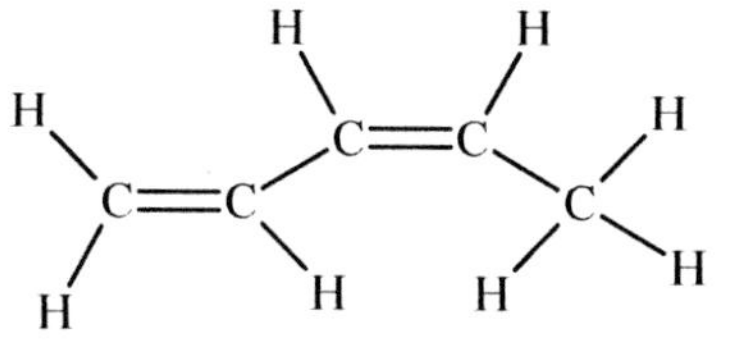

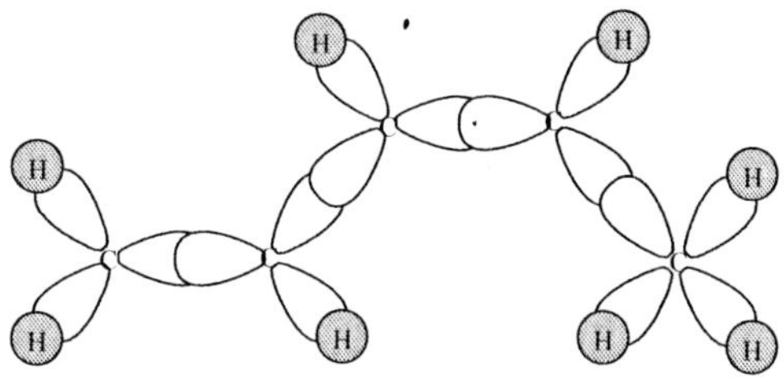

All carbon atoms are sp^2 hybridized except for the carbon on the far right side of the molecule, which is bonded to three hydrogen atoms and is sp^3 hybridized.

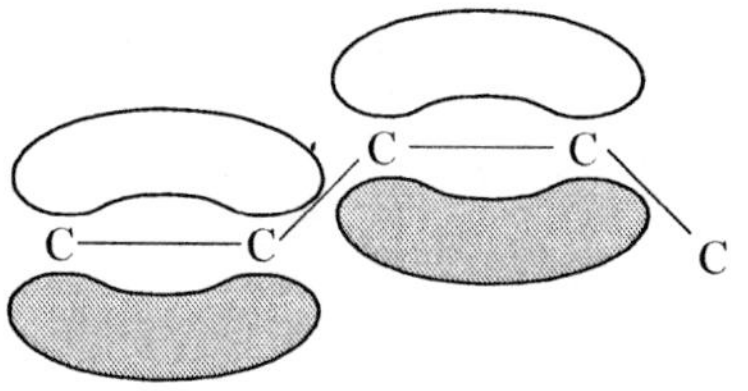

Pi bonds form the second bond in the C=C double bonds.

18.38 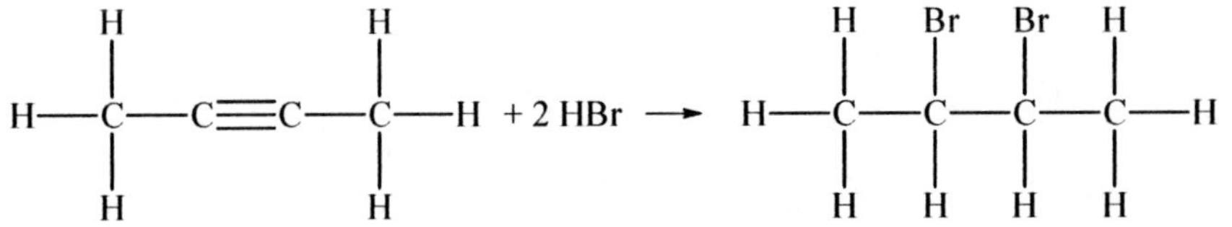

Addition reaction

18.40 (a) 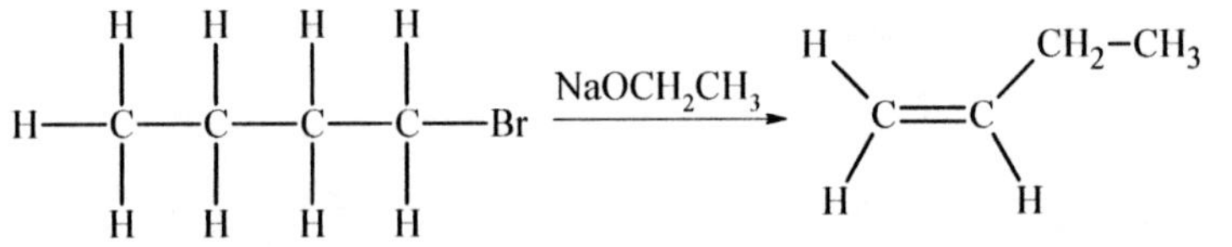

(b) 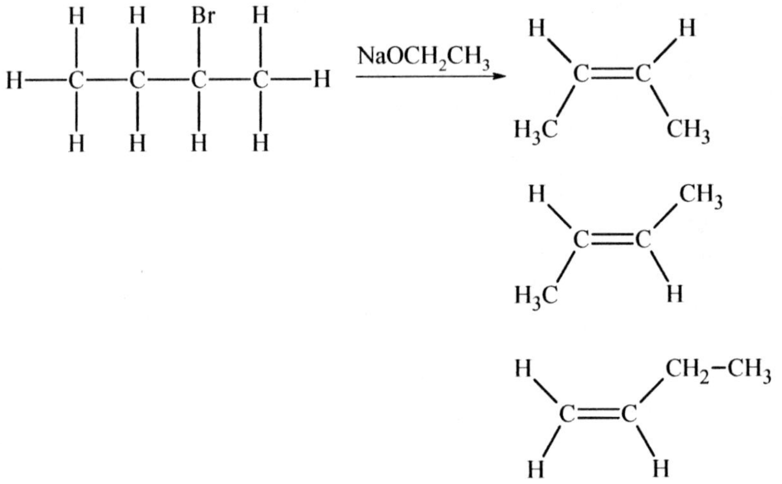

18.42 $C_2H_4 + HX \longrightarrow C_2H_5X$

We will break one H–X bond and form one C–H bond and one C–X bond. Using bond enthalpies:

Halogen	Cl	Br	I
H—X bond breakage $(kJ \cdot mol^{-1})$	$+431$	$+366$	$+299$
C—H bond formation $(kJ \cdot mol^{-1})$	-412	-412	-412
C—X bond formation $(kJ \cdot mol^{-1})$	-338	-276	-238
Total $(kJ \cdot mol^{-1})$	-319	-322	-351

A general trend in the exothermicity of these reactions is not as obvious as in Exercise 18.41. There are two opposing factors involved. Although it requires less energy to break an H—X bond as one descends the periodic table, there is also less energy released from the formation of a C—X bond.

18.44 (a) 2-ethly-4-methyl-1-propylbenzene

(b) 1-ethyl-4-propylbenzene

18.46 (a) 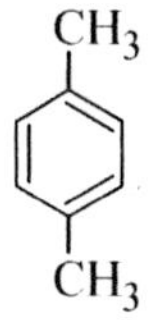 (b) 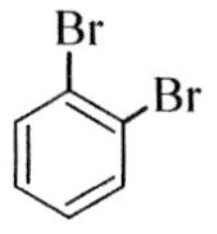 (c) 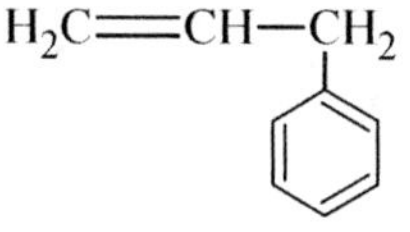 (d) 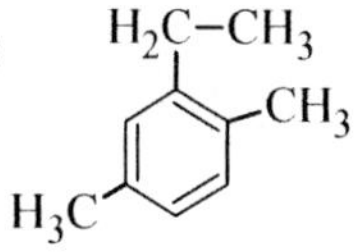

18.48 (a), (b)

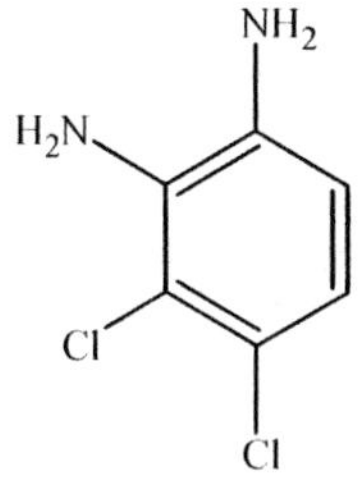

1,2-Diamino-3,4-dichlorobenzene

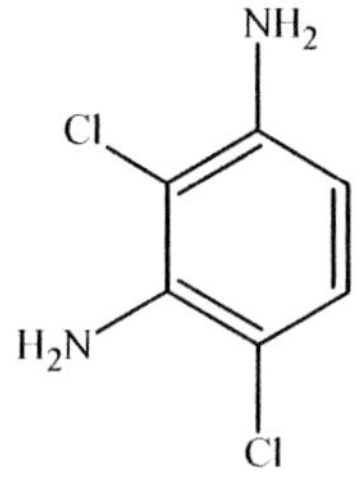

1,3-Diamino-2,4-dichlorobenzene

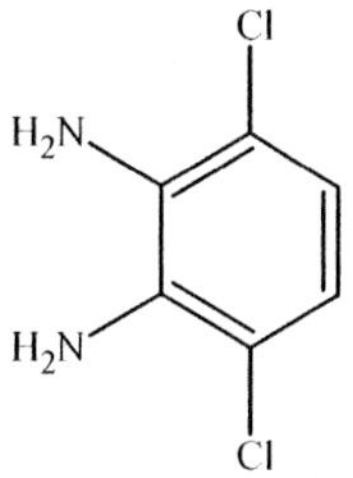

2,3-Diamino-1,4-dichlorobenzene

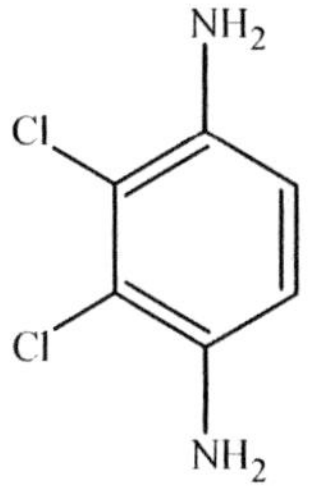

1,4-Diamino-2,3-dichlorobenzene

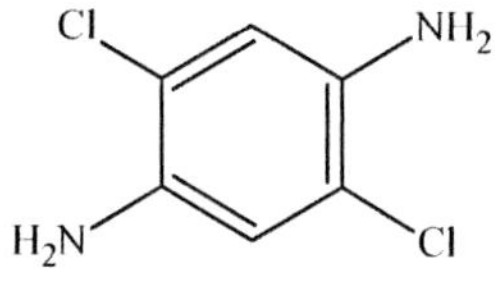

1,4-Diamino-2,5-dichlorobenzene

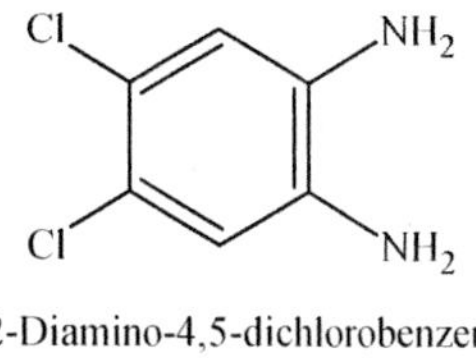

1,2-Diamino-4,5-dichlorobenzene

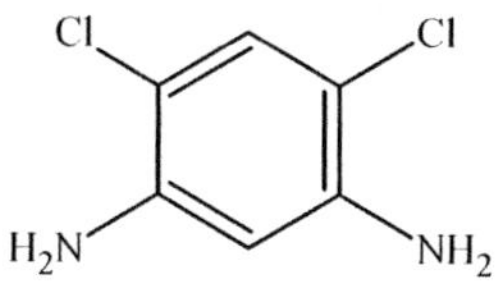

1,5-Diamino-2,4-dichlorobenzene

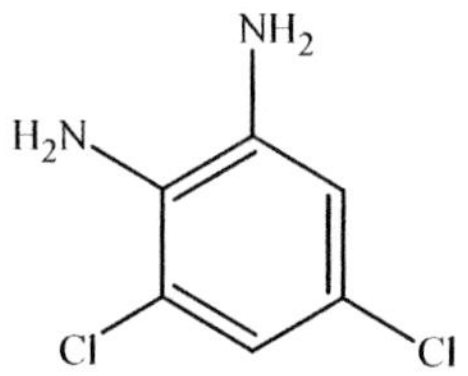

1,2-Diamino-3,5-dichlorobenzene

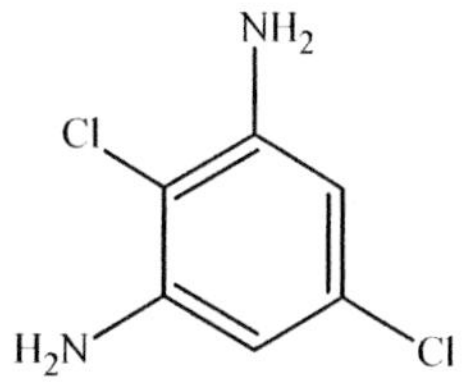

1,3-Diamino-2,5-dichlorobenzene

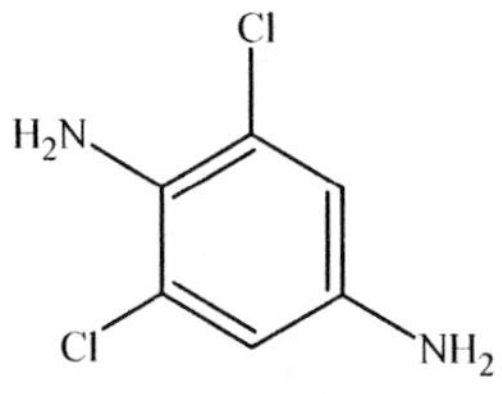

2,5-Diamino-1,3-dichlorobenzene

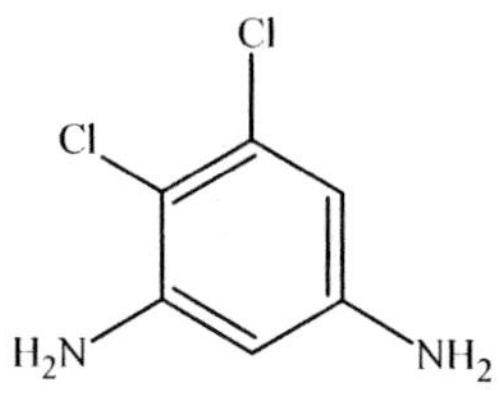

1,5-Diamino-2,3-dichlorobenzene

All these molecules will be at least slightly polar except for the 1,4-diamino-2,4-dichlorbenzene, which will be nonpolar.

18.50 $C_6H_5NO_2 + Br_2 \longrightarrow ?$

$C_6H_5Br + NO_2^+ \longrightarrow ?$

The two reactions will give different product distributions because the directing influence of the groups attached to the ring are different. A nitro group is a metadirecting substituent; a bromo group would be ortho- and para-directing. Thus, the dominant products in the nitration reaction would be a mixture of 1-bromo-2-nitrobenzene and 1-bromo-4-nitrobenzene, whereas for the bromination reaction, the product would be primarily 1-bromo-3-nitrobenzene.

18.52 Like the nitro group, the aldehyde group is electron-withdrawing and forms resonance structures that place positive charge on the ortho and para positions. This group is, therefore, a meta-directing substituent.

18.54 Coal is not a pure substance and, as a result, does not burn cleanly. Some types of coal produce considerable amounts of sulfur and nitrogen oxides, which contribute to air pollution. The burning of high-sulfur coal contributed very much to the environmental damage in many of the eastern European nations, such as the former East Germany. This damage persists to this day. Coal is also not as easy to transport as gasoline because it is a solid rather than a liquid or gas. Liquids or gases can be placed in fuel tanks and pipelines.

18.56 Ethanol is a renewable fuel source, which may be produced from biomass (fermentation reactions). Currently, ethanol is more expensive to produce than normal components of gasoline, but a strong push for its use comes from the agricultural industry, which, of course, would benefit by the fermentation of biomass to produce automotive fuel. The requirement to use ethanol in gasoline mixtures is a controversial issue in the United States, where there are strong lobbies both for and against its use.

18.58 (a) four tetrahedral sp^3-hybrid orbitals with σ-bonds

(b) three trigonal planar sp^2-hybrid orbitals with σ-bonds and one unhybridized p-orbital with a π-bond

(c) two linear sp-hybrid orbitals with σ-bonds and two unhybridized p-orbitals with two π-bonds

18.60 (a) $CH_3CH{=}CHCH_2CH_3 + H_2 \xrightarrow{\text{Ni}} CH_3CH_2CH_2CH_2CH_3$, addition reaction

(b) $CH_2{=}CHCH_3 + HCl \longrightarrow CH_3CHClCH_3$ and an isomer, $CH_2ClCH_2CH_3$, addition reaction

18.62 (a) The reaction is: $C_3H_6(g)$ (propene) + $H_2(g)$ → $C_3H_8(g)$ (propane). To calculate the equilibrium constant, first calculate the free energy of reaction: $\Delta G_r = (-23.49\ \text{kJ}\cdot\text{mol}^{-1}) - (-32.82\ \text{kJ}\cdot\text{mol}^{-1}) = 9.33\ \text{kJ}\cdot\text{mol}^{-1}$.

The equilibrium constant is then:

$$K = e^{\frac{-\Delta G_r}{RT}} = e^{\frac{-9330\ \text{J}\cdot\text{mol}^{-1}}{(8.3145\ \text{J}\cdot\text{K}^{-1}\cdot\text{mol}^{-1})(298\ \text{K})}} = 0.0232$$

(b) To determine how increasing the temperature will affect K, we must determine if the reaction is endothermic or exothermic:

$\Delta H_r = (-103.85\ \text{kJ}\cdot\text{mol}^{-1}) - (-84.68\ \text{kJ}\cdot\text{mol}^{-1}) = -19.17\ \text{kJ}\cdot\text{mol}^{-1}$.

The negative enthalpy of reaction indicates an exothermic reaction. Applying Le Chatelier's principle, we would expect that increasing the reaction temperature would, therefore, favor reactants and decrease K.

18.64 (a) Because the double bone in *trans*-3-hexene should be planar, the torsion angle should be 0°. (b) The measured values from the Web site files are *trans*-3-hexene, 0°; cyclohexene, 8.8°; cyclohexane, 59.8°. The presence of the ring structure prevents the double bond in cyclohexene from being perfectly planar. There is a distortion of about 8.8° from planarity as a result of the geometrical constraints on the molecule.

(c) The cyclohexane ring allows the molecule to adopt normal bond angles (109.5°) for all the sp^3-hybridized carbon atoms. While the sp^2-hybridized carbon atoms in the double bond of cyclohexene show a slight distortion from the ideal bond angles (about 121–122°), so do the sp^3-hybridized carbon atoms immediately adjacent to the double bond (angles are about 112°).

18.66 (a) 1-pentene; (b) *cis*-3-methyl-2-pentene, *trans*-3-methyl-2-pentene; (c) 2,4-dimethyl-2-pentene; (d) 5-methyl-2-heptyne; (e) 5-methyl-3-heptyne

18.68 (a) 2-methypropane; (2) 2-methylbutane; (c) *iso*- is added as a prefix to denote the isomer of a straight chain hydrocarbon in which the end CH_3 group has been moved from the end of the chain to a position one carbon unit inward. The systematic name for isohexane will thus be 2-methylpentane.

18.70 (a) 1 mol of hexene will react with 1 mol of H_2 to form one mole of hexane:

$$C_6H_{10} + H_2 \rightarrow C_6H_{12}$$

The volume of one mole of H_2 at 298 K and 1.00 atm is:

$$V = \frac{n \cdot R \cdot T}{P} = \frac{(1\text{ mol})(0.082057\text{ L}\cdot\text{atm}\cdot\text{K}^{-1}\cdot\text{mol}^{-1})(298\text{ K})}{1\text{ atm}} = 22.5\text{ L}$$

Three moles of H_2 are required to completely react with one mole of benzene:

$$C_6H_6 + 3H_2 \rightarrow C_6H_{12}$$

Therefore, the volume of H_2 needed is three times that needed to react with one mole of hexene or 67.5 L

(b) For the reaction of hexene with H_2:

Break 1 C=C bond:	$+612\ kJ\cdot mol^{-1}$
Break 1 H—H bond:	$+436\ kJ\cdot mol^{-1}$
Form 1 C—C bond:	$-348\ kJ\cdot mol^{-1}$
Form 2 C—H bonds:	$2(-412\ kJ\cdot mol^{-1})$
Total:	$-124\ kJ\cdot mol^{-1}$

For the reaction of benzene with H_2:

Break 6 C⋯C bonds:	$6(+518\ kJ\cdot mol^{-1})$
Break 3 H—H bond:	$3(+436\ kJ\cdot mol^{-1})$
Form 6 C—C bonds:	$6(-348\ kJ\cdot mol^{-1})$
Form 6 C—H bonds:	$6(-412\ kJ\cdot mol^{-1})$
Total:	$-144\ kJ\cdot mol^{-1}$

Hydrogenation of benzene is not three times more exothermic than the hydrogenation of hexene due to resonance stabilization of the C-C bonds in benzene. The resonance stabilization of benzene makes the carbon-carbon bonds more stable, reducing the overall amount of energy released during the reaction.

18.72 amount (moles) of $H = 6.75\ g\ H_2O \times \frac{1\ mol\ H_2O}{18.02\ g\ H_2O} \times \frac{2\ mol\ H}{1\ mol\ H_2O}$

$= 0.749\ mol\ H$

amount (moles) of

$$C = 17.55\ g\ CO_2 \times \frac{1\ mol\ CO_2}{44.01\ g\ CO_2} \times \frac{1\ mol\ C}{1\ mol\ CO_2} = 0.399\ mol\ C$$

$$\frac{0.749\ mol\ H}{0.399\ mol\ C} = 1.88 \frac{mol\ H}{mol\ C} \approx \frac{15\ mol\ H}{8\ mol\ C}$$

The empirical formula is C_8H_{15}. The molecular formula could be $C_{16}H_{30}$, which is of the type C_nH_{2n-2}, corresponding to an alkyne or a dialkene.

18.74 (a) Ethane can form no chiral products. (b) Propane can form no chiral products. (c) 2-Bromobutane is chiral. (d) 2-Bromopentane is chiral.

18.76 Assume a 100 g sample.

$$\text{amount (moles) of C} = 90\ \text{g C} \times \frac{1\ \text{mol C}}{12.01\ \text{g C}} = 7.5\ \text{mol C}$$

$$\text{amount (moles) of H} = 10\ \text{g H} \times \frac{1\ \text{mol H}}{1.01\ \text{g H}} = 9.9\ \text{mol H}$$

$$\frac{9.9\ \text{mol H}}{7.5\ \text{mol C}} = \frac{1.3\ \text{mol H}}{1\ \text{mol C}} \approx \frac{4\ \text{mol H}}{3\ \text{mol C}}$$

The empirical formula is C_3H_4, which has a molar mass of $40\ \text{g} \cdot \text{mol}^{-1}$. Because $40.06\ \text{g} \cdot \text{mol}^{-1}$ is the molar mass of the compound, the molecular formula is C_3H_4 as well. The compound could be a cyclic alkene, dialkene, or alkyne. See the structures of the possible isomers below.

$$\text{amount (moles) of } C_3H_4 = \frac{1.46\ \text{g}}{40.06\ \text{g.mol}^{-1}} = 3.6 \times 10^{-2}\ \text{mol}$$

$$\text{amount (moles) of } H_2 = n_{H_2} = \frac{PV}{RT} = \frac{1.0\ \text{atm} \times 1.600\ \text{L}}{0.082\ 06\ \text{L} \cdot \text{atm} \cdot \text{K}^{-1} \cdot \text{mol}^{-1} \times 273\ \text{K}}$$

$$= 7.2 \times 10^{-2}\ \text{mol}$$

$$\text{Thus } \frac{n_{H_2}}{n_{C_3H_4}} = \frac{7.2 \times 10^{-2}\ \text{mol}}{3.6 \times 10^{-2}\ \text{mol}} = 2$$

There are two π-bonds across which H_2 can add. This is consistent with the compound's being an alkyne or a dialkene. No definite identification of the hydrocarbon can be made on the basis of the data provided. The three possible isomers consistent with the formula C_3H_4 are

(1) HC≡C—CH_3

(2) $H_2C{=}C{=}CH_2$

(3) 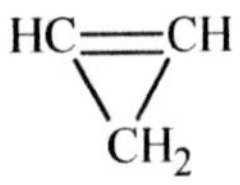

Structure (3) is inconsistent with the hydrogenation data.

18.78 The question here is twofold: (1) Which ring will be nitrated? and (2) Which position of that ring will be the site of electrophilic substitution?

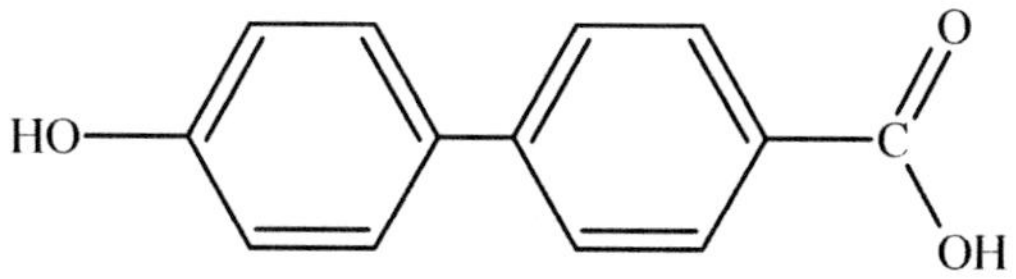

The OH group is an activating function because it donates electrons to the aromatic ring, whereas the COOH group is electron-withdrawing and, therefore, deactivating. We would thus expect the ring with the OH group attached to be the one that is substituted. Because OH is an ortho-, para-directing group, we would expect the substitution to occur at one of those sites. Because the para position is already substituted, this leaves only the position ortho to the OH group for the substitution. The expected product is, therefore,

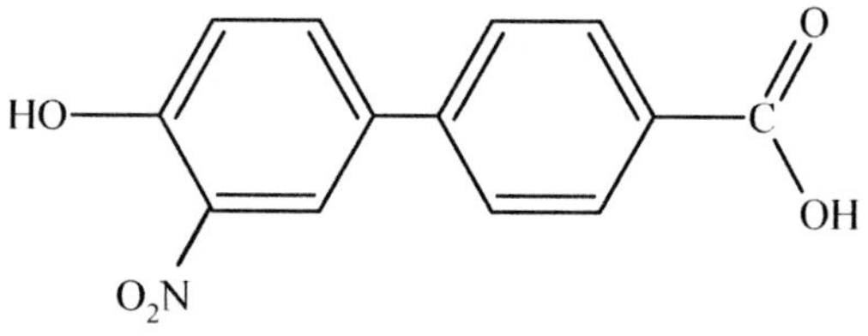

18.80 (a) There are four carbon atoms in the molecule that each have four substituents attached. Each of these carbon atoms is labeled with an asterisk. Each of these carbons has two possible configurations.

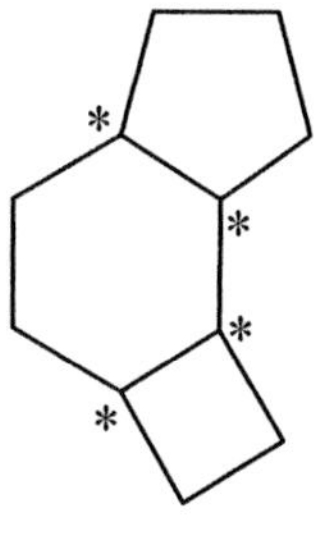

(b) Because each chiral carbon atom has two possible configurations, there will be $2 \times 2 \times 2 \times 2 = 16$ different configurations. These are shown below. This can be seen readily if we label the configuration at each carbon atom. In organic chemistry, conventions have been developed that use the symbols R and S to designate the two different configurations possible at a particular chiral carbon atom. It is not important here to understand the conventions associated with each configuration; they are simply provided as labels to show what the possible combinations are. Thus, if we label each chiral carbon atom as R or S, we have the following possible configurations:

RRRR	RRRS	RRSR	RSRR
RRSS	RSSR	RSRS	
RSSS			
SRRR	SRRS	SRSR	SSRR
SSRS	SSSR	SRSS	
SSSS			

(c) Yes. In fact, the 16 isomers can be divided into eight pairs of enantiomers. If we use the R,S designations, we find that isomers that are completely opposite in configuration are mirror images of each other. Thus, RRRR is the mirror image of SSSS, RSRS is the mirror image of SRSR, etc.

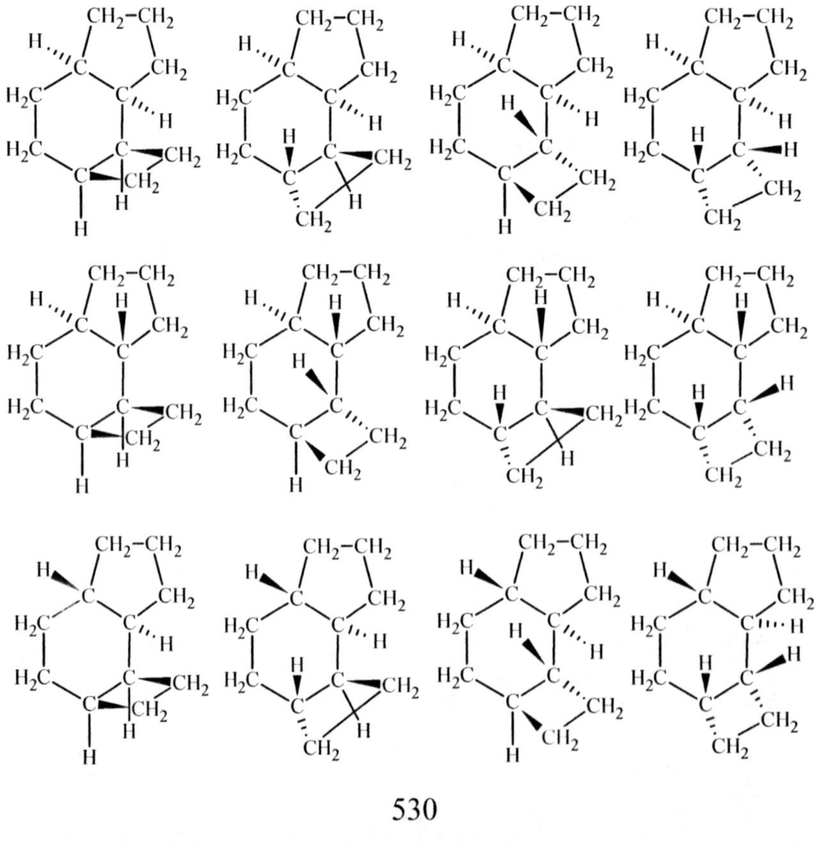

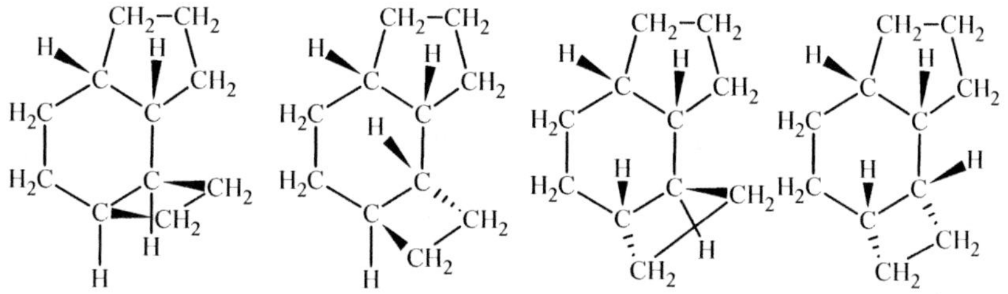

18.82 (a) 2,4,6-trinitrotoluene. (b) The methyl group on the toluene directs electrophilic aromatic substitution toward the meta and para positions (see section 18.9) because the methyl group is slightly electron-donating. A nitro group, however, is electron-withdrawing and will tend to cause electrophilic substitution meta to its position. The result is that, once a nitro group has substituted onto the ring, it reinforces the directing influence of the methyl group. For example, in 2-nitrotoluene (meta-nitrotoluene) the methyl group will cause preferential substitution at the 4 position. The position is also the meta position relative to the nitro group, so that group enhances substitution at this location as well.

18.84 Because D has twice the mass of H, the parent ion for D_2O will occur at $(2 \times 2\ \text{u}) + 16\ \text{u} = 20\ \text{u}$, rather than at 18 u. Although H_2O shows a peak at this mass, that peak is very small due to the natural probability of finding two D atoms or one ^{18}O atom in the water molecule. For D_2O, this peak will be the largest peak. A small peak at 22 u will be observed for $D_2{}^{18}O$, as will peaks at 18 u for $D^{16}O$ and at 16 u for ^{16}O. A peak for $D^{18}O$ would overlap with a peak for $D_2{}^{16}O$, but they will not be exactly the same; a high resolution mass spectrometer may be able to resolve them. Because D_2O easily absorbs water, one might also find signals associated with normal isotopic abundance H_2O.

18.86 This reaction is a classic electrophilic substitution reaction, so we would expect the product to be one of the possible chloromethylbenezenes. If we calculate the parent ion mass expected for $C_6H_4CH_3Cl^+$, we find that Cl has two isotopes in a roughly 3:1 ratio (^{35}Cl, ^{37}Cl, respectively). The parent ion will thus show two peaks in roughly a 3:1 ratio. The masses of these peaks will be $(7 \times 12 \text{ u}) + (7 \times 1 \text{ u}) + 35 \text{ u} = 126 \text{ u}$ and $(7 \times 12 \text{ u}) + (7 \times 1 \text{ u}) + 37 \text{ u} = 128 \text{ u}$, the latter peak being about 1/3 the size of the former. The peaks at 113 u and 111 u represent the parent ion minus the methyl group (15 u). The formula of this fragment will be $C_6H_4Cl^+$. As with the parent ion, these two peaks will be roughly in a 3:1 ratio. The peak at 91 u represents the loss of the Cl atom from the parent ion with a formula of $C_6H_4CH_3^+$. Note that because this ion does not contain Cl, it does not have a companion peak with which to form a 3:1 intensity ratio.

CHAPTER 19

ORGANIC CHEMISTRY II: FUNCTIONAL GROUPS

19.2 (a) R-O-R; (b) R-CO-R; (c) R-CO-OR; (d) R-CHO

19.4 (a) aldehyde; (b) carboxylic acid; (c) amide; (d) alcohol

19.6 (a) 3-bromo-1-propene;. (b) 1,1-difluoro-1,3-butadiene; (c) 1-bromo-2-pentyne; (d) 1,1,1-trichloro-2-butanone

19.8 (a) $CH_3CH(CH_3)CH_2OH$, primary alcohol; (b) $CH_3CH(OH)CH_3$, secondary alcohol;

(c) 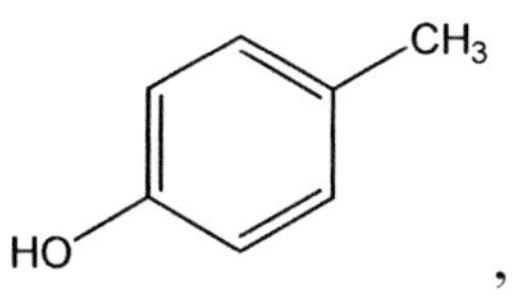

, phenol (d) 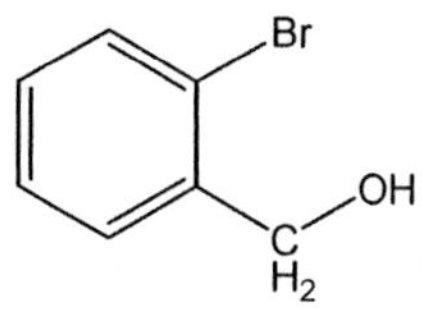

, primary alcohol

19.10 (a) $CH_3CH_2OCH_2CH_3$; (b) $CH_3CH_2CH_2CH_2OCH_2CH_2CH_2CH_3$; (c) $CH_3OCH_2CH_2CH_3$

19.12 (a) ethyl propyl ether; (b) butyl methyl ether; (c) dipropyl ether

19.14 (a) aldehyde, propanal; (b) aldehyde, 4-methylbenzaldehyde; (c) ketone, 4-ethyl-2-hexanone

19.16

(a) 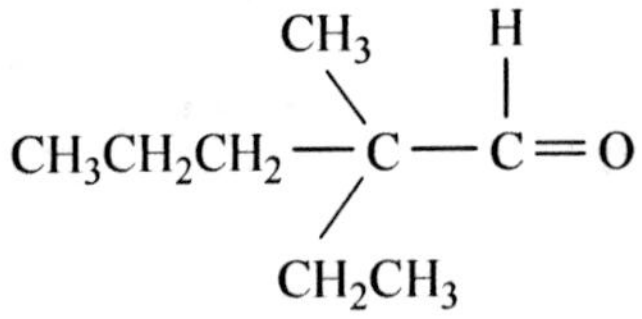

(b) 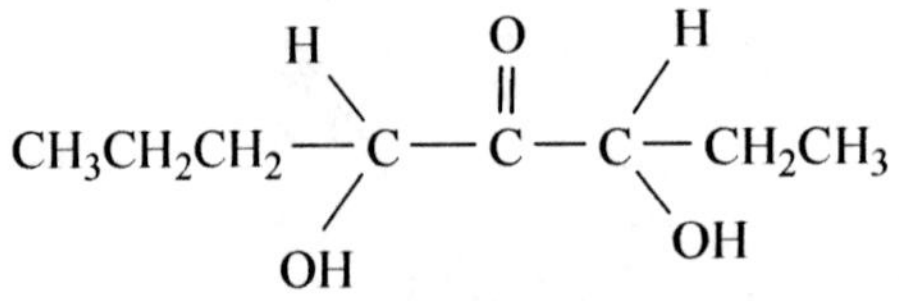

(c) 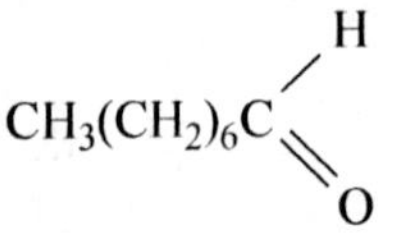

19.18 (a) propanoic acid; (b) 2, 2- dichloroethanoic acid; (c) nonanoic acid

19.20

(a) 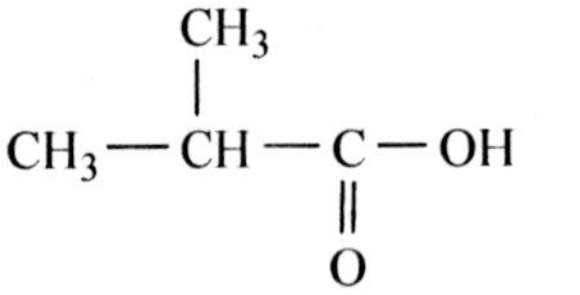

(b) 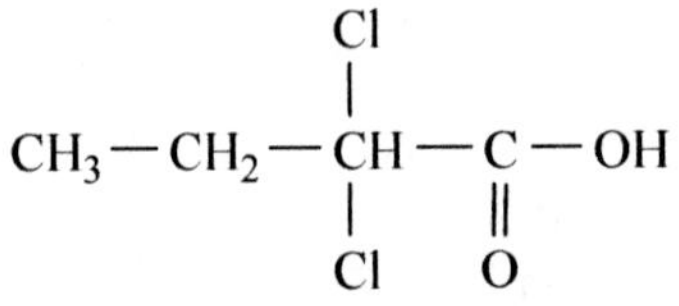

(c) 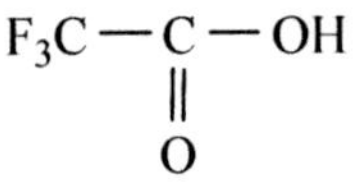

(d) 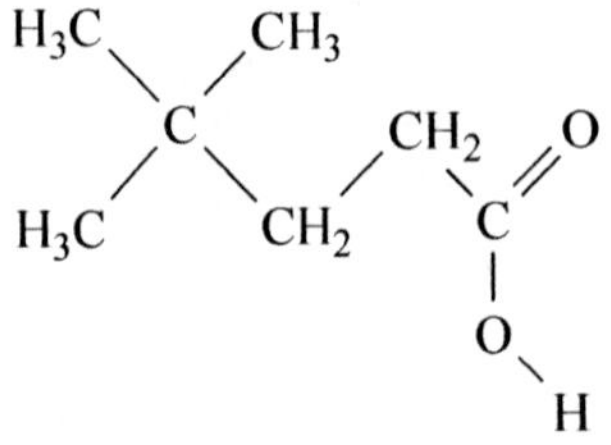

19.22 (a) propylamine; (b) tetraethylammonium ion; (c) *p*-chloroaniline or 4-chloroaniline

19.24

(a)

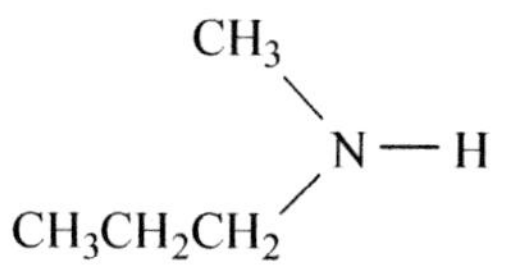

(b)

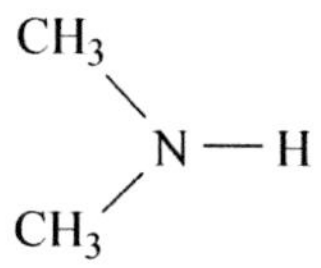

(c)

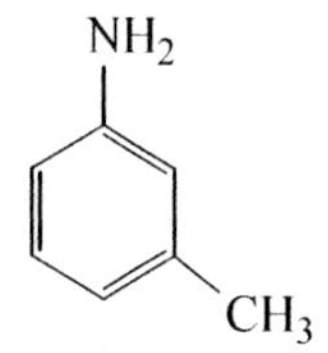

19.26 (a), (b) and (d) may function as nucleophiles. All these molecules have lone pairs of electrons that will be attracted to a positively charged carbon center. NH_4^+ does not have lone pairs of electrons and is positively charged, making it less likely to function as a nucleophile. If it is deprotonated first to give NH_3, however, it could function as a nucleophile.

19.28 (a) CH_3OH

(b) $CH_3CH(OH)CH_3$

(c) $CH_3(CH_2)_3CH(CH_3)CH(OH)(CH_2)_3CH_3$

These reactions are accomplished with an oxidizing agent, such as $Na_2Cr_2O_7$, in an acidified solution.

19.30

(a)

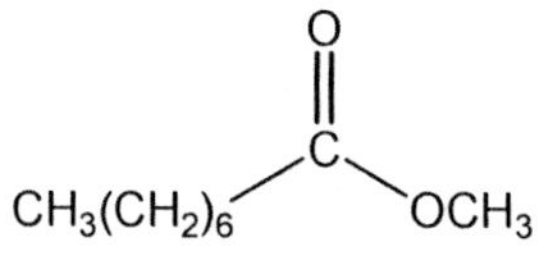

(b)

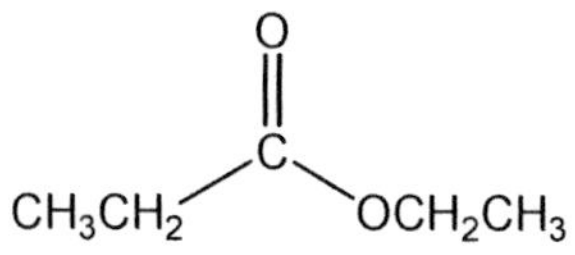

(c)

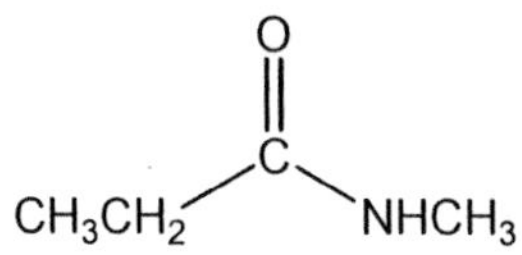

(d)

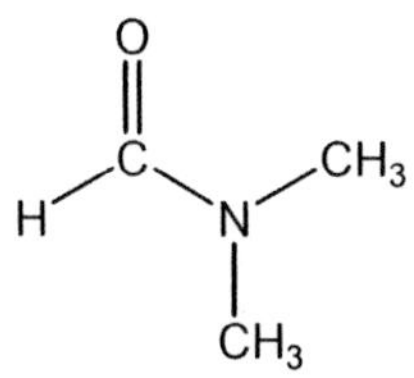

19.32 (1) $CH_3CH_2CH_2OH \xrightarrow{\text{base (aq)}}$ (dissolves)

(2) $CH_3(CH_2)_3CH_3 \xrightarrow{\text{base (aq)}}$ (does not dissolve)

(3) $CH_3COOH \xrightarrow{\text{base (aq)}}$ (dissolves)

(4) $CH_3CH_2CH_2OH \xrightarrow{\text{indicator}}$ no color change

(5) $CH_3COOH \xrightarrow{\text{indicator}}$ color change to acidic

(1), (2), and (3) distinguish pentane from the others. (4) and (5) distinguish propanol from ethanoic acid.

19.34 The weakest base is methylamine, followed by dimethylamine. The strongest base is diethylamine. The basicity of amines increases with increased substitution on the nitrogen atom. In acting as a base the nitrogen atom develops a positive charge, which is stabilized by the electron-donating alkyl groups. Hence, the primary amine, methylamine, is a weaker base than the other two secondary amines. Diethylamine is a stronger base than dimethylamine because the larger ethyl group is a better electron donor than the smaller methyl group.

19.36

(a) tetrafluoroethene: $CF_2{=}CF_2$

$$—CF_2—CF_2—CF_2—CF_2—CF_2—CF_2—$$

(b) phenylethene (styrene):

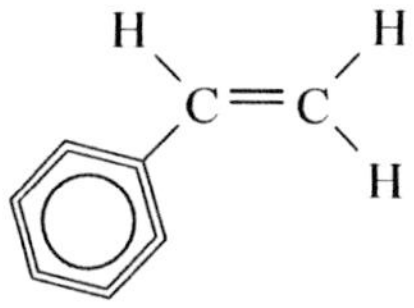

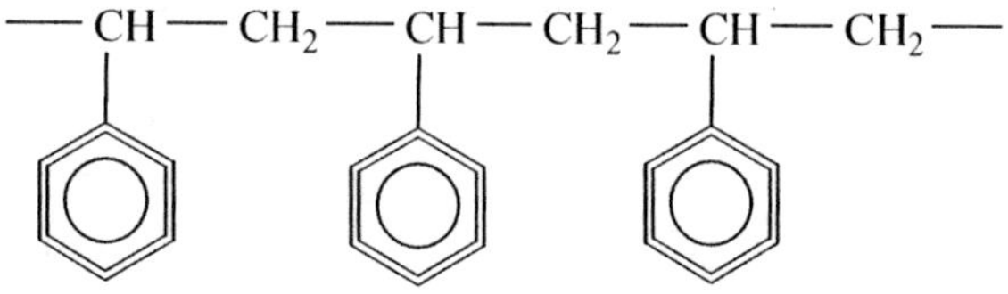

(c) $CH_3CH{=}CHCH_3$

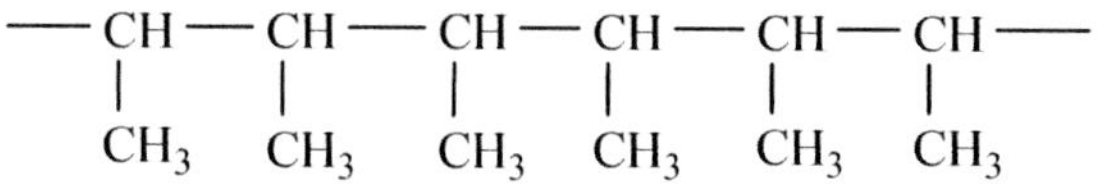

19.38

(a) $HO—C(CH_3)_2—COOH$ (b)

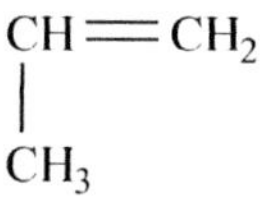

(c) NH_2CH_2COOH

19.40

(a)

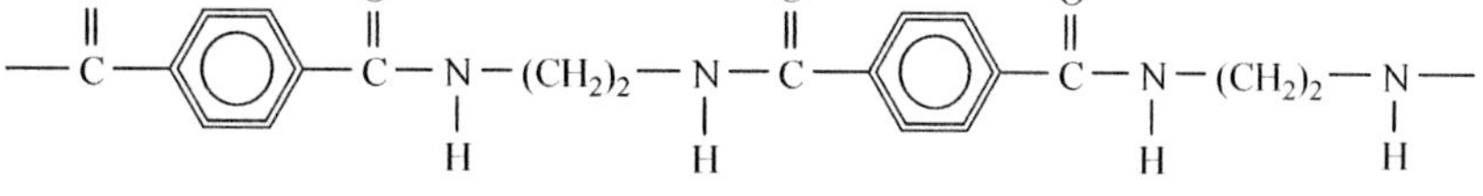

(b)

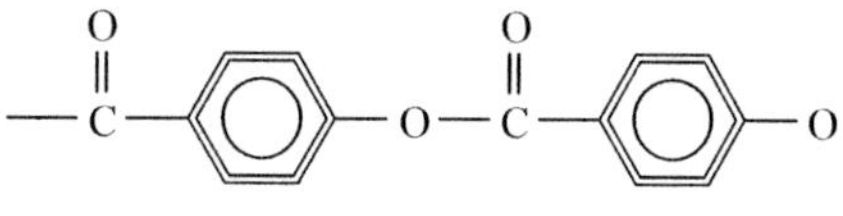

19.42 The regular alternation of the placement of the substituents in a syndiotactic polymer (as opposed to the random attachments of substituents in an atactic polymer) permits the molecules to fit together snugly, forming a highly crystalline and dense material.

19.44 Random copolymer.

19.46 The more polar functional groups present, the greater the intermolecular forces. Greater intermolecular forces result in (a) higher softening point, (b) greater viscosity, and (c) greater mechanical strength.

19.48 When polymer chains are cross-linked, the polymer does not soften as much as it otherwise would as the temperature is raised. It is also more resistant to deformation on stretching, because the cross-links pull it back. However, extensive cross linking can produce a rigid network that resists stretching.

19.50 (a)

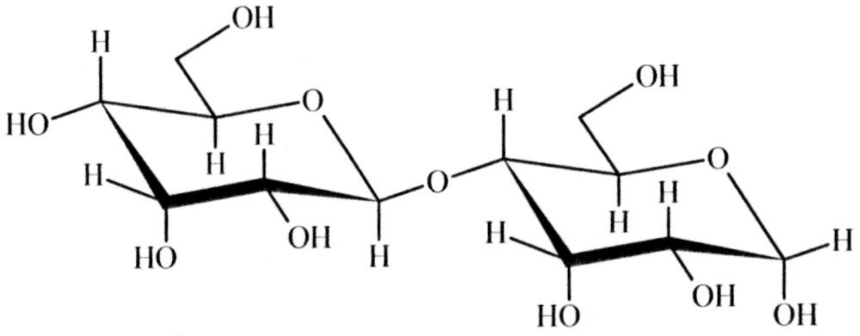

(b) The functional group formed acetal linkage; (c) the polymerization is a condensation reaction.

19.52 Glycine, alanine, phenylalanine, valine, leucine, isoleucine, tryptophan, and methionine contan nonpolar side groups

19.54

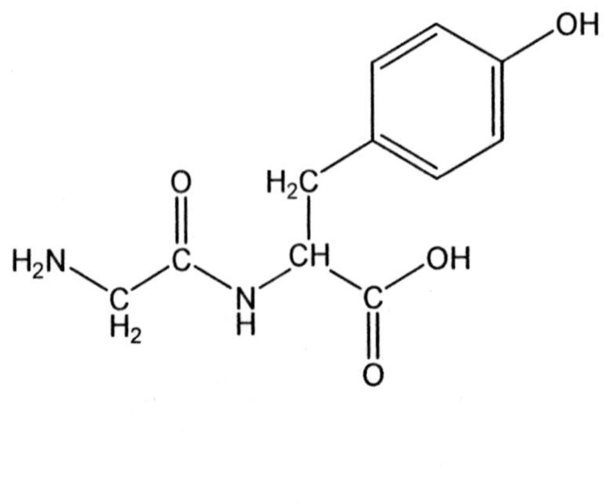

19.56 (a) The functional groups include primary, secondary, and tertiary amines, as well as a carboxylic acid function. Histidine is an essential amino acid. There is one chiral carbon atom. It is labeled by an asterisk (*) in the following structure.

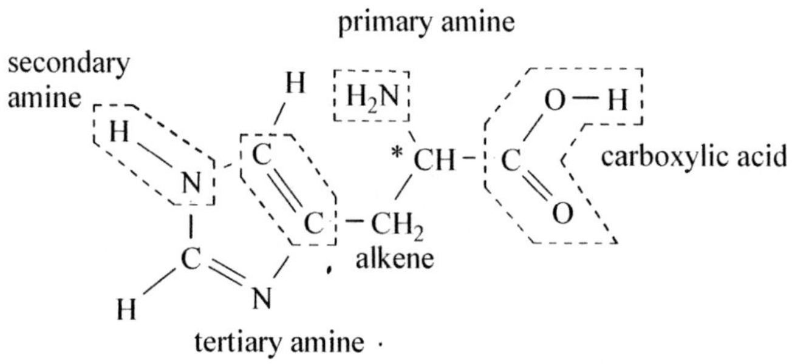

19.58

(a)

G	G	A	T	C	T	C	A	G
|	|	|	|	|	|	|	|	|
C	C	T	A	G	A	G	T	C

(b)

G	T	A	G	C	C	T	G	T
|	|	|	|	|	|	|	|	|
C	A	T	C	G	G	A	C	A

19.60 (a) $C_3H_7NO_2S$ (b) $C_5H_5N_5$ (c) $C_9H_{10}O_2$

---

* Chiral atoms are marked with an asterisk.

19.62 All the functional groups are circled in the figures.

(a)

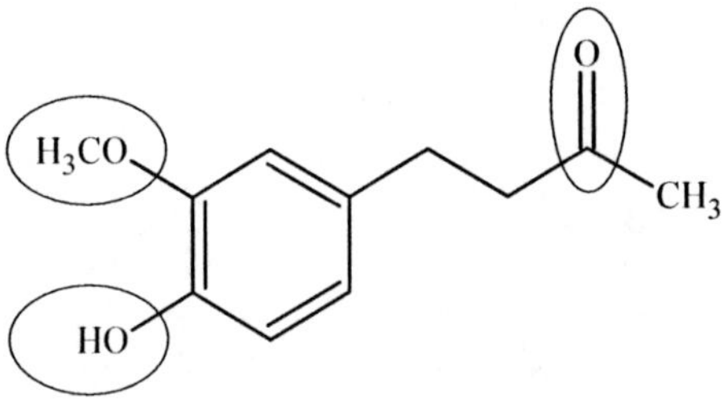

ether ($-OCH_3$), alcohol ($-OH$), ketone 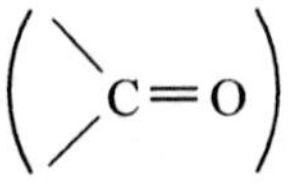 aromatic ring 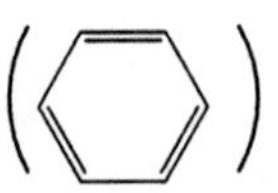

(b)

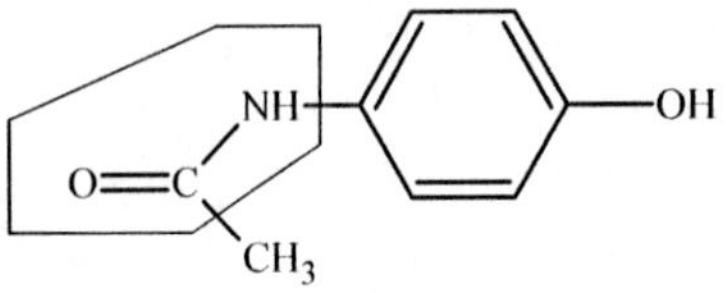

amide $\left(R-\overset{\overset{\displaystyle O}{\|}}{C}-\underset{\underset{\displaystyle H}{|}}{N}-R'\right)$, phenol ($-OH$), aromatic ring 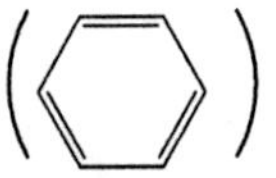

(c)

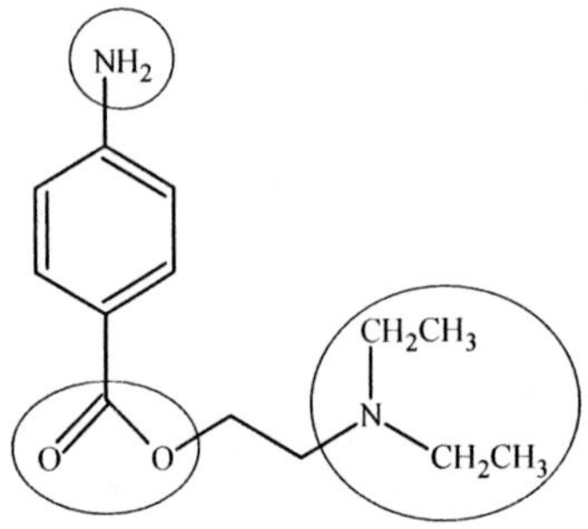

amine ($-NH_2$), ester 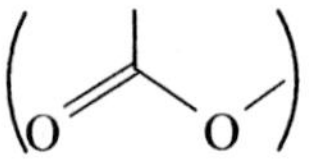, tertiary amine $\left(-H_2C-N\begin{matrix} CH_2CH_3 \\ CH_2CH_3 \end{matrix}\right)$

19.64

(a) alkene and alcohol

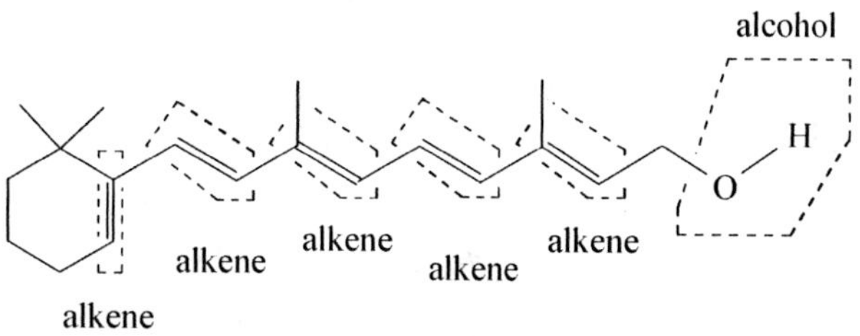

(b) ether, alcohol

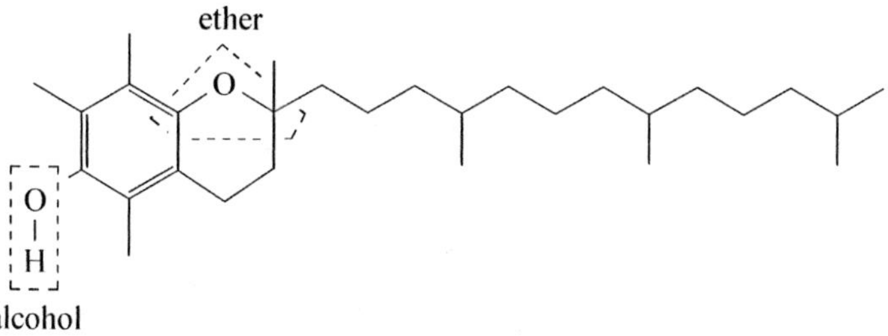

(c)

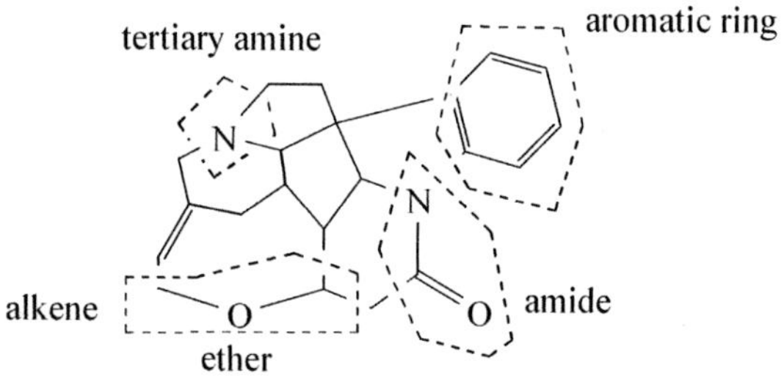

19.66

(a)

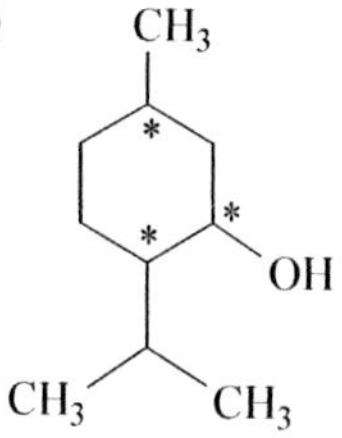

(b)

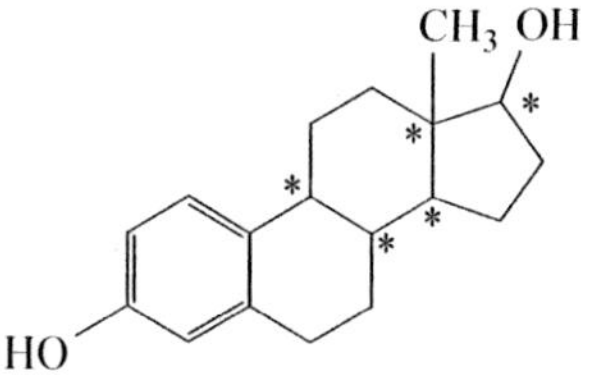

---

*Indicates the chiral carbon atoms.

19.68

(a)

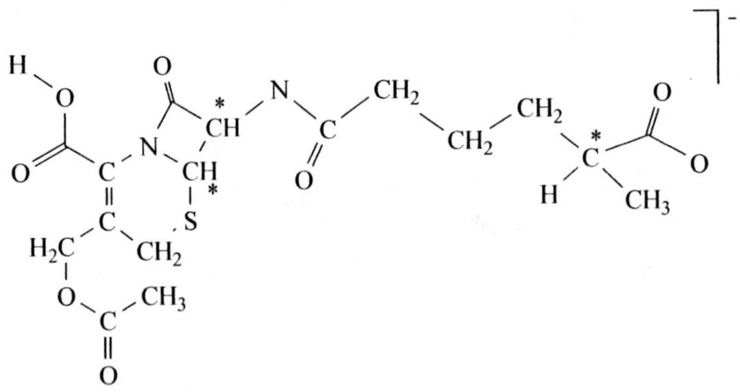

(b)

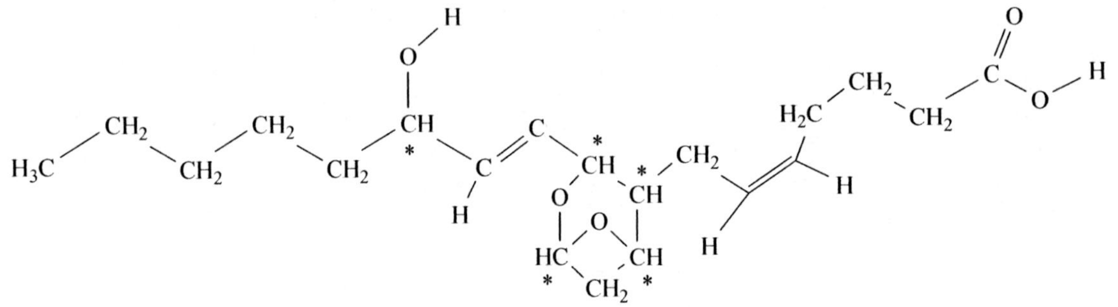

19.70 (a)

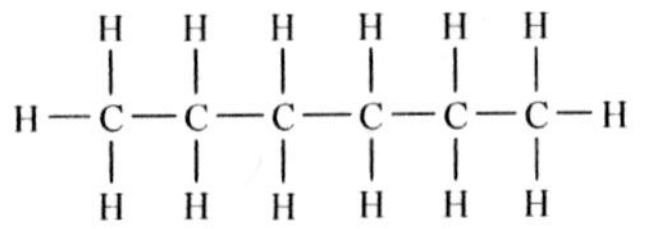

hexane
alkane

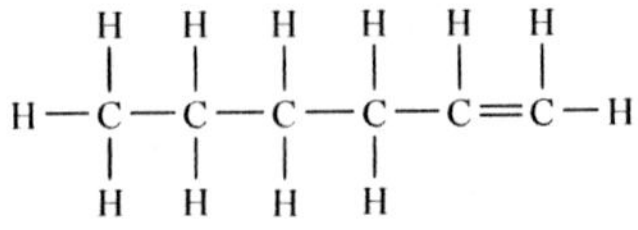

1-hexene
alkene

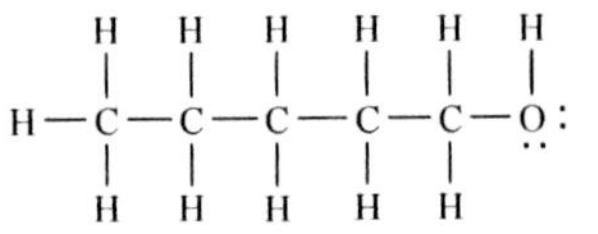

1-pentanol
alcohol

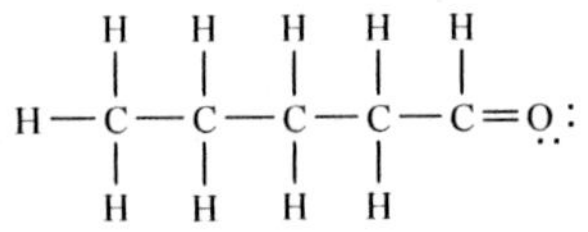

pentanal
aldehyde

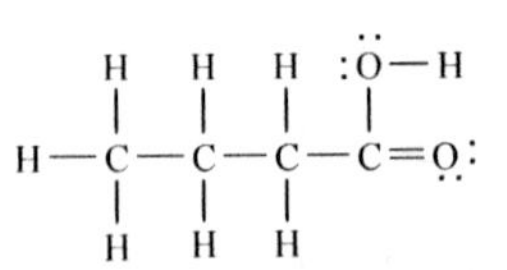

butanoic acid
carboxylic acid

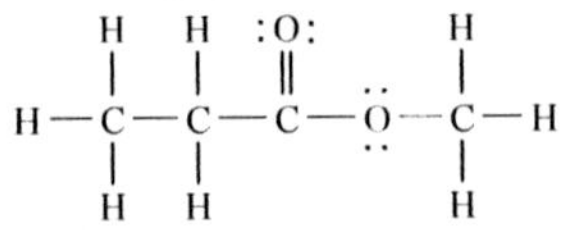

methyl propanoate
ester

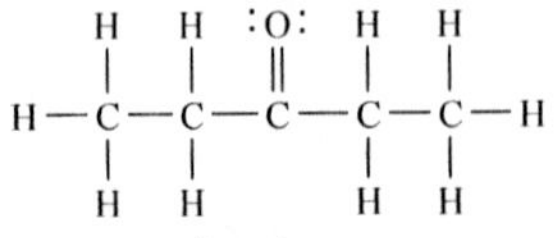

3-pentanone
ketone

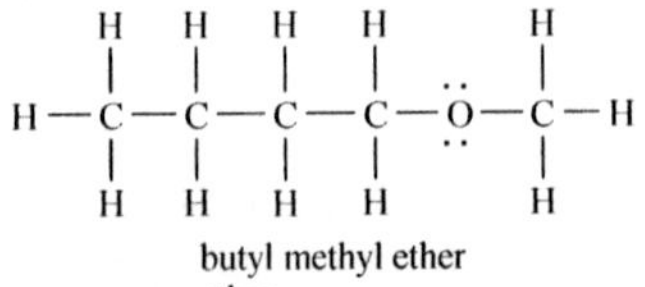

butyl methyl ether
ether

(b) 1-pentanol and butyl methyl ether are isomers. Pentanal and 3-pentanone are isomers. Butanoic acid and methyl propanoate are isomers. None of these molecules are chiral.

(c) Hexane – London forces

1-hexene – London forces

1-pentanol – hydrogen bonding, diplole-dipole interactions, London forces

Pentanal – dipole-dipole interactions, London forces

butanoic acid – hydrogen bonding, London forces, dipole-dipole interactions

methyl propanoate – London forces, dipole-dipole interactions

3-pentanone – London forces, dipole-dipole interactions

butyl methyl ether – London forces, weak dipole-dipole forces

(d) We can classify the molecules in the list into three categories based upon their intermolecular interactions: (1) those with London forces only, (2) those with London forces and dipole-dipole interactions, and (3) those with London forces, dipole-dipole interactions, and hydrogen bonding. The boiling point should increase with the increase in the types of intermolecular forces available, so compounds of type (1) will boil at a lower temperature than compounds of type (2), which will in turn boil at a lower temperature than compounds of type (3).

Type (1) molecules: hexane and 1-hexene

Type (2) molecules: pentanal, methyl propanoate, 3-pentanone, and butyl methyl ether

Type (3) molecules: pentanol and butanoic acid.

Within these classifications, it is more difficult to order the molecules; however, we can make some generalizations. The C=O in an aldehyde or a ketone will have a larger dipole moment than the C—O—C linkage in an ether, because of the geometry around the O atom. Thus we expect the ether to boil at a lower temperature. Among the molecules of Type (3), carboxylic acids generally exhibit stronger H-bonding than alcohols, so we would expect butanoic acid to boil at a higher temperature than pentanol.

Compound	Observed Boiling Point (°C)
hexane	69
1-hexene	64
pentanol	119–120
butanoic acid	103
methyl propanoate	79
3-pentanone	102
butyl methyl ether	70–71

19.72

(a)

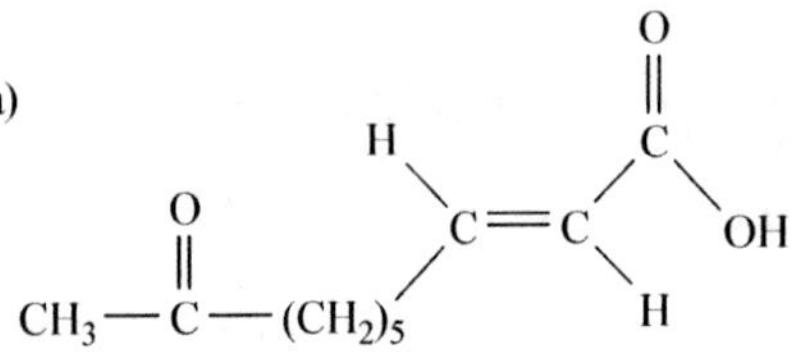

(b)

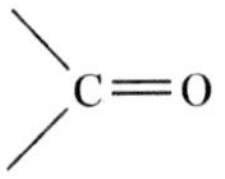

carbonyl group, ketone

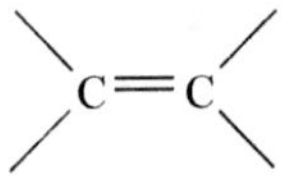

alkene

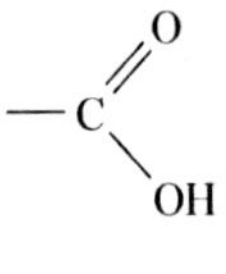

carboxylic acid

19.74 (a) condensation; (b) Because the sodium hydroxide is concentrated, the elimination reaction should dominate; however, nucleophilic substitution may also occur. (c) nucleophilic substitution; (d) condensation; (e) electrophilic substitution

19.76 Polymers generally do not have definite molecular masses because there is no fixed point at which the chain-lengthening process will cease. The chain stops growing due to lack of nearby monomer units of the appropriate kind or a lack of properly oriented smaller polymeric aggregates.

A polymer is, in a sense, not a pure compound, bur rather a mixture of similar compounds of different chain length. There is no fixed molar mass, only an average molar mass. Because there is no one unique compound, there is no one unique melting point, rather a range of melting points. Thus, there is no sharp transition between solid and liquid, and we say the solid softens rather than melts.

19.78

1,4-di(hydroxymethyl)benzene

terephthalic acid

19.80 (a)

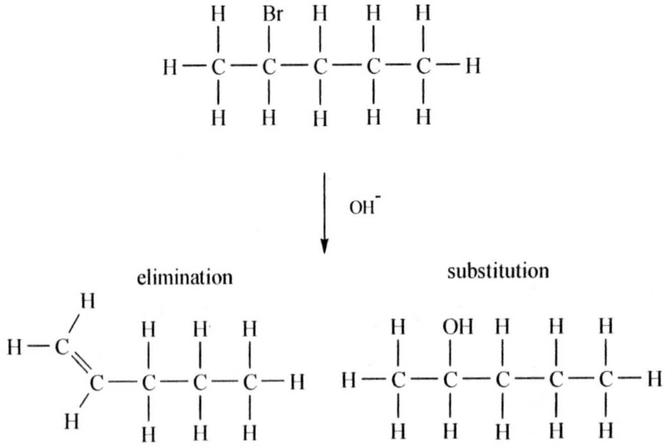

trans

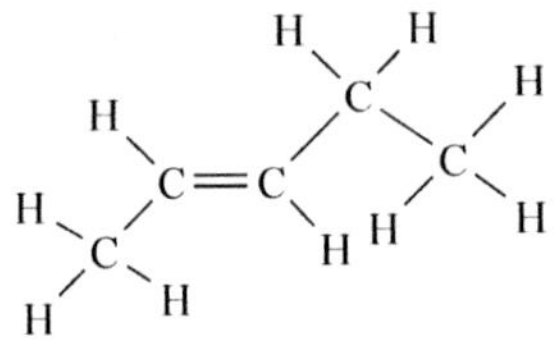

cis

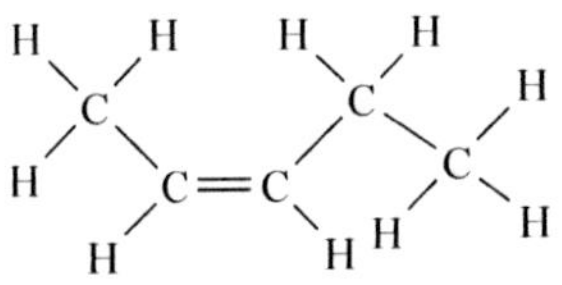

(b) Substitution will be favored over elimination if the hydroxide solution is dilute.

19.82 The nitro group in general is an electron-withdrawing group and would normally be expected to enhance the acidity of the phenol, simply by pulling electron density away from the phenol oxygen atom. This happens by two processes. One is the simple inductive effect of having the electronegative groups attached to the aromatic ring. The second effect is through the resonance interaction of the nitro group with the aromatic ring, which reinforces the resonance interaction with the oxygen atom for the o- and p-isomers. The m-isomer, however, does not have a reinforcing resonance interaction. The appropriate resonance forms are shown below:

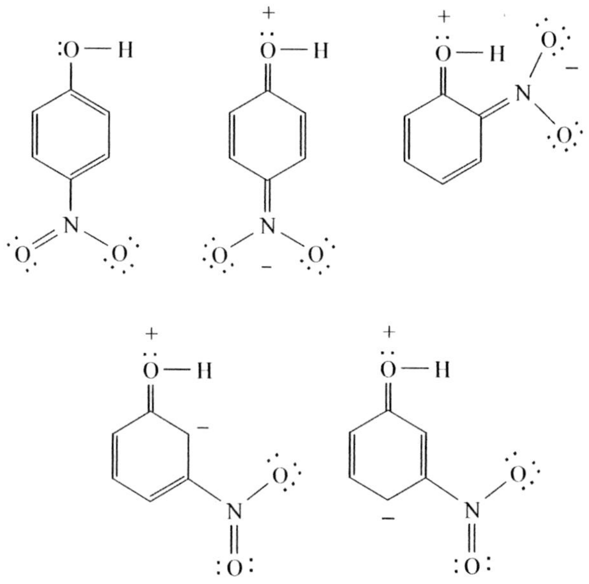

19.84 There is one nucleotide every 340 pm, or 3.40×10^{-10} m. Over the course of 0.299 m, there will be 8.79×10^{8} nucleotides. Each nucleotide has a molar mass of 625 g/mol on average, so the molar mass of the DNA molecule will be 5.50×10^{11} g/mol.

19.86 (a) The Fe^{2+} ions exit the ferritin protein through the "three-fold channel," as this channel is lined with the polar amino acids aspartate and glutamate. This hydrophilic channel allows Fe^{2+} ions, complexed with water molecules, to exit the protein. The "four-fold channel" is lined with hydrophilic leucine, which prevents the water-soluble ion from exiting through this route.

19.88 (a) The nitrogen atoms as drawn are sp^3 hybridized. (b) The polymer absorbs light due to the long chain of alternating single bonds and double bonds in the structure.

19.90 (a)

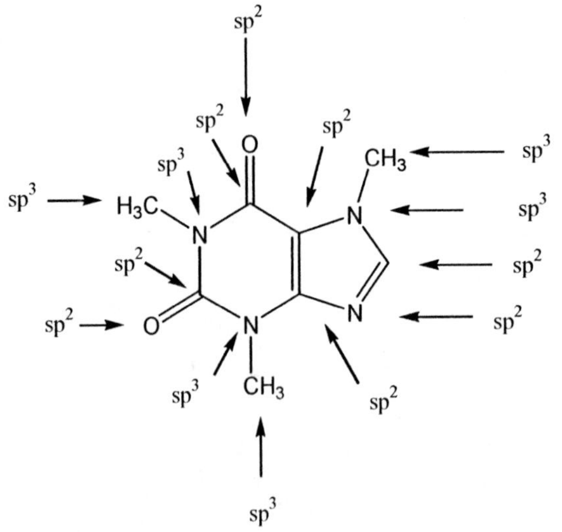

(b) Bond angles for sp^3-hybridized atoms should be around 109.5°. Bond angles for sp^2-hybridized atoms should be 120°.

19.92 Two peaks with relative overall intensities 3:1. The larger peak is due to the six methyl protons and is split into three lines with intensities in the ratio 1:2:1. The smaller peak is due to the two methylene protons and is split into seven lines with relative intensities 1:6:15:20:15:6:1.

19.94 The structures of the two compounds are shown below.

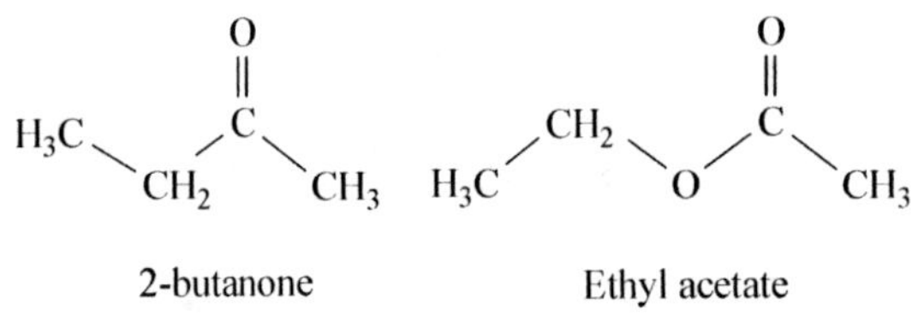

Both molecules contain an ethyl and a methyl group. The quartets are assigned to the CH_2 groups of the ethyl unit. They have an intensity of two, and the fine structure arises from coupling to the three protons on the adjacent methyl group. Similarly, the triplet peaks in each spectrum correspond to the CH_3 part of the ethyl unit. They have an intensity of three and are triplets, owing to coupling to the two CH_2 protons. The peaks that have no coupling arise from the isolated methyl groups. The major difference between the two spectra is the chemical shift, especially for the CH_2 groups. The CH_2 group of ethyl acetate is found farthest downfield because it is attached directly to an electronegative oxygen atom that acts to pull electron density away from the atoms attached to it. The C=O group is also somewhat electron-withdrawing but not nearly so much as an oxygen atom. The CH_3 parts of the ethyl groups are not much influenced by the change in molecular structure because they are removed from the electronegative center and the influence is not felt as strongly. Similarly, the isolated CH_3 groups are both attached to carbonyl functions and occur at similar chemical shifts.

19.96 ^{31}P, ^{19}F, ^{13}C, ^{15}N, ^{29}Si, ^{103}Rh, ^{107}Ag, ^{109}Ag, ^{77}Se, ^{111}Cd, ^{113}Cd, ^{117}Sn, ^{119}Sn, ^{125}Te, ^{207}Pb, ^{203}Tl, ^{205}Tl, ^{129}Xe, ^{183}W, ^{195}Pt, ^{187}Os, ^{89}Y, etc.

19.98 All the H atoms in benzene are equivalent and so there is only one peak observed, which contains no fine structure due to H—H interactions. The structures of the dichlorobenzenes are given below. The H atoms are labeled to show which ones are the same.

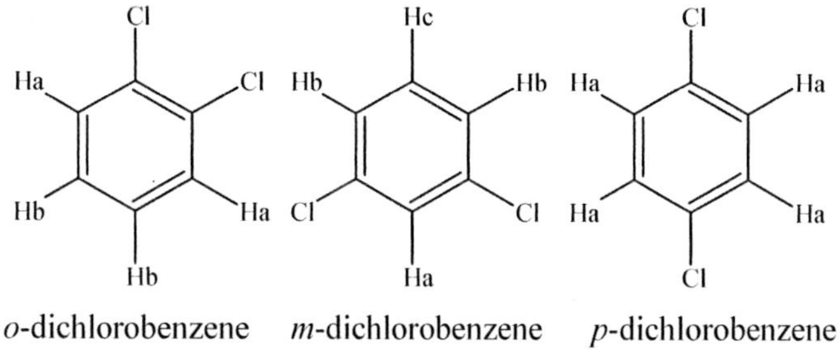

o-dichlorobenzene *m*-dichlorobenzene *p*-dichlorobenzene

As can be seen from the diagrams, *o*-dichlorobenzene will have two peaks in a 1:1 ratio, *m*-dichlorobenzene will have three signals in a 1:2:1 ratio, and *p*-dichlorobenzene will have only one signal. This alone is enough to distinguish between the three isomers. Coupling of H atoms attached to aromatic rings to other H atoms attached to the same ring is sometimes not simple. The degree of splitting of the signals is often small and may not be resolvable unless one is using a spectrometer with a very large magnetic field strength. For the molecules in question, it is possible to say definitively that all the ^1H atoms of *p*-dichlorobenzene are equivalent and will show no splitting. The H_a and H_b atoms of *o*-dichlorobenzene will couple to each other, but the pattern will be more complicated. An even more complicated fine structure is expected for the *m*-dichlorobenzene, in which there are three different types of protons attached to the aromatic ring.